Pollution in a Promised L

"And I brought you into a plentiful land to eat the fruit and the goodness thereof, and when you came, you defiled my land and turned my heritage into an abomination."

Jeremiah 2:7

Pollution in a Promised Land

An Environmental History of Israel

ALON TAL

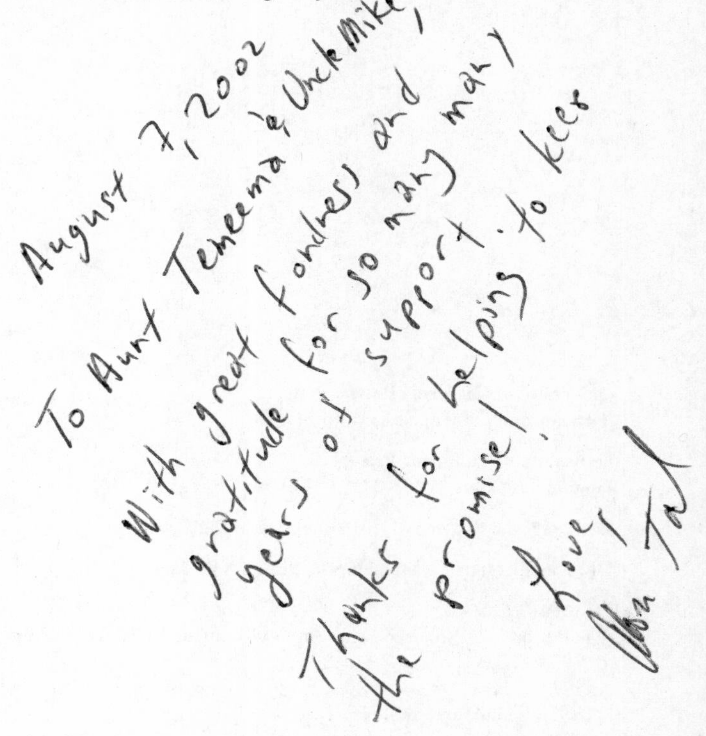

University of California Press
BERKELEY LOS ANGELES LONDON

University of California Press
Berkeley and Los Angeles, California

University of California Press, Ltd.
London, England

Library of Congress Cataloging-in-Publication Data

Tal, Alon, 1960—.
Pollution in a Promised Land: An Environmental History of Israel / Alon Tal.
p. cm.
Includes bibliographical references and index.
ISBN 0-520-22442-6 (cloth: alk. paper).—ISBN 0-520-23428-6 (pbk.: alk. paper)
1.Pollution—Israel. 2. Israel—Environmental conditions. 3. Environmental protection—Israel. 4. Environmental policy—Israel.
I. Pollution in a Promised Land

TD187.5.175 T35 2002
363.7'02'095694—dc21 2001040658

Manufactured in Canada

10 09 08 07 06 05 04 03 02
10 9 8 7 6 5 4 3 2 1

The paper used in this publication meets the minimum requirements of ANSI/NISO Z39.48-1992(R 1997) (*Permanence of Paper*). ♾

For Robyn, Mika, Hadas, and Zoe

"Whose home I placed in the Arava,
 dwelling amid the salt flats,
Amused by the multitudes in the city,
 oblivious to the demands of a taskmaster,
Roaming the mountains for pastures
 and seeking every green thing."

Job 39:6–8

Contents

Illustrations

Acknowledgments

Many people have assisted in putting this book together. Well over a hundred people agreed to be interviewed or to provide information, despite their busy schedules. I am extremely grateful for their openness. I especially appreciate the many talented individuals who helped me by reviewing different sections of the manuscript, in particular, Rochelle Adam, Azariah Alon, Dror Amir, Dror Avisar, Gideon Biger, Valerie Brachya, Michael Cohen, Noah Efron, Brock Evans, Michael Graber, Noam Gressel, Yerahmiel Kaplan, Rene Karschon, Arthur Kessler, Uri Marinov, Stuart Masters, Danny Orenstein, Uzi Paz, David Pargament, Yisrael Peleg, Berry Pinshow, Aviva Rabinovich, Oren Rosenthal, Ruth Rotenberg, Menahem Sachs, Yoav Sagi, Uriel Safriel, David Saltzman, Dianne Saxe, Stuart Schoenfeld, Eilon Schwartz, Benny Shalmon, Uri Shanas, Hillel Shuval, Bill Slott, Emily Silverman, Josef Tamir, Gil Troy, Danny Yoffe, Reuven Yosef, Amotz Zahavi, and Dan Zaslavsky.

No one could enjoy better editorial support than that provided by Technion professor Dr. Yonina Rosenthal, my marvelous and assiduous mother. While she was not always successful in getting me to downsize gratuitous ten-dollar words, she brought her wisdom and considerable technical writing skills to the task, and it shows. As always, my father, Dr. David Tal, provided constant encouragement, fervent opinions, and understanding about the technical aspects of environmental issues. All of those mentioned above offered thoughtful suggestions and flagged many mistakes. Of course they could not catch them all, and those remaining are my own.

The Conanima Foundation's support is gratefully acknowledged. While it shuns recognition, many of us know just how much this institution has contributed over the years to Israel's environmental progress. Dr. A. Joshua

Sherman, a superb historian in his own right, offered encouragement and friendship throughout. Three books were exceptionally valuable, and my indebtedness to them is reflected in the notes: Ofer Regev's 1993 history of Israel's Society for the Protection of Nature, *Forty Years of Blossoming;* Uzi Paz's 1981 history of the Nature Reserves Authority, *The Land of the Gazelle and the Ibex;* and Simcha Blass's anecdotal water history *Water in Strife and Action*. In addition, Shoshana Gabbay's *Environment in Israel* series remains the best collection of facts and figures on the subject in any language.

I cannot say enough about the faculty and staff of the Otago University Law School, whose Kiwi hospitality was truly "good as gold" even prior to our arrival in Dunedin. Dean Richard Sutton's support and Dr. Jim Allan's good humor (and even better jump shot) helped me keep the project on pace. During my stay in New Zealand, Dalit Ucitel was a resourceful and devoted research assistant back in Tel Aviv. More recently, Arava Institute students Shirly Riceman and Jennifer Lorenzen offered excellent and cheerful assistance.

Three remarkable organizations also have my undying gratitude. First, I salute Adam Teva V'din, the Israel Union for Environmental Defense, in Tel Aviv, and its devoted staff, with whom I worked from 1990 to 1997. I shall always be especially grateful to its first three workers: Irit Sappir, Tirtseh Keinan, and Ruth Yaffe, as well as the eight magnificent founding board members, headed by the ever brilliant and supportive Eilon Schwartz. They have been companions and colleagues of mine for almost a decade. Much of what appears in this book we learned together. In addition, kudos and gratitude to the staff and faculty at the Arava Institute for Environmental Studies at Kibbutz Ketura, to Miriam Ben Yosef, Cecil Rimer, and Sharon Benheim, who have been partners in the endeavor from the beginning. Special thanks got to Dr. Noam Gressel, who covered for me there so effectively during my Kiwi voyage and subsequently remained a wonderful friend and "boss." During the past two years I have been privileged to work with Life and Environment, the umbrella group for Israel's environmental organizations. Josef Tamir, the organization's founder, and his granddaughter, Orr Karassin, its present director, have taught me a great deal about Israel's environmental history and about its future.

I am of course indebted to the University of California Press and its staff. My gratitude goes to Howard Boyer, who believed in a very prolix manuscript, brought it on board, and helped me start the painful process of slimming it down. Doris Kretschmer has been a most thoughtful editor who has shepherded this book masterfully through its many phases, along with her very able assistants. Nicole Stephenson was always especially positive and ready to lend a hand. Special thanks go to the extremely talented team at Publication Services in Champaign, Illinois: Jerome Colburn, senior editor; Kay Suleiman, copy editor; Louise Toft, who handled the logistics and offered much encouragement; and the production staff.

For permission to use Reuven Rubin's evocative painting on the cover, I would like to thank Carmela Rubin (curator and director) and Shira Naftali of the Rubin Museum Foundation, as well as my friend, public transportation advocate Rachel Horam, for facilitating the connection. Many institutions and individuals graciously granted rights to print the photographs that appear in the book, and I am grateful to them all.

Without my wife, Robyn's, encouragement, support, and comments, the book would not have happened. She knows how much I admire and depend on her. My sisters and brother, Gabriella Tal, Aliza Stark, and Oren Rosenthal, were there for me during the vicissitudes of the book's preparation, as they always are. Bill Slott's wit and wisdom have been precious to me since I was nine, but never more so than during the past two years. I should close by thanking Mika and Hadas Tal, who over the years never complained about their father's frequent absences. For months they faithfully monitored the manuscript's progress during dinnertime conversations in New Zealand. More recently, Zoe has joined us, and her intermittent cooperation and constant good spirits helped me to complete the editorial revisions. Perhaps, Zoe, when you are old enough to get through this book, you will understand why I liked reading *The Lorax* to you so much.

A. T.

Preface

On a technical level, I have been preparing this book for the past five years. In fact, the project represents the culmination of some eighteen years of work in the field of environmental protection in Israel. By training and trade I am not a historian. Yet, I hope that *Pollution in a Promised Land* offers a fair, engaging, and well-documented description of Israel's environmental experience.

On a subject this broad, I am well aware of the highly selective nature of the editorial process. There are large areas about which I might have written and did not. For example, regional environmental issues are not reviewed in great detail because I believe they have been of limited historical importance. Issues such as radiation, hazardous materials, noise, and consumption may be underrepresented. And a future edition will certainly have to discuss the growing concern about cellular phone technologies and antennas. A recurring theme in the book is that limited space forces hard choices. The text would undoubtedly look quite different if it had been written by someone who had worked for industry for the past two decades or grown up as a Bedouin in the Negev.

It is well to set my biases on the table at the outset for readers so they can weigh the perspective of the narrator as they read this story. Two personal and professional passions informed almost every stage of writing. The first, my "green" bias, was probably formed somewhere between the friendly woods of North Carolina and many dinner table conversations with my family as I was growing up, the son of an analytical chemist with an expertise in pollution monitoring, who cared about his work. This perspective crystallized during the course of work in government environmental agencies in Israel and the United States and more recently in public-interest environmental advocacy. As an environmentalist, I am concerned that human beings today are pushing this planet too hard. I believe that Israel offers an extreme example of this phenomenon. I also retain a healthy suspicion toward manufacturers and developers, having seen how the profit motive can twist people's perceptions and their relationship to the environment. My visceral revulsion at the destruction of lovely corners of the planet, in particular in the ancient land of Israel, seems to grow stronger with the years. It is far too ingrained to be purged in the name of "objectivity," especially in a venture stretching on as long as the present one. So this prejudice begins at the very start of this book and is manifested throughout the chapters.

At the same time, certain principles have remained axiomatic with me as a Zionist since visiting Israel at the age of twelve in 1972. Given the history of the first half of the twentieth century, I believe that a Jewish state in the Land of Israel constitutes a moral imperative. The Zionist tradition has always encouraged free and open criticism, and like all Israelis I have more than my share of gripes and suggestions. But ultimately I believe that Israel remains a remarkable country—a wonderful, even inspirational, place to live and raise children. The country's achievements during the past fifty years, by any impartial standard, appear to me to be breathtaking. When the more mystical of the locals speak of the nation and its progress in terms of "miracles," it resonates.

The trouble is that Zionist and environmental viewpoints can clash when they leave the realm of theory and meet in a planning commission hearing or in an emission standards debate. Although I do not believe that these two ideologies are mutually exclusive, they have not yet found a healthy equilibrium or even a clear basis for coexistence. This book represents my own attempt to reach some harmony between the two impulses, or at least a more sober understanding of why they are at odds. In this regard, the writing process has been an edifying and encouraging endeavor. The one thing Zionists and environmentalists clearly share is a fundamental optimism and a belief that people can do better.

As I mentioned, I am not a professional historian. To me, my amateur status seemed like a liberating factor rather than a liability. Nonfiction can be hard to read, especially a book of this length and potential density. It was important to me that Israel's environmental history be accessible, digestible, and, whenever possible, fun to read. On the whole, I prefer a good anecdote to scholarly commentary or analysis. On the scale ranging from "scholarship" to "journalism," therefore, I often intentionally leaned in the latter direction, while censoring "juicy" tidbits that crossed the line from relevant history to gossip. I have tried to make a presentation that is scientifically and factually precise. To compensate for the popular style and the sheer chronological and substantive breadth, I have given in to a lawyer's and academic's weakness for citations and footnotes.

The text also reflects my lack of enthusiasm for "debunking" the past or painting it with the cynicism that seems to characterize the thinking of many so-called post-Zionist revisionist historians. This world is full of heroes. One finds disproportionately large clusters of them among environmentalists, and I have been privileged to meet many myself. In reviewing the manuscript, one friend gently advised that I beware of hagiography, and indeed I have tried to tone down the hyperbole without masking my admiration. Given the rather brief opportunities for personal profiles that a book like this offers, it seemed appropriate to emphasize the positive. If Israel's environmental movement might benefit from a little less self-righteousness, it certainly could also stand to be a little more collegial. I hope this book contributes to such a culture of ingenuousness and appreciation and will be delighted if its positive tone sparks a "revisionist" critique.

A word or two about the book's narrative perspective. History books belong to the realm of the third person. In the present context, however, there may be a few places where I departed from this convention. I drew many of my recent examples from personal experience or direct involvement, especially in Chapter 11, when documenting the emergence of Adam Teva V'din, the Israel Union for Environmental Defense. As founding director and later chairman of the organization, I relied heavily in this section on my own memories and observations. Friends there advised me that it would be unnatural to expunge myself from the text, and perhaps they were right. Nonetheless, writing with a first-person voice seemed even more awkward. In any event, I did not want to claim undeservedly disproportionate recognition for the work done by the talented staff during the period of my affiliation.

Finally, this book helped serve a personal need. My awareness of Israel's environmental experience really began in 1980, after I moved to Israel and joined the army. Everything before that time remained something of a "black box." Pieces of the past would pop up, but I lacked a comprehensive picture of where the country had come from environmentally. That made it a little harder to know where it should go. After eight years of acting locally, it seemed wise to take some time and try to think more globally.

So, armed with transcripts from dozens of interviews and a couple hundred kilograms of books and materials, I set off for New Zealand. It was there, at the University of Otago, that for eight months I wrote the first drafts of the book's twelve chapters. No place could be further from Israel's environmental reality. It never stops raining in New Zealand, much of the countryside is uninhabited wilderness, and ecologically, there are no natural predators—biological or political. But the contrast and the calm of this wonderful land proved to offer an incomparable perch for gaining a perspective on Israel's situation.

There I learned that the environmental movement in Israel has a truly rich history: It has its elders; it has its poetry and songs; it has had its failures and successes. One can even begin to identify a unique local ecological ideology—a curious amalgam of romantic, ruralist, pantheist, Western, and, in many contexts, Jewish beliefs. Recently I spoke on the subject of environmental heritage with my friend and teacher (and first boss), Rutti Rotenberg, head of the legal department at Israel's Ministry of Environment. She told me how shocked she was to discover the growing number of coworkers who had never heard of Uri Marinov, the controversial first director-general at the Ministry. But, then, where might they go to learn of their roots? I hope this book will remedy the situation. More important, I hope this book will help all people who care about the Land of Israel to better understand the origins (and the magnitude) of present environmental challenges. Perhaps such an understanding can contribute to greater action and commitment to healing this ancient yet new, fertile yet arid, violent yet holy, and polluted yet promising land.

A.T.
Kibbutz Ketura, April 2002

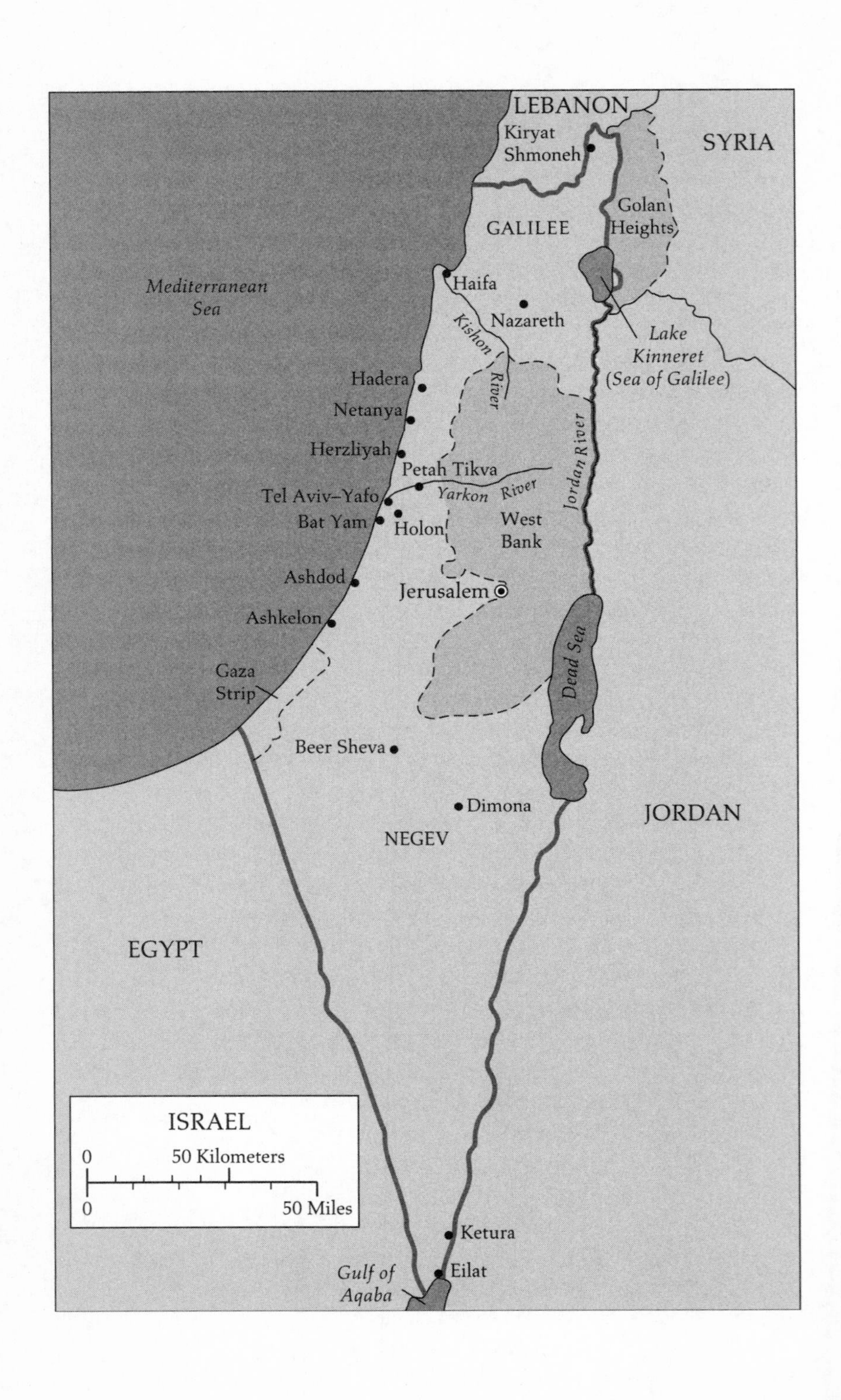

LEBANON
Kiryat Shmoneh
SYRIA
Golan Heights
GALILEE
Mediterranean Sea
Haifa
Nazareth
Kishon River
Lake Kinneret (Sea of Galilee)
Hadera
Netanya
Herzliyah
Petah Tikva
Jordan River
Tel Aviv–Yafo
Yarkon River
Bat Yam
Holon
West Bank
Ashdod
Jerusalem
Ashkelon
Dead Sea
Gaza Strip
Beer Sheva
Dimona
JORDAN
NEGEV
EGYPT
ISRAEL
0
50 Kilometers
0
50 Miles
Ketura
Gulf of Aqaba
Eilat

1 The Pathology of a Polluted River

An Introduction to Israel's Environmental Crises

The bizarre clinical symptoms bewildered even the seasoned veterans of the emergency room medical team. Although the Israeli doctors were well accustomed to grisly disaster situations, no prior experience prepared them for what they encountered on that warm summer evening of July 14, 1997, at the Beilinson Hospital in Petah Tikva.

The athletes had been making their way across the Yarkon River, the second of fifty-six national delegations marching into the Ramat Gan Stadium for the opening ceremonies of the Maccabiah Games, often called the "Jewish Olympics." To the dismay of organizers and athletes alike, the newly (and shoddily) constructed bridge collapsed. Unable to withstand the weight, the beams snapped. The athletes, packed together in parallel rows of six, toppled eight meters into the waters below.[1]

Only one, Gregory Small, a thirty-seven-year-old bowler from Sydney, reached the hospital dead on arrival, apparently as a result of injuries sustained during the fall. Once they were pulled out of the river by rescue teams (Figure 1), the other sixty-six athletes seemed to be on track for a reasonable recovery. Shallow enough to stand in (1.6 meters under the bridge), the Yarkon River hardly seemed to pose a serious health hazard.[2] The term "river" in fact is a misnomer. Like almost all Israeli rivers, the Yarkon is more precisely a stream—generally a languid one, especially in July after four months without rain. Most of the injuries seemed routine. But then something went wrong.

Within an hour or two of arriving, patients who initially tested healthy for respiratory function were tottering on the edge of asphyxia. Some unidentified organism was devouring their respiratory systems and pulmonary blood vessels. "We inserted a tube into the heart," explained Dr. Patrick Surkin, who ran the intensive care unit at Tel Aviv Medical Center. "We saw that the heart muscle had been directly damaged. There was a lack of air supply and a steady drop in blood pressure. These aren't the symptoms in drowning cases."[3] When news reports around the world hit the stands the next morning, of the 373 members of the Australian delegation, thirty-five remained hospitalized, seven in critical condition.[4]

Sensing that he was witnessing a toxic incident, Dr. Surkin immediately sent a physician on a motorbike to the site of the accident to take water samples. By midnight he roused David Pargament, the burly head of the Yarkon River Authority, to hear whether there was cause for particular suspicion regarding toxic discharges into the water. Pargament explained that thirty-six hours prior to the collapse, the River Authority had sprayed the waters with MLO (mosquito larvicide oil), a combination of jet fuel and oil, to prevent mosquito infestation. Subsequent lab reports found no traces of the substance.[5]

THE MACCABIAH GAMES, 1997

It took some time for the Israeli press to pick up the disaster's environmental angle. Initial attention remained focused on the criminal liability of the engineers and the construction company as well as on the controversial decision to continue with the festive opening ceremony in the face of the massive injuries.[6] On July 16, for example, the *Jerusalem Post* published a scorching editorial entitled "A Sense of Shame." It did not even refer to the pollution that was the actual cause of the Australians' medical problems.[7]

The public's response seemed a mix of embarrassment and wistful regret that yet another symbol of Zionist achievement was marred by incompetence and negligence. The opening ceremonies were especially staged to celebrate the hundredth anniversary of the first Zionist Congress, which launched the movement that ultimately produced the State of Israel. The Maccabiah Games were created to be a centerpiece of this "Zionist revolution." Established in 1932 as the Jewish Olympics and held every four years in Israel, the Games offered Jews around the world the opportunity to join together and shed the age-old stereotype of the frail and intellectual Jew through the demonstration of athletic prowess. By 1997 the Maccabiah was billed as the third largest sporting event in the world, with fifty-three hundred participants from around the world and thirty-eight sporting events ranging from ice hockey to rugby.[8] Many Israelis cynically perceived the games as more of a tourist opportunity than a serious sporting competition, a cynicism reflected in the relatively modest press coverage of the sporting events; for most citizens, though, the Maccabiah still held sentimental significance.

The notorious bridge was specially built, as one is every four years, to herald the dramatic entrance of the athletes into Israel's largest stadium. As the subsequent investigation revealed, the construction of the bridge was riddled by a litany of irregularities, presumably to reduce costs. Ultimately, the head of the Maccabiah Committee, the contracting engineer, and the construction companies were convicted of negligence.[9] But no expense had been spared on the opening ceremonies themselves. There were hundreds of dancers, high-tech sound and light displays, and the obligatory torch lighting by the Israeli basketball legend Mickey Berkovich. Ironically, the opening ceremonies were the only part of the games to receive live national television coverage. And so it was that households tuning in from around the country saw a show that bounced back and forth between chaotic rescue efforts along the river and festive performances in the stadium. The juxtaposition captured a schizophrenic mix of the tragedy and euphoria that have characterized so much of Israel's emotional history.

SASHA ELTERMAN'S NIGHTMARE

Even six months later, the story would not go away. The weekend magazine of Israel's most widely circulated daily, *Yediot Ahronot*, featured the most high-profile victim, Sasha Elterman, on the eve of her sixteenth birthday. Sasha, a dark-haired, lanky fifteen-year-old from Sydney, was a particularly promising tennis player with a training regime that included ten kilometers of daily running and dozens of laps at a swimming pool each day. It was her first trip to Israel, "the homeland," and she was moved by her visit to Jerusalem, the Dead Sea, and key local attractions with her parents prior to the games.

"On the evening of the opening ceremonies, I stood together with all the delegation and it was our turn to move forward. We started to walk," she recalled.

> I truly remember that one of the kids that stood by me said: "There's no way we're going to go on that shaky thing." I didn't even manage to answer him or think about the sentence, when I heard a shout, and immediately after, I fell into the water. There were people on top of me, and I was stuck in the mud and I couldn't get myself up above the water. From the stories of the other athletes, they pulled me from the Yarkon unconscious. After treatment in the ambulance they restored me to consciousness, and I remember they wanted to cut my clothes to treat me. With my last bit of strength I told them: "Don't cut my Maccabiah uniform."[10]

Sasha lapsed into a coma within hours and woke up four days later in the hospital with her entire body aching. Little could she know that her nightmare was just beginning. For six weeks she languished in the Shneider's Children's Hospital in Petah Tikva, waiting for a diagnosis of her condition that would lead to curative therapy. With her lungs barely functioning, biopsies were sent to local and American laboratories. Only when *Pseudallescheria boydii* infection was identified in the lungs of Warren Zeins, the fourth Maccabiah fatality, was there a clear direction for treatment.[11]

This was hardly good news—of the twenty-six diagnosed cases of *P. boydii* infection in the world, only four have survived.[12] *P. boydii*, a relatively ubiquitous fungus, is an opportunistic organism always ready to exploit a weakness in the body's defenses. Lung transplant recipients are at special risk because of their immunocompromised status. The fungi tend to colonize in the lungs or pleural cavities, producing a pneumonia that kills cells in its wake. The microbes and the deadly pus they spawn spread readily, disseminating to the brain, kidney, heart, and thyroid.[13]

Six months after the disaster, Sasha had been through three hundred X-rays and eighteen operations (thirteen of which were brain surgeries). She had lines inserted to bring food and medicine directly to her stomach, and three tubes for draining the excess fluids in her head, which had been putting pressure on her brain.[14] Her permanent lung function was 60 percent of what it was before the accident, and she was plagued with periodic convulsions.[15] Eventually, she recovered sufficiently to carry the Olympic torch a few hundred meters during the opening of the games in Sydney.

The outpouring of concern by Israel's public, including a visit by Nava Barak, the Prime Minister's wife, along with the ultimate conviction of the defendants for negligent homicide, may have ameliorated the family's frustration and sense of loss. But the Yarkon River's toxic touch was largely irreversible. Sasha's mother, Rosie Elterman, spoke candidly: "She won't be what she was before. She was extremely intelligent, skipped a grade, was an outstanding athlete, a quick thinker. And now, all that I want is that she should be average, just an ordinary girl. I dream that she may one day have a normal life."[16]

ENVIRONMENTAL DISBELIEF AND DENIAL

Sasha and her fellow athletes' pain, as well as at least three of the four Maccabiah fatalities, were caused by pollution. "How did the Yarkon River come to be so polluted?" and "What is being done to stop it?" are questions that barely seemed to interest the victims or the concerned public. Indeed, initially even environmentalists were confused about what actually happened. When activists from the environmental group

GreenAction organized a small protest about Yarkon water quality days after the accident, those who noticed saw it as insensitive opportunism. Early news reports in Israel implied that the victims suffered from trauma due to the fall.

As Dror Avisar, hydrologist at the activist organization Adam Teva V'din (the Israel Union for Environmental Defense), acknowledged in an interview a month after the disaster: "Even we who are constantly dealing with Israel's severe water pollution were surprised by the deaths in the Yarkon. No one anticipated that a person that falls into the water would die from it."[17]

Part of the confusion can be attributed to the Israeli government's environmental agencies, who were eager to avoid blame when the public was clamoring to find a scapegoat. The Ministry of the Environment was slow to acknowledge the environmental aspect of the tragedy. First, it denied the toxic exposures; later, its director general, Nehama Ronen, thought it inappropriate to capitalize on human misery. Months after the tragedy, Ronen maintained a highly defensive position: "Look, the Yarkon has been a place that people want to run from for over thirty years. The second that people came looking for someone to blame, suddenly, we at the Ministry of the Environment are the guilty party."[18] Even in informal settings, David Pargament, director of the Yarkon River Authority, still takes this line: "You have to remember that ultimately the Maccabiah incident was a mechanical disaster. A bridge collapsed. It's not like Love Canal. The people had no business reaching the water there."[19]

Eventually, pollution's role as *a causa sine qua non* in the Maccabiah deaths could not be denied. On August 18, 1997, an "Intermediate Report" submitted by the Water and Streams Division of the Ministry of the Environment reported a veritable cocktail of chemicals around the area of the bridge. The results were based on six separate analytical tests performed in Israeli laboratories and at the Global Geochemistry Laboratories in California. Although fecal bacteria appeared low in the river, *Pseudomonas* bacteria were found in high concentrations of ten per hundred milliliters. (*Pseudomonas* were isolated in the phlegm of Warren Zeins, the fourth victim of the incident.) Oils, heavy metals, petroleum hydrocarbons, and siloxane were all found in the samples.[20]

There was much quibbling over the proximate causes of death. Yoel Margalit, director of the Center for Biological Control at Ben-Gurion University in Beer Sheva, went on record blaming the MLO. Nehama Ronen responded that this just served to frighten the public, and threatened legal action.[21] The absence of a single "smoking gun" among the chemicals revealed in laboratory results has led most scientists to assign some role to the river's sludge: The collapse of the bridge temporarily disturbed the greasy mud created by years of suspended solids settling onto the river floor. This virulent anaerobic cocktail was temporarily stirred up and led to the toxic exposures.[22]

With the submission of the intermediate report, headlines began to appear attributing the death of the athletes to the contamination in the Yarkon. Uri Minglegreen, the chief scientist at the Ministry, described to reporters the synergistic effect caused by the mixing of a variety of hazardous chemicals.[23] "These pollutants are sufficiently toxic and there's no need to look for any additional mystery substances," he explained.

THE DEATH OF A RIVER

The Yarkon River had not always been contaminated. Until the 1950s, a remote risk of bilharzia (schistosomiasis) was still the only "environmental" concern for the children who swam and fished its rushing waters.[24] The river's 28-kilometer run to the Mediterranean at Tel Aviv's northern border still resembled this description from an 1891 travel guide: "The Yarkon River leaves from the base of a small hill called Ras El Ayin. Many other springs and rivers flow into it until it becomes a roaring river that zig zags until falling to the sea. This is one of the largest rivers in the land, and its power turns mills, and small fish can be caught in it."[25] Biologists would later discover that the river boasts a species of small fish, *lavnun ha-Yarkon,* that exists nowhere else in the world. Even today, a rich variety of native plants remain along the banks. When the little nutria (a furry rodent) was brought to Israel in an unsuccessful fur enterprise near the river, the animals soon escaped and took to the warm climate, joining the local fauna.

Historically, the Yarkon also enjoys a rich and varied history. The strong flow and the thick wetland vegetation surrounding it created a substantial land barrier. This gave the river considerable military importance. Whoever controlled the narrow land passage between Migdal Zedek and the Fort of Antipater dominated the surrounding area and a major north-south route. Over the past millennia, a series of military contests ensued around the river's narrow headwaters, as so often is the case for such strategic locations.[26]

But the Rosh ha-Ayin springs were tapped in 1956, and most of the Yarkon's natural 220 million cubic meters of annual flow (220 billion liters) was diverted to the south of Israel as part of a national irrigation program. The river quickly lost its vitality.[27] Were it not for the allocations to upstream farmers, the river would have become completely moribund. The untreated sewage of the many municipalities that make up the Central Israeli Dan and Sharon region replaced the clean natural spring waters bubbling out of the Mountain Aquifer. These wastes were augmented by chemicals and detergents from factories in the many industrial zones that lay along the flood plain, discharges from the many solid waste dumps in the watershed, and runoff, carrying residues of oils, from roads and industrial debris.[28]

It took little time for Israelis to grow used to the new stench-filled and stagnant reality. It was just another annoyance of daily life. A line in the popular song "Only in Israel" from the 1960s summed up what was already the common perception about the perennially polluted river: *Ha-Yarkon tamid yarok*, "The Yarkon is always green."[29]

In fact, the river is not uniform, breaking into three separate sections:

- The upper Yarkon, which has been returned to a fairly clean state
- The central (and largest section), which receives municipal wastewater and effluents at various levels of treatment
- The lower four kilometers, where seawater flows in to fill the vacuum and tidal fluctuations ensure regular replacement of the saline estuary

Even if authorities had been interested in addressing the water quality of the Yarkon, its complex pollution profile makes it a particularly challenging resource. The river's watershed contains some eighteen hundred square kilometers stretching from cities in northern Samaria, such as Nablus, to the sea. Controlling for all the runoff that eventually reaches the Yarkon is a daunting task, today requiring international cooperation.

Raw sewage coming from the Kaneh River in the West Bank, for instance, reaches the stream with a biological oxygen demand (BOD) reading of 250 milligrams per liter. The Israeli standard for this parameter, which measures the amount of organic waste that steals oxygen from aquatic systems, is 20.[30] The Yarkon River master plan, based on the river's sensitivity, recommends that entering sewage be treated to a BOD level of 10, or 25 times lower than the concentrations that actually reach the river.[31] The Kaneh is just one example of many pollution sources. In addition, the Ayalon, Shiloh, and Hadarim tributaries carry their own sewage loadings.

The range of pollutants is also enormous. The small factories and sewage discharges whose effluents indirectly reach the tributaries make up only the urban part of the pollution portfolio. Farmers near the banks contribute a range of agricultural pollutants, including pesticides and animal wastes, through diffused (nonpoint) runoff, which may have been responsible for the *P. boydii* that harmed and killed the athletes.[32] In the most densely populated area of the country, where land prices are highest, there simply is not much land available to install the biological processes required for treatment.

Once the heart of the river became polluted, it lost its biological integrity and the ecological law of "unintended consequences" set in. Pest control offers one example. Since the British brought the Gambusia fish (*Gambusia affinis*) to Palestine in 1924 as a remedy for the malaria problem, these fish have been employed for mosquito control,[33] but the central Yarkon pollutants are too virulent even for these hardy mosquito eaters. Without fish in the waters, mosquito populations are not curbed by a natural predator and quickly became an urban nuisance. This is particularly aggravating for Tel Aviv residents, who like to leave their windows open and frequently do not have screens.

The light spraying of MLO, a thin sheen of oil on the surface, was selected as the "lesser evil" of the available control options.[34] Even if it affects only the mosquito population (a dubious assertion to be sure), it is not always possible to apply the MLO from a boat, so access roads along the river are required. This in turn disturbs vegetation that is supposed to grow back as part of reclamation efforts. The exposed land hastens erosion, silting, and turbidity. Treating the symptoms rather than the causes of a sick river has limitations.

REVIVING THE YARKON

After twenty-five years of total dormancy, the Streams Authority Law was finally given some life with the establishment of a Yarkon Streams Authority in 1988. (It took the better part of another decade for the next River Authority to be declared.) The idea behind the 1965 statute was to introduce a watershedwide approach to river reclamation in Israel. The responsibilities of the Authority include pollution prevention enforcement, planning, licensing, and drainage. These responsibilities extend beyond the meager 20 meters of land on each river bank over which the Authority exerts direct control.

David Pargament took over the Yarkon River Authority after one too many photogenic "fish kills" hit the Israeli press and left top brass at the Environmental Ministry dissatisfied with his predecessor.[35] Pargament, a surprise appointment, strikes an unconventional figure in Israel's environmental bureaucracy. A physically imposing man, his long beard and ponytail (still red), preference for jeans, and American education give him a rugged cowboylike persona.

Selected for the Yarkon River Authority position, after working as a relatively aggressive nature reserve field inspector, Pargament had all the requisite ecological qualifications, enforcement orientation, and rapport with environmental groups to make him a relatively popular figure. He downplayed expectations after his appointment, explaining that much of the reported improvement in the river's quality was exaggerated. The recent spate of fish kills was largely due to the fact that previously there had simply been no fish in the river at all.[36] No one expected miracles, but they did want progress. And they got it.

A comprehensive master plan was drafted to preserve the river. Pargament, with the help of a broad coalition, got the strategy passed in principle through the labyrinth of planning procedures and the affected city councils. The operational objectives that the plan set for itself included drainage, ecological rehabilitation, preserving the "green lungs" for the most populated area in Israel, and changing the Yarkon's image "so that the public perceived it as a front yard rather than a backyard."[37]

With the personal interest of a concerned, Tel Aviv–based Minister of the Environment and his director general, the River Authority's budget grew. The restoration of the upper seven kilometers made great progress, becoming one of the better-kept recreational secrets of the greater Tel Aviv area. Many of the recalcitrant cities, such as Kfar Saba and Ramat ha-Sharon, began to make a serious commitment to sewage treatment upgrading.

Even before this, the Yarkon rehabilitation efforts were already hailed as a success story by Israel's Ministry of the Environment. A 1994 propaganda piece reads:

> The success of the rehabilitation program is already evident in the return of flora and fauna to a restored seven-kilometer stretch, in the development of boating and fishing areas, and in the eradication of mosquitoes using biological control. A few kilometers upstream, near Petah Tikvah, the National Parks Authority officially inaugurated the Mekorot ha-Yarkon (Sources of the Yarkon) National Park in October 1993. The park includes historic sites, a pastoral atmosphere, and riverbank vegetation with public access.[38]

Reconciling this rosy picture with the poisoned athletes is not simple. However, in small countries such as Israel it is not uncommon to have pristine natural enclaves adjacent to significant contamination.

RESPONDING TO THE YARKON CRISIS

At a senior staff meeting of Israel's Ministry of the Environment, days after the Maccabiah Games, the importance of a cogent government response was raised, and opportunities created by the tragedy were also pointed out. Historically, environmental disasters have spawned quantum progress. The "London Fog" killed 4000 people in 1952 but ultimately led to England's Clean Air Act in 1956 and the subsequent dramatic improvement in London's ambient air quality.[39] The toxic tribulations of residents in New York State's "Love Canal" neighborhood produced an enormous nationwide investment in hazardous chemicals cleanup in the United States through the establishment of Superfund.[40]

But the response to the crisis from Israel's Ministry of the Environment was remarkably sluggish. It is not clear whether the hesitation stemmed from not wanting to exploit the misery of the victims, or not wanting to implicitly acknowledge responsibility for the pollution problem, or perhaps, just not knowing what to do. At a formal level, sewage reaching the Yarkon River had been allowed by the Ministry of Health and Israel's Water Commissioner, which might have served to temper the virulence of these agencies' responses.[41]

By the end of the summer, the Environmental Ministry had changed its tune. The international media's focus on the Yarkon and related pollution stories were becoming embarrassing, with top periodicals running exposes like *Time*'s "The Filthy Holy Land."[42] Lack of clear information led to wild claims, such as an erroneous theory appearing in the Australian press that the river contained nuclear wastes. There was a need to muster some sort of response to the disaster. Environmental Minister Rafael Eitan, a former general who typically paid more attention to his second portfolio at the Ministry of Agriculture, eventually went on the record demanding 15 million shekels from the Finance Ministry for special Yarkon rehabilitation.[43]

Later, the Environmental Ministry announced an enforcement campaign against Yarkon polluters. But it was unsuccessful in identifying any point source polluters beyond the sewage from the cities of Ramat ha-Sharon, Kfar Saba, and Hod ha-Sharon.[44] No legal actions against polluters were ever filed.[45] Of the requested funds, only five million shekels were ultimately approved for conducting general exploratory research about the contents of the river's sludge. Director General Ronen explained: "It's a good start. Before you take out the sludge you have to know how and when. We've commissioned a report from some Dutch experts. At least Holland has reclaimed a few rivers and actually might know what they are doing. I just got back from visiting England and they explained to me that river reclamation is a matter that takes twenty years. Until the sewage flow is stopped, it's basically a hopeless issue."[46]

The follow-through was even less impressive. It took some three years for the modest public funds to be transferred by the Treasury to support an assortment of Yarkon-related research initiatives and projects.[47] Four years after the catastrophe, a thirteen-million-dollar compensation package was finally closed for Sasha Elterman and the other victims.[48] Baruch Weber, Deputy Director of the Ministry's Water and Streams Division, offers a pessimistic prognosis for the Ministry's efforts. "The problem is the legal standards for sewage treatment.[49] They may be acceptable as a generic effluent criterion, but if we're talking about a recreational resource, they just aren't relevant. With the sewage treatment facilities in place, chemicals can slip through the treatment process and will continue the mass destruction of aquatic life."[50] Indeed, the Ministry of the Environment recommends effluent standards that are twice as stringent.[51]

Hydrologist Dr. Dror Avisar is also pessimistic about the adequacy of the remedial measures pledged. The problem, he argues, will ultimately be maintaining the new sewage treatment plants. Municipalities are not always ready to pay for the expertise necessary to keep the level of treatment high. Even if, hypothetically, they are treated to a BOD level of 5 or 10 milligrams per liter and the Yarkon flow was entirely based on high-quality effluent, there is the problem of periodic "overflows." These are inevitable because of maintenance requirements or equipment breakdown. "If these overflows happen five times a year (a reasonable level of operation based on present experience from Tel Aviv), multiply it by seven cities. The river will die. It just can't handle this level of direct raw sewage discharge. That's after you've cleaned the water, brought back the fish, and rehabilitated the river. Without the lands for an overflow backup settlement pond, all the efforts will be for naught."[52] Rachel Adam, an attorney at the Ministry of the Environment specializing in water quality, agrees that zero discharge should be the ultimate goal. All treated sewage should be utilized by agriculture.[53]

As Israel's Water Commissioner for most of the 1960s and 1970s, Menahem Kantor was a central figure in national water policy. He has reached an even more discouraging conclusion: "There's no choice but to dry up the rivers of Israel. We don't have the groundwater available to dilute the streams. If you want a river full of water, it creates an illusion. The public thinks it can swim in this water even though it's wastewater. It creates the disaster that we saw at the Maccabiah."[54]

The Ministry of Health was much less hesitant in its response from the start. It took protective measures. On September 22, Shalom Goldburger, the Ministry's chief environmental health engineer, recommended a ban on all boating and fishing (and, of course, swimming) on the lower stretch of the Yarkon. The decision was based on water quality monitoring in the Tel Aviv section from April to August 1997 that showed clear violation of water quality standards. This effectively shut down a range of recreational activities that had been part of the Tel Aviv experience for years.[55]

Following the assassination of Prime Minister Yitzhak Rabin, the spontaneous creation of memorial sites through the lighting of traditional mourning *(yahrtzeit)* candles at the sites of tragedies emerged as a ritual of Israeli youth culture. All signs of the bridge were removed, but months after the event, the Maccabiah site became a shrine. Candles were scattered on the concrete blocks adjacent to the river bed. The death of four healthy athletes and the untold suffering of many survivors may not have inspired any meaningful policy changes, but it left in its wake a wave of graffiti on the blocks with sketches of fish skeletons and an ominous engraving: "This pollution kills."[56]

A RIVER AS AN ENVIRONMENTAL INDICATOR

Pollution killing athletes in a highly publicized international sporting event should constitute a watershed event in the environmental history of a nation. Unfortunately, this was not the case with the 1997 Maccabiah tragedy. The upshot of the events on the Yarkon from the perspective of the Israeli environmental experience is discouraging. Officially, boating has been banned, and roughly one million dollars in additional funds were allocated for theoretical research about pollution characterization and reclamation strategies. David Pargament at the Yarkon River Authority put together a long list of restoration projects that were ready to go if they could only get funded. However, his agency's budget actually went down in real terms during the eighteen months following the disaster.[57] Only lip service to Israel's rivers flowed freely, while the Yarkon's putrid waters continued to meander to the sea.[58] The message was one of chronic neglect, evasion, impotence, and resignation.

Most alarming is how closely the Yarkon River fiasco fits into an established pattern, as part of a long series of inauspicious ecological delinquencies. Israel's environmental history is marked with a disheartening number of environmental disasters. Epidemiological estimates suggest that over a thousand Israelis die each year from the fine particles in the country's ambient air.[59] Toxic exposures are almost certainly more hazardous to Bedouins living adjacent to the Ramat Hovav industrial zone and hazardous-waste disposal site. Only a fraction of the country's coral reef has survived the uncontrolled expansion of the city of Eilat. Scattered around Israel's towns and cities there may be thousands of "brownfields"—polluted and abandoned tracts of land, too contaminated for development. A considerable portion of the landscape of Israel, which for millennia inspired prophets and pilgrims alike, lies decimated by careless development and sprawl.

The Yarkon disaster is not even the most severe instance of water contamination. A decade earlier, some 10,000 residents of the Krayot district, north of Haifa, became sick, hundreds were hospitalized, and two died after drinking water contaminated by the sewage of the nearby city of Shfaram.[60] The laundry list of hazards goes on and on. As environmental experts were quick to point out, the Yarkon is not the most polluted river in Israel: The Kishon River, a receptacle for the acidic effluents of Haifa's petrochemical industry for decades, stands uncontested for this dubious distinction. When an epidemic of cancer among veterans of the Israeli Navy's elite commando frogman unit who trained in its waters was confirmed by a high-profile government inquiry in 2001, the river's notoriety was unsurpassed in the public's mind.

Israel's ecological record is particularly tragic because its environment is so unique. Despite its diminutive size (smaller than Denmark), Israel has an almost unparalleled combination of biological, climatic, and geological diversity. Sitting at the crossroads of three continents, Israel has as many indigenous bird species as are found in all of Europe, and many more types of bats. After thousands of years of human settlement, the sheer density of Israel's archaeological treasures is unmatched. Its status as a holy land makes its preservation of particular importance to Christians, Moslems, and Baha'is, as well as Jews.

Amos Keinan, the acerbic journalist and author, pronounced at a 1993 symposium held by the Society for Protection of Nature in Israel: "Two thousand years of conquerors didn't cause to the land of Israel a fraction of the damage produced by a century of Zionist settlement." At first glance it might seem that the Zionist quest for Jewish national rejuvenation through aggressive economic development has produced an appalling environmental legacy. The 1997 Maccabiah is but one small chapter. Any environmental history of Israel should have massive ecological deterioration as its point of departure and overall context.

But Keinan's indictment is also somewhat banal. There is hardly a nation on earth where the ecological destruction wrought by the twentieth century has not exceeded all previous anthropogenic harm throughout human history. Moreover, the magnitude of the damage is less important than understanding how environmental conditions deteriorated, if we are to ever begin the process of restoration. Understanding the Zionist vision is part of the diagnosis. How it might be transformed into a more environmentally friendly form must be part of the cure.

It is also important to recognize the many ways in which the Third Jewish Commonwealth has been a blessing to the land of Israel. Israel's environmental profile is far more complex than a one-dimensional, pessimistic snapshot would imply. Alongside the story of pollution and pernicious development is a parallel narrative, which offers hope. It is a story of unprecedented afforestation in a semiarid climate, of relatively extensive nature reserves, of comprehensive planning, of legislation, of solar innovation, of world leadership in water conservation, and even of enforcement against polluters.

There have been dramatic reductions in concentrations of sulfur dioxide in the cities.[61] The concentration of tar on Israel's Mediterranean beaches dropped more than a hundredfold, from 3.6 kilograms per front meter to 20 grams per front meter, during the 1970s and 1980s.[62] Even the Yarkon waters *are* cleaner today than they were ten years ago. Such changes do not happen by themselves. Hence, Israel's environmental history is also the story of a unique movement that, despite the growing population and economic pressures, often prevails in preserving its promised land.

Israeli environmentalists rarely refer to a formal environmental movement. The press frequently lumps together "the Greens," but in fact ecosolidarity is rare. The aphorism "two Jews—three opinions," which explains so much backbiting within Israeli society, certainly characterizes environmentalists.[63]

Ironically, many of these organizations, as well as other government and nongovernment environmental institutions, have attracted international admiration for their achievements. Their individual histories are as unique as the idiosyncratic country they work to preserve. Their stories are also the only way to truly understand Israel's environmental experience. While the Yarkon incident is in a sense a reflection of their shortcomings and collective failure, there are other cases in which the environment emerges triumphant. Understanding when and why these groups were effective, and when they were not, is a crucial part of Israel's environmental puzzle.

It is also instructive to consider the relationship between Israeli society and its environment. Increasingly, the environmental crisis is described as a crisis of values. An environmental history of Israel must consider the cultural paradox that continues to baffle activists and professionals alike. A country that justifiably boasts a national ethos for hiking, nature education, and a visceral identification with "the land," remains strangely detached from the ongoing ecological damage.

In the past, Israelis could be likened to the proverbial frog and the pot of boiling water. Al Gore conjures up this well-known metaphor in his book *Earth in the Balance*.[64] Reportedly, frogs have a very crude internal thermostat. If you throw them in a pot of boiling water, they will hop out, recognizing that their lives are in danger. But if they are put in a pot of lukewarm water that is slowly heated to the boiling point, they will boil and die, unable to detect the gradual increase in temperature. For its first forty years, as environmental conditions in Israel worsened, the public was seemingly unable to detect the deterioration.

The tremendous media attention, government sponsored publicity, and impact of pollution itself over the past ten years, however, has left little room for doubt in the Israeli public's mind that the environment is in serious trouble. *Yediot Ahronot's* New Year's "Survey of the Nation" on Rosh Hashanah in 1997 reported that environmental degradation was considered the single greatest impediment to quality of life.[65] A 1994 poll conducted by the Ministry of the Environment three years before the Maccabiah disaster showed that 63 percent of Israelis perceived the nation's rivers to be "an environmental nuisance."[66] And yet politically, the environment remains a "nonissue."

WHY STUDY ENVIRONMENTAL HISTORY?

No Israeli man of letters is more associated with environmental protection than Yizhar Smilansky. His novels under the pen name "S. Yizhar" may frequently be controversial, but are universally admired for their masterful descriptions of a lovely land. Smilansky is not just an observer. While serving in Israel's Knesset during the 1960s, he was largely responsible for passing landmark legislation that established a nature reserve system. Forty years later, however, he questions whether there is much utility for environmentalists in studying earlier periods in the country's history:

> When A.D. Gordon wrote about nature at the turn of the century, we were maybe 40,000 people and the few pioneers were scattered like seeds. In those days, it could take more than a day to get from Tel Aviv to Degania; today it can be done in an hour and a half. It has become a suburb of Tel Aviv. . . . Television and newspapers have made everybody urban. In my wildest dreams, I would never have imagined we'd be five million people in Israel. In this context, the Zionist ideology of those days isn't one you can actually apply. The children of farmers may feel close to the land, but they can't make a living unless they hire Thai workers to do the work for them. Whoever wants to find nature at the Kinneret today gets stuck in traffic. There's nothing left of that world.[67]

A century of astonishing technological development and population growth has changed the landscape (political and natural) in most countries beyond recognition. International influence is profound. Yet national environmental realities continue to be diverse, a reflection of the dreams, failures, and triumphs of the people and their land. Israel is also the product of myths and passions. While Israeli political leaders may rarely get outside of office buildings, they reflect the landscapes and ideals of their youth, which, despite Smilansky's intuitive pessimism, largely endure.

Israelis rarely wave the banner of Zionism. Except for politicians and school teachers, a Zionist label today is not a particularly trendy one. This cynicism is more a healthy response to the self-righteous, bombastic rhetoric of the nation's founding fathers than an actual reflection of Israeli patriotism. Youth are volunteering to challenging combat units in greater numbers than ever. Israelis continue to prefer to tune their radios to Hebrew folk and pop music than imported international tunes.[68] Even expatriates who live outside the borders of their homeland tend to be more passionate about what goes on in the Knesset in Jerusalem than in the national legislatures of their adopted countries. Also, even Israeli citizens who choose not to define themselves as Zionists still carry with them Zionist values and assumptions that inform their perspectives about life and the environment.

For example, Israelis take it for granted that the government owns 92 percent of the land, and that those who dwell there may only lease for periods of 49 years. The importance of water delivery to semiarid regions, the edifying effects of hiking, the desirability of large families, an intolerance toward high unemployment—all these are manifestations of a Zionist perspective that continues to shape the country's environmental conditions. Thus the source of Israel's pollution problem will be found beyond atmospheric or groundwater chemistry. These are mere symptoms. Rather, the problem begins with an ideology that was a reaction to European anti-Semitism during the late nineteenth century and blossomed into a pragmatic, national endeavor in the soils of Palestine. Israelis may have been the players who acted out the rebirth of a Third Jewish Commonwealth. One hundred years of unyielding Zionist determination and achievement unwittingly wrote the ecological script. While Zionism gave birth to a political and economic culture that often exacerbated the environmental impact of modernity, there are cases where it also provided the philosophical foundations for breathtaking ecological gains.

To understand Israel's present environmental problems, therefore, one must know its past—both the physical and the intellectual realities. One must also be familiar with the many efforts and organizations that sought to make the Zionist pursuit a more gentle one for the land, resources, and creatures of Israel. This book describes both, considering the local ecological challenges alongside the institutions and people who tried to influence Israel's checkered environmental history. Diverse in orientation, they share a common passion and patriotism. (It is not surprising that environmental leaders were primarily Ashkenazic men. The astonishingly high percentage of those who are or were members of kibbutzim is less predictable.) The book concludes with a closer look at some of the critical issues that will need to be tackled if Israel is to begin to move in a sustainable direction. Zionism, which has been part of the problem in the past, must also evolve.

Jewish tradition describes the past as a point of departure, as in navigating a boat on open waters. It enables one to guide one's course as the trip unfolds. If one never looks back, one quickly becomes lost.[69] A litany of avoidable environmental blunders in Israel's past suggests that by not studying history, the country has indeed repeated it. The tragedy wrought by the polluted Yarkon River will certainly be repeated in other, more pernicious forms if Israel's environmental challenges are not addressed. Priorities and perspectives must change. If the ingenuity, determination, and emotional power of the Zionist dream is at the heart of Israeli environmental problems, it is also true that a newly modernized, environmentally sensitive Zionism has the power to solve them. The Zionist view of the natural world and how it was manifested in pre-State Israel, therefore, offers a natural starting point to begin Israel's environmental history.

2 Reclaiming a Homeland
Zionism's Mixed Ecological Message

THE ZIONIST IDEA

To the extent that "history is the polemics of the victor," the environmental history of Israel in this century can be told through Zionist eyes. During the twentieth century, the Jewish nationalist movement and the state it established dominated the activities that most influence landscape, natural resources, human health, and the many creatures of the land. In some areas, this influence was quite formidable even before the military results of 1948. Moreover, the cultural legacy of Zionism and the attitudes of its leaders may hold the key to comprehending future ecological outcomes. Therefore, it is crucial to understand Zionism in order to understand "why."

Zionism has its spiritual origins in traditional Jewish beliefs and practices. Three times daily, prayers express the longing to return to the promised land. This yearning is particularly acute when the destruction of the Temples and the origins of exile are commemorated each year on the Ninth of Av. Seasonally based Jewish festivals, reflecting the weather of the ancestral homeland, often make little sense in European and American climates.

During the centuries of exile, the land of Israel, *Eretz Yisrael,* itself took on mythical dimensions. Jews around the globe, suffering endless cycles of persecution, relief, and again persecution, found comfort in visions that reflected local aesthetics and bore little resemblance to the Middle East of the nineteenth century. The poet Micah Joseph Lebensohn offered a typically green idealization of the Holy Land:

> Once in a leafy tree, there was my home.
> Torn from a swaying branch friendless I roam.
> Plucked from the joyous green that gave me birth,
> What is my life to me and of what worth?[1]

When Theodor Herzl wrote *The Jewish State: An Attempt at a Modern Solution to the Jewish Problem* in 1896, he was only one of a long line of Jewish intellectuals advocating a political solution to the chronic problem of European anti-Semitism. Yet the disappointment left by false messiahs, the growing hope for assimilation in an enlightened Europe, and a general sense of impotence made previous calls to "return to Zion" during the 1800s seem like wistful fantasies. Herzl's book and his energetic follow-up, on the other hand, were a spark that lit latent Jewish aspirations, translating spiritual longings into concrete political action. A year later Herzl presided over the First Zionist Congress, where representatives from around the world adopted his program to "create a home for the Jewish people in Palestine."[2]

Already the first wave of pioneering settlers, or *aliyah,* from Europe had arrived in Palestine. There would be five identifiable *aliyot* before World War II. The first immigrants brought with them little more than a fierce ideological commitment "to build and be built" (as the folk song went) in their ancient homeland. From the start, there were many schools of Zionist thought, and even Herzl could not squelch the incessant philosophical and political squabbling. But it was a diversity born more of passion than of pettiness.

The majority of Zionists who moved to Palestine shared a visceral rejection of the Jewish condition in the Diaspora: not just the chronic physical vulnerability and political feebleness but also the pervasive mentality of alienation. After two thousand years of wandering among inhospitable hosts in dozens of nations around the earth, Zionism envisioned Jews once again putting down roots in a single land, the land of their forefathers. In today's rhetoric, the impulse was to reclaim their status as "indigenous people."

The Labor Zionists, who arrived between 1904 and 1923 in the Second and Third Aliyot, sought to redefine Jewish identity from what they perceived as the rabbinic distortion it had been given in exile. This cohort's Labor Zionist perspective is particularly important. Although the Second and Third Aliyot comprised fewer than 10,000 people, their socialist viewpoint soon came to control key institutions of the Jewish settler population (the *Yishuv*), thereby dominating cultural and political life in Israel for the next seventy years. Many of these Jewish East Europeans identified with Tolstoy's idealization of peasants' connection to "Mother Russia" and with political leftists who saw in them the most promising revolutionaries.

Yet, the Zionists' revolution was not directed against their entire heritage; the settlers actually sought to reclaim "the Land" by returning to "the Book." The Bible held the key for many Zionist pioneers in their search for a more authentically Jewish or Hebrew identity. David Ben-Gurion was representative of his generation when he wrote: "I believe that the inspiration of the Bible sustained us, returned us to the land, and created the state. . . . All of my humanitarian and Jewish values I drew from the Bible."[3]

The land of Israel was the other inspiration for producing a renewed Jewish identity. Many settlers shared a biblical view that sees the land of Israel as imbued with human characteristics. It is not a neutral setting. The land enjoys the right to rest periodically and is a player in the drama between the people of Israel and their God. Pastoral images from the Scriptures envisioning "every man under his fig tree" gave the land a role to play in the drama of spiritual redemption.[4] Zionists had come home to redeem her and be redeemed. The Zionist poet Saul Tchernichovsky's oft-quoted aphorism, "Man is an image of his homeland's landscape," was more than a slogan. After kissing the holy ground upon arrival, immigrant pioneers set out to become acquainted with their motherland and thereby to discover their own identity. Here again, the Bible offered the ultimate road map in this spiritual journey. Among the more powerful rituals adopted by the early Zionist settlers were their hikes throughout the countryside, which they called "discovering the landscape through the Bible."[5]

It is hard to say just how much the Bible actually influenced a Zionist ecological viewpoint. The ecological merits of the biblical perspective are disputed.[6] The Scriptures do mandate ecologically protective practices, such as prohibition of cruelty to animals and limits on the destruction of fruit trees during wartime.[7] But others seize on Adam's anthropocentric dominion over all the other creatures as the West's first slide down the slippery slope toward environmental oblivion. Indeed, the same Scriptures condone slavery, animal sacrifice, and execution of witches, all of which mercifully remained alien to the new Zionist culture. Ultimately, any actual ecological message of the Bible, and the Bible's influence over the ideology and policies of the Yishuv, were certainly colored by other contemporary philosophical trends, especially Romanticism.

THE ROMANTICS

Until recently, scholars did not systematically consider the early Zionist perspectives on environmental issues; the definitive study on the subject remains to be written. This left many Israelis with the impression that classical Zionism constituted the "antithesis" to a sound ecological ideology.[8]

The enormous ideological variation within the Zionist tradition makes blanket categorizations impossible. Many point to the writings of Aaron David Gordon, the Romantic "prophet" of the leftist labor movement, to support the argument that early Zionism was in fact ecologically progressive. Gordon (1856–1922) came to Israel at the age of forty-eight. He quickly captured the imagination of the younger pioneers through his diligence and mystical belief that human liberation could only come through manual labor.

Today, Gordon might be typed a New Age eccentric: a vegetarian with a long flowing beard. He was essentially apolitical, maintaining that only immediate personal deeds could lead to individual salvation. For Gordon, working the soil assumed a cosmic significance, binding man not only to nature but to the great All. "It is clear that man-as-man always needs to be among nature," he wrote. "For nature is, for a man who feels and knows, truly what water is for a fish. Not just something to look at. Man's very soul is in need of it."[9] While considered a visionary, Gordon was by no means a marginal personality. By the time of his death, mainstream Zionist institutions hailed Gordon as a cross between the Baal Shem Tov (the legendary Hasidic rabbi) and Tolstoy.[10]

If Gordon was the prophet of the Romantic school in Zionism, "Rachel the Poet" (Rachel Blubstein) was the psalmist of the Second Aliyah. Her lyrical, frequently haunting poems of unrequited love and lonely landscapes drew many metaphors from Lake Kinneret near her home. Chronically infirm and frail, she would die young of tuberculosis in 1931, but not before leaving a rich body of work—many paeans that expressed a generation's devotion to its new homeland. For example, these lyrics became popularized with a Naomi Shemer melody:

> The Golan Heights are over there, stretch your hand and touch them.
> They order you to halt with silent confidence. . .
> There is a low palm tree on the lake's shore, its hair disheveled as a naughty baby. . .
> Even if I lose my fortune, broken, lose my way, and my heart became a foreign beacon,
> How could I forget, how I could I forget the kindness of my youth?[11]

Of course, the very heights to which they aspire make all Romantics vulnerable to a painful fall. Zionist Romantics are no exception. Tel Aviv University's Izhak Schnell finds a great ambivalence toward the natural world among the so-called Romantic settlers of the Second and Third Aliyot. Examining diary entries and letters, Schnell sees a pattern of initial euphoria and subsequent disillusionment. In support of his view, he cites the diaries of pioneers like Rachel Yana'it Ben-Zvi, who was probably the most prominent feminist of her generation in the Yishuv: the first female member of the Jewish defense organization Ha-Shomer, a leftist political leader and essayist, a cofounder of the Gymnasia High School in Jerusalem, and the wife of Izhak Ben-Zvi, Israel's second president. Soon after arriving in the country at the turn of the century, she writes of the Judean desert: "What natural fortunes are these mountains concealing? I am mainly attracted to the spirit of the prophets that are hidden between their cliffs rather than to their beauty." Then, days later she is seized by anxiety and despairs at the land's desolation.[12]

Other pioneers describe the relentless heat, the thorns, the stone-filled fields, and the sense of alienation caused by an unfamiliar and unfriendly climate and the increasingly hostile Palestinian Arabs. Yet, as in many relationships that become troubled, the original love prevailed.

The Romantic school had a natural ally in the National Religious Zionists of the period. Beginning with the mystical writings on national redemption of the Belgian Sephardic rabbi Judah Alkalai in the mid 1800s, these Orthodox calls for settlement actually preceded Herzl's. By design, they were a less politically powerful faction than other competing parties. Yet, the unswerving reverence for the land of Israel proved highly influential in keeping the Zionist movement focused on Palestine as its geographical destination. In their eyes, everything in the land truly was holy.

The central figure of religious Zionism during the first half of the century was Avraham Kook, the venerated chief rabbi of the Yishuv. A man of unique intellectual and spiritual faculties, his tolerance and indeed affection for the nonbelieving agricultural pioneers set the tone for the National Religious communities in Israel for many years. Kook was comfortable with the biblical mandate that granted dominion over the natural world. Kook explained that the intention was not the "dominion of a tyrant who deals harshly with his people and servants" but rather a dominion comparable to that of God "whose mercy extends to all creation."[13] For many of Kook's followers, vegetarianism was ideological. Innumerable anecdotes describe his deep compassion for a natural world that God had created and that man had no right to destroy.[14] Perhaps the most famous tale involves his sudden rebuke of a follower for thoughtlessly picking a wildflower, fifty years before the issue became a rallying cry for the incipient environmental movement in Israel. The rabbi explained: "You know the teaching of the Sages, that (there) is not a single blade of grass below, here on earth, which does not have a heavenly force telling it 'Grow'!

Every sprout and leaf says something, conveys some meaning. Every stone whispers some inner hidden message in the silence."[15]

The political philosopher Avner De-Shalit feels that a Romantic perspective such as Gordon's is more precisely defined as "ruralism." Ruralism is a moral glorification of rural or country life and a rejection of urbanism, not only in the purely ecological sense (e.g., as a source of pollution), but also as an inferior moral condition or even a state of decadence. Only when the massive wave of refugees from Hitler's Europe created new economic pressures did Zionism deviate from this aggrandizement of country life. De-Shalit argues that there were three major ideological phases in Zionist history: the initial vision of romantic ruralism, replaced in the 1930s by a development ethos, and modern environmentalism, a synthesis of the previous two that has only recently begun to capture Israeli hearts and minds.[16] Progression through these three stages was in his view ineluctable.

Interestingly, most of the academics who analyze this subject are themselves native-born Israelis ("Sabras") and describe the pioneers' views with an appropriate sense of historical detachment. Yet the ambivalence and anxiety described by commentators like De-Shalit and Schnell are immediately recognizable to the immigrants to Israel who still constitute about 50 percent of Israel's population.

Eilon Schwartz emerged as one of Israel's profound figures in environmental education during the 1990s, after playing a leadership role in the American Zionist movement. Schwartz criticizes De-Shalit's dialectical perspective as simplistic. Zionism, he argues, has always been a mirror of general trends existing in the West. At the turn of the century, Europe was torn between a rationalist tradition and a Romantic one. This tension is firmly entrenched in Zionist theory. Zionism in its original form was an expression of Romanticism and its general veneration of the uncorrupted natural world. At the same time its practitioners were influenced by a rationalist view with the implicit belief in the human ability to control (and improve upon) nature. This conflict existed not only between identifiable ideological camps but also within the complex psychologies of the individual settlers.[17]

David Ben-Gurion is representative of the competing philosophical impulses among settlers during the early aliyot. On the one hand, because Palestine needed builders, as a teenager he wished to delay immigration until he could acquire an engineering degree. (His formal studies in Warsaw never materialized, either because of limitations on Jewish enrollment in Polish universities or possibly because of a romance that drew him back to his native Plonsk.)[18] Yet, the would-be engineer wrote his father about the existential mysteries of plowing the land of Israel: "This soil that stands revealed in all its magic, and in the splendor of its hues, is it not itself a dream?"[19] This dynamic tension was to continue throughout his political career. During the 1960s, Ben-Gurion would deliver an impassioned speech in Israel's Knesset extolling nature and its preservation.[20] At the same time he personified a school of Zionism that embraced the Baconian model of science and human ascendancy.[21]

The Romantic tradition in early Zionism may have become a minority view in the operational decisions of the Yishuv. But, according to Schwartz, it wielded a special influence because it was the "heart and soul" of the movement from which the rationalistic camp drew its inspiration. Undoubtedly, the ruralist impulse remained very strong. Even though most Israelis were urban dwellers themselves, a 1949 public opinion poll indicated that almost all felt new immigrants should be directed to agricultural settlements. Half the respondents felt the immigrants should be forced to move there.[22]

Accordingly, the Romantic stream's passion for a harmonious relationship with the land remained consistent throughout the century and was relatively unaffected by the economic exigencies of any given period. The Society for Protection of Nature, which became Israel's largest membership organization in the 1950s, is an authentic expression of the Romantic tradition. Schwartz argues that the two competing paradigms continue to merge in individual psychologies. For instance, many of Israel's environmental scientists are fundamentally Romanticists whose love of nature inspired them to pursue biological studies. At some point in academia they "bought into" the trappings of a more rationalist approach, but their hearts remained with the land.[23] Similarly, one can see Ben-Gurion and the secular Zionism he represented as advancing Romantic objectives through instrumental rationalism, in the same way that contemporary religious fundamentalists enthusiastically utilize the latest technologies.[24]

THE RATIONALISTS

Uri Marinov, who headed Israel's fledgling Environmental Protection Service and later its Environmental Ministry, is fond of telling a story about his father, who in his day was an influential Labor Zionist leader. Whenever they would pass by polluting factories and the younger Marinov would criticize the emissions, his father would respond: "It's good to see smoke coming out of the stacks. It's a sign that they're working."

Those who had responsibility for the Zionist endeavor faced enormous pressures to find work for the hundreds of thousands (and eventually millions) of Jews who joined them in the task of creating a state. Just as the aspiration to return to the land was wrapped in ideological significance, so was the development of a modern industrial economy.[25]

Development per se need not be synonymous with environmental devastation. However, many argue that Zionist development has always been of the particularly aggressive, environmentally unsustainable variety. In both academic and casual discussions about Zionism and the environment, the texts of folk songs from the 1920s and 1930s are invariably invoked to authenticate a given perspective. None is cited more frequently than the homage to urbanization and construction penned by the Yishuv's most topical poet and commentator, Natan Alterman: "We shall build you, beloved country . . . and beautify you. . . . We shall cover you with a robe of concrete and cement." Countless other poems, songs, essays, and plays from the period celebrate the heroic human domination of a recalcitrant land that must be subjugated.

Attempts to attribute this aggressive development ethos to the Mapai, or socialist wing of Zionism, ring false, despite its control of the Yishuv prior to the State. The economically ambitious, rational viewpoint was typical of an entire generation. Yitzhak Shamir, a lifelong right-wing, "revisionist" Zionist, agreed about very little with his leftist contemporaries and rivals, but the environment lies outside the realm of their ideological controversy. As prime minister, Shamir oversaw the formation of a Ministry of the Environment, but in retrospect was decidedly unsentimental on the subject. "They talk about clean air and natural resources and that's all very important. But on the other side, there is development. I mean, why have we come here anyway? To bring the Jewish people here back to the land of Israel. To do this we need development. Ultimately, in the name of development, I am willing to sacrifice anything."[26]

Avram Burg, the thoughtful speaker of Israel's Knesset, cites an innate aggressiveness in the Hebrew language when it refers to the relationship between man and land. For Burg, the words evoke a model of chauvinistic male domination over females. "Conquering the land," the modern Hebrew rhetoric for settlement, is adopted by Israeli Don Juans when they speak of conquering a woman. An owner, or *baal,* of land has the same title as "husband," which also means fornicate (one of the ways that a man can legally acquire a wife).[27] Others have noted that even the seemingly innocuous expression for geographical expertise, *yediat ha-aretz,* technically means "knowing the land," which in turn connotes an act of sexual possession, "knowing" a woman.[28] There are of course other ways to wield the Hebrew language on the subjects of land and nature. Some argue, for example, that the uniqueness of Gordon's alternative style and use of maternal metaphors represents something akin to an ecofeminist approach.[29]

In any case, most experts would agree that hairsplitting over the monolithic character of a very heterogeneous political movement is silly. They also would agree that there was a decline in the practical influence of the Romantic approach to nature in the period following World War I. The explanation may be more demographic than philosophical. The immigrants arriving in the Fourth and Fifth Aliyot (1923–1939) were mercantile, professional, and decidedly more urban in their orientation than their predecessors.[30] Although much is made of Nazi Germany's contribution to immigration, the 1924 anti-Semitic economic restrictions imposed by the Polish government a decade earlier led scores of bourgeois Jews to liquidate their assets and move on. With the United States already imposing immigration quotas, many immigrants brought their middle-class values to the new settlements of the Middle East. Besides their money, they carried an appreciation for formal education, fine arts, and material comfort. The urban immigrants were also less enamored of bucolic landscapes and felt little need to discover their essence as human beings in a direct relationship with the soil and the natural history of the Holy Land. However, one-dimensional caricatures are inappropriate. These refugees from Poland and Germany were not dispassionate about the wonders of the new countryside, but they also found beauty among civilization's buildings.

"The old-time Revisionists were city people," writes the novelist Amos Oz.

> They didn't travel to the village. The smell of the manure, the fragrance of the hay was not to their noses' tastes. My grandfather lived in the land of Israel forty-five years and never was in the Galilee or went south to the Negev. . . . But the land of Israel he loved with all his soul, and he wrote love poems in her honor (in Russian). He loved it as it was revealed to him from his window. A piece of stone, mountain's edge, the sunsets of summertime, the light of a geranium in the courtyard, the lizard on the stone wall, and two or three birds whose names he never bothered to learn on the branch of a tree whose name he never learned.[31]

DESPERATELY SEEKING HOME

There is a rich sociological literature about the mixed emotional profile that pioneer immigrants have concerning their new surroundings: awe, alienation, inspiration, hostility, and, of course, anxiety.[32] The Zionist pioneers shared many of these feelings. At the same time, theirs was a uniquely Jewish pioneering experience. The very word for immigrants in Hebrew—*olim*, or "they that have ascended"—suggests a privileged status.

For many first-generation Israelis, moving to the proverbial "homeland" is the culmination of a lifelong personal dream. For example, Adam Werbach, elected president of the largest U.S. environmental activist group, the Sierra Club, in 1996, is Jewish but never a particularly active Zionist. Yet when the organization decided to install an environmental hotline using 1-800-HOMELAND as a toll-free number, he felt peculiar, because his automatic association with the word was Israel.[33]

The very depth of the immigrants' aspiration created complex psychological dynamics. On the one hand, they had an intellectual sense of coming home to the well-known landscape of the Bible, the cradle of Jewish history. On the other hand, particularly for the European immigrants who made up the bulk of the Yishuv, little about the land was truly familiar. Coming to Israel therefore was more than simply learning to appreciate the sparseness in local flora, different species of animals, scorching summers, unfamiliar smells, and an entirely new aesthetic. It also involved a host of less glorified transitions: diarrhea from the unreliable food and water supply, disease-carrying insects, a peculiar diet, and, of course, a foreign language. Part and parcel of the "Promised Land" travel package were malaria, the threat of violence and theft from neighboring Arabs, and abject poverty. Hunger was also part of the aliyah experience. Many who had left behind their disapproving families in acts of youthful rebellion also carried enormous anxiety about falling short of the self-righteously vaunted ideals they had preached.

For the majority, the disparity between the dream and the reality of daily life was too great. David Ben-Gurion, for example, estimated that 90 percent of the original Second Aliyah immigrants left Israel, unable to overcome the enormous challenges of adaptation.[34]

A nagging nostalgia often left even the ostensibly successful Olim uncertain: Was the decision to come to Israel final, or might it be more sensible to return to the lands of their birth, which constantly beckoned? One apocryphal story tells of a cabinet meeting opened by Prime Minister Levi Eshkol, who wryly asked his fellow ministers: "*Nu*, when can we finish the job here already so we can go back home to Russia?" Even the unshakable Ben-Gurion struggled occasionally with mixed emotions.[35]

For many, a key element of this nostalgia was the lush scenery of Eastern Europe. One way to overcome the sense of alienation and the resulting cognitive dissonance was to transform nature into a more hospitable, or at least more familiar, form. Cultural manifestations of settlers' desire to transform local conditions are expressed in the fundamentally European styles of music, art, dance, theater, and literature of the Yishuv. Efforts to integrate local, Middle Eastern culture were only symbolic and peripheral. Physical expressions of this dynamic were just as common; it is not by chance that Tel Aviv bears more resemblance to Europe than to Jaffa or even West Jerusalem. The most insulting description that Meir Dizengoff could imagine for his beloved Tel Aviv was for it to be labeled a Levantine city.[36] The commitment to landscaping and lawns within the new kibbutzim, giving these communities a European "public garden" feel, was not coincidental. Richard Kaufman, a German Jewish architect with landscaping experience from Europe, was commissioned to design the new settlements.[37] Kaufman did not hesitate in applying his foreign perspective. Those who worked with him during the 1930s remember little more than lip service to the idea of integrating the land's natural layout into the designs of new settlements.[38]

The pre-State slogan for agricultural settlement, "Conquering the Wilderness," implied combat, with Zionists battling on behalf of an imported vision of natural beauty. While the pioneers may have perceived themselves as liberating or returning the Holy Land to its former glory, they could not mistake the exogenous geographic origin of the eucalyptus trees and conifers they brought to the task. Moreover, although their burning desire to "know" *Eretz Yisrael* was sincere, not all immigrants had the tools, free time, or transportation to do so. It is not surprising, then, that the ecological perspectives appearing in the writing and discussions of first-generation settlers lack the authenticity that comes from specificity. Acquiring the biological, geological, and zoological intimacy of natives is no easy task. Many of the settlers were informed by what they had read, and the myths upon which they were nurtured abroad, rather than by the actual ecological reality of the land.[39] Perhaps the most telling critique of Romantic Zionist environmental views comes from zoologist Heinrich Mendelssohn. Now considered the father of ecological science in Israel, Mendelssohn arrived in Palestine in the 1930s, when Gordon's mythological status was at its apex,[40] yet he cannot suppress a condescending smile when Gordon's view of nature is mentioned: "Gordon was completely unrealistic, a total Romantic. He didn't know the first thing about the natural world."[41]

Toward Indigenousness

It may have been their own sense of inadequacy, as well as old-fashioned patriotism, that made first-generation pioneers so committed to teaching their children about the land of Israel. The Hebrew word for "homeland,"

Muledet, literally means " birthplace" but connotes homeland, like the Russian *Rodina,* "Motherland."[42] Indeed, comfort with the Middle Eastern environment influenced the ecological outlook of the Zionist second generation, who were in fact born of the land. It is fascinating to read the impressions of the children of First Aliyah settlers and compare them with their contemporaries, European immigrants from the Second Aliyah.

"As a will, bequeathed for future generations, great is the work and the concern that man wrapped up his inheritance, the estate that he received from nature." Thus one of the first Sabras, Avshalom Feinberg, described his impressions in 1914, while traveling in the north of Israel. "The entire way, the Druze and the Christians (especially the Druze) show the requisite virtues to be our teachers, teach us the practical art of loving our land (an art that exists with us, unfortunately, as only a theoretical sprout)."[43]

Familiarity with the land of Israel and its natural wonders was not colored by political affiliation; it was considered a virtue by Zionists from all backgrounds. Declared a compulsory topic in 1923 by the Yishuv's central committee, a series of Hebrew *Muledet* textbooks were printed in Hebrew in the 1920s and 1930s.[44] Teaching nature studies became a particularly prestigious area of instruction. Azariah Alon, one of Israel's most eminent conservation leaders for fifty years, ascribes much of his success in lobbying government ministers and politicians to his prestigious status as a teacher of nature at Kibbutz Beit ha-Shita in the Harod Valley.[45]

When Alon reflects on the origins of his own environmental consciousness, roaming freely as a child at Kibbutz Kfar Yehezkel in the 1920s, it almost conjures up a Rousseau-like state of nature.[46] Physical proximity to nature was then supported by classroom theory and later supplemented by extensive field trips, or *tiyulim,* at every stage.[47] This focus, particularly during the early grades, continues to this day, leading to the common Israeli phenomenon of immigrant parents learning not only Hebrew grammar but the names of plants, birds, and insects from their children when they return from kindergarten.

The initial meeting between children of the Yishuv and nature was broadened geographically and elevated to even greater prestige within the youth-movement culture that was such an essential part of the children's socialization experience. Even today, much of the appeal of Israeli youth movements for the hundreds of thousands of participating adolescents is the opportunity to leave the confining framework of the city and enjoy the freedom that nature imparts. With vacations spent backpacking or at work camps on kibbutzim, the idealization of nonurban natural living became a key component of the psychological makeup of many adolescents. It helps explain the well-known Israeli obsession for hiking and international travel.

This mixing of natural and national impulses before the establishment of the State reached a peak in the training of the secret Haganah militia, especially in the elite Palmach units, where field skills had particular utility.[48] The intense Romantic idealization of nature produced a culture that sociologist Oz Almog defines as "pantheistic," with reverence for nature serving as nothing less than a secular religion. These values are reflected in the eulogies to the fallen fighters of the period, where sensitivity and competence about the natural world were the grounds for the highest praise. Almog quotes a number of these memorial books whose tributes are both moving and distinctively nonmilitaristic. "More than all of us, Moshe was a child of nature, in the deepest meaning of the word. Powerful senses that felt the soul of nature and its most subtle movements. Deep love and connection to plant, to bird, to stone, to the landscape, to the land."[49] Immigrant pioneers did not honor their dead friends with such accolades.

AN ISRAELI ENVIRONMENTAL ETHIC

When considering the course of Israel's environmental history, the intellectual is no less important than the physical profile. The Yishuv succeeded in leaving a distinct environmental ethic as an integral part of the Jewish national culture it reinvented.

It is sometimes argued that Israel's environmental problems stem from a failure of Zionism. According to this view, Israelis still carry with them a conscious or subconscious alienation from public spaces which they inherited from the Diaspora. While in exile, the Jew felt unsafe when venturing out into the street or the forest, and thus the home became a fortress—clean, safe, and totally cut off from the natural world. The well-known phenomenon of Israeli apartment buildings, with their filthy corridors and stairwells but immaculate personal dwellings, supposedly reflects this mentality.[50]

It is impossible to define a single Zionist environmental perspective that emerged from the first half of the twentieth century. Frightened alienation, however, hardly seems to characterize the thinking of the first Israelis at midcentury. Diverse and conflicting ways of thinking about the natural world frequently broke down along philosophical and generational lines. There was no shortage of inconsistencies, because cultures—like people—are best defined by their contradictions. The sundry intellectual traditions that colored Israeli and particularly Sabra environmental values may have marveled, embraced, or attacked the land, but nobody hid from it.

The ideology of the young country was a peculiar amalgam of Tolstoyan agrarianism and sturdy faith in enlightened rationalism and technocracy—beliefs that had already largely vanished from Europe.[51] Steeped in such competing and simplistic paradigms, decision makers lacked both the theoretical tools and the sense of caution needed to understand and confront the environmental neglect whose consequences would soon come home to roost. At the same time, Zionist education and the Yishuv experience unquestionably produced a substantial group of Israelis with a unique indigenous sensitivity to their natural world.

Whether this ecological consciousness was pantheistic, Romantic, or distinctively Jewish, the passion for intimacy with the land of Israel continues to influence much of Israel's population. It explains why the Society for the Protection of Nature quickly became the largest membership organization in the country. One hears this consciousness in folk songs. One sees it when looking for a vacant stretch of beach at Lake Kinneret, or for an empty trail during an Israeli holiday; everybody is out trying to get in touch with the land. Even when competing policy considerations such as security or job creation for immigrants trump environmental sentiments, the commitment to preserving the landscape of Israel remains a widely held ethos.

Israeli indigenousness is reflected in lay citizens' mastery of local flora, fauna, and geography alongside that of specialist scholars. Meir Shalev is a prize-winning Israeli author whose intense interactions with nature during his childhood are a constant theme in his work. Of the many honors bestowed on him, none thrilled him more than an award from the Israel Entomological Society, in recognition of the meticulous and fascinating descriptions of insect activity in his best-selling novel *The Blue Mountain*.[52] Shalev contends that, although he has visited many scenic lands around the world, he still continues to find the Mediterranean landscape more compelling and personally moving than other, objectively more magnificent, ones.[53]

Meron Benvenisti, a noted politician/historian, echoes this sentiment: "This land is part of me and I am part of it. My American friends laugh when I tell them that the flowers and trees in Central Park seem fake to me."[54] Benvenisti, a product of a sabra education, boasts that even as an amateur, he was so familiar with the land that in 1959 he dictated from memory complete hiking guides for a five-day walking trip from Beer Sheva to Masada and a seven-day walking trip to Eilat. Later he would become the leftist deputy mayor of Jerusalem under Teddy Kollek. This political experience undoubtedly influences Benvenisti's belief that there is a darker side to the traditional Zionist reverence for nature.[55] He argues that, as with other Romantic movements, naturalism provides intellectual justification for chauvinistic attitudes in modern Israel:

> Our sensitivity is to things. We are obsessed with the landscape. The Palestinian Arabs who dwell in it are viewed as part of its natural features—a kind of fauna: objects not subjects. *Muledet* textbooks are full of Romantic descriptions of the Arabs, their customs and folklore are always perceived as an integral part of the scenery, never as a legitimate entity in their own right. . . . (I)n the wake of the 1967 war, all this added momentum to the growing religiomessianic attachment to Eretz Yisrael.[56]

Benvenisti's perspective is interesting but counterintuitive. The Labor Zionist culture that produced this fascination for nature has been more conciliatory to Palestinian interests than to competing Zionist schools. Right-wing *Gush Emunim* (Bloc of the Faithful) settlers in the West Bank may play on the tradition of *Muledet,* but its followers are primarily driven by an orthodox messianic zeal. For many Israelis, green interests are often (and not without some empirical justification[57]) equated with leftist ones. Moreover, commitment to nature is an area of cooperation between Israelis and Arabs that offers a modest basis for reconciliation and meaningful cooperation. To the extent that objectification and chauvinism toward non-Jewish minorities exist, it is a rational, albeit regrettable, response to a century of mutual suspicions, conflict, and violence.

The Israeli ecological perspective shares little with the deep ecology philosophy that holds appeal for many environmentalists in Europe and North America. To begin with, Israelis have the physical constraints of living in very crowded conditions. Even the most venerated local environmental leaders perceive idyllic, "ecologically pure" lifestyles as unrealistic. "I don't recommend that anyone return to the stage before civilization," warns Azariah Alon. "If anyone wants to, it's their problem. I think it will be very hard to do it in this country. And if there is room for ten people like this, the eleventh such person will have to ask himself, Where is there room for me?"[58] With the human imprint everywhere, "pristine" is invariably a misnomer for Israeli landscapes.

Rather than eschewing technology, Israeli environmentalists for the most part welcome it. Technical resourcefulness holds the key to improving environmental quality in such a crowded, high-consumption country. There is broad support for waste-to-energy incineration, desalination, high-speed rail, and extended tunnels to save scenic countryside from ugly highways. Thus, Israeli environmental ideology is also a product of the rationalist side of Zionism. There is an unmistakable pragmatism that creeps into the Romantic views of most Israeli intellectuals associated with green causes. Amos Oz, Israel's most famous novelist, perhaps best captures the paradoxical environmental ethos that Zionism ultimately produced:

> And now it is my turn for a terrible confession. I object to nature preservation. The very ideal of "preservation" is not acceptable in almost any area of life. We have not come into this world to protect or preserve any given thing, mitzvah, nature, or cultural heritage. . . . We have not inherited a museum, to patiently wipe off the dust from its displays or to polish the glass. . . . Nature also is not a museum. One is allowed to touch, allowed to move, to draw closer, to change, and to leave our stamp. . . . Touch the stone. Touch the animal. Touch your fellow man. On one condition. How to touch? . . . 'on one leg,' and in a word, I would say: 'with love.'[59]

This love, shared by so many Israelis for their land, was the prime weapon that Israel's environmental movement would carry in the many trials ahead.

3 Palestine's Environment, 1900–1949

Prelude to Disaster or Benign Half-Century?

There may be no landscape in Israel that inspires as much national pride as that of the Jezreel Valley. This vast triangular patchwork of field crops and orchards is encircled by rounded hills with names like Tabor and Gilboa that evoke a majestic biblical past. The fertile soil is dark and heavy, rich in organic matter,[1] yielding a cornucopia of produce that celebrates a successful partnership between humans and this goodly land. Natan Alterman's lullaby from the 1930s remains a popular standard even today:

> The sea of grains sway. The song of the flocks rings out.
> This is my land and her fields. This is the Valley of Jezreel.

One can hardly think of scenery more quintessentially Israeli and precious to the national heritage.

Yet the Jezreel Valley at the start of the twentieth century was an undeveloped swamp[2] described as "barren and boring, a miserable plain without a tree or river . . . , all in ruins."[3] In 1905, Gertrude Bell, the American diplomat/traveler, described mud so deep that the mules fell and the donkeys almost vanished in it: "'Lord,' said the mule driver, 'One cannot even see the mules' ears.'"[4] It was not until 1924 that the Zionist settlers completed the arduous two-year task of converting the wetlands to the arable farmland of today.[5]

To frame the history of the Jezreel Valley and the other results of pre-State Zionist activity in modern environmental terms is at best a highly subjective venture. The pioneering Jews sought to "redeem" their promised land, and change it they did. By 1948, the 8 percent of Palestine's soil acquired by the Yishuv bore little resemblance to the lands when they were purchased. The transformation, however, did not necessarily leave the land less healthy.

After almost four hundred years of relative equilibrium under Ottoman rule, the first half of the twentieth century was a tumultuous period for the land of Israel.[6] The population increased more than fourfold, from roughly 400,000 people to 1.8 million (prior to the massive Arab migration of 1948).[7] Violence was common. The Turkish defeat in World War I ushered in British colonial rule. Habitual Arab rioting and later Jewish armed resistance hastened an unceremonious British departure that in turn sparked Israel's sanguinary War of Independence. Twentieth-century technology also left its inevitable stamp with the introduction of factories, electricity, motor vehicles, telephone infrastructure, airports, and the cultivation of hundreds of thousands of acres. Over three hundred new Jewish settlements sprang up during this time, providing the operational base for the new State of Israel.

What long-term effect did the vicissitudes of the first half of the twentieth century actually have on the land of Israel and its natural resources? A dispassionate analysis within an environmental history requires objective criteria. Three factors—population, affluence and technology—are typically associated with deterioration in environmental quality.[8] These factors will drive the discussion about the environmental consequences of Zionism and parallel activities in pre-State Palestine. Their derivation also requires a cursory understanding of the economy that blossomed during these years.

Judged by objective criteria, the impact of these ecologically turbulent years was limited in scope and not necessarily negative. Yet the antecedents of Israel's modern environmental problems can often be found in the initiatives and achievements of this remarkable period.

PALESTINE CIRCA 1900: THE MANY MEANINGS OF "DESOLATION"

The early Zionist settlers spoke of their enterprise as a "return to Zion." But to what sort of land did the Zionist settlers "return"? No matter how the borders are cut in Israel (or Palestine, the fourth-century Roman name for their province), it has always been small, never much larger than twenty thousand square kilometers. (This is one-fiftieth the size of Egypt, one-thirteenth of New Zealand, or roughly the size of New Jersey.) Yet as the first British High Commissioner, Herbert Samuel, wrote, "While only the size of a small province, the land has the variety of a continent." Over 2600 species of plant life are indigenous to Israel, with hundreds of bird, reptile, and mammal species.[9] The ancients recognized only two seasons:

rainy and dry. During the seven months when rain does fall, the northern hills and mountains are blessed by as much as 800 millimeters of precipitation. The southern desert receives no more than 30 millimeters.

Biblical and talmudic citations suggest that most of the land in Israel was forested in ancient days, with some twenty species of trees.[10] For instance, the southern Galilee was known as the "woods of nations," and Absalom's ill-fated encounter with trees occurred in the "Ephraim forest of Gilead." Josephus, the Jewish Roman historian, wrote of the general plethora of horticulture in the first century B.C.: "For this land is a blooming garden, and in it grow a great quantity of precious and extremely beautiful fruit trees of all kinds."

The land of Israel did not benefit appreciably from its relatively low population density during the following millennia. As was the case in many other Middle Eastern and Mediterranean countries, the years of human abuse, predatory military campaigns, and overgrazing from the Roman conquest onward were manifested in soil erosion and deforestation. Similarly, some two millennia earler, the Habur plains of northern Mesopotamia had deteriorated to desert, and its rich local ecosystem disappeared.[11] In his 1997 bestseller, *Guns, Germs, and Steel*, Jared Diamond describes the mechanics of this environmental degradation:

> With the tree and grass cover removed, erosion proceeded and valleys silted up, while irrigation agriculture in the low-rainfall environment led to salt accumulation. . . . Thus, Fertile Crescent and eastern Mediterranean societies had the misfortune to arise in an ecologically fragile environment. They committed ecological suicide by destroying their own resource base.[12]

Palestine was part of this Levantine phenomenon.

The question of how barren the land of Israel was prior to Zionist settlement has become highly politicized in the "tit-for-tat" debates between pro-Israeli and pro-Arab camps. Environmentalists now question the appropriateness of such loaded terms as "barren" and "desolate," given the remarkable underlying biodiversity. Indeed, this critique lies behind the historic tension between the foresters of the Jewish National Fund and Israeli conservation activists.

Undoubtedly the Zionist movement's perception of a sparsely populated "Holy Land" at the turn of the century was not without justification. The countryside was not crowded. It was not empty, either. The first census taken by the Ottomans in 1882 revealed a collection of small villages, with Jerusalem's population reaching 30,000, Haifa's 6000, Jaffa's 10,000,

Hebron's 10,000, and Safed's 7500.[13] The most precise estimates suggest that by 1900 there were 350,000 predominantly Muslim inhabitants settled in the land, and some 20,000 nomadic Bedouin.[14]

Arab historians and representatives generally present a far more positive description of the land prior to the twentieth century.[15] These writers justifiably resent the tendency of ninteenth-century chroniclers to caricaturize the native population as ignorant, fanatical, violent, or lazy, or, even worse, to ignore them completely.[16] Travel reports greatly influenced the perspective of the European Zionist visionaries. Their apprehension of Palestinian geography culminated in Yisrael Zangvil's 1894 slogan "a land without a people, for a people without a land." This view was at the core of the hostility felt by native Arabs, who did not regard themselves as nonentities and who saw the Zionist settlers as Jewish "colonialists."

Arab scholars, however, fail to offer a substantive basis for dismissing the preponderance of geographical descriptions chronicled by pilgrims and travelers[17] and the jejune desolation revealed in early aerial photographs. Israeli experts estimate that a mere one hundred square kilometers of natural woodlands still survived in Palestine by the turn of the century,[18] increasingly segmented into small, isolated groves.[19] Winter heating needs and the local glass trade's steady demand for charcoal quickly led to the further decimation of the remaining forests between Hebron and Tekoah.[20] Grazing by goats was particularly pernicious, because it stymied regeneration.[21] By the time the Turkish army retreated in World War I, the soldiers' relentless search for firewood and ties for a railroad to the front erased all sign of the trees that had covered the coastal Sharon region.[22]

No account of the land's condition in the nineteenth century is likely to be quoted more often or is more entertaining than Mark Twain's irreverent 1867 travelogue-turned-bestseller—*The Innocents Abroad*. While prone to exaggeration, the narrative betrays his disappointment at the neglected condition of the Holy Land: "If all the poetry and nonsense that has been discharged upon the fountains and the bland scenery of this region were collected in a book, it would make a most valuable volume to burn."[23]

Much of the Middle East is arid or semiarid, so forests should have looked different from those to which Western travelers were accustomed. In the absence of summer rains, trees are shorter and distribution is naturally not as dense, making the term "forest" in the local context something of a misnomer. "Woods" may be a more appropriate noun; indeed, the original 1920 British conservation legislation was entitled the Woods and Forest Ordinance.[24] The "cedars of Lebanon" and the "oaks of the Bashan" probably never covered a lush countryside as the biblical narrative might suggest.[25] But a diverse Mediterranean flora flourished, even if trees rarely exceeded a height of three or four meters. Human activity was responsible for the disappearance of most of this vegetation.

A variety of rationalizations are put forward for what is probably a textbook case of the unregulated "commons" abuse. Explanations include the abject poverty and lack of alternative fuel sources, the impotence of the indigenous "feudal" *fellaheen* (small farmers and farm workers) to undertake conservation initiatives, Ottoman taxes on trees (which actually led to episodes where healthy groves were uprooted to avoid the levies), and overgrazing.[26] After the excesses of World War I, it was always easy to blame the Turks. (For instance, one-third of Palestine's estimated 300,000 dunams—75,000 acres—of productive olive groves were devoured by the Ottoman military machine.[27]) In any event, the Turkish Forest Law, purported plans by the sultan to bring French forestry expertise to the Ottoman Empire during the 1860s, and specific afforestation plans around Jerusalem did not amount to much.[28]

Poor stewardship was not limited to forestry and soil. During the Ottoman rule, hunting was completely unregulated. When German religious settlers came to Ottoman Palestine in 1868, they brought modern European firearms with them. It did not take long for these high-precision weapons to become accessible to the native Arab population, in particular the Bedouin. World War I also greatly increased the availability of rifles. The consequences were swift in coming. Many animals were hunted to extinction, among them the Syrian bear, the fallow deer, and the crocodile.[29] The ostrich, the cheetah, and the wild ass would soon follow suit. In 1912 the first of the Jewish zoologists, Yisrael Aharoni, managed to buy a two-year-old roe deer from Bedouin friends for study and preservation. Ironically, it turned out to be the very last one to be born in the wilds of Palestine.[30]

By world standards of the period, Palestine of the nineteenth century was a depressed region. There were no intercity roads until 1869. That year, the sixteen-hour horse ride from Jaffa to Jerusalem was shortened by improving the old pilgrim trail to honor Austro-Hungarian Kaiser Franz Joseph's visit. Three years later, a Jerusalem-Nablus road was added.[31] The famed train to Jerusalem (brokered by Yosef Navon, the grandfather of the future Israeli President Yitzhak Navon) became operational only in 1892.[32] Jerusalem installed the country's only municipal water system the previous year. It would take another sixty years for a proper sewage system to be built,[33] and for proper sewage treatment, another century. When Theodor Herzl visited Israel in 1898, his diary recorded the lack of trees and the pervasive stench of human wastes in Jerusalem.[34]

While greater prosperity and substantial population growth characterized the final days of Ottoman rule, it was quickly offset by the heavy burdens imposed by World War I. The Jewish population alone dropped from 85,000 in 1914 to 65,000 by 1918.[35] Hence, the British acquired a land with only 700,000 inhabitants, practically no automobiles,[36] and no industrial activity to speak of.

This was not for lack of trying. Baron de Rothschild, the wealthy French-Jewish patron of the incipient Jewish settlements, valiantly supported economic diversification, investing in several industrial ventures. Although a few initiatives, such as his Rishon L'Tzion wineries, were successful, a silk-thread spinnery at Rosh Pina and perfume plant in Yesod Hamaalah failed. A factory in Tantura, on the Mediterranean coast to the south of Haifa, was established in 1893 to supply glass to the nascent wineries. It, too, soon closed down, presumably due to technological deficiencies rather than to the competency of its manager Meir Dizengoff, who later became the first mayor of Tel Aviv.[37] The Baron's efforts paralleled previous efforts by Moses Montefiore, the British philanthropist, who had set up weaving and sewing workshops that quickly floundered.

In 1890, 70 percent of the people in Palestine remained engaged in primitive subsistence agriculture. Even in the Jewish sector, technology was antiquated. For example, in 1900 sixteen of twenty orange groves in the vicinity of Petah Tikva were irrigated by animals pulling waterwheels rather than by then state-of-the-art steam-powered pumps.[38]

But a letter of November 2, 1917, from Arthur James Balfour, the British Secretary for Foreign Affairs, to Lord Rothschild would change all that. The Secretary explained that "His Majesty's Government view with favour the establishment in Palestine of a national home for the Jewish people" pledging "its best endeavours to facilitate the achievement of this goal." The 117-word Balfour Declaration gave the Zionist movement the ticket it needed to ride and changed the course of history.

The British Mandate

It is almost impossible not to divide a systematic environmental analysis into three spheres. Arab, Jews, and the British colonial rulers ostensibly lived together during the Mandate, but in fact were fully sequestered in their separate worlds, meeting only superficially. Speaking three distinct languages, living in three different cultural milieus, and developing three separate economies, in the best of times they had an uneasy coexistence.[39]

Jews remember the British Mandatory government primarily for reneging on the Balfour Declaration's commitment to Jewish national aspirations and for its brutal actions to limit Jewish immigration during the Holocaust. Arabs have their own bitter memories about a mendacious policy of "equal obligation," behind which they saw a consistently pro-Jewish bias. But if the land of Israel itself could speak, it would surely recall the twentieth century infrastructure introduced by the British.

The Ottomans had left behind a tiny network of poorly maintained roads, irregular mail service, a few train lines, telegraph wires, and a small port in Jaffa. This was the condition of the country at the beginning of the British Mandate. In contrast, the airports, military bases, telephone service, modern port, and oil refineries in Haifa and the extensive highway network that the State of Israel inherited thirty years later provided the core of the country's infrastructure for years thereafter. The remarkable economic growth that enabled a tenfold increase in the Jewish population and the doubling of the Arab sector during the Mandate was made possible by the groundwork laid by the colonial government.

Historian A. Joshua Sherman describes this postwar effort:

> Often frustrated in their civil tasks, British officers were able to mobilize their energies in the non-political work of relief and rehabilitation: units of the Royal Engineers built roads, dug drainage systems and wells, worked to restore the shattered railways and brought in medical supplies and foodstuffs, largely from Egypt for the population. Officers restored the Palestine economy by establishing a Palestine currency pegged to the Egyptian system and helping the local farmers to cultivate their neglected fields.[40]

The motives behind British investment were not entirely magnanimous. Mandate policy was based on self-interest, with the stated goals of avoiding any burden on British taxpayers, promoting British exports, helping British firms expand, and supporting a capitalistic, local economy.[41] Despite the considerable disruption caused by the anti-Zionist Arab riots of the 1920s and the full-fledged revolt during 1936–1939, the British military managed to maintain a relatively stable business climate and thus achieve these ends. The GNP quickly swelled as the rudiments of a modern industrial economy, replete with polluting emissions and effluents, took hold.

The Mandatory government showed surprisingly little favoritism toward English interests in awarding concessions to develop Palestine's limited natural resources. Despite the objections of Conservative English politicians, Pinhas Rutenberg, a Ukrainian Jew, was granted rights to build a Palestinian electrical system (outside of Jerusalem).[42] No ordinary Ukrainian Jew, Rutenberg had served in the unlikely position of police chief for Alexander Kerensky during the brief provisional government in

Russia that separated the overthrow of the Czar from the Bolshevik takeover. Rutenberg liked to intimate that had he been given the green light to carry out a few surgical assassinations, the entire Bolshevik revolution could have been prevented.[43]

Having become a persona non grata in Soviet Russia, Rutenberg turned his formidable energies to Zionism.[44] In Israel Rutenberg will always be primarily associated with electricity. After raising a quarter of the one million Palestinian pounds necessary for the project, he commenced work on the first electrical station at Naharayim (or "two rivers"), which was located on the Yarmuk and the Jordan Rivers. The design selected required a total revision in the natural hydrology of the area. Four years after work began, the station began to generate power. Physically the plant sat in an artificial lake, fed by a diverted Jordan River that was rechanneled eastward into the Yarmuk River.

Rutenberg built additional facilities in Tel Aviv, Haifa, and Tiberias, so that by the 1930s electricity was no longer the limiting factor for the new cities' industries and irrigation projects. As the former two facilities expanded, they posed increasingly severe public health hazards, eventually galvanizing activists and spurring environmental campaigns and litigation in Haifa and Tel Aviv.

In addition to producing electricity, the British expanded the Ottoman train system. This was complicated by the fact that the Turks had used a narrow-gauge track system, whereas the British relied on standard gauge. Following World War I, some of the Turkish tracks were converted, and several hundred kilometers of new lines were laid. To provide a reliable supply line to the front, this network was expanded during World War II deep into Lebanon, with through trains to Egypt.[45] It was the British roads (built, of course, by both Arabs and Jews) that really affected the local landscape. By the time the Mandate departed, two-lane highways crisscrossed the land, linking the Galilee, the Negev, and the agricultural valleys to the center of the country, even though Arab ambushes frequently made them unsafe.

The British also bequeathed to the new State a comprehensive legal infrastructure. Despite a tenfold increase in population, Israel still clings to the Mandate's three-tiered court system and many of its laws.[46] Notwithstanding an avowed policy to avoid disturbing local life and custom, many of the Turkish laws and norms were changed during the course of the Mandate. The Government promulgated important ordinances, and British common-law precedent guided the courts. Some changes had important environmental implications. While Parliament had not yet enacted "modern environmental statutes" in England, traditional limitations on public and private nuisances were brought to Palestine through principles of British tort law that were later distilled into the Civil Wrong Ordinance of 1944.[47] Indeed, seven brief provisions of the Ordinance, summarizing British nuisance law, were the basis for most environmental litigation in Israel until the 1990s.

The reliance on private and civil law to control nuisances and hazards was a reflection of British philosophy of the day concerning the environment. The British perceived pollution to be essentially a local problem, best addressed by municipal authorities or dissatisfied neighbors. The Municipal Corporations Ordinance of 1936 required each town to appoint a municipal sanitary engineer or sanitary inspector to oversee such key issues as the quality of drinking water, construction of sewage systems, and prevention of sanitation hazards.[48] The Public Health Ordinance of 1940[49] contained many far-reaching provisions for the setting of centralized standards in areas such as drinking water and nuisances. These laws provided the groundwork for basic sanitation in the Jewish cities. They did little, however, to create a centralized authority that could both assist and regulate the largely unequipped local personnel. Environmental enforcement was not yet a salient concept. Although the Penal Code of 1936[50] included prohibitions against intentionally polluting air and water, these offenses were not on the Mandate police's priority list.

In general, sanitation efforts by the Jerusalem government focused on preventing major epidemics such as plague and smallpox. Environment and sanitation received only secondary interest and resources. What did take place was almost entirely due to Yehudah "Louis" Kantor. Kantor was an American sanitary engineer who came to Palestine as part of a Hadassah delegation following World War I. He was immediately put to work by the military Occupied Enemy Territory Administration (OETA) authority to create a sewage system for Jerusalem. Impressed, the authority's leaders made him director of the government's Sanitation Department. He served in this capacity from 1920 until his death in 1933 at the age of 49.[51] Notwithstanding his dedication, most rural communities during the Mandate had no indoor plumbing, and the majority of cities and villages relied on cesspools or septic tanks for waste disposal.[52]

Unlike sanitation, natural resources were considered a legitimate area of involvement for the central government. The Mandate promulgated ordinances pertaining to fishing,[53] forestry,[54] and wild animals.[55] All these laws survive in amended forms, fifty years after the Mandate's cessation, with varying degrees of implementation.

Water, considered the scarcest resource of all in Palestine, was not addressed in a comprehensive manner until 1940, when the Palestine Order of Council of 1922 was amended.[56] The High Commissioner was granted rights to surface waters, "holding them in trust" for all of Palestine. Implicitly, the "trust" canceled the private rights to streams that had been granted under the Ottoman Mejelle law. The Commissioner was also empowered to enact ordinances stipulating the beneficial use of all water sources, including groundwater.

During the last ten years of the Mandate, water allocation became a subject of major contention between the Yishuv and the Mandate government. By this stage, Zionists viewed all British regulatory efforts with suspicion, as part of an overall British strategy to stymie Jewish development.[57] Although a water commissioner was appointed under the 1944 edict, plans for redistributing water allocation in controlled areas were never carried out. Apparently the Jewish agricultural lobby was already a formidable political force with which to be reckoned.[58]

As the legislative agenda suggests, the Mandate's hydrological concerns were limited primarily to issues surrounding allocation. Water quality was not something that warranted much attention.[59] By allowing municipal authorities to determine the location of their sewage outfall, towns were in effect encouraged to convert streams and wadis into carriers of waste.[60] For many cities and towns, this dynamic continued into the 1990s.

Ironically the Mandatory government's policies that exerted the greatest influence on the Palestine environment were those involving broader geopolitical considerations. A series of administrations could not seem to make up their minds between the Balfour Declaration's commitment to the Jews and the Mandatory Government's desire to appease the much larger Arab world. The resulting flip-flops in London's policy on settlement were quickly felt on the ground in Palestine. Quotas on Jewish immigration in white papers, as well as restrictions such as the 1940 Land Regulations, which limited Jewish land acquisition to 5 percent of Mandatory Palestine, were the valves that turned population growth and development on and off.

World War II shifted the economy of Palestine into overdrive. The British encouraged the Yishuv to mobilize for the war effort. By 1943, a full 63 percent of the total Jewish workforce was involved in occupations directly connected to defense needs.[61] Factories produced everything from boots and uniforms to machine and weapon parts.[62] The amount of cultivated land increased by 70 percent, and twelve hundred new Jewish factories for military-related products were up and running by 1945, an absolute increase of 60 percent over antebellum levels. This remarkable expansion occurred despite the fact that 136,000 Jewish-Palestinian men and women enlisted in the National Service within five days after registration opened at the War's outbreak.[63]

CONSERVATION POLICY DURING THE MANDATE

If pollution control and water quality were nonissues for the Palestine government, the British did bring modern ideas about land management and conservation to the Middle East. Soon after Allenby's armies assumed control of the country, a stringent ordinance was passed protecting woodlands. The law banned a variety of activities in forests, including unpermitted removal or cutting of timber, extraction of resin, removing stones or minerals, and starting a fire "without due caution." Most important, grazing was banned, as well as any passage by domestic livestock without a permit.[64] Violations could bring fines or up to six months imprisonment. In 1924 the Mandate passed a Hunting Ordinance to protect the animals on these lands.[65]

Upon assuming the Mandate, the British conducted a survey of lands showing that 622,370 dunams could still be classified as indigenous forest—roughly 2 percent of the lands in Palestine.[66] Even this is a somewhat inflated number. As much as 30 percent of the territory was defined in later surveys as closed vegetation—impenetrable thickets, accessible to goats but too dense and inhospitable for cattle- or sheep-grazing.[67] Another 50 percent of the lands had only sparse trees, and these rarely exceeded a few meters in height. But the British recognized that even these areas were quickly shrinking, because of the local Arab population's tendency to seek firewood or new lands for grazing and tillage.

The Forest Ordinance established a Department of Forestry, headed by a Forest Officer, in the Ministry of Agriculture. The designation of protected forests limited logging and restricted other harmful activities (e.g., fires). The British Mandate government would eventually declare forty such areas, covering over 800,000 dunams.[68] Under the ordinance, eighteen types of trees were declared to be protected throughout the country and could be felled only with a permit from the Forest Officer. Although British motivation may not have been environmental, the policy preserved many scenic areas and enabled certain woodlands, such as those in the Carmel and Meron mountains, to rebound ecologically.[69] These served as the physical and cultural foundation upon which Israel's Nature Reserves Authority and the JNF (Jewish National Fund) Forestry Division could later build.[70]

The British rationale behind their forestry policy is summed up by Amihu Goor, a Yale and Oxford graduate and one of the few Palestinian Jews to reach the pinnacle of the Mandate's civil service, serving in the capacity of Conservator of Forests for the Mandate:

> In practice all local villagers are permitted to graze and cut, but not to cultivate, so that no new claims to ownership based on cultivation are allowed to arise. As a result, with the minimum of disturbance to the life of the villages, the rights of the state have been safeguarded and the state still has the chance of afforesting and developing these uncultivated lands at some future date when circumstances permit.[71]

Goor's summary does not mention the afforestation conducted by the Mandatory Government. Twenty million trees were planted on 53,500 dunams of reserve lands. The Government distributed eleven million more trees for local initiatives. Most of these were grown in a network of government nurseries. Government efforts dwarfed the parallel afforestation efforts of the Jewish National Fund, involving 4.5 times more land and 6.5 times as many trees.[72] The principal goal of British forest policy, however, was to halt destruction of existing stock. A significant percentage of the trees planted died, primarily as a result of pest infestation.

Because they were designed to constrain them, not surprisingly, the reserves were unpopular with the local Arab populations. Arson was a common form of protest. Years later, during the Palestinian Intifada of the 1980s and 1990s, this phenomenon would destroy scores of trees.[73]

A fascinating and little-known aspect of British conservation involved the protection of wildflowers. Goor, as a senior official in the Forestry Department, in 1930 requested from the head of the Botany Department at the Hebrew University of Jerusalem a list of plants that might benefit government protection. Goor also explored the idea of adopting 1929 German wildflower protection regulations.[74] The Hebrew University responded with a series of suggestions, including a blanket prohibition on vegetation in the Dead Sea region, protection of the Tabor oak tree, and a ban on picking various wildflowers that were already beginning to disappear. The recommendations were transformed into a 1931 regulation forbidding the picking of flowers in forest reserves.

The British record in protecting local wildlife was far less impressive. The hunting law reflected a lack of familiarity with local species, particularly of those most requiring protection.[75] Once a hunting license was granted, there were practically no limitations on the types of animals that could be shot. Some of the rarest indigenous mammals disappeared during the course of the Mandate. The ordinance allowed hunters to shoot two gazelles a year, but very quickly the number of licensed hunters exceeded the number of gazelles surviving in the Palestine wilds.[76] Hardy ibex, with their spiraled horns, were slaughtered when they gathered around the water holes in the evening after leaving the safety of their rocky desert perches. Over 30,000 gazelles were shot between 1918 and 1935 (often to be used in a soup popular among Arab residents). Even today after fifty years of stringent Israeli protection, these populations have not returned to their original levels.

The feeble implementation of conservation policy was not without critics. A few Jewish academics stepped forward to fill what today would be called a watchdog function. Professor Heinrich Mendelssohn was one of them. When he was barely into his teens, Mendelssohn decided that he was interested in learning about natural history and animals. His parents would not hear of it—in Berlin, gifted Jewish children were expected to study medicine. As a compromise however, they allowed him to try a double major with zoology at Humboldt University. By then Mendelssohn was already an active Zionist. A copy of Leo Pinsker's classic Zionist treatise *Autoemancipation* had fallen into his hands at age twelve, changing his life forever. Only when he got into a fist fight with Nazi thugs in 1933 did his parents relent and give their blessings to his aliyah. In Israel he eventually assumed the role of scientific guru for the budding conservation movement.

At the age of 90, Mendelssohn still maintains remarkable recall over events during the Mandate period. He argues that not only was the Hunting Ordinance inadequate—it was never enforced, despite the many Arab policemen hired to this end. In the 1930s, already enjoying the respect accorded academics, Mendelssohn received an audience with the responsible British authorities in the Ministry of Agriculture. There he complained about the widespread slaughter of the gazelle taking place in the reserves. The patent bureaucratic response was that police or inspectors at the reserves were responsible for enforcement. The British attitude was: "'Original inhabitants have the right to hunt whatever they please'— which meant, 'Jew stand aside, don't interfere.'"[77]

On his honeymoon in 1945, taking his wife hiking in Nahal Amud between Safed and the Sea of Galilee, Mendelssohn bumped into a group of Druze men from the village of Mrar who were out for a day of hunting. Not short on courage, the diminutive Mendelssohn scolded the armed Arabs for breaking the law. Mendelssohn recalls, "They knew very well they shouldn't be hunting and didn't take it too badly." In fact the jovial hunters must have been charmed by the chutzpah of the bespectacled professor. "Soon thereafter they made a special trip to come visit me in Tel Aviv so we could talk about nature."[78]

Perhaps because Jewish dietary laws permit the slaughter of animals only when they are in captivity, hunting is alien to traditional Jewish culture. Like the biblical matriarch Rebecca, who favored Jacob over Esau, Jewish sentiment generally sees hunters as brutish and barbaric. (Not surprisingly, Josephus reports that the non-Jewish King Herod was a rabid hunting enthusiast.) For the most part, the Zionist revolution did not try to reinvent this part of Jewish identity; hunting was predominantly an Arab pastime during the Mandate. There were more than a few exceptions though. Few know about a Jewish hunting club, based in Jerusalem, that convened periodically during the Mandate. The wealthy Jerusalemites would take a bus to the Huleh Valley, bringing along local Arabs who would venture into the swamps to fetch the birds that fell into the cold winter waters. These elitists, however, were perceived by the Yishuv as pathetic mimics of the Gentile aristocracy, and with the founding of the State, the club fell into obscurity.[79]

SUBSISTENCE PERSISTS: THE AGRARIAN ARAB ECONOMY

Generalizing about the indigenous Arab population of Palestine during this period is difficult, as it was truly a heterogeneous community. In 1880 Laurence Oliphant, the British diplomat-anthropologist, identified nine different ethnic and religious groups in the farm villages around Haifa alone.[80] As Arab immigration ballooned during the British Mandate, it only increased the diversity. Most of the Sunni Muslim majority worked as small farmers, or fellaheen. In contrast, 10 percent of the Arabic speaking community identified by the 1931 census was Christian, and of this group only one in seven worked in agriculture. A rapidly growing community of some sixty thousand Bedouin kept a distance from the permanent population, who feared these unpredictable nomads (see Figure 2). Arab villages were prudently sited on the crests of hills for reasons of self-defense.[81]

With the improvement in personal security during the twentieth century, Arab villages spread out across the rural lands of Palestine, particularly in the coastal region, where Jewish activity was most intense. The proportion of land they controlled did not grow, despite calls in the Arab press as early as 1913 to buy state-owned lands before they were snatched by the Zionists. As land prices rose, there was little effort—and ultimately, insufficient funds—for the systematic expansion of Arab landholdings and agricultural activity.

In practice, the new century did little to change the traditional, subsistence farming practices of the locals. Arab entrepreneurs such as Beirut-based Ibrahim Sursuk, who for the early years of the century hired thousands of laborers and even dabbled in swamp draining to establish a modern cotton industry, suffered heavy losses.[82] The Arab fellaheen were only marginally influenced by the new technologies introduced by the European Jewish settlers, such as the more sophisticated thresher, or the "Jewish plow," which replaced the traditional single-nail plow by the end of the Ottoman period.[83] Retention of fellah farming methods probably had more to do with financial constraints and the absence of formal extension programs (despite the educational program of the Mandate, by World War II rural illiteracy among men was still 70 percent[84]) than with ideology.

Today Palestinian agricultural experts are quick to point out the environmental benefits of the old fellah techniques.[85] The 1930 report of the Simpson Commission about conditions in Palestine (which was very negative toward Zionist aspirations) reflected similar admiration; it spared no praise in its description of the diligence of indigenous Arab farming.[86] The market was less sympathetic.

The technology gap led to economic disadvantage for the fellaheen, as agriculture shifted from subsistence farming to cash crops. While far below Egyptian (not to mention European) levels, Jewish yields for cereal crops such as wheat and barley were more than twice those of the Arab farmers.[87] In particular in the lucrative citrus branch, the Arabs' mule-driven pumps could not provide the necessary irrigation for the plantations that expanded along with the ever-growing citrus demand from Great Britain.

With the rural population in the Arab sector doubling between 1922 and 1944 (from 375,000 to 734,000) and the available lands shrinking in the face of Zionist acquisition, subdivision led to further subdivision.[88] Reforms in the Mandate tax policy were designed to ease the burden on fellaheen, ultimately reducing their contribution to 2 percent of national collection.[89] It was not enough, however; taxes continued to exacerbate the destitution of the debt-ridden peasants. Sentimental and even environmental benefits notwithstanding, the old fellah methods of farming could not compete.

For a brief period, an exception to this trend, owing to government intervention, was the raising of olive trees. To assist the beleaguered fellaheen after Turkish plundering, the British offered olive trees at subsidized prices. Mandate officials believed that the effect of the policy would be beneficial to the land, because it necessitated terracing and other soil conservation practices.[90] Some one hundred thousand new dunams within the Arab sector were planted during the late 1930s. The new olives were typically bought by the soap manufacturers centered around Nablus. Previously, these presses and factories had imported olives from Syria as a raw material.[91] It takes eight years for olives to begin giving fruit and twenty years to reach full production levels. Eventually, however, supply exceeded demand, and the market became glutted. Actual revenues from olive production remained a distant third behind citrus and wheat, leaving the financial profile of the fellah unchanged.[92]

The 1920s and 1930s witnessed an exodus of fellaheen from the farm to jobs in British public-works projects, Jewish industrial and agricultural ventures, and limited new Arab enterprises.[93] Already at the turn of the century, one thousand Arabs worked for the handful of Jewish farmers in Zichron Yaakov. Typical Arab wages in the late 1930s in Palestine ranged from 150 to 600 mills (25 U.S. cents to one and a half U.S. dollars) a day. While this may seem exploitive, it was more than double the going rate in Syria and three times the wages in Iraq.[94] In fact, the high salaries made Palestine a magnet for workers from the entire Arab world, much like Kuwait in the 1960s and 1970s. By 1935 large shantytowns sprung up outside Haifa and Jaffa, the cities experiencing the most dramatic growth during this time. Relative to other parts of the region, residents enjoyed better health care and a higher quality of life. For instance, the infant mortality rate among Palestinian Arabs dropped by almost 100 percent between 1927 and 1940.[95]

The spectacular growth in the Arab population during the Mandate, therefore, was not reflected in an expansion and diversification in the Arab industrial base or even a proportionate rise in Arab agricultural productivity. The soap produced from the local olive oil in Nablus, the cotton woven in Gaza, and the glass produced in the workshops of Hebron remained the best-known of the commodities manufactured in Arab cities.[96] Manufacturing was typically carried out in small workshops rather than large industrial plants. An interesting inventory of Arab industrial activity in the Mandate can be found in the records of the Abandoned Property Commission, which appropriated the workshops and plants left behind by the fleeing Palestinian Arab population. In the city of Lod, this included shops for producing buttons, ice, sausages, pasta, and soft drinks.[97] These processes were not without environmental side effects. Environmental authorities in the Galilee today, for instance, express concern about the organic sludge produced from the small, traditional olive-oil production units in the Arab community. Their impact is modest, however.

Although 64 percent of Palestinian Arabs still lived off the land at the end of the Mandate,[98] they had become more dependent on the burgeoning Jewish sector. Some left the country after losing their jobs during the 1936–1939 revolts (or simply fled because of the wave of internal assassinations that left five hundred Arabs dead).[99] The British Royal Commission of 1936 (known as the Peel Commission), however, reported that the large importation of Jewish capital into Palestine in general had a "fructifying effect" on the economic life of the entire country and the expansion of Arab industry and citriculture.[100] Yet it was not an empowering influence.

Although the Arab sector may not have been a major source of pollution, some of their activities were ecologically problematic. Naturalists are justifiably critical of Palestinian Arabs and Bedouin both for overgrazing rangelands and for unrestrained hunting during the Ottoman and British regimes. (Ironically with the drop in numbers of livestock, and of their carcasses due to natural death, after the Arab exodus, local scavenging birds of prey were denied an estimated six thousand tons of food and suffered accordingly!)[101] It is difficult to find precise figures regarding sanitation and sewage services in the Arab sector, but clearly these also fell behind those of Jewish settlements at the time. Nevertheless, since Arabs lacked the capital for large industrial and agricultural initiatives, the overall environmental impact of their activities was minimal. When the dust settled after the 1948 war, the number of Arab residents in Israel had dropped to an astonishing 156,000. With a tiny population, modest technologies and a low standard of living, Arab communities contributed little to the overall pollution burden inherited by the new State of Israel.

THE ZIONIST AGRICULTURAL REVOLUTION

In 1878 Yoel Moshe Solomon, an eccentric, third-generation ultraorthodox, Jerusalem Jew, took along two naïve Hungarian immigrants to buy swampland on the banks of the Yarkon River. Like other Jews of the period who began building homes outside the cramped quarters of the Old City, Solomon felt claustrophobic and sought a more natural existence.[102] The resulting farm village, Petah Tikva, was the first of a string of agricultural settlements associated with modern Zionism and the First Aliyah. The settlement was eventually abandoned by the original group, who threw up a white flag after being ravaged by malaria. When Baron Edmond de Rothschild contributed financial resources to the recently arrived farmers of Israel Belkind's "Bilu" project (1882), it started an alliance that would launch a cultural and agricultural revolution. New Jewish settlers would return to Petah Tikva and stay.

With the departure of the oppressive Ottoman administration after World War I, the Jewish agricultural revolution took off. In the twenty years that followed, average three-year yields skyrocketed, from 11,000 to 205,000 tons.[103] Between the years 1926 and 1944, for example, potato production increased from 821 tons to 35,000 tons per year.[104] The paramount importance of agriculture was not a subject of debate within the Yishuv. General, Labor, Revisionist, and Religious Zionists all shared an ideological fervor about transforming, or "redeeming," the land of Israel for agricultural production. As it turned out, rapid agricultural expansion was an expedient and labor-intensive economic strategy for the Yishuv. Its real significance, however, was geopolitical, solidifying Jewish control over large tracts of land.

Citrus production led the way. Nimbly responding to England's unquenchable demand for Jaffa oranges, Jews invested seventy million dollars in groves, expanding production sevenfold between 1918 and 1938.[105] Although citrus trees only occupied 4 percent of Palestine's eight million hectares of cultivated lands, the total value of production (3.6 million pounds sterling) exceeded the value of all other agricultural products,[106] generating 80 percent of Palestine's export revenues.[107] The increase was largely due to synthetic fertilizers. The profitability of citrus justified the high costs of imported fertilizers, whose use jumped from 1077 tons in 1922 to a peak of 14,698 in 1937, before settling at around 10,000 tons during World War II.[108] Eventually fertilizers would become the primary source of the nitrate that pollutes the groundwater under the western coastal plain. Prior to Israel's independence, application rates were low, and contamination was unknown, with nitrate concentrations of 0 to 10 parts per million, or "background levels," during the 1930s.[109]

With monies supplied primarily by the Jewish National Fund, land was systematically purchased on a regional basis and converted to agricultural uses. The popular slogan "A dunam here, a dunam there" produced one settlement campaign after another. After the successful "redemption" of the Jezreel Valley, the reclamation of the marshy lands of Emeq Hefer that linked Haifa and the Tel Aviv region became the top Zionist priority. A string of settlements in the Beit Shean Valley was next. The particularly inhospitable swamps of the Zvulon Valley, directly north of Haifa, were soon purchased during this period. It would take some time before area kibbutzim would overcome the ferocious mosquitoes there. In the 1940s it was on to the Negev. Although British land decrees slowed progress, the official *Palestine Statistical Abstract* indicates the steady, almost geometric, increase in Jewish agricultural activity.

Many of the settlements that emerged were set up in the paramilitary "stockade and tower" operations of the 1930s. Overnight, a kibbutz with a defensible wall around it would burst into existence. The system was designed to create quick demographic facts on lands purchased by the Jewish National Fund, preempting British evictions and Arab violence. There is little doubt that the remarkable expansion of agriculture in the Yishuv was possible only through these aggressive tactics.

The "stockade and tower" approach, though, was also inherently myopic and led to mistakes. The paradigm promoting the "creation of facts on the ground" would live on as a macho modus operandi for the State of Israel. Beyond many environmental disasters, this "just do it" and fait accompli ethos would lead to scores of misadventures, from the ill-fated 1982 Lebanon War to the prohibitively expensive production of the Lavi jet.

Ra'anan Weitz tried to introduce theories of regional planning, in particular in the area of water resources, when as a young upstart he joined the Jewish Agency's planning department in the 1930s. (The Jewish Agency served as the de facto government of the Yishuv.) Weitz returned home to Palestine after formal training in Italy. He was more than just another junior staffer, being the son of a leading figure in the JNF. Weitz quickly came to sit on key committees, but just as quickly came to recognize that without a supportive government, integrated, or "sustainable," development as it is known today, was impossible.[110] The political realities ensured that planning by exigency would prevail in the Yishuv.

By the 1930s, farming in Eretz Yisrael conformed to a Western monoculture approach taught by JNF agronomists and the pioneering "hachsharot," or preparatory farms, in Europe. The original farmers of the First Aliyah generation sought to learn from the fellaheen and imitated many of their techniques.[111] Aaron Aharonson, the premier agronomist in the Yishuv during the early years of the century, shared this inclination. Founder of Palestine's first agricultural research station, in Atlit, Aharonson was widely admired locally and abroad for discovering the biblical genotype of wheat and for leading the national effort to control locusts. (Aharonson also headed the legendary NILI spy ring, which worked during World War I on behalf of the British.) Notwithstanding a modest formal training, which did not go beyond high school, he received offers for professorial positions in several California universities. Fluent in Arabic and with many Arab friends, Aharonson was deeply committed to understanding the scientific basis of the Arabs' traditional agricultural practices in order to develop expanded applications with modern technologies.

After Aharonson's untimely death in a 1922 plane crash, the agricultural establishment of the Yishuv never shared his interest and admiration for indigenous fellah agriculture. With his departure, the foremost agricultural authority became Professor Yitzhak Volcani, a Lithuanian-born agronomist, who had founded the Zionist Executive's agricultural research station in Rehovoth in 1921.[112] Volcani was convinced that emulating the fellaheen was a sure formula for economic stagnation.[113] Rather, progress could be linked only to what he called the "mixed farm," with intense irrigation, European plows (later tractors), and diverse produce.

The paramount agricultural challenge involved water. Between 1924 and 1938, Zionist colonizing agencies dug 548 wells and almost as many canal systems to tap springs and streams.[114] The Yishuv did not have experienced water engineers and operated largely in a vacuum, with little assistance from the Mandate. Slowly proficiency improved.[115] The functional "hydroautonomy" within the Yishuv needed to be coordinated. The result was the establishment of the Mekorot Company in 1937 by a consortium of the four leading Yishuv development institutions: Keren ha-Yesod, the JNF, the Palestine Land Development Corporation, and the Nir Corporation.[116] Mekorot was charged with planning, operating, and administering the companies that supplied water for irrigation and household needs. Today Mekorot maintains its status as the national water utility.

Tapping available water resources and utilizing all the arable land for farming changed the landscape dramatically. No single ecological system felt these changes more than the wetlands of Palestine. At the start of the twentieth century, some 180,000 dunams were categorized as swamps and marshes. For millennia they had been part of the local landscape. Even the Romans had made unsuccessful efforts to drain them. With mosquitoes making the vicinity virtually uninhabitable, Arab effendi landowners were all too happy to unload the swamplands onto the eager Zionist Europeans who could offer cash up front.[117] As a result, little was left of the wetlands by midcentury. The draining of the Huleh swamp during the 1950s was just the final stage in a reclamation process that paralleled geometric agricultural expansion.

Criticism of agricultural practices during this time needs to be tempered by an awareness of the state of the art in the early science of soil conservation as well as the economic conditions prevailing in Jewish settlements. The tenacity of Zionist farm communities is reflected by their finding time and energy for culture, infrastructure, and ideology, in addition to soil conservation measures, while their communal farms tottered on the verge of starvation. During the first half of the century, the pioneers lived well below today's poverty lines.

SUSTAINABILITY

From an environmental perspective, the Yishuv's intense agriculture changed the landscape. But was the change necessarily negative? From the perspective of conservation, the most outspoken champion of Zionist agricultural achievements was probably a Christian American—Walter Clay Lowdermilk. Arriving in the Middle East in 1938 as part of a United States Department of Agriculture fact-finding mission, Lowdermilk brought with him a remarkably advanced "ecological" perspective for the period. A world-renowned soil scientist, Lowdermilk was sent to assess what might be learned about conservation practices in ancient lands such as Tunisia, Algeria, Morocco, and Egypt. The picture he saw was bleak.

His impressions from Palestine, detailed in his widely read 236-page book, *Palestine, Land of Promise,* offer a radical contrast. Lowdermilk was as critical of erosive practices of local fellaheen as he was inspired by the stewardship of 550,000 Jews over 6 percent of Palestine's mandated lands.[118] His clearly documented record challenges ecological revisionist aspersions about pre-State Zionist farming:

> Along with the records of decay in the Holy Land we found a thoroughgoing effort to restore the ancient fertility of the long-neglected soil. This effort is the most remarkable we have seen while studying land use in twenty-four countries. It is being made by Jewish settlers who fled to Palestine from the hatreds and persecutions of Europe. We were astonished to find about three hundred colonies defying great hardships and applying the principles of cooperation and soil conservation to the old Land of Israel. . . . Here in one corner of the vast Near East, thoroughgoing work is in progress to rebuild the fertility of land instead of condemning it by neglect to further destruction and decay.[119]

One can argue that Lowdermilk's account is a biased one. Indeed, he was so impressed by his fact-finding visit that in 1951 he agreed to return and give "a year of volunteer service to the young little nation." This stretched on for many more years, and he went on to found the Faculty of Agricultural Engineering at the Technion Institute of Technology in Haifa.[120]

A less flattering view of the agriculture in the Yishuv is offered by Professor Said Assaf. He perceives traditional Palestinian farming as environmentally superior, with its terracing; legume planting on shallow, nitrogen-poor, hilly soils; and minimal irrigation. Indeed Assaf claims that the small proportion of cultivated lands in the West Bank under irrigation (still only 5 percent) remains a virtue. The region simply does not have the water resources to support the type of agriculture established by the European Zionists.[121]

Today there is considerable concern about agricultural pollution in Israel. Yet it is hard to draft a compelling ecological indictment of farmers during the pre-State period. Efforts were made to combat erosion,[122] synthetic pesticides were not utilized,[123] and fertilizers were limited to citrus orchards in a very limited area. Notwithstanding Zionist efforts, only a fraction of the replenishable water supply was utilized.[124] Reliance on irrigation in a semiarid region is hardly a compelling argument, given irrigation's central role in growing much of the food required to feed the planet. Even in tropical areas like Florida, farmers are grateful for Israeli drip-irrigation technology that was born of this dependence.

Whether or not the intensive farming spawned by the Zionist enterprise was a desirable transformation for the land is a question that depends on one's philosophical and aesthetic point of view. Historically, however, there was nothing new about efforts to bring large, even mountainous areas under cultivation. According to one study, man-made terraces occupy a remarkable 56 percent of the mountains surrounding Jerusalem.[125] Finally, even today, Israel's environmental agencies and organizations tend to be extremely charitable toward the agricultural sector, despite the contribution of farms to groundwater contamination. This may be related to the very high percentage of former kibbutz members in leadership positions. Presumably they share a visceral identification with a bucolic landscape.

But even the most outspoken antiagricultural Green activists see farmers as a lesser evil than industrial polluters. The most-favored-sector status of the kibbutz and the moshav (semicollective farms) constrained the sprawl of many Israeli cities, forcing them to plan more efficiently. Farming has its own ecological problems, but it does not impose nearly the damage to wildlife habitats that cities do. Indeed, preserving the enormous percentage of Israeli land under cultivation from insatiable real-estate developers has become a top priority today for Israel's environmental movement.

ANTECEDENTS OF HEAVY INDUSTRY

Industrial development was a central part of the Zionist agenda. By 1931 the new census showed as many Jews employed in factories and small workshops as in agriculture.[126] The influx of "capitalist" immigrants from Germany (and the estimated 63 million pounds sterling they brought with them) spurred economic expansion, in particular in metal trades, textiles, and chemicals. Between 1930 and 1937, the number of industrial firms increased from 6000 to 14,000, with a corresponding jump in the "industrial" work force, from 19,000 to 55,000. During this period electricity usage in the Yishuv increased sevenfold.[127] Four years into World War II, industrial production reached new highs, as output increased 75 percent.

As in the Arab sector, few Jewish manufacturing ventures at midcentury involved heavy industries. There were several notable exceptions, however, where more-polluting industrial activity took place. The environmental impacts of these factories became problematic after they expanded during the 1950s and 1960s. Most notable was the industrial infrastructure (Figure 3) that sprang up to meet the postwar construction and immigration housing needs.

In 1909, sixty wealthy families left the city of Jaffa to build homes on its northern sand dunes. Like the Jerusalem Jews who escaped from the old city thirty years earlier, they too were looking for a moral, pastoral life.[128] They gave their settlement Nahum Sokolow's freely translated Hebrew name for Herzl's novel *Altneuland;* twenty-five years later, Tel Aviv was the most populous city in the land.[129] Its Municipality's Technical Department reported that during the late 1920s building increased geometrically, from 13,000 square meters in 1927, to 29,000 in 1929, to 44,000 square meters in 1930.[130] In absolute terms, the housing needs of the Arab population were even greater during this period. The new factories that provided building materials employed 6 to 7 percent of the Jewish laborers in the country.[131] One was the dusty Nesher Cement plant, established in 1919 on the outskirts of Haifa. Nesher prospered so much that it eventually established a monopoly on cement production in Israel. The particulate emissions from its smokestacks would become sufficiently severe during the 1960s and 1970s to spark the first real grassroots movement for air quality in Haifa.

Many obstacles stymied heavy industry in the Yishuv. One threshold problem involved the availability of capital. Early in the century Herzl encouraged Zalman Levontin to establish the Anglo-Israel Bank, which he did in 1903.[132] The lion's share of official Zionist investment, however, went into land acquisition and reclamation rather than manufacturing. Another problem was Article 18 of the Mandate, which stipulated that there should "be no discrimination in Palestine against goods originating in or destined for any of the countries which are members of the League of Nations." The inability to protect fragile startup ventures from international imports constituted a major drawback for entrepreneurs.

Although the Mandatory government offered no capital to subsidize early heavy industries, the "concessions" that they granted were often sufficient to spur private foreign investment. Private Jewish money invested in Palestine during the Mandate reached one hundred million pounds sterling, four times the level of "communal Jewish investment."[133] The oil refineries in Haifa, the two potash plants along the Dead Sea, and, of course, electrical power stations, all enjoyed special arrangements with the Mandatory government. In retrospect, the tradition of government patronage toward monopolistic industries proved to be an unfortunate part of the Mandate's environmental legacy. The extraterritorial, privileged status of these corporations engendered a dismissive attitude toward regulatory authority. For years after the establishment of Israel, this status hindered effective government control of industrial pollution.

Bureaucracy also slowed Zionist industrial progress. Moshe Novomeski arrived in Israel in 1920, eager to put his training in engineering and experience in Siberian gold mining to work. The Dead Sea—the lowest place on earth (and at times one of the hottest)—is a far cry from Siberia, and yet exploiting the mineral-rich lake quickly became an obsession for Novomeski. British enthusiasm for the endeavor was lukewarm. It took nine years of sustained efforts by Novomeski, including overcoming opposition in the British Parliament, to win a complicated tender. Only then was it possible to attract American Jewish investors and to commence mining.[134]

Within two years, the twin facilities of the Palestine Potash Company employed four hundred workers and exported thousands of tons of potash. By the end of the 1930s, annual sales would reach half a million Palestinian pounds.[135] The northern Kalia plant was destroyed during the War of Independence. The southern Sodom plant, however, survived to become a true industrial behemoth. Its present capacity of two million tons per year has had an enormous environmental impact.[136] The mining operations have obliterated any remnant of the natural landscape in the southern Dead Sea area. No less important, for many years the facility mined the bromine for one-third of the world's methyl bromide, a highly toxic pesticide responsible for destroying one-tenth of the stratospheric ozone layer.[137]

MALARIA

There is no better example of Zionist intervention to improve the land than the successful war waged against malaria, and Israelis are proud of it.[138] From a strictly environmental health perspective, the dramatic drop in the incidence of this disease was a remarkable achievement. Malaria had been part of the life in the Holy Land from time immemorial. Many biblical commentators believe that it was malaria that so intimidated Moses' spies, who fearfully described "a land that devours its inhabitants."[139]

Malaria is caused by protozoan parasites of the genus *Plasmodium* that are carried by female anopheline mosquitoes. (Male mosquitoes feed only on plant juices and consequently do not transmit the disease.)[140] The parasites develop in the gut of the mosquito and are passed on in its saliva when it takes a new blood meal. The parasites are then carried by the victim's blood to the liver, where they invade the cells and multiply. It takes approximately two weeks for the parasite to return to the blood and penetrate the red cells, producing the well-known symptoms of fever, shivering, pain in the joints, and headache, as the red cells are broken down. Long-term damage to vital organs, as in cerebral malaria, where infected red cells obstruct the blood vessels in the brain, can cause death.[141]

Morbidity and mortality data are available primarily from the Jewish sector in Palestine, where the initial research and prevention efforts took place. The data reflect a scourge of staggering proportions. The first published surveys in 1912 showed that some 40 to 80 percent of Jerusalem schoolchildren had malarial symptoms. It was simply assumed that malaria was part of the Palestine experience—almost an initiation rite. Mosquitoes were the real winners of World War I, with 40 percent of returning Turkish soldiers infected with malaria and 28,500 recorded cases among the British army. Few immigrants were spared during their first year. In 1920 there was a rate of 355 cases per 1000 people among Jewish residents of Palestine, causing a total of 526 fatalities.[142] Malaria was the number one public-health enemy, with a relative risk of death sixty times greater than that posed by traffic accidents in Israel today.

The first signs of progress in fighting malaria were made by Aaron Aharonson and Dr. Hillel Yaffe at Aharonson's Atlit agricultural research center. Their combined strategy involved draining breeding grounds, netting windows and spraying oils on surface waters to exterminate larvae. The measures reduced incidence from 380 cases between August and November of 1910 to 39 for the same period in 1911. This pilot intervention raised hopes that malaria need not be a fact of life in the Promised Land after all. The following year, Yaffe brought these techniques to Kibbutz Mishmar ha-Yarden and Yisod ha-Maaleh.[143]

Although the Mandatory government was technically responsible for the problem, Jewish initiative was behind the broad-based public health activities. U.S. Supreme Court Justice Louis D. Brandeis, during his 1919 visit to Israel, posited that the Yishuv had no prospects for surviving and developing unless malaria were eradicated (Brandeis also made an anonymous donation to support malaria research and control efforts, but the donation somehow enjoyed widespread publicity).[144]

The Hadassah Medical Organization was among the key scientific players. Its founder, Henrietta Szold, published an illustrated brochure on the subject in 1921 that was much admired by professionals in the field. At that time the organization sponsored a pilot program in the agricultural settlements of Migdal, Kinneret, and Yavanael. The Hadassah scientists sampled the blood of every resident, distributed quinine as a preventative measure and systematically identified puddles and possible nesting grounds, for oil spraying (Figure 4). Here again, within a year, malaria rates dropped by 80 percent.[145] Not surprisingly, when the results became known, other Jewish communities demanded similar interventions.

It was then that the Joint Distribution Committee (an international Jewish philanthropy) proposed to the British authorities a national initiative to assist the Jewish and Arab sectors alike. The Joint, as it is still called, funded antimalarial measures, while the Mandate government offered official endorsement. Pragmatism rather than generosity may well have motivated the scope of the Jewish effort in Arab villages. For the first time the ecological realities of the country's small size sank in. Local efforts were not enough. Research revealed that the *Anopheles* enjoyed a range of up to 3.5 kilometers, but that the *A. sacharvoi*, common in the Huleh region, could fly 15 kilometers. This explained the seasonal outbreak in Rosh Pina. By 1923 substantial antimalarial initiatives were up and running in fourteen regions of the country. In 1939 an English commission praised the Yishuv's antimalaria contribution to public health in the Arab sector. By 1946 the population had grown 200 percent since 1920, but the incidence of malaria had dropped to only twelve thousand reported cases.[146]

That was the year that DDT was introduced for control at an experimental level. It signaled the final stage in the battle against malaria. The pesticide's impact was precipitous. There were only 1172 cases reported in 1948. The general chaos that accompanied the 1948 War of Independence interrupted the meticulous drainage work that lay at the heart of much of the antimalarial success, but it was resumed with the creation of the State. With the expanded indoor and outdoor spraying of DDT, incidence dropped to eight to ten cases a year, enabling the Ministry of Health to close down its malaria department in 1962.

Some environmentalists point to the ecological price of the initiative: massive habitat destruction as well as reliance on a persistent and ecologically insidious chemical. Public-health practitioners, as well as the families of the two million people around the world who even today are lost each year to the disease, would see it differently. In this dilemma of competing environmental values, it is not surprising that Zionists chose an anthropocentric, people-first approach. In this way, Israel is no different from other Western and developing nations.

URBAN ENVIRONMENTAL CONCERNS

Because most of the industry that developed in the towns and cities of the Yishuv was light and involved only small tradesmen, the levels of pollution generated were marginal by present standards. It is wrong, however, to assume that the Yishuv was so busy "creating a state" that it had no awareness at all about environmental issues. Urban environmental concerns certainly existed, but they fell into a generic "quality of life," or nuisance, category. In retrospect, this approach made sense, for pollution did not yet pose a serious risk to public health.

The mayor of Tel Aviv, Meir Dizengoff, addressed the key environmental problems of the city in his 1934 pamphlet, *Tel Aviv and Its Life Styles*. In this remarkably sanctimonious harangue, printed and widely distributed by the municipal government, Dizengoff exhorted his city's residents to mend their antisocial behavior. The text is not expressed in the nomenclature of today's environmental rhetoric, yet the mayor of Tel Aviv today, Ron Huldai (who has carefully groomed his own image as a Green mayor), would certainly feel comfortable with the message.

There are those who see Israel's litter problem as a recent phenomenon, caused by such factors as the introduction of plastic packaging, the paucity of trash cans (owing to security concerns), or even the arrival of non-European immigrants. But litter was just the first misdemeanor on the first Jewish city's environmental rap sheet:

> The city has one hundred sixty workers who clean the streets. But even if we add more to them, will they be able to clean up a city with 100,000 people that pollute nonstop? In contrast, it would greatly influence cleanliness in the city if every resident understood that it is forbidden to litter the streets with fruit peels and paper, intended for the special waste bins and cans. It is prohibited to pour polluted water into public spaces, and it is prohibited to leave building materials on the sidewalks. . . . It's the obligation of every civilized person to keep himself and his environment clean.[147]

Dizengoff went on to denounce the chronic noise pollution, habitual smoking of cigarettes in public venues (despite signs forbidding it), and the lack of respect for trees and flowers in public parks, as well as chronic tardiness and immodest dress by the younger generation. Presciently he also dedicates several paragraphs to the environmental hazards of transportation. His claim, however, that Tel Aviv traffic is twelve and a half times greater than that in other international cities with a similar population does not reference any supporting data.[148]

It is interesting to note other environmental issues that might have been on a Tel Aviv mayor's agenda but were not. The wells of the city, for instance, had already begun to show signs of increasing salinity, as seawater rushed in to fill the vacuum caused by overpumping. Not long after the creation of the State they were closed, owing to high chlorine levels.[149] Similarly, the fine sand, or "zifzif," on the beaches was rapidly disappearing, mined to supply concrete for the construction of the many new buildings. When legislation would finally be introduced banning the mining of beach sands, very little of the generous natural deposits was left. Natural-resource constraints, however, were not part of the Yishuv's environmental consciousness.

The urban environmental perspective received massive reinforcements during the course of the 1930s from the Fifth Aliyah. Dirt, disorder, and a lack of general hygiene were a particular affront to the German-Jewish refugees, known locally as "Yekkes," who fled Nazi anti-Semitism. It would take many years, however, for these values to resonate with the more senior members of the Yishuv, who mocked the newcomers' accents, promptness, orderliness, and passion for classical music. As with almost every immigrant group, it took years for the Yekkes to feel as if they belonged.

Shimon "Sigfried" Kanovich is a good representative of the environmental orientation of the Yekkes, of the fastest-growing community in Palestine during the 1930s. A pediatrician and psychologist of some renown, Kanovich moved to Israel from Berlin in 1933. There he got involved in Tel Aviv politics, helped found the Progressive Party, and eventually served in the Knesset. But his outsider status prevented him from bringing sanitation and civility to his adopted city. He shared his alienation with his diary: "We were neither here nor there."[150] Three decades later, in 1961, he was able to draft and push through the first air and noise pollution statute as a member of Israel's Knesset. For years it was dismissed as a foolish parliamentary exercise and even spawned a pejorative expression: a "Kanovich Law" (a well-meaning, but unenforceable statute). Kanovich died five months after the legislation's passage, in the heat of the 1962 Knesset campaign, but he would exert an influence beyond the grave. His Abatement of Nuisances Law eventually became the primary tool for implementing Israeli air and noise pollution abatement policies.

It was the urban environment that produced the few political figures in Israel who adopted environmentalism as a cause. The most famous, Josef Tamir, earned a reputation as an environmental maverick in the Knesset during the 1970s.[151] When asked what brought him to the issue, he recalls his early childhood memories of the forests and rushing rivers of the Ukraine. Then there were his many class trips around Palestine, such as a particularly memorable visit to the Dead Sea in 1925. Tamir, however, was a city dweller. During the 1940s he served as Director General of the Petah Tikva rural area local council. His efforts to maintain the town's rustic character brought the political nature of environmental issues home to him, forty years before anyone thought of Green parties.

THE YISHUV'S INCIPIENT ENVIRONMENTAL MOVEMENT

For all intents and purposes, there were no focused environmental activism or formal Green organization in the Yishuv prior to or during the Mandate period. Nonetheless, numerous researchers and teachers in the area of biology, zoology, botany, and nature studies are generally recognized as the forebears of Israel's environmental movement.[152] For many years, their interaction was informal and unorganized. However, with the establishment of the professional journal *Ha-Teva v'ha-Aretz* (*Nature and the Land*) in 1931, communication between the natural scientists of the Yishuv improved. The agronomist Ariyeh Feldman founded and edited the journal. It offered an accessible forum for academics and teachers, as well as amateur nature lovers, to report findings on subjects related to natural history. And today for those interested in nature, reading back issues from the 1930s is still fascinating. Despite the chaos and disruption of the period, for twenty-eight years, until his death, Feldman did not miss a single monthly issue.[153] Today the periodical lives on in a more glossy format as *Land and Nature*.

Even before publication of the journal, Alexander Eig's 1926 pamphlet *Additions to the Knowledge of Plants in the Land of Israel* contained a remarkably prophetic analysis about disappearing habitats and the threat to Palestinian biodiversity:

> Already today, if one of the botanists from the past century rose from the grave, he would not recognize entire regions. Those interested in the nature of the Land and its fate must get organized into an association for

> the purpose of preservation of nature. One of its primary tasks would be to keep a constant watch about all that concerns the plants of Israel.[154]

Eig would die an untimely death at the age of 44 but not before amassing an astonishing list of achievements, which included founding a Department of Botany and establishing the Botanical Gardens at the Hebrew University, and researching and writing numerous publications, including the first comprehensive taxonomic guide for names of plants in Palestine.[155] He did not live to implement his call for ecological activism, and it probably went unnoticed by most of his scientific colleagues. His writing, however, appears to have moved the Mandate forester Amihu Goor to initiate a wildflower protection program. Indeed, it was Eig who prepared the proposal of protected plants.

Similar calls to save the Palestinian landscape were heard in the 1930s in Hebrew architecture and planning journals.[156] These were lone voices with no organizational follow-through. Nonetheless, the network of researchers and teachers with a passion for nature and the outdoors continued to grow. With the establishment of the Biological Pedagogical Institute in Tel Aviv in 1931, nature lovers in the Yishuv had a nerve center. Many consider its founder, Yehoshua Margolin, to be the greatest environmental educator of his day.

Prior to coming to Israel, Margolin had taught nature in Kiev. It took no time for him to take over the subject at Mikveh Yisrael, the oldest agricultural school in the Yishuv, located south of Tel Aviv. There "Uncle Yehoshua" was always leading his adoring students into the field. Some of the results can still be seen in an impressive taxidermal collection.[157] As luck would have it, a former pupil from Kiev, Shoshana Parsitz, oversaw education for the city of Tel Aviv. Parsitz had fond recollections of her favorite class from school days, and when the old Central Synagogue with its 750 square meters of adjacent land moved to its present, more prominent location, she offered Margolin the old site on 12 Yehudah ha-Levi Street (today it is a parking lot). Moreover, she found funding for a budget: the then-astronomical sum of one hundred Palestinian pounds a year.[158]

Margolin had no trouble attracting top young academics such as Heinrich Mendelssohn and Alexander Barash to the Biological Pedagogical Institute. Beyond formal and informal teaching, the Institute sponsored a youth group called The Young Naturalists, who would hike throughout Israel, bringing back samples from their travels. National Societies of Botany and Zoology were established, with the Institute serving as their base. The former synagogue lot quickly became a small zoo, home to a variety of species that had been collected by amateur enthusiasts. Until his death in 1947, Margolin taught hundreds of young people, and they became the new generation of nature teachers. Among them was Azariah Alon, who would eventually bring his encyclopedic knowledge of Israel's land, nature, and history to the new nation through a weekly radio show and many books. Although these incipient conservationists might today be categorized as "public-interest scientists," most in fact were driven by the Zionist Romanticism of the period.

TROUBLE AHEAD

In general, there is a tendency to frame impressions about early Zionist environmental crimes and misdemeanors in the light of contemporary ecological knowledge. Even by such standards, however, environmental conditions in the land of Israel were in decidedly reasonable shape in 1949. After the industrial boom of World War II in particular, the newly founded Jewish state was much less agrarian than Palestine had been at the start of the century, but air pollution was still moderate enough that a modern nomenclature had not entered the vernacular. (Twelve years later, when Shimon Kanovich proposed his nuisance law, he would refer to it as "harmful smoke.") The prevailing agricultural practices would be defined today as low-impact, if not organic, for pesticides and synthetic fertilizers were not heavily used in mainstream Israeli agriculture until the 1950s.

The rivers were largely unpolluted. At midcentury, a plentiful flow of 220,000 billion liters of water a year easily diluted whatever sewage reached the Yarkon River, and Haifa children swam in the Kishon. Groundwater contamination was for the most part unstudied in the hydrology literature (retrospective scientific studies suggest that nitrate levels during the 1930s were quite low, averaging only two milligrams per liter).[159] Only in the 1950s did the first tests show unusually high concentrations of chlorine and nitrates, today's primary pollutants. There were only six thousand privately owned cars on the roads, the equivalent of three weeks' nationwide car sales for 1997.

Many positive environmental achievements from the period deserve honorable mention. The blight of malaria had essentially been removed. A system of protected forests had been established. A practical train system was in place that could take people from Jerusalem to Europe or Africa. Sanitation had improved dramatically, and outbreaks of diseases such as cholera and typhoid were increasingly rare. Primitive but important environmental laws existed, and rudimentary government institutions and offices, including a water commissioner, were in place. It is not surprising that some Israeli experts feel that Palestine was better off environmentally as a result of the British Mandate.[160]

The three key factors that affect the magnitude of environmental damage—technology, affluence, and population—showed no cause for alarm. Technologies, for the most part, still reflected a nonindustrial economy or the craftsmanship of light industry. By Western standards, both Jewish and Arab populations were decidedly poor, even if they were more prosperous than their neighbors in Arab countries. The level of personal consumption was low, and there was a culture of frugality (for example, toothpaste containers were recycled in Tel Aviv during World War II). But most important, population pressure was not yet felt. Whether their own panic, the urging of Arab leaders, or premeditated Zionist harassment was the cause,[161] the exodus of more than seven hundred thousand Arabs in 1948[162] left the land of Israel with demographic levels close to those that General Allenby had found when he conquered Palestine in 1918. From the inception of the Mandate, there were arguments between Zionist and British authorities about the ultimate carrying capacity of the land of Israel and how many immigrants could be absorbed.[163] By the 1940s, however, even conservative estimates were in the millions.[164]

Although the Yishuv settlement proliferation was considerable, Zionist mythology tends to exaggerate the extent of pre-State development. For instance, thirteen thousand dunams of JNF woodlands, at most, were planted prior to 1948, much less than a single park such as the Yatir forest today.[165] Israel's coastline remained largely untouched. The little construction that did approach the shore was limited to one- or two-story structures. Aerial photographs show that only fifteen million square meters were urban or paved at that time (less than 8 percent of today's level), leaving 93 percent of the land undeveloped or under cultivation.[166]

But this snapshot does not tell the whole story. Judging the first half of the twentieth century by the environmental consequences of the second half yields a much tougher verdict. Of course, such a standard is an unfair basis for criticism. No nation on earth at the time had premonitions about the long-term persistence of pesticides, groundwater contamination, air pollution, ozone depletion, endocrine disruption, or the host of other ecological maladies that are of concern today. The more pressing concerns of food, shelter, the Nazi Holocaust, and political independence provided sufficient distraction. With the wisdom of hindsight, it is easy to identify worrisome phenomena and environmental trends in Palestine between 1900 and 1949. These would reach disastrous proportions when fueled by the new nation's ambitious agenda for development and by a phenomenal population increase—roughly one million people each decade.

The Jordan River offers a good example of such early warning signs. Pinhas Rutenberg broke a psychological barrier when he built his Naharayim power plant during the 1920s. Diversions and manipulations would continue until 1993, when the last natural segment of the Jordan River was redirected to establish a small hydroelectric facility at Kibbutz Kfar ha-Nassi.[167] The Jordan River had never been as "deep" and "wide" as the old American spiritual depicted it, but Israeli agriculture's titanic thirst would eventually reduce the river's flow to a salty trickle. (In 1900 the estimated annual inflow of water into the Dead Sea was 1.2 billion cubic meters. By 1940, this had dropped to 900 million cubic meters. Twenty years after the State was established, it was down to 810 million cubic meters, and by 1985, only 125 million cubic meters—a tenth of the original amount.[168]) Much of the River's little remaining water comes from diverted saline springs in the Kinneret Lake. Current water agreements between Israel and Jordan plan to tap even more. It is little wonder that the Dead Sea, the River's ultimate destination, is beginning to dry up.

Many other modern hazards originated in the early half of the century. Pipes were laid in many towns to collect sewage, but the sewage was then released untreated into the nearest river. In the case of Tiberias and Safed, the pipes reached Lake Kinneret.[169] DDT would not be completely phased out until the spring of 1997, after pressure from environmental groups and affected farming communities. Dizengoff never did solve Tel Aviv's litter problem. Intensive agriculture would lead to massive overpumping and contamination of Israel's Coastal Aquifer from nitrates and salinity. An aggressive policy to encourage population growth remains an integral part of Israeli culture to this day.

The period also left behind certain values among Israeli leaders that were at the heart of key environmental problems. Zionist tradition, for instance, perceives agriculture as inherently virtuous. Many of Israel's first political leaders spent formative years behind a plow. Some, if not most, of their best friends really *were* farmers. Acquiescence to an agricultural lobby was not so much a political expediency as the moral thing to do. The fact that the lobby represented a myopic sector, insensitive to the long-term impacts of high-input, water-intensive agriculture, failed to penetrate Israeli politicians' conceptual universe.

While the economic infrastructure of the Yishuv had few severely polluting factories, industry and the employment it provided formed the engine that sustained immigrant absorption. With the removal of British restrictions, the trickle of immigrants became a flood. Factories and production became even more highly venerated. The tradition of formal monopolistic concessions to protect industrial interests would be expanded to include various forms of environmental immunity. The socialist tradition, which even the right-wing Likud chose not to jettison when it came to power in 1976, exacerbated this phenomenon. For much of Israel's first fifty years, the most severely polluting industries were government-owned corporations—the electric company, Israel Chemicals, and the oil refineries. Similarly the Israeli army and defense establishment remain the ultimate sacred cows. Israel's justifiable commitment to security has proved convenient for officers and managers in the vast consortium of military industries and army bases.[170] Above reproach, even today they are exempted from key environmental statutes such as Israel's National Parks and Nature Reserves law.[171]

The more charitable acknowledge the remarkable accomplishments of the Yishuv, which had not yet spoiled the solid environmental indicators existing at midcentury. It is also true, however, that the very zeal with which the settlers took vengeance on the mosquitoes spelled trouble for the land of Israel once it was harnessed by an energetic young state.

4 The Forest's Many Shades of Green

"Before it did anything else, the first thing that the Zionist movement did was to set up an official conservation agency—the Jewish National Fund—and start planting trees." So begins Uri Marinov's stock introductory presentation about Israel and the environment. Marinov was the top environmental public servant in Israel's government during the 1970s and 1980s. His admiration for the JNF is genuine. "No other national movement made land restoration and forestry the centerpiece of its operational activities."[1]

Aviva Rabinovich, well into her seventies and still one of Israel's most provocative ecologists, presents a very different picture. As the first Chief Scientist at Israel's Nature Reserves Authority from 1970 until 1988, she became appalled at the practices of the Jewish National Fund: "They bring in bulldozers and, if this doesn't work, pneumatic hammers to destroy the stone, and all to plant a few pines. Wildflowers, tulips, and anemones—don't they have any rights? These people are my friends. They are good people. But who gave them the right to destroy?"[2]

No institution in Israel's environmental community has inspired as much controversy and passionate disagreement as the JNF. Some see it as a picture of ecological innovation, a model to the world of the human potential to reverse deforestation and desertification. Others can see only the disastrous results of JNF imposition of monotonous European scenery on an unreceptive land—the "pine deserts where only rats can survive."[3] In any case, now that forests cover over 10 percent of the lands in the northern half of Israel, all agree that the JNF changed the face of the country.

Judgment on the merits or defects of JNF activities, and the beauty of the 220 million trees it has planted, ultimately depends on the environmental values and aesthetics of the beholder. Advocates tend to exaggerate early environmental inclinations. Critics sanctimoniously impose the wisdom of hindsight, with little sensitivity to the ecological and historical milieu (and constraints) in which the organization operated. As is usually the case, the truth lies somewhere in the middle. The history of the JNF in fact shows a dynamic organization that rewrote its mission statement several times before arriving at its present identity as an "ecological agency." It evolved in response to the vicissitudes and challenges of the Zionist movement. Israel's forests therefore largely reflect the historical exigencies of Zionism, including the JNF's institutional development and the visions of a few passionate leaders.

THE BIRTH OF THE JEWISH NATIONAL FUND

While Zionist lore associates JNF and forestry with Theodor Herzl's vision, the minutes of the First Congress of the World Zionist Organization suggest otherwise. In fact, the charismatic chairman tried to squelch the proposal for a Jewish National Fund. Only after rival factions forced the issue did Herzl relent and bring the motion before the Congress, just prior to adjournment. Zvi Herman Schapira, a bespectacled, bearded mathematics professor and opponent of Herzl's political Zionism, was the man who actually designed and presented the plan to create a Fund for "acquisition and cultivation of land in Palestine."[4]

Schapira was a bona fide genius. Born in Lithuania in 1840, he could read Hebrew by the age of two and a half and was already studying the inscrutable Talmud at the age of four. By the time he was eight, his teachers accepted him as an equal, and he had mastered Maimonides' *Guide for the Perplexed.* While still in his early twenties, Rabbi Schapira stumbled on a book about mathematics that captured his potent intellectual curiosity. After serving for three years as a rabbi and yeshiva (religious school) director in the Lithuanian town of Birtushla to fulfill a promise to his father,[5] he set off for a Western education in Germany at the age of twenty-eight.[6]

It was a bold move. With no knowledge of even elementary German, Schapira endured poverty so great that he once stole a loaf of bread to avoid starvation. Once his formal studies began, though, it took him only two years to complete a doctorate in mathematics at the University of Heidelberg. Although he quickly attained the title of professor, it was not a paying position. The heavily accented lecturer was not a popular teacher and had to supplement his family's income by watchmaking.[7]

By the time of the First Zionist Congress in 1897, Schapira had been involved in Chovevei Tzion (the Lovers of Zion network of Zionist groups) for twenty years. When he submitted his idea for a national fund to buy land in Palestine to the organization's 1884 conference in Katowice, the proposal went nowhere. But Schapira was persistent and raised the issue again thirteen years later. The presentation was brief, in deference to the unenthusiastic Herzl's request that Schapira keep his comments short.[8]

The specifics called for the raising of ten million pounds sterling to be invested in a fund, with the interest going to Zionist settlement. Perhaps the most distinctive component of the proposal was the stipulation that "acquired territory shall be inalienable and cannot be sold even to individual Jews; it can only be leased for a period of forty-nine years maximum, according to regulations yet to be devised."[9] This policy of long-term leasing remains the cornerstone of Israeli land policy to this day.

Given the diffidence of the Congress, a compromise motion was accepted that set up a committee to consider the fund as well as a proposed Jewish bank. Eight months later, Schapira died of pneumonia during a visit to Cologne. It would take four more Zionist Congresses for Herzl to change his mind and throw his weight behind a scaled-down version of the proposal.

On December 29, 1901, at the end of a rousing speech, Herzl called for a revote: "Do you want us to start a Jewish National Fund immediately—yes or no?" he thundered, and the frenzied delegates responded with calls of support. "You can, if you wish to, delay its establishment for another two years or even until the coming of the Messiah." And the obliging crowd answered: "No. No."[10]

A SLOW START

The first donation to the JNF was made by Johan Kremenezky, a Viennese electrical engineer and industrialist, who had taken on Schapira's role among the delegates as chief lobbyist for the Fund. A month later, at the age of fifty-one, Kremenezky was appointed to head the newly created Jewish National Fund, or Keren Kayemet L'Yisrael (KKL), as it is called in Hebrew.[11] Quickly he instituted the three fundraising gimmicks that have been associated with the JNF's remarkable revenues ever since:

- The "Gold Book" recognizes important events in contributors' lives.
- JNF stamps, depicting famous sites and personalities, are either collected or used decoratively in correspondence (the stamps were

even briefly recognized by the Austrian post in 1909 and by the fledgling state of Israel before its post office was organized).

- The "Blue Boxes," in use to this day, tapped into centuries of Jewish experience; the so-called "Rabbi Meir Ba'al ha-Ness charity box" had been a fixture in many observant homes in Europe, collecting coins to support indigent religious brethren in Palestine. Donations to the new JNF blue boxes, however, went for land redemption rather than personal handouts.[12]

In fact, the effect of these three fund-raising devices was more symbolic than practical. As with most nonprofit organizations, the majority of the JNF budget has always come from "big donors"[13] or, with the inception of the State, from the payment of user fees on JNF lands.[14] The increasingly ubiquitous JNF blue box, however, raised the consciousness of Jewish communities around the world regarding the Fund and its mission. Nothing was more effective than the blue box in giving Jews around the world a sense of involvement with the Yishuv and in making Zionist settlement synonymous with overall Jewish aspirations.

The operational mission of the JNF remained amorphous, however. As early as 1896, Herzl's diary records his friend Kremenezky's enthusiasm for a "national forestry association to plant ten million trees throughout the country" (along with his desire to develop a chemical industry alongside the Dead Sea and hydroelectric plants on the Jordan River).[15] But in retrospect, forestry was a peripheral part of Kremenezky's agenda and remained so in the JNF at large until the State of Israel was established. Moreover, prior to World War I, JNF soil reclamation was limited to removing stones at Ben Shemen before planting olive trees there and to draining a modest plot near the Kinneret.

During its first twenty years, purchasing land and bankrolling agricultural activity were the centerpieces of the JNF's diverse activities. Other bodies, such as the Jewish Colonization Association (JCA) and the Palestine Jewish Colonization Association (PICA), were already active in this realm, sponsored by the great patrons of Jewish settlement, Baron Edmond de Rothschild and his son James.[16] In 1903 the JNF acquired its first parcel—200 dunams (about 50 acres) in the new settlement of Hadera—followed by larger parcels in Kfar Hitim and Hulda. Slowly, the JNF (and the world Zionist movement it represented) became the key player in this sphere.

Finding sellers and cutting a reasonable deal were the greatest hurdles that the JNF faced in those days; Zionist activity, after all, was hardly popular among local Arab landowners. The Fund's success in acquiring one million dunams (100,000 hectares) prior to 1948 was largely attributable to one of the most colorful characters in JNF and Zionist history—Yehoshua Hankin.

Arriving as a teenager in Ottoman Palestine in 1882, Hankin acquired a fluency and familiarity with Arabs and their business practices by helping his father in the family's Jaffa store. When, at age twenty-five, he stumbled on a seller of a ten-thousand-dunam tract, he realized he had a unique gift as a real-estate broker. In a moment of mystical acuity, Hankin took an oath dedicating his life to "redeeming the land." For sixty years he stuck to this sacred mission, largely at the behest of the JNF, who eventually became his best client.[17] Zionist legends abound describing this eccentric lone ranger with flowing hair and unruly beard. Fearlessly riding the backcountry on his horse, armed only with Ottoman land titles and sacks of silver, Hankin would go anywhere, any time, to buy yet another parcel of the total six hundred thousand dunams of Palestine he would purchase during his lifetime. During slow periods in the real-estate market, he had to rely on his devoted wife Olga's earnings as a midwife, but the couple never veered from their common quest.[18] In a fittingly peculiar postscript, their graves overlooking one of Hankin's biggest purchases, the Jezreel Valley, are now a popular destination for women with fertility problems, who come to pray for the childless couple's intervention.[19]

Despite heroic individual efforts, from an organizational standpoint the Jewish National Fund was all over the place during its first twenty years of operation: With only moderate success, it dabbled in tree planting, took in orphans from pogroms, subsidized high schools, and even bought the buildings for Israel's national art academy. Limited by a modest budget, it could ill afford such a diffuse strategy. After World War I, however, new leadership would get the organization on track.

THE IRON MAN TAKES THE HELM

Before the State of Israel was established, the JNF was largely associated with the dominating personality of Menahem Ussishkin, its chairman from 1922 to 1941. He took a small, poorly defined, and mismanaged charity and formed a JNF in his own image: pragmatic, aggressive, committed to Jewish labor, open to Zionists of all political persuasions, and relentless in the pursuit of "land redemption."[20] Born in Russia in 1863, Ussishkin, because of his family's affluence and influence, was able to live and study in Moscow, a city closed to most Jews during the Czars' rule. Despite his integration into Russian society, Ussishkin's supplementary education at home left him with a strong sense of Jewish identification. The pogroms of 1882 had a profound impact on him, and at the age of nineteen he became a devoted member of Hibbat Tzion, precursor of the world Zionist movement.[21]

As one of the few young leaders with the means and time to dedicate himself fully to Zionist causes, Ussishkin became something of a protégé to many early Zionist visionaries, such as Leo Pinsker and Moshe Lilenbloom, while still maintaining close ties with religious Zionist leaders. Indeed, part of the stature he enjoyed when he finally moved to Israel was linked to the Yishuv's perception of him as the "last of the first generation." From the perspective of the young pioneers, Ussishkin had walked among the legends whose writings had shaped their personal ideologies and destinies.

By the time of the First Zionist Congress, after fifteen years of activism, Ussishkin was well on his way to becoming the central figure in the powerful Russian Zionist movement. In fact, over a period of forty-five years, until his death in 1941, he would attend every Zionist Congress except the sixth, during which he was visiting Palestine.[22] A pragmatist who believed that the collective impact of small acts of settlement was the most likely road to Jewish liberation, he was instinctively suspicious of the grandiose political solutions espoused by the newcomer Herzl. Although he made his peace with the political Zionists, his consolidation of power in Chovevei Tzion during the next twenty years earned him nicknames such as "Dictator of Russian Zionism."[23] Once he was in Palestine, the epithet would soften, and he would merely be called the "man of iron," although the British, who found his imperious demeanor insufferable, dubbed him "Czar Menahem."[24] *Our Program,* his treatise, proved to be highly influential for many settlers of the Second Aliyah.

Ussishkin arrived in Palestine following World War I, at age fifty-eight, a stage of life when many immigrants would be finishing their careers. But with his fierce energy, signature gray beard, and historic credentials, he became a central figure in Yishuv politics. Indeed, he was a natural choice to assume the JNF leadership from the unpopular Dutch industrialist Nehemia Di Lieme,[25] who had taken on the job when the JNF offices moved to neutral Holland during World War I. Perhaps his background in insurance made him naturally cautious, opposing the purchase of the Jezreel Valley properties. But he was completely out of touch with the optimistic spirit in the Zionist movement following World War I and failed to attain widespread support in the movement. Di Lieme argued against the Jezreel purchase on procedural grounds (the Palestine representatives had never received formal approval from JNF headquarters in London), as well as for financial reasons (the price was outrageous, given JNF's modest resources). His bureaucratic perspective left him looking feckless alongside the passionate Ussishkin.[26]

The contrast between the two men could not have been greater. Di Lieme was Western; Ussishkin was Eastern European. Di Lieme was a newcomer to Zionism; Ussishkin had been in the thick of things for forty years. Di Lieme was analytical; Ussishkin, emotional. Di Lieme was fiscally conservative; Ussishkin, bold and optimistic. Di Lieme wanted the JNF to focus on the urban sector, where the danger of land speculation was greatest, whereas Ussishkin was a ruralist. Di Lieme saw Europe as the key playing field, whereas Ussishkin established the JNF offices in Jerusalem. Di Lieme's JNF board operated in German, but Ussishkin's minutes were entirely in Hebrew.

In short, by 1922 Ussishkin was the right man to translate the promise of the Balfour Declaration into the territorial basis for the third Jewish commonwealth: Average land purchases during his tenure as the head of JNF were forty thousand dunams a year, reaching as high as one hundred thousand by the time of his death.[27] Above all he was pragmatic. In the 1920s, for example, middle-class Zionist representatives were fearful that JNF national ownership would discourage private initiative, while Labor Zionists took the extreme view that private ownership by Jews should be suspended in Palestine. Ussishkin wisely took the middle road, arguing that areas that could be used for citrus should be left for private entrepreneurs, but that the JNF should redeem land that private enterprise would be hesitant to buy.[28]

In retrospect, Ussishkin kept his message simple,[29] with only two issues on his agenda. The first was fundamentally political: broadening the land base for Jewish settlement while creating footholds in regions that lacked Jewish presence. The second was cultural: outreach and fund-raising to bring Diaspora Jewry to the Zionist cause. Trees were perceived as a biological declaration of Jewish sovereignty; a forest's merits were primarily evaluated by its contribution to geopolitical facts, establishing borders de jure under the arcane Turkish land laws as well as marking out property lines de facto.

Many who worked with Ussishkin found him intimidating, and he used his overpowering style to squelch opposition and get things done. At the same time, old-timers in Israel remember fondly that, unlike most of his contemporaries, he kept his speeches short.[30] Ussishkin's stubbornness and frugality were legendary. He would personally review expense accounts from JNF agents around the world, canceling claims he felt were excessive. Paradoxically he was largely uninvolved in the details of what actually went on regarding the purchase and management of JNF lands, preferring instead to deal with the "big picture."[31] Ussishkin did not bring a tremendous passion for forestry to his new job. Future JNF Director General Shimon Ben Shemesh's first ten years at the Jerusalem headquarters coincided with Ussishkin's final ten years. "Ussishkin definitely didn't know from trees," Ben Shemesh recalled. "He was an urban sort. He didn't think about those things but rather focused on making money and taking care of the immigrants or getting the funds to start the University."[32]

Ussishkin was quite talented in this most important funtion—fundraising. In the annals of Zionist fund-raising, undoubtedly Golda Meir's visit to the United States in 1947 (to fund arms for the new State) ranks as the top achievement. Ussishkin's earlier 1927 visit to Canada, however, would also make the "top ten." Canada was home to 150,000 Jews, many of whom had recently arrived from Russia, where Ussishkin was something of a Zionist celebrity. The packed crowds who came to hear him speak learned of the opportunity to buy the thirty-thousand-dunam Hefer Valley and thereby link Haifa and Tel Aviv. The Canadians delivered: Ussishkin would sail home with $309,000 in checks for the next major JNF purchase after the Jezreel Valley.[33]

When Ussishkin died of a heart attack on October 10, 1941, the entire Jewish world mourned him. The collection of obituaries from the Jewish press around the world, kept at the JNF archives, is remarkable.[34] The ubiquitous JNF blue boxes made Ussishkin the best-known Zionist leader of his day, with the possible exception of Chaim Weizmann. His death was the top story from Melbourne to Montreal.

But by then Ussishkin's work was largely over. The advent of World War II and the tough British restriction on Jewish land acquisition outside of the Negev finally slowed JNF's relentless land acquisition. Meanwhile, however, the JNF under Ussishkin had bought close to a million dunams of land that literally shaped the borders for the new state.[35] (Zionist publications wistfully speak of a JNF deal that almost went through in Samaria in the 1940s and could have led to Jewish sovereignty there as well.)

THE FIRST BUDS OF A FOREST SERVICE

Menahem Ussishkin's lukewarm commitment to trees was typical of early Zionist leaders. Herzl, having witnessed "the barrenness of the land" in 1900 during his only visit to Palestine, seemed to support anything that offered some shade. In his diary he called for the planting of ten million trees in Palestine.[36] Yet given the many emergencies facing the Zionist movement, initiatives in this area remained symbolic. An interesting Herzlian anecdote foreshadowed future controversies surrounding Zionist forestry. During this visit Herzl participated in a tree-planting ceremony outside of Jerusalem. After his death, the cypress tree he planted was mysteriously uprooted, allegedly by anti-Zionist orthodox Jews who found his secular viewpoint blasphemous. But the Zionists got the last word, planting another two hundred saplings on the site with much fanfare.[37]

The forests envisioned by the European Zionists were decidedly European in nature. Herzl embraced forests as a Zionist issue only after discussions with Otto Warburg, a professor of botany and an expert in African and Asian flora. (Warburg was also a politician who ultimately became the third president of the Zionist Organization.) Warburg confirmed that planting trees would improve the country's climate and economy, and he recommended a decidedly non-Middle Eastern planting mix of African, Asian, and Australian fruit trees.[38]

The impulse to plant trees went beyond pragmatic and political considerations of preventing erosion and asserting Jewish ownership. There was a deeper psychological component to the JNF vision of land redemption. The Jewish immigrants saw the treeless land as more than ugly; they saw it as abandoned and awaiting a redeemer. Trees not only transformed the landscape into something more familiar and hospitable—the woods also evoked the freedom of the European settlers' youth. They were a source of spiritual renewal, a validating biological symbol of their hopes for a Jewish and Hebrew cultural renaissance.

Although forestry held a respectable place in the JNF's general ideology, it made for better rhetoric than policy. In terms of budget and organizational energy, it remained a low policy priority. For example, soon after he was elected chairman in 1922, Ussishkin presented to the JNF an organizational strategy that barely mentioned forestry. Trees were considered by Ussishkin to be primarily sentimental, at best an exigency required because under the old Turkish law the JNF had to demonstrate possession of purchased land if it were to maintain ownership. Ussishkin's strategic plan for the JNF cited forestry only in passing, as a tool for garnering small donations from communities that wanted to name forests after themselves.[39] By 1948 trees covered only about twelve thousand dunams, or 1 percent of all JNF lands.[40]

Perhaps the strongest proof that forestry was not a serious part of the original JNF mission was the establishment of a separate "Olive Tree

Donations" fund in 1903 by the Sixth Zionist Congress. Donors could purchase a tree for six deutsche marks. For Otto Warburg, who spearheaded the initiative, both their longevity and the economic potential made olive trees the natural choice. Soon this fund was merged with the JNF, but it would take a full decade for the JNF to launch an afforestation project. Herzl's death in 1904, at the age of forty-four, provided the impetus (and fund-raising opportunity) for the JNF's first memorial forest east of Jaffa. The project was a dreadful failure, however.[41] Ravaged by the violence and anarchy of World War I, the wooded area had dropped from 1400 to 173 dunams, with only some fourteen thousand trees left standing.[42]

In fact, some afforestation efforts in the Yishuv that preceded the establishment of the JNF were highly successful. For example, Jamal Pasha, Ottoman commander of the Jaffa region and later of the entire colony, was hardly a friend of Jewish settlement, yet he had a begrudging respect for their afforestation achievements. After visiting Rishon L'Tzion in the early 1900s, he was so impressed by the ability of groves planted there to contain the Mediterranean sand dunes that he granted the settlers rights to continue their efforts until they reached the sea. This decision infuriated the local Arab populations, who frequently responded by uprooting the plantings even though their own lands stood to benefit from the soil conservation efforts.[43]

The eucalyptus emerged as the Yishuv's tree of choice during this period. Prior to 1920, some 78 percent of all JNF trees were eucalyptus.[44] The first eucalyptus seeds were sent to Palestine in the 1880s from Australia. Several characteristics soon led to their popularity. First, they are fast-growing trees. In six years they can grow twenty meters tall with a one-meter trunk radius. In 1939, fifty-five years after they were first planted, the first eucalyptus trees at the Mikveh Yisrael agricultural school were close to fifty meters in height. Second, eucalyptus was seen as the most effective biological assistant in overall swamp-reclamation efforts. Their thirsty root system enables eucalyptus trees to flourish in wetlands, and they have been used around the world to dry swamps.[45]

In 1900 eucalyptus trees were first introduced on a large scale by Baron Edmond de Rothschild's development agency, PICA, as a swamp-draining measure in Hadera. No forestry work was more thankless.[46] The business of planting eucalyptus was so nasty that one of Rothschild's agents decided to compensate the foresters by giving them each a bottle of cognac per day as a perk. Morale quickly improved, and although workers began to arrive at work completely drunk, this did not seem to affect the productivity of their labor.[47]

Soon the eucalyptus spread out far beyond the original towns of Petah Tikva and Zichron Yaakov. The tree was usually planted in the springtime in communities and settlements throughout the Yishuv. So common was the tree that Arabs began to call it *shajarat al-Yahud,* or "the Jews' tree." The greatest enthusiasts harbored illusions of the eucalyptus providing the basis for a local lumber industry; a crate factory even sprang up briefly in Jaffa that relied solely on eucalyptus wood.[48] It was only logical that the JNF would continue this trend when it assumed the role as chief forester of the Yishuv. During the 1920s, however, the tree fell out of favor. Between 1920 and 1925, 53 percent of all JNF trees planted were eucalyptus. A year later the rate dropped to 32 percent, and by 1930 JNF hardly planted them at all.[49] What were the reasons for the disenchantment? Fifty years later, ecologists around the world came to loathe the Australian tree that had immigrated so enthusiastically. Eucalyptus trees were seen as adversely affecting the soil, water cycle, wildlife, biodiversity, fire cycle, and local vegetation.[50] The JNF's reasons for discontinuation, however, were more prosaic. There were fewer wetlands that needed to be drained.[51]

THE SCIENCE AND SOCIOLOGY OF EARLY JNF AFFORESTATION

When the JNF returned to its afforestation work at the Herzl Forest after the Great War, it no longer planted fruit trees. As it began a more systematic forestry effort, cypress, tamarisk, acacia, casuarina (beefwood), and, ultimately, pine replaced the eucalyptus. Chaim Blass, who began his long career in the JNF afforestation department during the Ussishkin period, explains the rationale behind that era's approach:

> Foresting really began with the first kibbutzim that took on projects of a few hundred dunams with JNF money. There were two ideological goals to the initiative. First, to help the economies of the kibbutz. But there was also a practical element: holding the lands, so that they wouldn't revert to Arab hands. And tree planting was a good way to achieve it. First of all, British law protected trees, which provided us with some legitimacy. And there was no activity that could hold land as cheaply as forests. Just a year or two's work and the trees really didn't need any more help.[52]

For the Jewish settlers who actually did the planting, trees indeed meant steady paying jobs. Aesthetics, however, offered a motivational bonus, although not always enough to overcome the drudgery of the work. One of the kibbutz members who made a living planting JNF trees in 1921 was a recently arrived American immigrant named Golda Meirson, who would later shorten her last name to Meir. She found the job miserable, but in retrospect was extremely proud that both she and her trees survived.[53]

The JNF afforestation policy prior to 1948 was devised by Akiva Ettinger, one of the JNF's first scientists. Ettinger had a Ph.D. in agronomy and had studied in Germany. He was sent by Baron Maurice de Hirsch to travel around the world to look for appropriate farmlands for Jewish settlement. He began work at the JNF after it relocated to The Hague during World War I. When he later moved to Palestine, he brought with him clear ideas about how forestry should be conducted there: "Planting many fast-growing trees that do not require long-term maintenance, and at the same time planting on lands that cannot be utilized for agriculture other than forestry, such as rocky lands, swamps, and moving sand dunes." [54] No one played a greater role in implementing Ettinger's practical but ultimately controversial JNF strategy than his assistant and successor, Yosef Weitz.

THE FATHER OF THE FORESTS

When Yosef Weitz was appointed head of the JNF lands and afforestation department in 1932, afforestation received a great boost. Weitz brought a forester's bias to the job, although the new position involved general responsibility for land acquisition and settlement. He would remain the key player at the JNF until his death forty years later and would deserve his nickname, "the father of the forests" (Figure 6). Indeed, probably no single sector of Israel's environment was influenced by a single individual as much as Israeli forests were by Yosef Weitz. Weitz was a truly prolific writer, so we know much about him. His personal diaries stretch on for five volumes, and his definitive 1970 history, *Forestry and Afforestation in Israel,* is almost autobiographical.

Forestry was not Weitz's first love. He had grown up around wood, because his father was in the lumber business, but when he left his small town in Russia for Palestine at the age of eighteen, he was a typical Second Aliyah farmer. Steeped in the culture and politics of Labor Zionism, he started as an agricultural worker in Rehovoth in 1908, moved on to Hadera, and eventually reached the Galilee. There, at Sejira, he was made a farm manager owing to his formidable organizational skills and work ethic. He was also very smart, and, like many of his generation who never attended university, he displayed extraordinary autodidactic talents. His collected papers in the JNF archives in Jerusalem reveal all sorts of surprises, including a 1919 academic journal article that he wrote in French about sand dune reclamation techniques. A few months after this publication appeared, the JNF came calling (Figure 7).

When Akiva Ettinger, then the JNF's chief agronomist, asked Weitz to join his staff, the immigrant farm manager was already 31 with no formal training in forestry. But he was a quick study. Weitz's longtime personal assistant, Shimon Ben Shemesh, believes that most of his boss's strategies, as well as his conceptual approach to afforestation, were taken directly from Ettinger. Yet Weitz gets credit for several JNF innovations—such as importing the first date palms from Egypt to the Kinneret and to Degania and bringing carob scions from Cyprus for grafting.[55]

Weitz's career was a marathon run during which he rarely slowed down. Throughout the fifty years he worked at the JNF, photographs of him show the same erect figure with wire spectacles, thinning hair, dark mustache, enormous walking stick, and very few smiles. The first foresters who worked under him during the 1930s speak of him adoringly as a father figure,[56] but Weitz's public persona was strictly authoritarian. In the early days of the new nation, Professor Heinrich Mendelssohn came to talk with Weitz about setting aside some lands in the Huleh swamp as a nature reserve before the JNF drained the land. "He started screaming at me and banging on the table. 'You want the land for the animals and plants, and if it was up to you, there'd be none left for the Jews. You're an *ocher Yisrael* (an enemy of Israel),' he yelled and then threw me out of his office. I remember he called me an *ocher Yisrael,* because my Hebrew has never been that great, and I had to look the expression up in the dictionary."[57]

Weitz was never comfortable as an office bureaucrat. His diaries are filled with the frustration of a caged bird, and he sounds truly happy only when recounting his numerous trips across the country. Environmentalists who passionately opposed his approach to development still retained great respect for his familiarity with the specifics of local geography.[58] A field visit by Weitz would begin at 6:00 A.M., with workers joining him at his home for a full breakfast that always included an omelet.[59] Often he brought his wife or one of his three sons on visits. Ben Shemesh fondly remembers Weitz's inclination to convene departmental meetings under the shade of the trees.[60] Despite his reputation for stubbornness, however, Weitz eventually would make his peace with the environmentalists and show remarkable intellectual openness.

During the War of Independence, Weitz was the leading figure on the three-member "Transfer Committee," which sought to expedite the departure of local Arabs, and which operated at first with tacit and later formal approval of the prime minister.[61] As a result, some historians have painted a one-dimensional picture of Weitz as a callous, Arab-hating official.[62] That view, however, ignores the historical circumstances as well as Weitz's tireless efforts to offer Arabs compensation for lost lands. As an indication of his actual position on the Palestinian issue, he opposed occupation of the West Bank after 1967 and even refused to attend dedication ceremonies when old JNF settlements at Gush Etzion were resurrected.[63]

It is a much more sentimental Weitz that emerges from his diary. Recalling his first serious afforestation effort with stone pine trees (*Pinus pinea* L.) in the Herzl Forest at Huldah in 1922, he wrote:

> One morning, I'm coming up the road from the train, filled with blissful expectations of "the forest." I climb, skipping up the incline to see my soft, newly nurtured darlings. But then they came into view and my eyes turned black. The green saplings had turned brown, except for a few isolated ones here and there, and from the heights I could hear the ridicule of the Angel of Death. I fall to my knees and gently touch the saplings. Needles crumble and fall, leaving only a few orphaned needles behind. I dig out all the soil around the saplings to uncover the mystery of their sudden demise, and I don't find anything but a broken heart and tears in my eyes. I was ashamed of myself. My heart asked remorsefully: "What will I show Ettinger?" and total failure encompassed and followed me.[64]

Weitz not only had to overcome grief from lost trees: His youngest and dearest son, Yehiam, was killed in a 1946 Palmach demolition attack. In a posthumously published journal entry he writes: "I don't have the courage to go out on the street, to come to the office, to look at the everyday world. Only the plants in the garden look at me honestly. It is only with them that I can talk."[65] Still, Weitz would bounce back and lead the JNF through its most ecologically controversial period.

THE JERUSALEM PINE COMES OF AGE

In absolute terms, before 1948 JNF forestry was overshadowed by British efforts.[66] Nonetheless, this period has enormous conceptual significance. In a sense, the thirty years of the Mandate served as a pilot run during which the JNF established the strategy that would accompany it during its most intensive period of afforestation after Israel's independence. It was during the 1930s that the Aleppo (or Hebrew, or "Jerusalem") pine (*P. halepensis*) became the dominant tree in JNF forests. Its emergence was swift. At the start of the 1920s, JNF forestry projects were initiated at eight sites. The foresters had no clear preference for any given tree type, eucalyptus experience notwithstanding. They experimented with a variety of species and combinations. On three mountainous locations (Ben Shemen, Huldah, and Kiryat Anavim) the stone pine was favored. The five wetland and sand dune planting locations (Beer Tuviah, Rishon L'Tzion, Merchaviah, Kinneret, and Degania) featured eucalyptus and tamarisk.

By the time work commenced at the larger Balfour Forest near Kibbutz Ginnegar and Mishmar ha-Emeq in 1929, needle-leafed conifers, in particular pine trees, had captured JNF hearts and minds. For the JNF foresters, the results looked like what a forest should be. The trees seemed adaptable to various climates and seasonal rain patterns. Most of all, they grew quickly. By 1936, forest plantings were integrated in the creation of new settlements such as ha-Zorea, Chanita, Ein ha-Shofet, and Kfar Choresh. In 1939, before World War II slowed afforestation efforts, over half of the JNF forests were located in the general vicinity of the Jezreel Valley.[67] All were planted in Jerusalem pine, with little use of stone and brutia (*P. brutia*) pine species.[68] Many veteran JNF foresters argue that it was available terrain rather than doctrine that was behind the shift in inclination to pine trees.[69] But as is often the case, exigency can turn into ideology.

The scientific debate about the indigenousness of the Aleppo or Jerusalem pine species has not been entirely resolved. For years the prevailing view was that of the noted botanist Michael Zohari, who saw the tree as the proverbial *etz ha-shemen*, or "tree of oil" from the Bible. He emphatically argued that even where no signs survive today, before man's intervention the hills of Israel were covered with pine trees. The tree might even be climax flora (the fully developed natural vegetation when ecological equilibrium is reached with climate and soil) in areas from Zfat to Samaria.[70] This view is now disputed. Based on genetic and enzyme analysis, later researchers would show that the brutia pine was in fact the dominant local variety. Their tests suggest that the Jerusalem pine is predominantly a North African species whose seeds were imported in ancient days and planted in isolated locations.[71]

How in twentieth-century Israel did the Jerusalem pine attain this "most favored tree" status? (In 1994, two decades after its official fall from grace, this species of pine tree still constituted more than half of the forested lands in Israel.[72]) In an extremely thorough study of JNF planting records, Nili Lipshitz and Gideon Biger of Tel Aviv University document the sudden ascendance of the Jerusalem pine, which came to dominate JNF afforestation for fifty years. As early as 1918 Akiva Ettinger reported the first JNF plantings of Jerusalem pine. They were only 4 percent of the trees planted in the Herzl Forest, hardly noticeable when half of the groves were eucalyptus.[73] Of this original pine stand, only 13 percent survived World

War I. In future comparisons, however, Jerusalem pines consistently outperformed the oak trees planted between 1926 and 1929 at Kiryat Anavim, as well as the carobs and other species in the Balfour and Mishmar ha-Emeq Forests. Yosef Weitz began to take note.

When the Jerusalem pine successfully replaced his failed crop of stone pines at the Huldah site, Weitz was sure he had found a winner. The percentage of Jerusalem pines planted skyrocketed; by 1926 they constituted more than 50 percent of the 69,335 JNF trees planted in Palestine. After 1930 this percentage would increase to 86 percent, and peak at 98 percent in 1934.[74] What was the appeal of this particular conifer? A 1936 article by Weitz, "Forest Trees in the Land of Israel: The Jerusalem Pine," spared no praise of this botanical wonder, revealing the JNF thinking that prevailed until the 1970s:

> For one, it adapts to different climates: from the Jordan Valley, the wasteland receives its shade. In Jericho and Degania you will meet it. In the coastal plains and in the Sharon it will flourish, and on mountains eight hundred to one thousand meters above sea level. For another, it does not discriminate according to soil type. It is happy to blossom in sandy and organic soils alike, and even on rocks it sends its roots to explode them and grab hold. It finds soils rich in lime to be most pleasant, so it can be planted in the most desolate places in the land. And, finally, it expands and grows quickly.[75]

With Weitz overseeing every detail, an entire JNF science was developed for planting pine trees, as foresters moved up the learning curve and corrected flaws in their silviculture technique for the species.

JNF pine forests in this period were perhaps most distinct for their high density of planting. Forests around the world rarely have more than one thousand trees per hectare (one hundred per dunam), leaving ten square meters per tree. In contrast, even at the end of the 1960s JNF tree density typically reached four to six hundred per dunam.[76] High-density plantings were thought to offer a visually satisfying rapid green cover, inhibit competing vegetation, direct tree energy into the main trunk rather than the branches, and ensure sufficient numbers of surviving trees.[77]

The real reason for the density, however, was the JNF vision of an Israeli timber industry. Intensive growth might make forests impassable for hikers, but would increase their profitability as tree farms.[78] The silviculture life-cycle theory adopted by the JNF advocated continuous pruning until trees reached the age of fifteen. More important, thinning was supposed to be done ten years after the planting of saplings and to be repeated every five or six years until the end of the forest's rotation, at about age fifty. By that time, tree density might be as low as twenty-five to thirty trees per dunam.[79] Yet this schedule was rarely met, because of personnel constraints. The result was unnaturally straight stands of weakened trees that were vulnerable to fire, prone to collapse in storms, and highly susceptible to drought.[80]

Although the planting menu favored by British government foresters during the Mandate was more diverse than that of the JNF, roughly half of the thirty-one million trees the British planted between 1920 and 1948 were also pine. Unlike the JNF, the government workforce was almost entirely Arab. But the British also saw forestry as an important step in fighting soil erosion, stabilizing sand dunes, and providing timber for the local economy. The fact that the British also preferred *P. halepensis* was interpreted for many years as a vote of confidence in JNF forestry methods.[81]

NATURAL AND UNNATURAL ENEMIES

Jerusalem pine trees have a natural life expectancy of eighty to one hundred years; if irrigated, they may live to be 150. In fact, only a fraction survived that long. The pine forests of both the British and the JNF were under a constant state of siege prior to 1948. Arson and other forms of what Mandate officials called "political sabotage" were a constant problem for the JNF; the Jerusalem pine's flammable sap literally added fuel to the proverbial flames. During the Arab Revolt of 1936, vandalism became so violent that JNF planting efforts became unsafe. This led to a strategy of pairing a young forest with an adjacent kibbutz that could protect it. The Arabs saw the forests as easy military targets, and, in fact, the young forests were used to conceal the illegal bunkers and clandestine military training of the Haganah and Palmach.

In retrospect, the Arab disturbances proved counterproductive. The JNF defiantly planted one million saplings between 1936 and 1938, almost double previous afforestation efforts.[82] The arson doubly backfired, much as the Intifada would fifty years later: While foresters became more vigilant, JNF fund-raisers were quite adept at exploiting the destruction of helpless trees. The JNF budget expanded considerably in the late 1930s, as a sympathetic Jewish public around the world dug into their pockets to fight the flames with record donations.[83]

But the greatest threat to the young forests was not from fires but from insects—in particular an aphid known locally as the Jerusalem pine blast (*Matsucoccus josephi*). Lipshitz and Biger's research suggests that as early as 1933 the first definitive scientific identification of the blasts was made on the Carmel.[84] Damage to trees from this persistent airborne pest includes both acute and chronic injuries. The crawlers love everything about Jerusalem pines, settling on all parts of the tree that lie above ground. The preferred sites are the partially smooth stem sections with scaly bark in mature trees and the buds and the base of the fast-growing shoots. The pests dry out the branches and in acute cases cause the death of the tree within months. When damage is chronic, the process can last for decades, as limbs slowly dry and atrophy.[85]

In 1935, with a 70 percent loss in the Mishmar ha-Emeq forest, JNF forestry officials could no longer ignore the scourge. A formal scientific survey revealed that, in contrast, the stone pines in the forest were unharmed.[86] Several of the Yishuv's top botanists, including Hebrew University professors Hillel Oppenheimer and Fritz Bodenheimer, were enlisted in the research efforts. (Oppenheimer's site visits to the JNF's northern forests were coordinated by Sharon Weitz, Yosef's middle son, who had already joined his father's afforestation department and would eventually run it.) The chief government entomologist, Dr. Karl Shveig, identified other problematic pests such as the caterpillar *Evetria rhyacca buoliana* (nicknamed by Shveig the "Pine Fire"). But the biological detectives explicitly singled out the *Matsucoccus* blasts as the key guilty party.

There was little the entomologists could do to help, however, particularly once the upper branches of trees became infested. This scientific impotence was apparent in October 1938 when Professor Oppenheimer submitted his report to Weitz. The professor sounded a clear warning about the long-term implications of the pest problem: "After the disease has spread to such a great extent, I doubt if we can continue to successfully grow forests comprised only of densely planted Jerusalem pines."[87] The warning went unheeded. Although allowing for some diversity among its conifers, on the whole JNF managers remained obtuse for forty years. Planting policy eventually changed, but as recently as 1969 JNF literature still hailed the Jerusalem pine as its flagship species.[88]

A NEW INSTITUTIONAL IDENTITY

On March 10, 1949, two Israeli brigades took competing routes down through Israel's southern Arava to see who would be first to reach the police station at Um Rash Rash at the northern tip of the Red Sea. They both made it safely; not a shot was fired as Israel's southern border was thus forged. The armistice agreement with Egypt had been signed two weeks earlier on the island of Rhodes. The War of Independence was over, and the newly independent State of Israel woke to discover that it held over twenty million dunams of land—far more than that allocated under the UN-approved partition plan. With the exception of the Etzion block, which fell to Jordan early in the war, the borders included virtually all the JNF forests.

The creation of the State of Israel touched off something of an identity crisis for the Jewish National Fund. Military victories had expeditiously completed the primary mission for which the JNF had toiled so patiently and steadfastly: The land of Israel was under Jewish jurisdiction. What was left to be done? Even in the area of forestry, the JNF was no longer the only player. Israel's Ministry of Agriculture inherited the British forestry department, whose professional experience and woodlands exceeded those of the JNF. To complicate matters, the future of the JNF was linked to a broader issue: the relationship between its parent institution, the Zionist Organization/Jewish Agency, and the State of Israel it had created.

There was no real reason for concern, however. No one ever thought seriously about closing down the JNF. Despite tension with the JNF over Arab refugee policy, Aaron Zisling, the leftist Mapam party interim Minister of Agriculture, wrote to the JNF Directorate requesting that it continue its work.[89] Actual land purchases by the JNF, in fact, reached a new peak immediately after the War of Independence: On the block were the so-called Absentee Properties—lands abandoned by Arabs during the course of hostilities. The JNF believed that it, rather than the State, should hold these lands. First of all, it assumed that the Fund would be in a better position to pay for the "millions" in compensation that would ultimately be exacted. Second, given the instability of the new State politically, and Israel's unresolved demographic situation, the JNF felt better able to guarantee Jewish land ownership.[90] Yosef Weitz took the lead in negotiations with his old Second Aliyah comrades who now held power.

The new Israeli government was initially receptive to Weitz's overtures. As a result, JNF land holdings tripled between 1948 and 1954, from 936,000 to 3,396,333 dunams. The acquisition of such controversial lands remains an enormous source of bitterness and outrage for Arab-Israelis and Palestinians, who still see the JNF as representing the most imperialistic aspects of Zionism. As the diplomatic stalemate set in and Israel's map stabilized, however, the government reconsidered its position, and the real-estate sweepstakes promptly ceased. When James de Rothschild, Edmond's son, transferred the baron's remaining Israeli land holdings to the JNF in 1957 (including 3,000 dunams of woodlands),[91] this brought

JNF-owned territory close to its final level of 3.6 million dunams or 15 percent of the country's lands.[92]

Fundraising was reason enough for the state to preserve the organizational infrastructure of the JNF, a nongovernmental corporation with a nonpolitical persona, forty-eight national affiliates around the globe, and unparalleled name recognition among Jewish communities in the world. In retrospect, however, this was probably the least compelling reason to continue the JNF. After the creation of the State of Israel, the JNF was no longer dependent on donations from the Diaspora: In 1997, for instance, 83 percent of the its budget came from payments of Israeli leases.[93] United Jewish Appeal and Israel bonds would soon supplant the high profile of the JNF in Diaspora fund-raising.

The JNF scandal of 1996, however, offers an excellent example of how detached the Israeli JNF program has become from its Diaspora supporters since independence. That year, American JNF offices were rocked by journalist Yosef Abramowitz's claims of illegal diversion of funds to South American JNF offices and general mismanagement.[94] A subsequent probe revealed that despite its slogan that "Israel is our only business," only 21 percent of U.S. donations to the JNF were reaching Israel. The scandal resulted in the unceremonious resignation of the JNF's senior American management.[95] Before 1948 this sort of public-relations debacle would have been a major source of concern for the Yishuv, but now the story failed to crack the newspapers in Israel.

Of course, sentimental factors, as well as the natural tendency of bureaucracies to perpetuate themselves, helped keep the JNF alive after the establishment of the State. There was a substantive reason too, however. With colossal challenges facing a young nation in every conceivable sphere, the government of Israel was truly relieved to have an existing and reliable nongovernment agency to which it could delegate some of its functions. It would take a decade for the formalities to be worked out, but the new focus of the JNF was soon apparent: infrastructure and agricultural land development, with a particular emphasis on forestry. The JNF would maintain its status as an independent institution within the World Zionist Organization, working with the Jewish people in Israel and abroad to raise funds "for the redemption of the land from its barrenness."[96]

The JNF would require internal and external reorganization to adapt to this new role. The first formal institutional change by the JNF after independence was the establishment of a legal entity in Israel for the Jewish National Fund.[97] Since 1907, the JNF had been registered as an English company, Keren Kayemet L'Yisrael Ltd.[98] Now it was established as an

Israeli corporation: Keren Kayemet L'Yisrael. Working out a relationship between the JNF and its government counterparts in areas where activities overlapped was a thornier task. The functional division of labor in the area of afforestation was set forth in a 1949 proposal by Weitz: The government's Forestry Division would be involved in applied research into planting techniques, particularly in semiarid areas, and into industrial timber opportunities as well. In addition, the government would establish nurseries for subsidized distribution of saplings to citizens and for educational initiatives. The JNF, in turn, would be responsible for improving indigenous forests, afforesting hilly regions, stopping the spread of sand dunes, and planting windbreaks. In addition, the Israeli army formed a special tree-planting unit to shield sensitive roads from enemy cannon and rifle sights.[99]

As is often the case, sorting out responsibilities was easier on paper than in the field. Amihu Goor and his staff at the Ministry of Agriculture's Forestry Department were perceived by JNF leaders as tainted by their association with Mandate foresters. (Goor, whose sister had married the son of England's first High Commissioner to Palestine and who had completed degrees at Oxford and Yale, perceived himself as something of an aristocrat during the Mandate.[100] Those credentials did little to endear him to Yosef Weitz.) Coordination between the competing agencies fell apart, and the appointed liaison committee soon stopped meeting. The inefficiency associated with duplication of efforts set in.

The problem was not limited to civilian government foresters. JNF foresters saw the commander of the military's forestry division, General Akiva Atzmon, as unrealistic and suffering from perennial flights of fantasy.[101] It was not until 1959 that the government granted the JNF full authority over afforestation, and it was not until 1964 that the Ministry of Agriculture bowed out of the forestry business altogether, disbanding its department.[102] The supremacy of the Forestry Department (operating under JNF's Land Development Department) in related matters was uncontested at last and remains so today. This remarkable autonomy entrusted to a private agency has been criticized by environmentalists and developers alike. Ultimately the traditional autocratic approach was challenged in the Supreme Court, which in 2001 ruled that present JNF forestry procedures were inherently undemocratic and illegal.[103]

The final stage in the institutional evolution of the JNF came about in its 1961 covenant with the Israeli government and in a series of Knesset laws that formalized government land policy in Israel. The covenant was the culmination of a lengthy negotiation process that began when Ben-Gurion appointed a joint JNF-government committee in 1957 to consider issues of land ownership and development. By the time the covenant was signed on November 28, 1961, it was rendered a mere formality. The Knesset had already enacted a Basic Law for Lands some eighteen months earlier, which constituted a somewhat more binding statutory expression of the agreement. Of the covenant's principles, none had greater symbolic significance than the State's pronouncement that the land of Israel is owned by the Jewish people and must not be sold in perpetuity.[104]

The JNF entrusted the management of its lands to a newly established Israel Lands Administration, which also oversaw government holdings. In practice this means that leasehold agreements issued by the Lands Administration mirror the standard JNF contract prior to the covenant (for example, they run for 49 years with an option for renewal). Both entities maintain formal ownership of their lands. Supervision, however, is in the hands of the Israel Lands Administration, which is overseen by a council of six JNF and seven governmental representatives. The JNF retains the right to withdraw from the covenant if the Basic Law is repealed or amended without its approval.

It might have seemed that by transferring administrative authority for its lands to a government agency, the JNF struck a foolish deal, yielding its primary source of influence, but in retrospect the agreement turned out to be lucrative. The direct influence of the JNF was extended to the 92 percent of Israel's lands that were now State-administered. An Advisory Board to the Israel Lands Administration even has a JNF majority. Yosef Weitz was appointed as the first director of the Israel Lands Administration, during its first six years bringing his distinctively JNF perspective to the task. Most important, the JNF maintained the revenues from its considerable holdings. Had forestry been solely a governmental activity, its level of appropriations would be far smaller than the level it enjoys today. And despite being a nongovernmental corporation, the JNF maintains almost total independence in determining forest policy.

THE NEW STATE STARTS PLANTING

In a speech to the newly formed Israeli Defense Forces in 1949, Yosef Weitz described the state of the forests inherited by the JNF and his vision for the future. He classified 170,000 dunams of Israel's land as forest, covering less than 1 percent of the country and consisting of dispersed tree stands of modest dimensions. He called for a 1700 percent increase in the nation's forestland.[105]

Weitz would have no trouble adapting to the JNF's new organizational identity and tackling the challenge of developing a forestry policy for the new nation. He had been waiting almost thirty years to do just that. For the first time his ambitious visions of a forested country enjoyed substantial political backing, and at the highest level: from the seemingly omnipotent prime minister. Not so much Zionist ideals as massive unemployment sent JNF planting into high gear. Ben-Gurion's enthusiasm for forestry is best understood within the dynamic context of Israeli demography, his Bible-thumping rhetoric notwithstanding. Almost overnight the population had grown from six hundred thousand to one million people. At the end of 1949, one in ten Israelis languished in immigrant camps. The situation in the camps was more than just a drain on national resources; it was demoralizing both for a young country that failed to meet its own expectations and for disappointed and increasingly bitter immigrants.[106]

For many years JNF tree planting was inversely proportional to the national employment situation. For instance, in the early 1960s, when the country reached full employment, plantings tapered off and work became more capital intensive. Plantings waxed again during the recession of the mid-1960s.[107] The massive wave of Russian and Ethiopian immigration that began at end of the 1980s once again put job creation on the JNF agenda.[108]

Chaim Blass recalls one of the most famous forestry stories of the period from August of 1949: "Ben-Gurion had no patience. He wanted to settle the country in one day. He came to Weitz after the War and told him that he wanted to plant one billion trees over the next decade—one hundred million a year. Imagine Yosef Weitz's frustration."[109] In his diary, Weitz wondered whether the prime minister, who insisted on seeing the forests in political rather than botanical terms, had lost his mind.[110] Ben-Gurion was so concerned about unemployment that he pressured Weitz to plant in summertime, when the likelihood of success was lowest.[111]

Still, the JNF Forestry Department responded to the prime minister's marching orders with an unprecedented campaign.[112] Weitz, for his part, was strongly opposed to offering charity to new immigrants and saw inculcation of a work ethic as a crucial part of acculturation.[113] Dividing its operations into four geographic regions, the JNF set up work villages where immigrants were sent to reclaim and plant JNF lands. The work was highly regimented, almost military in style, and the immigrants had to meet daily production quotas.[114]

Forestry thus became the paramount JNF task, because it was the most labor intensive. By 1950, Weitz could report to Ben-Gurion the planting of a million eucalyptus and tamarisk seedlings for windbreaks and shelter belts in the Negev, with a 60 percent success rate. During the JNF fiscal year 5610 (1950/1951), 12,650 dunams were planted—four times the level of the previous year. In 1951/1952, the number increased to 56,400 dunams—five times the area that had been planted by JNF during its first fifty years![115] For the rest of the 1950s the number dropped to around twenty thousand dunams and six million trees per year.[116]

The JNF's accomplishments as an agricultural assistance agency during this period deserve mention along with its forestry work. The hundred new rural settlements established each year during the 1950s had a very tenuous economic basis. The contribution of JNF workers, who prepared some thirty thousand dunams a year for cultivation, was critical. The most ambitious of these projects, the controversial draining of the Huleh swamp, will be considered at length in the next chapter. This level of intensity eventually dissipated: During the 1960s the amount of soil reclamation declined by 50 percent, reflecting an 80 percent drop in new settlements.[117] But JNF aid found other forms, including the start of a poultry industry for financially beleaguered mountain settlements.[118]

It was during the critical post-independence period that the JNF's new institutional identity came together. Three factors elevated afforestation's status during this process: its ability to offer a quick fix for unemployment; the need for the JNF to find a new organizational niche in an independent Israel; and Yosef Weitz's personal proclivities. When these factors combined with the commitment of Israel's early leaders to a pioneering culture of land settlement, and their desire to import a European natural aesthetic, JNF's new functional role was cast. With time, a more mature and sophisticated ecological program would emerge. But its roots remained planted in the unique circumstances that characterized the incipient State of Israel.

THE NEW FORESTERS

Mordechai Ruach arrived in Israel from Egypt in 1949 at the start of the mass migration from Arab countries, and, like thousands of other immigrants from countries such as Morocco, Yemen, and Kurdistan, he found temporary employment with the JNF. The work was hard because it was designed to be labor intensive and light on capital. Workers used simple hoes, shovels, and manual drills. Israelis were typically cynical about the quality of the JNF's immigrant employees, whose work was subsidized by the Ministry of Labor. Chaim Blass still bristles at such sneers when recalling the dedication of these *olim* to the new forestry projects.[119] Ruach also remembers most clearly the zeal of fellow immigrant workers who, like the Second Aliyah pioneers fifty years earlier, saw planting as a form of personal redemption and a sacred rite. Newly arrived Yemenite workers would kiss the ground each time they finished planting a row.[120]

Because of the unprecedented magnitude of the planting, improvisation characterized much of the decision making in the field. To a certain extent this was unavoidable. Even under normal circumstances, rainfall can undermine the best-laid plans and afforestation schedules: A few weeks without rain in the winter can jeopardize a planting timetable, and plentiful showers can extend a season beyond expectations. Weather offered many surprises during that energetic period. During Haifa's freak snowstorm of 1950, for example, a desperate effort to shovel the snow off pine seedlings in the country's largest nursery proved hopeless. To the astonishment of the foresters, those trees left safely under the protective snow blanket survived, whereas the few exposed seedlings died.[121]

One failed experiment from the period involved carob trees. The carob is not mentioned in the Bible but figures prominently in the Mishna and Talmud, suggesting that it was imported from Yemen with the Nabateans after the destruction of the First Temple.[122] The carob tree has thick dark evergreen leaves that form a broad green crown. The black pods of the carob are rich in sugar and are sometimes used as a chocolate substitute. Although materials in the pods can be extracted for certain paints and glue, Weitz was most excited by their potential to replace imported fodder for cows and poultry. The tree requires deep soil where plowing is possible, but it can also survive in relatively high altitudes with little rainfall. JNF experiments indicated that one ton of carob fruit could be harvested from each dunam of trees. Twenty thousand dunams were planted alongside the conifers in the Judean hills and the Galilee. Then suddenly dairy farmers began complaining that the fruit inhibited milk production. The project was abandoned, but the trees remain.[123]

By this time, JNF planners were making a clear distinction between protective and productive forests. Along with their political, military, and aesthetic benefits, the former were designed to prevent erosion, stop sand dune movement, and mitigate the effects of dust storms. They were not designed to produce wood. Although most JNF officials argue that economic profit was never a key factor in planting strategies, some foresters, educated in Europe, clung to illusory hopes for a serious Israeli timber industry.[124] Even in the late 1980s, JNF planners had to provide assurances that new tree species could eventually be logged to support the elusive particle-board industry.[125]

Today, timber yields exceed one hundred thousand tons per year, showing a steady increase. Roughly 40 percent of the wood comes from thinning activities, another 40 percent comes from felling conifers, and the remaining 20 percent from eucalyptus trees.[126] This provides about 10 percent of domestic wood consumption. While this is not an insignificant level, logging no longer seems to influence JNF decision makers or supporters.

THE ENVIRONMENTAL CRITIQUE

The pace and magnitude of the postwar tree-planting blitz did not go unnoticed by the small, but vocal, environmental community in the nascent State. Without exception, Israeli environmentalists liked forests and the biodiversity they could support. Yet sterile rows of scrawny JNF pines, separated by fire lines into symmetrical matrices, left them with a cold sense of alienation rather than any renewed feeling of closeness to nature.[127] As the JNF brought mass-production forestry to greater and greater tracts of Israel's land, the results produced indignation.

In fact, the first organized environmental campaign in the State of Israel was directed against a JNF plan to establish a settlement on the site of a native oak forest. A coalition of activist zoologists and members of Kibbutz Alonim wrote Yosef Weitz: "We sacrificed strength and blood so that these ancient oaks wouldn't be destroyed by our Arab neighbors. There is no justification for these trees to be uprooted by the State of Israel." Weitz assented and ordered the plan revised to preserve the trees.[128] This case of accommodation was an exception that proved the rule. To begin with there was an ideological divide that was difficult to bridge. "A forest is not just trees but other plants, as well as a place for animals to live," railed conservation leader Azariah Alon. "The foresters refused to accept this: 'We're foresters—not zookeepers,' they said."[129]

Little enraged environmentalists more than JNF methods of planting. As the JNF acquired heavy equipment, tree plantings took on the sort of environmental insensitivity usually associated with Green caricaturizations of agribusiness. The attack was multipronged. First, fires were lit to erase any remnant of indigenous bushes, trees, and brush. Next, bulldozers were brought on to sweep away the debris; then plows prepared the soil for planting. Finally, pesticides ensured that the new pine seedlings would not be troubled by other undesirable biological activity. Environmentalists charged that the underlying soil inevitably suffered from the relentless onslaught, while the surrounding ecosystem was irreversibly knocked off balance.

Once the trees grew, their needles formed a highly acidic ground cover that decomposed very slowly. The result was a sterile forest bed inhospitable to additional undergrowth and to most animal populations. Environmentalists coined the term "the pine deserts" to describe them, seeing even humans as aliens among the crowded rows of skinny trees. Although Israelis make their way in droves to these forests during holidays, they tend to stay in the crowded picnic/playground *chanyonim* and rarely wander the forests themselves. Professor Mendelssohn rejects the term "forests" altogether for these sites: "The JNF planted pine orchards. A forest is an ecosystem that develops over thousands of years."[130]

In addition to the ecological critique, there was an aesthetic one. As Knesset Member Rachel Zabari complained in parliament, despite their ostensible contribution to security, the walls of JNF trees blocked scenic vistas.[131] JNF foresters generally lacked a sense of landscape. For instance, the same fire prevention lines that cut the woods into perpendicular farming cubes could have been designed to flow with the contours of the land.[132]

Another complaint that became more angry with time involved the use of chemicals. During the 1960s, the JNF foresters' reliance on pesticides grew. Environmentalists could not forgive the JNF for pressing on with Jerusalem pine monoculture, even after JNF's own scientific research suggested that the *Matsucoccus* aphids undermined its sustainability. The results could only be decimation or chemical dependence. Chemicals were used in the fight against each of the three primary adversaries of trees: weeds, fires, and pests. Until the trees' second year, when saplings are strong enough to withstand competitive flora, they need weeding twice in the springtime (weeds are less problematic during the dry summer months). The JNF's first forests were weeded by hoeing around the base of the trees, which subdued annual plants and reduced water loss to evaporation.[133] When the costs of labor increased, however, chemicals became the primary means of control. Foresters claim that their practices are fundamentally different from agricultural chemical dependence. Crop farmers spray on an ongoing basis, whereas foresters apply herbicides only once or twice at the start of a forest's life cycle, which can last sixty years or more.[134] Yet this argument does not cover the growing role of herbicides in forest fire prevention, where spraying is continuous (the ten-meter fire prevention lines separating stands of trees are useless if they fill up with opportunistic brush). The argument also fails to address the change in the composition and diversity of natural species caused by the chemical applications.

The JNF herbicide of choice for many years has been simazine. Selected because of its marginal impact on perennials, the chemical successfully prevents germination of annual plants. It is especially effective when applied at the start of winter, after the first rains. Unfortunately simazine is also a hazardous poison and is classified as a possible carcinogen. The U.S. Environmental Protection Agency moved to revoke the tolerance level for simazine in 1996 after studies indicated that the chemical caused cancer in female rats. Sheep and cattle are especially sensitive to it.[135] Seven European countries, including Germany, Holland, and Sweden, have imposed stringent restrictions on comparable chemicals. The chemical is highly stable, making it a potential groundwater contaminant in addition to a potential health threat to people and animals that frequent the forest. The chemical was first identified in Israel's rural aquifers during the 1970s. Subsequent studies identified residue levels in groundwater two to three days after spraying that were orders of magnitude higher than the 0.004-milligram-per-liter standard set in the United States.[136]

Spraying has also been directed at a variety of insects that are enemies of Israel's pines. For example, the pine processionary caterpillar is primarily controlled through the aerial spraying of endosulfan and diflubenzuron.[137] Here again, the ecological impacts of decades of spraying have not been studied but may be extremely severe. Endosulfan is a chlorinated hydrocarbon (and therefore highly persistent), designed to damage insects' central nervous systems. Human exposure to endosulfan can lead quickly to lack of coordination, gagging, vomiting, diarrhea, agitation, convulsions, and loss of consciousness. The chemical causes cell mutations, although there is as yet insufficient data to classify endosulfan as a carcinogen. Nonetheless, sheep, cows, and pigs that grazed in fields where the chemical was sprayed have gone blind, and there is much evidence indicating that birds and fish are particularly sensitive.[138]

In all fairness, it must be said that the JNF tries not to spray excessively. The quantities applied have been reduced, and foresters are at least directed to be more selective and spray only in critical situations.[139] For instance, although the Mediterranean pine shoot moth, a common Middle Eastern pest, attacks all species of pines, especially when they are young, it generally destroys only shoots and does not kill the trees. Hence the JNF does not spray against it.[140] And the damage that rats do to carob trees has no chemical panacea, so no spraying is attempted. And as for the pesky *Matsucoccus josephi* and the Jerusalem pines, years of research concluded that continuous, good-old-fashioned pruning probably offers the best defense.[141] Behold, the saw is mightier than the spray.

It is possible that Rachel Carson's 1961 polemic against pesticides, *Silent Spring,* influenced the Forestry Department. Since the 1960s, official JNF publications have bemoaned the side effects of spraying, and Israeli forest managers have called for biological alternatives, including the development of natural predator populations so as "not to upset the biological equilibrium in woodlands."[142] However, when one considers the ample budget of the JNF and its broad research agenda, these calls seem to be nothing more than lip service. Integrated pest management as part of an ecological forestry strategy was never an organizational priority, and calls for alternatives to simazine made today may carry no more force than those made thirty years earlier.[143]

Conflicts between the JNF and environmentalists sometimes spilled over into the personal realm and could grow ugly. Although many JNF foresters and environmental advocates were actually friends, some were not. Azariah Alon remembers almost coming to blows with Sharon Weitz when the second-generation Weitz was in charge of the Northern Region Forestry Department. The Gilboa Mountain, famed for its iris flowers, lies across the valley from Alon's kibbutz, Beit ha-Shita. When he saw JNF planters coming to burn the lands and all the local flowers with them, it was too much and he brought kibbutz members with him to stop the JNF work. Eventually they agreed to submit the matter to Sharon's father, Yosef.[144] The compromise, a stark border between the open nature reserve and the shaded pine forest, is symbolic of the clashing aesthetic and ecological perceptions of the period.

NEW LEAVES BEYOND THE PINES

Frequently the ideological fervor of the JNF left it deaf to environmental critiques, even when they were substantiated professionally. Novelist Meir Shalev relates a story from the 1950s when the JNF drained the Huleh swamp. A Dutch reclamation expert who was hired as a consultant warned that the peat in the ground could undermine the project. "Then the JNF hydrologist stood up, hit the table with his fist and declared, 'Our peat is Zionist peat. Our peat will not do damage.' As is known, the Dutch have much experience in the reclamation of land. But even they had not yet met land with a political conscience."[145]

Perhaps in spite of its critics, however, the JNF perspective eventually began to change. It was not philosophy, but biological realism, that led to the shift away from planting Jerusalem pines. Quite simply, the insatiable *M. josephi* blasts thrived, and trees succumbed in an uncontrollable epidemic. It became particularly dramatic when it snowed and weakened branches collapsed. Extensive damage to the Sha'ar ha-Gai forest on the

Jerusalem highway in the winter of 1972 was the most identifiable turning point, but the process had begun much earlier. The press clamored for a simple explanation, and of course there was none.

"I had to call a press conference because it became a politically sensitive issue," recalls JNF forester Chaim Blass. "The damaged trees were on the Jerusalem-Tel Aviv highways. Government ministers passed by them on their way to work and were bound to be confused by the reports. The *Jerusalem Post* threw up a headline: 'The Sha'ar ha-Gai Forest Is Dying,' but I explained that this is just a stage. There is undergrowth. And if you go see the forest, it really is still very much alive."[146]

But press conferences could not save the trees. The crisis, which coincided with Yosef Weitz's death, brought with it the end of an era. A 1984 internal report described 20 percent of JNF pine plantations sustaining damage nationwide, with extensive mortality and damage among mature trees. At the same time, natural stands were largely uninfested.[147] By the end of the 1990s *M. josephi* spread to the forests in the previously uninfected southern region, causing heavy damage.[148] The Forestry Department finally got the message.

During the 1970s, praise for Jerusalem pine trees in official JNF publications became more muted. Wood that Weitz had once called "the best of the pines that can be used for furniture and building materials" was now referred to as "inferior in quality when compared with other commercial timber pines, but its high resistance to drought led to its past popularity."[149] By 1973, the brutia pine was elevated as an equal to the Jerusalem pine in official JNF publications and called "best suited" for the country's hills.[150] In 1987, conifers (pines and cypress) still made up 65 percent of all forest trees in Israel, but Jerusalem pines were only 11 percent. (Eucalyptus were 8.5 percent and oaks 7.5 percent.) Today conifers cover less than half of newly planted lands.[151] During the same period, tree density also dropped dramatically, from roughly 300 per dunam to 120 to 150. With no reduction in the area of annual forest expansion, in the 1980s the JNF was planting three million saplings per year, as opposed to six million during the 1960s.[152]

Native trees suddenly made more sense to foresters. Evolution had granted them the requisite natural protection for local conditions. Trees that survive in arid climates generally are protected from grazing animals by a coat of thorns. For example, the local oak, *Quercus calliprinos,* only reaches a quarter of the size of European or North American species, but its tiny acorns have small thorns on the edges of their thick waxy leaves. Although the tree grows very slowly, its roots can penetrate hard limestone and dolomite. Today it is just one of the forty types of trees that the JNF mixes.

The new biodiversity can be seen as part of a natural, organizational, and botanical progression. Now in his 80s, Chaim Blass still represents the unrepentant first generation of Israeli pine foresters:

> The pines were the pioneers—it was the first stage, and that's how we saw it. At the time, beginning in the thirties, we had to be efficient. Bringing in a more diverse combination of local species is a logical next step. Today foresters come up to me and tell me, "Thanks to you, we have work," because there is little new land left that can be converted to forests. We had to grab as much as we could and create facts.

Dr. Yerahmiel Kaplan, one of the first formally trained forestry experts to join the JNF in 1945, is equally unapologetic:

> If I were religious I would thank G-d for the Jerusalem pine. It was the only appropriate tree at the time. It may be aggressive, sending roots in all directions and dominating other species. But this is what allows it to compete for water. The truth is that pine and eucalyptus were the only JNF trees to survive World War I. It was a clear sign of what would last in Israel.[153]

The younger generation of foresters basically concur with this perspective, citing the increased preservation value that planning commissions assign to JNF forests, regardless of their ecological merits.[154] In the debate over legitimate reclamation versus conservation, environmentalists may also tend to overstate their case or attack a JNF that no longer exists.

Policy analyst Danny Orenstein represents the new generation of educators who have spent time working at the JNF. He has stronger Green credentials than most environmentalist critics, holding a master's degree in desert ecology from Ben-Gurion University. "The ecological critique needs to remember the challenges that JNF faced at the different stages and avoid the 'Monday morning quarterback' phenomenon," he explains. "For example, with the creation of the State there was a political side to forestry, which required tree planting to delineate the borders. JNF was playing with a completely different set of priorities and directives back then."[155]

Pine forests may not be as desolate as Green rhetoric suggests. At the very least they produce the popular pine mushrooms. After rains subside, Israelis flock to these woods to pick the *Suillus granitulus* for their soups and omelets. Wildlife was never really a JNF priority, and there have been many raging arguments over the years with nature lovers who saw the woods as habitat rather than as timber production centers. The JNF invariably countered that ecologists had a tendency to exaggerate and overstate their case.

Indeed, pine forests probably do not provide a rich microhabitat for the variety of plants, insects, birds, and small mammals that are found in native brushlands. Yet even in a prototypic pine desert such as the Negev's Lahav Forest, JNF workers report rich populations of hyenas, wildcats, foxes, porcupines, and snakes that have managed to adapt. These animals may feed off the garbage of nearby kibbutzim and Bedouin settlements, retreating to the safety of the forest to sleep.[156] Unnatural perhaps, and certainly a nuisance to farmers—but Israel is filled with examples of wildlife successfully adapting to changes imposed by man.

New global considerations may ultimately provide the most compelling reason to alter the traditional suspicion that environmentalists have of conifers. Israeli ecologists have not yet internalized the meaning of the greenhouse effect and the potential value of increasing carbon storage. Humanity may no longer enjoy the luxury of taking an across-the-board hands-off approach to open spaces. If, for example, the aggressive anti-desertification tactics of the JNF were adopted in arid regions around the planet, it might, beyond food and fiber benefits, contribute to stabilized global carbon dioxide levels.[157]

BORN-AGAIN ECOLOGY

The Eshtaol nursery is located at the foot of the Judean hills just off the Tel Aviv-Jerusalem highway. Hidden behind the moshav of the same name lies a massive arboretum, headquarters and nerve center for the JNF Forestry Department. As one wanders through the thousands of rows of saplings, one searches in vain for a Jerusalem pine tree: Oaks, cypress, carobs abound, but there is not a Jerusalem pine tree on the lot.

The profusion of species reflects a new ecological perspective that began to "infiltrate" the JNF during the 1980s and 1990s. One important figure in the transition was Dr. Menahem Sachs. It was in 1981 that Sachs took a leave of absence as an ecological researcher at the Volcani Institute. A short, ebullient man with a goatee and a big smile, he was given the role of policy maker for the Forestry Department, and in 1994 he was appointed director. Besides his training in ecology, he brought with him an openness to change.

Sachs claims that the actual revolution in JNF afforestation can be traced to the sabbatical visit there of Imanuel Noy-Meir during the mid-1980s. Noy-Meir is a prominent professor in ecology at Hebrew University. Sachs told Noy-Meir to review all afforestation activities and "tell the JNF how to improve." Noy-Meir's conclusions, published in a little-known article entitled "An Ecological Viewpoint on Afforestation in Israel: Past and Future,"[158] appear to have had a huge impact. "If you hear us singing a new song in the JNF these days," says Sachs, "he wrote the libretto."[159]

Noy-Meir clarified that high tree density can actually shorten the life span of entire populations. Typically in nature there is a process of self-thinning, when smaller and weaker plants that cannot compete with the stronger ones die off. Then the stronger trees can exploit the remaining resources. However, in the absence of significant differences in population, all the plants will be equally weak and retarded. In times of crisis (drought or pests) they lack the strength to survive these challenges. When trees are scattered in a mixed stand, on the other hand, the spread of pests and pathogens is slowed.[160]

Noy-Meir's operational recommendations confirmed many of the environmentalists' positions:

> Past [JNF] afforestation policy and methods were suitable for their time and efficient in achieving the afforestation objectives of the period. However they created today's problems. . . . Nowadays, most of the inhabitants know and cherish the country's indigenous flora and fauna. In 'making the wasteland bloom' for the future, the important thing is not the number of hectares or trees which are planted annually, but rather the quality of the landscape and the environment as they are shaped by the various afforestation actions.[161]

This new ecological identity ostensibly goes beyond trees. The JNF's current strategic master plan and the promotional material with which it is marketed abroad are purely environmental and address areas other than forestry. For example, the sewage infrastructure so desperately needed to preserve water quality has suddenly become a hot item. The contribution of the JNF primarily comes at the delivery end, rather than in treatment; by the mid-1990s JNF heavy equipment had been used to create thirty-four ponds for storing recycled wastewater.[162] Stream restoration is another area where, with great fanfare, the JNF moved to assume a "leading role." Although the Ministry of the Environment likes to take credit for this national initiative, the JNF provided over 80 percent of the funding and has the heavy equipment and experience necessary to stabilize the gradients of the banks and plant the anchoring vegetation and protective foliage.[163]

This born-again environmental spirit infects the JNF educational apparatus. In the 1930s the JNF established the JNF Teachers' Council to disseminate its materials and message in the schools of the Yishuv. Today, with the backing of the Ministry of Education, it fields a staff of naturalist educators, tour guides, and a magazine entitled *Roots* that broadcasts largely pop-ecology "good news" from the JNF. A new Community and Forests Department was recently established. Perhaps the only Zionist youth movement to show any signs of real growth in the United States during the 1990s is that sponsored by the JNF, with a national student organization called "Eco-Zionism."

The ecological mission of the JNF received a major legislative boost with the government's approval of National Master Plan (or "TAMA") 22 on November 16, 1995.[164] As part of Israel's planning and building system, national master plans zone various areas of land for a different purposes. When the cabinet approves a national master plan, the land designations carry the force of law. Several dozen such plans have been approved since the system was enacted in 1965. These include nationwide blueprints for roads, power plants, mining, garbage disposal, coastal and tourist development, and, as of March 22, 1995, forests.

The JNF was one of three institutions charged with preparation of the Plan by the National Planning Commission during the 1970s. Yet its governmental partners, the Israel Lands Administration and the Interior Ministry's Planning Division, had no time for the venture. As the most interested party, the JNF ended up drawing up the proposal, not only setting the borders to forests, but determining the character of each stand of forest. The lack of a participating governmental ministry committed to forestry impaired the Master Plan's progress. It was not until the early 1980s that a draft forestry plan was ready, but the Minister of Agriculture's opposition froze all progress for a decade.[165] Only after JNF representatives resolved the outstanding areas of conflict with agricultural interests in the early 1990s was the plan ready to go through the tedious regional review process.

From both a legal and a substantive perspective, the fifteen brief sections in the seven-page plan were revolutionary. Under National Master Plan 22, the JNF is responsible for managing close to two million dunams—one tenth of the nation's lands. For a country with a population density as high as Israel's, the plan represents a serious official endorsement of forestry. Some six hundred thousand dunams have already been planted, and the plan calls for three hundred thousand more. The remaining lands falling within the forestry master plan are to remain as open spaces. By comparison, prior to the plan's approval, the JNF Forestry Department oversaw only eight hundred thousand dunams.[166] The master plan reflects the new ecological approach in the JNF. It recognizes seven different types of forests, from "natural forests for preservation" to "human-planted existing forests," and stipulates those areas in which indigenous flora must be maintained. Moreover, the plan's regulatory orienation for the first time allows the JNF to work "in a calmer atmosphere," whereas in the past its staff had to hurry so as not to lose land to developers.[167]

Old habits, however, are hard to break, and JNF foresters have not always been strict in sticking to prescriptions of the master plan. In June 1998, Adam Teva V'din (the Israel Union for Environmental Defense), a public-interest environmental group, filed a Supreme Court petition against the JNF, arguing that it was systematically destroying native woodlands in contravention to the zoning restrictions of the plan. Its most compelling arguments charged that the JNF was circumventing statutory procedures for public participation and oversight of forestry activities.[168] Three years later, in the legal equivalent of a technical knockout, the Supreme Court sided with the petitioners, frequently citing the vitriolic professional affidavit submitted by Dr. Aviva Rabinovich. Justice Mischa Cheshin wrote the unanimous decision that rejected the JNF's planting policies out of hand. The JNF's ongoing refusal to promulgate detailed forestry plans that could be reviewed by the public rendered their afforestation activities between the years 1996 and 2001 patently illegal.[169] Israel's press made some fuss about the fact that, for the first time, one environmental organization was suing another.[170]

Nor does having an area protected as a forest on paper ensure its preservation. Here the JNF is exploring a new role. In 1996, the Forestry Department introduced a watchdog system, capable of identifying development plans that are potentially harmful to forests.[171] During the first sixteen months of the computerized tracking system, 474 suspect plans came to light. The JNF, however, is hardly a radical organization. Only very modest efforts are made to get developers and government agencies to minimize damage on JNF lands. In a few isolated cases, the JNF has filed legal objections with planning authorities to stop lands designated as forests from being compromised. Ultimately, the political leadership of the JNF has consistently been disinclined to translate its considerable statutory responsibility into an operational commitment. The thirty-six-person JNF board, appointed by the full gamut of Zionist political factions, may well have been unaware of the enormous environmental challenges it was avoiding. Many environmentalists came to believe that JNF leadership was simply afraid to get entangled in a fight with powerful government and business interests. The growing number of reports highlighting corruption, and a damning comptroller's report alleging "dangerous and damaging mismanagement" by JNF chairmen, did little to change this impression.[172] This rift within the Green ranks is unfortunate. With virtually identical interests on conservation issues, activism and common campaigns hold great potential for healing the old rifts between environmentalists and the JNF.

UNRESOLVED CONFLICTS

In many areas environmentalists and JNF foresters have reached some understanding. At the same time, a clash of perceptions, objectives, and values continues to generate tensions and animosity. At the end of the day, the JNF remains a development agency whose raison d'être requires that it intervene in nature. While the degree of gentleness and sensitivity may vary, improving the environment is philosophically no different from redeeming the land. For example, the JNF is still in the road-construction business. This began as part of its contribution to the 1948 war effort, ensuring supply lines to the many besieged communities on JNF lands. Once their asphalt capabilities were recognized, however, JNF bulldozers were increasingly pressed into such service. Between 1948 and 1964 the JNF paved 1500 kilometers of roads, primarily to connect new and existing settlements.[173] By the 1990s, it passed the 6000-kilometer mark.[174]

Today drivers entering the newly expanded Mitzpeh Ramon highway at the Zichor junction are jolted by a bright green billboard proclaiming the central role of the JNF in the project. Ecologists might wonder why the JNF would want to boast about a road that cuts through one of the country's largest nature reserves. If the JNF really wants to stake its claim as an environmental agency, why is it not laying railroad tracks?

By the end of the century, the JNF Mechanical Equipment Division still had a budget of over fifty-five million shekels a year.[175] Its fleet of over a hundred pieces of heavy earth-moving equipment is replaced at a rate of roughly fifteen pieces per year.[176] In the same way that pacifists fear armament buildups, environmentalists know that the machinery is not there to collect dust.

Dr. Benny Shalmon has been on the cutting edge of conservation activity in Israel for almost thirty years and is among the most senior public-interest scientists in Israel. His moderate stance reflects a consensus among environmentalists: "I feel that the JNF has changed. They are much more sensitive, having learned that monocultures don't work—from the point of view of forestry, even disregarding ecology. They have moved beyond the conifers. The problem remains, however, that they have a lot of money and a lot of machinery. When they have an idea, they don't try it out gingerly and consider it after extensive experimentation. They work on a large scale and still lack the patience to really learn. For example, they still want to transform the Negev desert into a forest."[177]

Indeed, the appropriate approach toward Israel's deserts remains a heated area of controversy. Afforestation efforts in the desert regions began during the 1950s, a relatively wet period; yet years of low rainfall during the early 1960s led to less impressive results. JNF efforts moved north, to semiarid areas, where rainfall exceeded two hundred millimeters.[178] The resulting pine forests at Yatir and Lahav are among Israel's largest and are unprecedented for such a parched region. At the same time, "making the desert green" remained a Zionist axiom, and JNF research steadily progressed, exploring ways to grow trees in more arid zones. The JNF first began applied research on saline, arid soils in 1939 at Kibbutz Beit ha-Arava, north of the Dead Sea. During the 1980s, researchers associated with Ben-Gurion University's Sde Boqer campus began to develop the innovative techniques that today support JNF flagship antidesertification initiatives.

"Savannization" has become an important buzzword since 1986, when the JNF began planting and water harvesting on three experimental arid sites. Some thirty species of salt-resistant and water-efficient trees are scattered on water-enriched patches in the otherwise barren and crusted terrain. The landscaping reduces soil erosion and gullies, while deep infiltration of water into subsoils is enhanced by dividing the basin into runoff-collecting areas that are irrigated by a redirected flow from contributing lands.[179] The result looks nothing like a forest but is closer to an African savanna, with only ten trees per dunam. A remarkable assortment of 130 additional plant species and grasses takes advantage of the improved conditions in collecting areas to support seasonal grazing. The resulting positive feedback loop continues, as foraging animals, attracted to the shrubs, leave excrement that enhances the soil's organic content.[180]

Beyond savannization the other JNF innovation, limans (from the Greek for "pond of standing water"), involve smaller, isolated man-made oases. Research began in 1962 when rain was channeled into microcatchment basins between one and six dunams in size. Eucalyptus, acacia, tamarisk, and pine have all been successfully planted in these pool-like depressions. Even in areas south of Beer Sheva, having less than one hundred millimeters of rainfall, the trees at these sites survive the summer months based entirely on runoff from the surrounding watershed during the sporadic winter rainfalls. Limans have been planted by the hundreds, breaking up the desert landscape and serving as rest areas for motorists, army units in the field, and herds.[181]

The JNF promotes these projects as part of a larger development initiative called "Action Plan Negev." This plan is designed to provide employment and to improve the standard of living in Israel's last "frontier" region. Billions of shekels are ultimately to be invested in hothouse agriculture, fish ponds, orchards, olive plantations, and associated infrastructures. The afforestation that is to accompany the entire endeavor is suitably ambitious: the JNF talks about planting one million dunams.

Not every one sees this vision as a blessing: Making more water available can destroy the competitive advantage of arid-zone species in their interactions with species less adapted to the desert. The expansion of flora also contributes to a new distribution of animals in the region, with winners and losers. Among the winners are crested larks, chukars, and red foxes, whose ranges have expanded dramatically with the new plantings. Other populations, such as sand partridges, brown-necked ravens, and foxes besides the red fox, may suffer from the new competition.[182]

While the new savannas, such as the Sayeret Shaked Park, are fascinating as an ecological experiment, many find the resulting sparse landscape uninspiring or even unattractive. In fact, it has been argued that excessive afforestation may detract from the Negev's appeal as a tourist site.[183] Ecologist Noam Gressel argues that despite the increase in tree cover, the desert-adapted species provide little shade, and their low density offers little in terms of cooling for recreational purposes. It is even conceivable that, in the long term, the diversion of water may have a negative effect.

The savannas are designed to be a more sustainable model of afforestation, but are they? Part of the criticism of the conventional pine forests involves the inability of JNF pine saplings to survive outside of a nursery in semiarid zones. Many trees in the savannas are also incapable of naturally regenerating. While biodiversity is much greater on managed lands, there is nothing indigenous about many of the trees. The exotic eucalyptus are planted in savannas more than any other tree type: this implies that perpetual, intensive management will have to accompany these sites if they are to endure as botanical attractions. These issues all come back to a complex ethical question: Is the goal of these projects preservation of ecological process or content?

A related ideological and ecological debate that will not go away revolves around the legitimacy of JNF efforts to improve native woodlands, or so-called natural forests. These constitute roughly a quarter of the lands in Israel designated as forests, including some of the most interesting and highest-quality stands. Originally the issue focused on the 440,000 dunams declared by the British as forest reserves, but with the JNF's central role in overseeing National Master Plan 22, the zone of contention has widened.

For instance, the most characteristic of Israel's indigenous trees in these woodlands is the Mediterranean oak. It can reach considerable heights and offers extensive shade year-round when it grows unencumbered. Yet whenever it is cut or burned, it coppices, growing a great number of new branches, none of which becomes a new trunk. Each branch produces a few twigs and leaves on its upper part. Together they create an extremely thick bush, through which little sunlight can penetrate. Not only can grasses not grow below, but the tree forms a thicket that becomes impenetrable for grazing.[184] To what extent should the JNF be involved in improving such natural stands? The JNF afforestation policy has not changed in thirty years. Standard procedure involves a three-step, seven-year process for "improving" such environments by pruning low in the tree, transforming the natural landscape.[185]

Because the work in these stands is especially labor intensive, JNF progress is slow. In 1996, the JNF thinned three thousand dunams and constructed eight kilometers of forest roads on these lands. Environmental activists claim that the JNF still brings its herbicides and conifer bias to the task.[186] The real issue is aesthetic and philosophical, however. Under JNF policy, eventually all native bushes will be crafted into an improved, accessible form. Forests will end up more like gardens than wilderness—lovely to some people, inappropriate in the opinion of others. Although satisfaction levels of Israelis visiting JNF forests is extremely high, it is largely a social or recreational experience, much as urban dwellers enjoy their local parks: as pleasant places.[187] For more profound natural experiences, they prefer the authenticity of the southern desert wilderness.[188] Its new ecological sensitivities notwithstanding, today's JNF remains confident that it is on the right track with such policies.

"There is very little that is natural in Israel," explains forester Menahem Sachs. "There has been civilization here for the past ten thousand years. Man determined what is left. We burned the flora. We created the landscape. . . . As of 1995 there were only three hundred square meters of open space for every person in Israel. By international standards, this is nothing. In these circumstances, I believe that you have to manage the lands. If I work intelligently, I will make relatively few mistakes. But you can't take man out of the ecosystem. The lands will only deteriorate."[189]

TOWARD SUSTAINABILITY

Few expressions are as used and abused in today's environmental lingo as "sustainable development." First coined in 1987 by the United Nations' Brundtland Commission on Environment and Development,[190] the term offered a compromise that enabled developed and developing countries to declare a common strategy. In fact they hardly agreed on anything beyond the phrase that has remained sufficiently amorphous to enable even egregious polluters to wield it as a shibboleth. For many environmentalists "sustainable development" is just an oxymoron.[191]

As a development agency with environmental pretensions, the JNF has naturally drifted toward the ideology of sustainability. It is more than a rhetorical shift: the influence of the landscape architects and ecologists on the staff is undeniable. The JNF has a compelling operational definition for *ecologically* sustainable forests, based on Noy-Meir's work, although such sustainability in practice remains an elusive goal.[192]

The JNF's long-term, *institutional* sustainability, however, will require more than simply improving the biodiversity, indigenousness, pest resistance, and regenerative capacity of its trees. To begin with, its work must be based on broad-based enthusiasm across all sectors of Israeli society for trees and forestlands. At present, baseline support seems solid.[193] The outpouring of outrage and empathy for lost trees after a spate of politically motivated arson events validates JNF claims for the popularity and educational success of its current approach. Israelis, who rarely donate to environmental causes, were responsive to the media campaign calling for contributions to replace trees lost in a major fire in the Carmel Park. Based on the level of pledges, in 1993 economics professor Motti Schecter set the value of this forest alone at six hundred million shekels. [194] As the population of users grows, the value should rise accordingly. Yet, decision makers have yet to internalize this appreciation in formal and informal cost-benefit equations. By the time they do, it may be too late.

Demographic trends suggest that early in the next century the northern half of Israel will be twice as crowded as Holland, Europe's most densely populated country. Open spaces, already disappearing at an alarming rate, will be increasingly in demand for development. With the simultaneous development of new urban centers at Modi'in, Kiryat Sefer, Elad, and Shoham, the plains and Judean hills between Jerusalem and Tel Aviv are giving way to pavement and buildings. Trees are not sacrosanct and cannot seem to stem the tide: Forests surrounding these communities, such as the mature woods at Ben Shemen, are feeling the pressure.[195]

The survival of Israel's forests and open spaces may well depend on the ability of the JNF to define itself once again—this time as a much more aggressive advocate for the lands with which it has been entrusted. As environmentalists well know, modifying National Master Plans is not that difficult. What begins as a few isolated challenges to a protected region can quickly turn into a deluge of development. The JNF took its first baby steps toward assuming such a role by creating a computerized tracking system that attempts to protect lands designated as forests in Master Plan 22. Yet it has not begun to utilize its public-relations and political powers to meet the forces of development head-on. It is time to cash in some of the bulldozers for the attorneys, planners, and spin doctors required to win a fight that may go on indefinitely.

The ability of the JNF to provide high-quality recreational sites will also be crucial to forest preservation. If the people of Israel are to accept limitations on suburbanization and sprawl, they must believe that they are getting a fair return for their residential self-discipline. At the behest of the new Land Development Department director, Gidon Vitkon, in 2000 the JNF began to address these issues systematically through the creation of its Forest and Community Department. Run by Meir Barzilia, an anthropologist by training, the department intends to build coalitions of forestland users and supporters.[196] In a rare example of the potential for partnerships, the JNF and grass-roots environmental groups have begun to collaborate in local preservation efforts to save the Jerusalem Forest. Other campaigns, like the successful grass-roots protests that saved the Shaked Forest in 2000, prove that such partnerships can be victorious.

With revenues close to one billion shekels[197] (four times the Ministry of the Environment budget), the JNF has some of Israel's deepest pockets. Here again, if the public and politicians do not believe they are getting their money's worth, they will find a way to siphon away its funds. Recreation as the primary engine behind popular support for forestry is something the JNF management understood as early as the 1960s. As the trees grew taller, forests became a favorite destination for vacationers. Picnic sites, rest-room facilities, playgrounds, and grills were developed accordingly. Before terrorism damped the enthusiasm of Israeli picnickers, twelve million visitors a year availed themselves of JNF sites,[198] but present capacity is already inadequate. According to a survey of the Israeli population in the late 1980s, one-third of a dunam of recreation area should be available for each individual in order to provide a desirable quality of life.[199] Even the two million dunams provided under National Master Plan 22 is only half the amount needed to accommodate the ten million Israelis who will be seeking relief from their crowded cities in the first part of the new century. Moreover, as crucial habitats disappear, the JNF must also dedicate more thought to accommodating homeless wildlife populations.

The JNF needs to be responsive to the pulse of popular culture while preserving its own sense of ecological integrity. Some environmentalists resent the JNF for its unique, nonaccountable status. No other organization holds such government functions without attendant institutional oversight. Eventually, the Supreme Court called on the JNF to stop acting like a "State within the State."[200] There is little to suggest, however, that any of the Israeli government ministries would be more responsive to public and scientific scrutiny, should they be given authority in these areas.

The JNF still suffers much of the ossification that characterizes established and inflated Israeli bureaucracies. Even after a concerted effort to retire or release twelve hundred employees in 1998, it still employs nineteen hundred people. That same year, a political compromise in the Jewish Agency produced rotating JNF cochairmanships between long-time Likud and Labor party activists—which has proven to be a formula for paralysis and partisan political squabbling.[201] Over the years the JNF has attempted to remain open-minded, but no institution really enjoys criticism, any more than people do. In recent years, unfortunately, the Chairman and the politically partisan JNF board of directors, who ultimately determine funding priorities, have done little to implement external recommendations or indeed to follow the professional judgment of their senior ecological staff. New leadership with strong technical literacy in ecology and natural resource management is needed to instill within the JNF the environmental ethic, the nimbleness, and the humility required to reinvent itself in light of changing political and ecological realities.

Viewed over a tumultuous century of activities, ultimately, the JNF has proven that it is very capable of institutional change. There is little if any similarity between the Jewish National Fund created in Vienna and today's collection of geographers, landscape architects, foresters, and educators. Menahem Ussishkin and Zvi Herman Schapira might not understand the nuances of the current JNF's desert research agenda or the attractions offered in its recreation plans. Yet they would certainly identify with the underlying and continuing impulse. In biological terms, because of its willingness to evolve, the JNF not only has survived but in many ways has flourished.

Mistakes along the way must be weighed against the historical achievements of the JNF. Without the Jewish National Fund, it is unlikely that the Zionist movement would have created a sufficient demographic, agricultural, and economic base to launch a new nation. The amount of forests in Israel would be comparable to those in its neighbors—measured in tens of thousands of dunams instead of millions. The desert would be expanding rather than contracting. For most Israelis, the pine trees that soften the rocky hillsides and whose friendly green shade beckons are one of the intangibles that makes life in their demanding country a little sweeter.

It is time that environmentalists embrace the JNF as the powerful ally it should be in the larger and more important context of determining the ultimate fate of the land of Israel. JNF representatives at the Israel Lands Administration board have been unrelenting in fighting for continued public ownership of lands amidst growing pressures for privatization. The clear financial motives involved makes their position no less critical environmentally. At the same time, the JNF cannot rest on its laurels but needs to face its ultimate challenge, a challenge that is already knocking on the door: Will future generations see anything in Israel that resembles the land of the Bible, or will they simply swelter in a sprawl indistinguishable from southern California concrete? To be sure, if Israel's open spaces become dysfunctional because of mismanagement, society will have little use for them. The JNF must continue to apply its new ecological knowledge in managing a more diverse and stable network of woodlands. Yet the question very soon will not be which species of conifers should be planted, but whether any lands remain for forests at all. With its financial resources, educational infrastructure, legislative authority, and historical prestige, the JNF ought to be leading this fight.

5 The Emergence of an Israeli Environmental Movement

It was 1972. General Haim Bar-Lev had just shed his uniform to take a position in Israel's cabinet as Minister of Trade and Industry. A proposal from the Haifa-based Nesher Cement Company reached his desk calling for long-term quarrying rights on the northern ridge of the Carmel mountains in Haifa. Although the area had recently been declared a nature reserve, the Nesher experts explained that the quarry would really only remove the top forty-six meters from the Givat ha-Haganah peak. Then they would restore it to a condition even better than its present state. The minister was sympathetic; after all, his job was to support Israeli industry. He was also out of touch. A soldier since the inception of the State, Bar-Lev was unaware of the power that the Society for the Protection of Nature in Israel (SPNI) wielded in a democratic Israeli society.

"In 1972 I was eighteen years old and came to the SPNI rally in the Carmel," remembers anthropologist Danny Rabinowitz. "Ten thousand people joined to call for the quarry's cancellation. This was before the Yom Kippur War, when mass protests became part of Israeli political culture. It was a real moment in the history of Israeli rallies and demonstrations. "[1]

The rally—and SPNI General Secretary Azariah Alon's inspirational speech—did more than just electrify the participants: The cabinet appointed the Ministers of the Interior and Agriculture to join Bar-Lev on an official committee to assess the issue.[2] It became clear that regardless of its merits, the proposal was illegal and could move forward only after securing approval from several bodies, including the Knesset's Interior Committee. Given public sentiment about the Carmel Park, and given the SPNI's ability to galvanize that sentiment, such approval seemed unlikely.[3] An alternative quarrying site was found in the western Galilee near Tamra.[4]

During the pre-State years, it is not surprising that the Yishuv did not produce a serious environmental movement. With Jewish control of only 7 percent of the lands at the end of the Mandate, there was no clear avenue for public involvement in preservation issues. And of course the public was distracted by the business of national independence. Yet the speed with which the SPNI captured the Israeli public's support during the 1950s suggests that the organization answered a deep-seated need to stay in touch with and preserve a precious land.

By the end of its first decade, a young, indigent country, with no model of public-interest organizations, sported a bold and original public-interest Green advocate. For much of Israel's history, the Society for the Protection of Nature (or "the Society" as its members called it) was synonymous with the country's environmental movement.

Not all Israelis adore the SPNI. Some, for instance, do not share its empathy for nature and resent its perspective as self-righteous, elitist, and exaggerated. "If Michelangelo was born in the land of Israel (in the form of Michael Malachi) and wanted to immortalize David's image in marble," read one editorial, "the Society for the Protection of Nature would submit a Supreme Court petition against him, arguing that quarrying marble hurts the landscape. . . . These nature lovers oppose any change in the landscape, good or bad."[5]

But the organization is sufficiently venerated to make its iris logo a valuable marketing asset (many types of products brandish its endorsement, from cellular phones to credit cards). With tens of thousands of affiliates,[6] SPNI is the country's largest membership organization. Its literature claims that 20 percent of Israelis are involved in some aspect of its educational activities.[7] The SPNI's budget in 1999 was over 176 million shekels, and it employed over six hundred workers. This is roughly three times more personnel than the biggest environmental organizations in the United States, such as the Sierra Club or the National Audubon Society.[8] It has magazines, international tours, and camping stores and also oversees historic-building registration. In spite of its periodic disagreements with SPNI, Israel's government in 1980 granted the Society the country's highest honor—the Israel Prize.

A review of the Society's evolution touches on some of the most important environmental triumphs in the country's history. Yet what began as a nimble, aggressive, improvisational family has burgeoned into a bureaucracy that its greatest fans acknowledge can be oppressive. The organization's history can be divided somewhat arbitrarily into the periods before and after the SPNI's receipt of the Israel Prize. Each stage had its key personalities, culture, and dynamics and its defining environmental campaigns. The story of the SPNI is ultimately one of idealism and idealists who harnessed Israelis' visceral attachment to their homeland, the heart and soul of the Zionist conviction. How they transformed it into a sustainable institution for conservation is a remarkable tale.

A SWAMP IS LOST, BUT A SOCIETY IS BORN

Conventional wisdom holds that the Society for the Protection of Nature in Israel came as a response of activists to the JNF draining of the Huleh lake and wetlands during the 1950s. This is true but only indirectly. The Huleh campaign actually preceded the Society, but it had a significant role in shaping the perspective of the SPNI's founders. In fact the organizational genesis is the result of one of the most successful and durable partnerships in Israeli history: the team of Azariah Alon and Amotz Zahavi.

The two met after the War of Independence in the circle of devoted students who were associated with the Biological Institute in Tel Aviv. Zahavi (see Figure 8) worked as a bird guide at the affiliated Kibbutz Seminar Center, where Alon was studying. The attraction was less than obvious: Alon immigrated to Israel at age six, whereas Zahavi was a Sabra. Alon, already over thirty when they met, was a respected nature teacher from Kibbutz Beit ha-Shita. Zahavi, a city boy from Petah Tikva and aspiring graduate student in the sciences, was ten years younger. Alon was a gifted public speaker and a prolific writer. Zahavi was more technical and entrepreneurial. Despite their differences they shared a passion for Israel's natural world and a no-nonsense work ethic. Over the years their combined efforts would produce synergetic results.

Zahavi claims that starting the organization was Alon's idea. Alon counters that their ideas coalesced when the two shared a tent as part of a 1950 delegation that explored possible routes for a new road through the Negev.[9] Although there is a consensus that it was the young upstart Zahavi (rather than Alon) who actually got the ball rolling, they shared a mentor. "Amotz Zahavi is the father of the SPNI—but that makes Professor Mendelssohn the grandfather,"[10] concludes Uzi Paz, often considered the unofficial historian of nature protection in Israel. Fascinated by birds, Zahavi idolized Professor Heinrich Mendelssohn, who was the leading figure at Margolin's Biological Institute. After the War of Independence he began work as Professor Mendelssohn's research assistant.[11]

The outbreak of World War II and financial constraints had for twenty years delayed the JNF's ambitious settlement plans for the northern Galilee.[12] As its first major land reclamation project after Israel's independence, the JNF announced its intention to drain what it called the "Huleh Swamp" to make way for the agricultural cultivation of one hundred thousand dunams. To the JNF, the Huleh region in the northern Galilee was the epitome of forsaken land crying out for reclamation. Insects and disease had kept it nonproductive except for modest papyrus harvesting and some fishing. It was, however, an ecological treasure chest.[13] The valley held the greatest concentration of aquatic plants in the entire Near East, eighteen species of fish, and so rich a collection of migratory birds that, on one October day, fifty different species were sighted in a single place.[14]

Mendelssohn was already intimately familiar with the wetlands, having received the prestigious Bialik Prize for his ornithological research at the Huleh. In 1950 he sent his young aid Zahavi to continue surveying the bird population there. Accepting this project meant that he would have to delay his return to his university studies, which had been truncated by the War. Zahavi was concerned about the interruption but ultimately figured, "To hell with school. Soon there won't be any birds left to see."[15] With the help of a kibbutz fishing dinghy, Zahavi and his local volunteers overcame the ubiquitous leeches in the water to generate considerable new data.[16] (This impulsive decision to choose public interest science over the better-trod academic route would repeat itself during Zahavi's career and was one of the keys to the SPNI's success.)

Since his arrival in Palestine in the 1930s, Professor Mendelssohn had been unwilling to limit his zoological activities to research.[17] He was particularly outspoken on the issue of the Huleh reclamation, calling it a disaster for biodiversity.[18] Mendelssohn's was not the only skeptical voice in the scientific community.[19] In 1951 he formed a Nature Protection Committee of concerned scientists affiliated with the National Botanical and Zoological Societies. The Committee's work culminated in a 1953 proposal calling for the creation of *"reservatim"* in areas of the Huleh, the Carmel forests, and the Galilee.

The Committee was not just an academic advisory forum. Its members dug into their own pockets to fund Israel's first environmental media trip, a two-day guided visit for the press to the Upper Galilee to foster public awareness.[20] There was a strong case against draining the Huleh. The original rationale for the project was in fact no longer very compelling:

- Malaria was now under control because of the introduction of DDT.
- The additional reclaimed agricultural lands had little meaning for a country that was so sparsely populated.
- Alternative lands could be developed at a much cheaper price.[21]

With newspaper coverage and Mendelssohn's lobbying efforts, a reluctant Yosef Weitz at the JNF listened to their case. In 1951 he set up a special committee to consider the appeal. Simcha Blass, head of the Ministry of Agriculture Water Department, chaired the JNF committee. They heard all the experts, including twenty-three-year-old Amotz Zahavi's call for a four-thousand-dunam reserve—eight times the land his former teachers were requesting. For the JNF of the 1950s, the value of eradicating "swamps" was axiomatic. Thus it is hardly surprising that environmentalists lost this round.[22] By 1958 the complex of wetlands and lake was entirely drained. In the most massive reclamation project in Israel's history, the southern divider between the lake and the Kinneret was literally blown away. Local residents were stunned by the dramatically altered landscape.

Five years later, the SPNI's unstinting efforts pushed the JNF to stick to its promise to reflood 3100 dunams (apparently based on Zahavi's demands).[23] It is unclear whether Weitz was joking when he told Professor Mendelssohn that his motives for reflooding primarily were "to allow future generations to see how miserable conditions used to be here." Other signs suggest that his perspective may actually have changed. Publicly, at least, he was not happy with the decision. "Do you know how many families could make a living on the lands I'm giving you?" he roared.[24]

In either case, the Huleh reserve declaration in 1964 was an empty victory. In theory, the cumulative years of protest produced the country's first case of land designation for nature protection, but unfortunately the reserve bore little resemblance to the original wetlands. Under JNF management it got neither the water nor the infrastructure it needed. Only in the mid-1960s, when the Nature Reserve Authority took over its administration, did the 3100-dunam swamp begin to show signs of rehabilitation. Even though the ad hoc Huleh campaign was not immediately successful, it did serve to galvanize a core of Green activists and set the stage for the establishment of the Society for the Protection of Nature.

Zahavi and Alon did not start organizing from scratch. As Professor Mendelssohn's research assistant, Zahavi was encouraged to launch SPNI through the Joint Committee for Nature Protection, which was under the Zoological and Botanical Societies. After its first taste of action with the Huleh initiative, the Committee was ready for more. Despite his age and junior status as a graduate student, Zahavi was comfortable taking on the role of group organizer.[25]

In fact it took another ten months, until December 1953, to register the group formally as a nonprofit organization (or "Ottoman Association," as it was called under the old Turkish law still on the books). The group was restless, frustrated by its lack of influence and a sense of stagnation. When Zahavi raised funds for a half-time "secretary," Alon pushed the organization to go public.[26] In June 1954, during the Shavuot holidays, the "Society for the Protection of Nature" held its founding conference at the Oranim Kibbutz Seminar Center near Haifa. Typically Alon and Zahavi have different attendance estimates (70[27] versus 150[28]). They agree that it was a huge success. The crowd, composed predominantly of kibbutz members and teachers, left with a sense that they had witnessed a momentous beginning. In short, six years after Israel's establishment, it was quite clear that the State's priorities were not always consistent with those of nature. This new organization would put environmental interests first and try to teach the people and their leaders why they should too.

HITTING ITS STRIDE

After the initial fanfare, institutional follow-up proved feeble. Initially Zahavi believed that a university student working part time should be able to coordinate the Society's activities satisfactorily. After two or three were fired for poor performance, he saw that his expectations for the job were far greater than he had originally recognized.[29] Azariah Alon, however, was busy teaching in the Harod Valley, and Zahavi himself, committed to scientific research, left for a year of advanced study in London for much of 1955. Abraham "Boomi" Toran, a beloved but eccentric teacher from Kibbutz Mabarot, received a year's sabbatical from his responsibilities on the kibbutz to run the organization from Tel Aviv. Yet he lacked the connections and administrative savvy that the unknown fledgling Society needed.

When Zahavi returned from his studies in August of 1955, he found an ailing organization. Boomi Toran had to return to his kibbutz responsibilities. No budget had been generated, membership was still marginal, and a replacement for Boomi was nowhere in sight. The SPNI was on the verge of disappearing.[30] Reluctantly Zahavi made a three-year commitment to delay his doctoral research, a delay that would stretch on for fifteen more years.

Professor Mendelssohn was again the coconspirator who would revive the organization. These were the early days of Tel Aviv University. As a prestigious zoologist who had thrown his lot in with the school, Mendelssohn was well situated. By the end of 1956, the Society was still without financial resources, and Zahavi had no money of his own left to bankroll his work. Mendelssohn, who never tried to rein Zahavi in, agreed to an "outrageous" idea he proposed: He put Zahavi on the university payroll as a research assistant (presumably to catch animals in the wild) at the new zoological research facility.

An unprecedented membership campaign began. Later Zahavi would joke that he had no friends left after pressuring them all to pay SPNI dues. If so, he must have started with quite a few. Within three years, five thousand people would join the Society.[31] Newspapers offered free publicity in return for a column on nature preservation.[32] When Azariah Alon began appearing weekly on the country's only radio station in 1955, word got out. The simple mission of preserving the birds, trees, flowers, and vistas that the first Israelis had come to love resonated in the young, idealistic country. The SPNI served as an institutional outlet for the "pantheistic Sabra" who was raised on devotion to the natural world. And although political leaders may have found some of their demands annoying, they could not help but admire the young idealists as representing the best of the Zionist dream.

The role of the kibbutz movement in shaping the SPNI's institutional culture during the early days cannot be overemphasized.[33] The organization had a clear hierarchy but socially was classless. Professors mingled with farmers and high-school students. At the same time, as a group it was elitist. Just as the kibbutz of those days eschewed titles such as "Director," the organization was headed by a "Secretary" with its more egalitarian connotation. Its members were knowledgeable but decidedly not intellectuals.

The intermingling of the kibbutz with organizational norms had other implications. For instance, the Society resisted being run like a business for years and therefore was perennially in debt. The SPNI still does not offer bonuses for exceptional performance. Benny Shalmon, an SPNI scientist-guide for thirty years, believes that this might be one of the causes for the high turnover rate among the more talented employees.

Amotz Zahavi consulted with experts but made most of the key calls himself. From its inception, he led the SPNI on an aggressive, uncharted path to protect nature and natural resources. Decisions were pragmatic, and Zahavi would cut an imperfect deal if there was no better deal to be had. Yet, given the spirit of the times and the economic conditions in the young State, Zahavi and his crew were also remarkably uncompromising in pursuing their agenda. Urieh Ben Yisrael, an early SPNI worker, recalls a director of an immigrant camp in Beit Shemesh who came to him in tears. "He said that if we opposed the development of a local quarry, there would be unemployment and suffering there. As in all matters, Amotz was the last word, but he ruled that we were to continue to fight the quarry: 'The people in the camp will ultimately be taken care of. Nature won't be.'"[34]

BUILDING AN EMPIRE

Indefatigable as he was, Zahavi recognized from the outset that he could not do the job alone and needed soldiers in the field. The first group of SPNI employees (alternatively called "rangers"[35] or "regional coordinators"[36]) filled a quasi-government role, preventing illegal hunting and working with planning agencies to integrate ecological considerations. With no money, Zahavi could rely only on chutzpah to recruit workers. Still, by the late 1950s, the SPNI had personnel in the field in the Huleh, the western Galilee, Eilat, Ein Gedi, the Mediterranean coast, the Judean mountains, and Mount Meron.[37]

Unleashing this pack of brash, energetic idealists brought the message of conservation to the Israeli public. Over half of the country's citizens had recently immigrated and were therefore not well informed about the natural processes in their new land. The SPNI created a cadre of young activists who would form the core of Israel's environmental movement.

The experience of Uriel Safriel, one of the first SPNI "employees," was typical. Zahavi found the recently discharged soldier a collection of odd jobs, teaching nature in the schools, working as a porter in the Eilat port, and selling SPNI membership cards. Having patched together this motley array of salaries, Zahavi wished Safriel luck and sent him south.[38] Today Safriel heads Ben-Gurion University's prestigious Blaustein Institute for Desert Research. The eminent ecology professor cannot contain the twinkle in his eye when he remembers the general cacophony surrounding the quixotic crew that set out to protect Israel's natural world. In a young country with relatively few entrenched interests, the young rangers in fact accomplished a great deal. Safriel was supposed to focus on enforcing hunting laws and protecting the coral reef in Eilat, receiving an official appointment from the Ministry of Agriculture to this end. Like the other rangers, he was limited only by his own ingenuity:

> Luckily I found a sympathetic Nahal group (military-farming corps) that was patrolling along the border at Yotvata. I'd just hop on the command car and join them for the patrols and then continue to use it as my vehicle for inspections. Later a group of Bedouin were trying to smuggle firearms across the border. The soldiers caught them in an ambush and shot all their camels. All but one, which they gave to me. So I would ride it, conducting inspections up and down the coast, giving out fines to anyone I caught removing corals from the sea.[39]

Safriel also managed to make underwater observations and post signs protecting the rare doum palm trees. Although he sees his effort to fill up watering troughs for gazelles as well meaning but silly, he maintains that other initiatives were important. After he conducted a survey of the Eilat coastline, the city government accepted Safriel's recommendation for a reserve. The municipality built a primitive fence around the lands that today are home to Eilat's only official marine and coastal nature reserve. Zahavi found the time to come down and check up on his young ranger three times during this period.[40] Otherwise, Safriel and his replacement, Uzi Paz, operated entirely on their own during the 1950s.

The designation of Eilat's corals as "fish" is one of the more famous stories in the SPNI lore from the period; it offers a revealing insight about the level of collaboration between the organization and Israeli academia. Coral, of course, consists of a hardened secretion that provides an external skeleton for polyps—marine invertebrates and no fish at all. The problem was Israeli enthusiasm for the newly discovered exotic world of the Red Sea. People wanted to take the colorful corals home with them. When the SPNI looked for a regulatory angle to curtail the practice, all they could find was the old British fishing ordinance. The legal staff at the Ministry of Agriculture were suspicious, demanding a written affidavit from a zoologist that Eilat's coral were indeed fish. Zahavi approached Professor Shteinitz, a proper academic and German immigrant who ran Hebrew University's Biology Department. "You want me to sign that. You're insane," protested the distinguished biologist. Amotz told him: "Well, you decide then. Either you sign, or the corals will be destroyed."[41] Shteinitz signed, and the Ministry of Agriculture allowed regulation to begin.[42]

Even with Zahavi's ingenious manipulations, salaries for the growing staff imposed a heavy financial burden on the organization. It was during his early days as Secretary that the SPNI first came to rely on government funding to defray expenses. In 1956, at one of their weekly meetings, Zahavi told Alon that the organization was short four thousand lirot and was unable to pay the regional coordinators. The Minister of the Interior at the time was Israel Bar-Yehudah, a member of Kibbutz Yagur and from the same political faction as Azariah Alon. Alon took the matter up with his friend the Minister, and the money was passed on to the SPNI through the budgets of the Regional Councils funded by the Ministry of the Interior. (It was exactly for this sort of financial patronage that Minister of the Interior Ariyeh Deri was indicted in the 1990s, despite his claims that this was how Israeli politics had always worked.[43])

The same chronic financial woes pushed the SPNI beyond its original nature-protection mandate. As frequently happens, exigency soon became ideology. For instance, the organization entered the world of informal education and for-profit tour guiding only as a way to raise money. Prior to this, organized hikes were associated with sports and physical fitness. In 1956 Eli Ronen, an SPNI worker, discovered a transport company that could offer inexpensive weekend charters. The guided hikes around the country brought Israelis to little-known scenic areas and generated revenues for the Society.[44] The menu of SPNI trips soon included family options and eventually international travel. Most important, the Society began to make nature preservation and appreciation the centerpiece of the hikes, supplementing the public-school curriculum. So many SPNI workers were eventually employed in this area that education and hiking were quickly seen as an end in themselves. According to a study conducted by Technion researchers, by 1974 only 0.6 percent of the SPNI budget went to environmental protection activities.[45] In one of the great ironies of its history, in 1997 the SPNI was divided by claims that the management had abandoned its original mission of *education* for nature and Zionism.

From the SPNI's first decade, there may be no institutional innovation more associated with it than the field school. "Patent" rights in fact belong to another long-time preservation advocate: Yossi Feldman. Having just finished working on an experimental farm that recreated ancient Nabatean agriculture at Shivta, the restless, twenty-eight-year-old Feldman arrived at the Ein Gedi kibbutz, facing the Dead Sea and above the scenic desert streams and waterfalls. The kibbutz had recently moved into its permanent housing. Feldman immediately set his sights on its original wooden structures, now abandoned, to create an education and archaeology center for students.[46] He gained the rights to the complex and set about renovating. When Amotz Zahavi suggested a collaboration between Feldman's school and the SPNI in 1960, it seemed the natural thing to do. The partnership between the two men had been established earlier, when Zahavi brought a group of students to Ein Gedi to spend a couple of days clearing out a garbage dump from the entrance to the oasis at Nahal David.

Being close to the border, the field school had a strong military component as well as an archaeological focus. Yet the Ein Gedi facility pioneered some of the trademark field school activities, such as trail marking and hosting school groups. In the absence of any formal government presence, Feldman's outfit took over and set up the trails along the Nahal Arugot and Nahal David streams.[47] During the week-long Sukkot holiday of 1962, twenty-five thousand Israelis came out, taking advantage of the newly accessible oasis.[48]

Although the field school was ostensibly a marriage of convenience between an ambitious nature organization and a like-minded entrepreneur, the union had a third partner—the Ministry of Education. The initial support of ten thousand lira from the Ministry of Education for the Ein Gedi facility burgeoned when unexpected government sponsorship changed the scope of this and future field-school initiatives.

The field schools enabled SPNI to reach millions of Israeli schoolchildren over the years and present its nationalistic message of love and commitment to Israel's natural world and heritage. Yet they also blurred the Society's nongovernmental status. The very fact that in 1997, for the fourth time, Israel's State Comptroller audited SPNI activities and in particular its field schools implies that they had crossed the line and at least legally become an accountable government agency. This is due to the extent of the government subsidy for field schools, which amounted to roughly one-third of the SPNI's budget.[49]

Many of the guides of SPNI hikes and field schools became legends in their own right. Dr. Benny Shalmon, working out of the Eilat field school for close to thirty years, is one such figure. Holding a group spellbound for hours, he points out hidden features of the insects, rocks, plants, footprints, and droppings inside a hundred-meter radius of seemingly barren wasteland. Although his approach is decidedly secular, many people describe trips with him as a religious experience. Shalmon actually sees himself as a second-generation guide and remembers the thrill in the 1960s of finally being old enough to hike with SPNI's first full-time and most renowned guide, Shukah Ravek. Shukah, as he is known, is legendary for his endurance and apparent invincibility, surviving any number of falls, broken bones, and rappelling disasters. "So he limps a little bit, but you still can't keep up with him. I'm telling you he's indestructible," marvels Shalmon. Ravek has more than just remarkable stamina. He also has extensive knowledge about natural history. While hiking in Jordan, Shukah actually discovered a species of acacia that had not been identified since 1891.[50]

In 1968 the army began allowing women soldiers to work as guides at the SPNI field school at Sdeh Boqer; today over two hundred women soldiers serve as the SPNI's core educational staff at its schools. Although initially their professional training was haphazard, the present three-month course produces a competent cadre of young nature guides. "Ultimately, it's a very positive phenomenon," enthuses Benny Shalmon. "First of all, we really get the crème de la crème of Israel's youth. Also, at a time when combat positions for females were unthinkable, it allowed the army to offer women something that was intellectually more challenging than folding parachutes."[51]

From the 1960s, field schools provided intense educational experiences for Israel's youth and a training ground for a generation of environmental leaders. Danny Rabinowitz's experience as a guide in the Sinai mountains field school in the 1970s after his army service was a typically life-changing experience. "It was a breathtaking place, 150 kilometers from the nearest asphalt. Once a week a jeep brought food from the supermarket in

Eilat. Just twenty young people, and all we had to do was some guiding and ecological surveys. There was a sense that we were privileged."[52]

Field schools offered a compelling educational package, but the primary vehicle for delivering the SPNI message to the public was Azariah Alon, a one-man publicity machine. Paradoxically the founding partner who chose to stay on the farm became the household name.

Azariah Alon has been a member of Kibbutz Beit ha-Shita for sixty years. He always looks the part, showing up in shorts and sandals regardless of the formality of the occasion. Stocky, with thick, muscular legs that attest to the thousands of kilometers hiked, he is literally a walking nature encyclopedia. He thinks he may be the country's sole university lecturer with only a high-school diploma. In short, if Zahavi was the Society's organizational engine, Alon (see Figure 9) was its trumpet.[53] Thousands followed when he sounded the call.

THE EARLY BATTLES

Few campaigns are more associated with the early years of the SPNI than the battle to create the Carmel National Park. In the 1930s the Hachsharat Yishuv agency had subdivided lands in the Carmel mountains into hundreds of one-dunam tracts and sold them to Jewish workers as Haifa's newest suburb. Roads were even cut through the woodlands to expedite construction. Only the Arab Revolt of 1933–1936, World War II, and the War of Independence delayed development.[54] Hence, at the inception of the State, the Carmel was an anomaly: a scenic old-growth forest, overlooking the heart of Israel's residential coastal region and sloping down the hills to the Mediterranean plains.

Early national development plans drafted by the Ministry of the Interior envisioned a modest portion of the Carmel region as a park. In the initial discussions about the future of the Carmel, nature enthusiasts were too timid to ask for the entire area, but over time the SPNI grew bolder.[55] In 1962 developers began to move ahead with their plans; it was time for the organization to up the stakes. In contrast to the battle over the Huleh, which really involved only three or four academic activists, the SPNI brought out the masses for the first time to further its case for a Carmel park. It was Boomi Toran's idea to turn Tu Bishvat, the traditional birthday of the trees, into a series of solidarity hikes in the Carmel.[56]

The press coverage of the unprecedented "demonstration of thousands" was extensive. The cabinet responded with a call for a moratorium on all Carmel construction. This government involvement is the most likely source for apocryphal stories about Prime Minister David Ben-Gurion's role in promoting a Carmel Park legacy. In one form of the legend, when Ben-Gurion was taken to the top of the mountain, he looked upon the verdant landscape, and declared that the Carmel would remain a park for all future generations. Although Ben-Gurion was clearly sympathetic to Carmel conservation efforts, there is no record of such a trip,[57] and SPNI old-timers dismiss it.

Not until 1971 was the park's status resolved. Although the final borders fell short of the maximalist proposal, it was still a very good deal for nature.[58] It was also more than just a major victory that preserved a lovely corner of Israel—it was a defining campaign. The SPNI leadership immediately understood the breadth of public support and their new capacity to take on powerful development interests and win.

It was during this period that the SPNI's historical animosity toward the JNF's pine forests became increasingly influential. The campaigns of the 1960s also provided the opportunity for the next generation of SPNI leaders to cut their teeth. No campaign was more important than the efforts that led to the creation of a nature reserve on Mount Meron. The area held the largest old-growth Mediterranean forest in Israel. For almost a decade, the SPNI fought to stop military bases and Druze villages from encroaching on it. They even resisted the offer of a deluxe field school to be built in the heart of the reserve to replace the collection of trailers and dilapidated buses that housed the Meron Field School.[59] Eventually the area was declared the largest nature reserve north of Beer Sheva.

As head of the newly formed Meron Field School, Yoav Sagi was the key figure in these struggles. During his days as a military officer for the elite Druze commando "300" unit, he had been frequently stationed in the Negev. After returning to civilian life, he spent a year at his home at Moshav Muledet and then decided to go back to Eilat, where he met Amotz Zahavi. It was a meeting that changed his life, enabling him to "combine business and a hobby."[60] Zahavi found the twenty-four-year-old Sagi a job helping to oversee the ecological impacts of the National Water Carrier, then under construction. Thirty-nine years later Sagi remains a dominant figure in the organization, succeeding Azariah Alon as General Secretary in 1979 and in 1988 becoming the first full-time organizational chairman of the SPNI.[61]

The 1960s were a period when the organization fully defined its identity as an independent advocate for nature. For instance, in 1963 the many years of lobbying led to a prohibition of mining "zifzif," Mediterranean beach sands. When the sands disappeared, giant sea turtles had no place to lay their eggs. That was also the year that saw the passage of the National

Parks and Nature Reserves Law. Soon thereafter, the SPNI embarked on its most famous campaign of the 1960s: a protracted effort to stop the picking of Israel's wildflowers. (This joint venture with the SPNI's new partners in the Nature Reserves Authority is detailed in Chapter 6.)

It was the 1970s, however, that have been called the organization's "golden years,"[62] during which it won a string of key victories for nature. A sampling of the better-publicized triumphs gives a sense of the organization's enormous influence. In August 1971 Azariah Alon published a position paper calling for a new power station to be moved from its proposed location at the base of Nahal Taninim, or Crocodile Stream.[63] Although no longer a home to crocodiles, it is one of the few streams outside of the Galilee that remains relatively clean. A letter-writing campaign ensued. In April 1972, five hundred SPNI activists turned out for a protest at the site; the resulting publicity far exceeded the actual size of the crowd.[64] As a result of the lobbying, a committee headed by Professor Moshe Hill was appointed to reconsider the issue. The committee ultimately suggested moving the facility to its present site at Nahal Hadera, which was already heavily polluted from the discharges of Israel's only paper mill.[65] On June 6, 1972, the National Planning Council approved the new location.

In 1974 after almost a decade of hands-on work in the Galilee, based out of the Meron Field School, Yoav Sagi returned from his studies and established the SPNI's Nature Protection Department (see Figure 10). As he perceived the situation, the emergence of a strong, governmental Nature Reserves Authority had diverted the SPNI from its original mission of nature preservation. When a disagreement over the excessive tapping of the Dan streams with the Nature Reserves Authority blew up into a full-fledged controversy, he became convinced that the SPNI could not afford to limit itself to education. (After the SPNI lobbied the rank-and-file members of northern Galilee kibbutzim and eliciting a feasible alternative from government experts, its proposal for a modest "eastern pipeline" was ultimately accepted. With the pipes for a central line already in the field, it was the first of many last-minute victories.[66])

Among the first things he did as he pushed the organization into watchdog mode was to appoint a staff member from every field school to take personal responsibility for nature preservation in the surrounding area. The theory was to avoid centralized control from the SPNI's headquarters in Tel Aviv, whose perspective might be colored by complicated organizational considerations. Rather, people in the field, who presumably remain truer to the values in question, should drive the Department's agenda.[67] Sagi wanted a rebellious rather than a complacent crew of workers.

The organizational shift was reflected in a series of campaigns between 1974 and 1988, and according to Sagi's count sixteen out of eighteen were successful. One of the first initiatives from this period was the campaign to save the *Bulbusim* and the surrounding area near the so-called "Hor Har" mountain in the Negev. Bulbusim are midsized boulders, remarkably rounded, that bounced hundreds of kilometers until finally resting like statues on the desert floor. Guides from the Hatzeva Field School noticed surveying work in the area and uncovered a plan to run railroad tracks through the area to support a phosphate factory in Nahal Tsin. Once again, the Nature Reserves Authority was not inclined to stop the factory's expansion,[68] so the SPNI took on the issue single-handedly. In December 1975, a thousand members made the trip down to the remote site to protest. When the factory managers displayed a surprising alacrity to explore alternatives, the Nature Reserves Authority decided that it was an important site to preserve after all. In 1980 a new nature reserve was declared at Hor Har, which saved the Bulbusim.[69]

Yossi Leshem has been one of the most colorful SPNI leaders since the 1970s. During the past twenty years no one has had greater success in raising public awareness about any environmental issue than he has in his campaign to protect birds, in particular birds of prey. "He did for birds what Azariah Alon did for wildflowers" is the common comparison. With a Ph.D. in zoology, Leshem fits perfectly the idealized SPNI stereotype of the "scientist in sandals"[70]—with one exception. Leshem is an Orthodox Jew in a predominantly secular organization.

Azariah Alon, who served as the General Secretary of the SPNI when Leshem first began working there, has never hidden his distaste for the religious status quo on which Israel's Orthodox parties insist. For instance, whenever the issue of sustainable transportation arises, he invariably blames public transport's decline on the religious limitations placed on Sabbath travel. But Leshem quickly gained Alon's respect as well as that of the rest of the nonreligious management of SPNI. Leshem rose through the ranks as a guide, field-school director, head of the Nature Protection Department, and ultimately Director, from 1991 to 1996.[71]

Leshem (see Figure 11) has a gift for engaging celebrities. It began when he was hitchhiking home after being expelled from his yeshiva high-school field trip, only to bump into retired Prime Minister Ben-Gurion, with whom he spent the day. Later, as the director of the SPNI, he brought the Dalai Lama to the SPNI's fortieth anniversary celebrations. This was after years of excuses by Israel's Ministry of Foreign Affairs, which feared an official visit, on account of Chinese sensitivities.[72] More than once he flew international jazz artist Paul Winter in a glider to accompany birds, inspiring him to play a concert for ten thousand nature enthusiasts. Leshem enlisted U.S. Vice President Al Gore in support of his global bird-tracking project for schoolchildren.[73] He drafted Israel's president, Ezer Weitzman, to support any number of SPNI initiatives. He also remains the only SPNI figure to forge meaningful connections with environmentalists in the Arab world.

This "chutzpah" characterized his tenure as head of the Nature Protection Department during the second half of the 1970s, after he succeeded Yoav Sagi. Among the most remarkable success stories of his tenure was the campaign to save the Um Zafah forest. The eight-hundred-dunam tract near Ramallah was one of only two pristine woodlands remaining in the West Bank. Declared a forest reserve in 1927, it was a popular resort stop for the senior British officials who expanded its natural growth with cypress and pine plantings. The forest holds the largest natural growth of Jerusalem pines and catalpas in the area, nineteen mushroom types, and countless wildflowers and birds, as well as unique reptiles. Local legend holds that Adam and Eve settled here after their expulsion from the Garden of Eden. Except for some fire damage done by Iraqi troops who camped there during the 1948 War, the forest was well preserved.[74]

The late 1970s were a period when Ariel Sharon, then serving as Minister of Agriculture, aggressively established Jewish settlements throughout the West Bank. A group of Herzliyah residents asked to move to the Um Zafah area. The ministerial committee for settlement was favorably inclined, and Tomi Leitersdorf, a noted Israeli planner, drew up blueprints for a village with one hundred wooden houses in the heart of the forest.[75]

The Nature Reserves Authority, ostensibly independent, was still beholden to the Minister of Agriculture, who appointed its director and oversaw the agency. Reluctantly the Authority did not oppose the settlement plan, which was quickly approved. Land in the forest was appropriated; building was soon to begin. It seemed a done deal. Leshem first heard about the rapid decision while attending a jointly sponsored conference with the Nature Reserves Authority. Impulsively he chewed out the Authority director for capitulation and set out on a quixotic campaign.[76]

Leshem collected "intelligence" from friends in the army, who confirmed that the settlement had all the necessary approvals. There were no legal loopholes. Friends in academia, on the other hand, confirmed the uniqueness of the forest and helped him put together a publicity packet replete with supporting expert botanical and zoological opinions. Leshem began a frenzied lobbying campaign; he shuttled scientists, politicians, and reporters to the site in his car. Within two weeks, 150 scientists publicly supported the SPNI's opposition to the settlement. A group from the prestigious Technion Institute took the uncharacteristically radical step of writing a letter of protest to Prime Minister Begin.[77]

With West Bank settlement an extremely controversial political issue, it was important that the SPNI's campaign not be misinterpreted. Friends at the Settlement Department helped Leshem find an alternative West Bank site near Nahal Shiloh that was already designated for settlement. When his sources discovered that the Herzliyah settlers were actually white-collar types who were not planning on moving at all but only sought summer cottages in the forest, Leshem and Sagi approached them, to no avail.

To defuse the political issue, Leshem especially sought out backers among right-wing politicians. Knesset members Moshe Shamir and Geulah Cohen were convinced and publicly opposed the settlement.[78] The press was fully mobilized; cartoonists were called into service. Among the many editorials to save the forest was an essay by songwriter Naomi Shemer, a known supporter of West Bank settlement.[79] To preserve the apolitical character of the opposition, when leftist Peace Now activists tried to hitchhike onto the campaign, Yoav Sagi kicked them out of demonstrations.[80]

On June 1, 1979, Sharon relented. Only months earlier he had refused to meet with the SPNI about this issue, but in the face of public pressure he suddenly announced his cancellation of the project. Despite its remote and sensitive West Bank location, the following day three thousand people came to the Um Zafah forest for a demonstration, only to discover a victory party. In 1981 the forest's status was officially elevated to nature reserve.

The campaign was indicative of the SPNI's lack of hesitation about the occupied territories. It set out to protect nature wherever it could. The integration of the Um Zafah forest into Israel's nature reserve system contributed to the SPNI's reputation at the time as a right-wing organization. "It is no coincidence that many West Bank and Golan settlements have, as a central feature, a field school set up by the SPNI," charges Meron Benvenisti. "As instructional centers for Muledet, they were in operation well before the settlements themselves and in some cases served as a guise for projected settlements."[81] Azariah Alon was director of the SPNI during the period that followed the Six-Day War. His own inclinations favoring a "Greater Land of Israel" were well-known.[82]

Yossi Leshem still lives at the Gilo Field School, one of the SPNI facilities located over the Green Line in internationally disputed territory. "It's nonsense in retrospect to say that they built the field schools in order to create new settlements," he responds. "At most, the thinking was that if you could bring 150 kids to a site, it created a greater sense of presence. The SPNI has always been apolitical."[83] To be sure, not only does the SPNI's historical connection with kibbutzim counter such a political generalization, but so does its membership, which comes primarily from middle-class suburban Israelis, who typically vote for left-of-center parties.[84] Few Israeli organizations have been as successful in reaching out to Israel's Arab citizens. The Society also had no compunction about filing a Supreme Court action to stop West Bank settlement when it threatened to compromise ecological interests in the Nofim region of Samaria.[85] Nonetheless, for years the New Israel Fund, a liberal philanthropic foundation, refused to donate to the SPNI because of its activities in "occupied territories."

The campaigns of the 1970s and 1980s were intense, creative, openly political (but not partisan), and conducted with an unselfconscious sense of abandon. They could also rely on the Zionist idealism that pervaded mainstream institutions during the country's first thirty years. Owing to state socialism, in most cases the enemy was either a government initiative or a government-owned industry. Developers had a sense of public duty to which it was possible to appeal. The SPNI leadership was in the right social clique to approach the ruling Labor government and could package its demands in the rhetoric of national interest. At the same time, the Nature Reserves Authority enjoyed its most aggressive and popular period, making it an invaluable SPNI partner. (The Um Zafah case to a certain extent marked the end of an era.) Most important, the country was not yet as crowded as it was to become. Alternatives still existed.

MIDDLE AGE

Danny Rabinowitz brought home a dream when he returned in 1982 from studying in London. The terms of his fellowship committed him to working for the SPNI, and after his years in the field he was happy to be stationed at the Tel Aviv headquarters. Yet he turned down Yoav Sagi's lucrative offer to run the Nature Protection Department. A gifted writer, Rabinowitz wished to transform the SPNI's modest printing department into an environmental publishing house à la Sierra Club Books in America. Things did not work out as he hoped. After the intimacy of a Sinai field school, Rabinowitz found the general atmosphere at headquarters stifling and was surprised by the rather impersonal relations among

SPNI leaders; for instance, the staff did not really socialize after hours. He also became frustrated professionally by the lack of support and the petty territorial thinking. "All my ideas were squelched by the head of the economics department," he recalls ruefully. "When my time was up, I moved on."[86] Today Rabinowitz is a tenured professor at Tel Aviv University.

The SPNI in the 1980s and 1990s had little in common with the loose band of volunteers that Amotz Zahavi conscripted. Naturally, the organizational culture changed, and not always with his approval. "I hate the conferences today where they bring in singers and create a 'happening.' If I were the Director, it wouldn't happen. In my day gatherings were simple. Nature was enough."[87]

But there are far more profound differences than just the evening entertainment at educational conferences. The SPNI's main offices near Tel Aviv's old Central Bus Station in Tel Aviv are vast, housing the administration needed to run a company with hundreds of employees. The directors' "thumbnail" synopsis of annual activities runs to sixty pages. Departments include student, youth, and school-age education; thirteen information and scientific research centers about caves, mammals, amphibians, plants, birds, and reptiles; museums; local branches; Arab affairs; publications; computers; construction; historic preservation; membership; security and safety; trail marking; domestic and international tours; insurance; vehicles; and the SPNI's own rabbi.[88] On the third floor of the building across the street from headquarters, on the other hand, sits the SPNI Environment and Nature Protection Department with a nationwide staff that has never exceeded fifteen people.[89]

This organizational sprawl can be traced in part to a well-meaning openness to fresh initiatives. The organization has always tried to modify and broaden its orientation as new issues emerged. For instance, in 1984 it lobbied the Knesset to established a Public Council for the Preservation of Monuments and Sites.[90] When the Knesset came back and asked the SPNI to place it under its aegis, the leadership was receptive.[91]

One key area into which the organization did not successfully expand was "the environment." In SPNI lingo, this generic term connotes urban or pollution problems. Azariah Alon tried to push the organization's agenda in this direction, and in 1988 the SPNI executive secretariat formally decided to dedicate more resources to environmental issues.[92] The interest in urban problems was to some extent a response to the growing number of SPNI branches that attached greater priority to their own immediate environmental health problems than to those of wolf populations in the Golan Heights. Indeed many of the grassroots groups that sprung up in Israel during the 1980s and 1990s were the work of citizens who had either left the SPNI or established parallel frameworks.[93]

Yet to address pollution issues effectively required training different from that of typical SPNI activists. Besides biologists, the only in-house professionals the SPNI retained were regional planners, who naturally came back to issues involving land use and conservation. Frequently the SPNI lacked the technical literacy to understand, much less participate in, national debates about pollution and environmental health, rendering the organization irrelevant.

Azariah Alon points out that only in the area of biological sciences can the SPNI depend on serious professional assistance from volunteers. Otherwise, it has to buy consulting expertise,[94] and this has drawbacks. The fact that the SPNI's legal advisers remain external and charge per case cannot help but temper organizational eagerness for initiating litigation.

During his tenure as chairman, Yoav Sagi was straightforward about the SPNI's insular organizational strategy, which eschewed the importing of senior staff from outside the organization. Leaders needed to be groomed from within, he believed.[95] Eitan Gidalizon may be the best example of this model. Gidalizon led the organization for most of the 1990s as either Deputy Director or Director. (As a sign of the times, during the 1990s the SPNI's "Secretary" became the "Director.") A personable kibbutznik who commuted to work in Tel Aviv by plane from the Galilee, Gidalizon had a long history in the organization and was among the legendary team in the Sinai field schools during the 1970s. But he was also a trained urban and regional planner whose master's thesis on the qualitative ranking of open spaces has been influential in professional circles.

Planning was perhaps the one area where the SPNI made a major commitment to capacity building and personnel. Sagi invested heavily in developing in-house expertise through the DESHE framework he set up and runs (the name is an acronym for the Hebrew phrase "the image of the land"). DESHE is "a think tank for integrating construction and preservation of open landscape values." The group attempts to bring influential planners together from academia and government to devise sustainable strategies on a range of physical planning issues.[96] In retrospect these efforts succeeded in changing the paradigms among both government and NGO environmentalists, convincing them of the paramount importance of open spaces on the national agenda.

The idea of open spaces, however, remains an amorphous geographical concept, not an activist agenda. In selecting its initiatives, the SPNI has occasionally allowed aesthetics to trump ecological considerations. For example, the organization actually opposed the mandatory installation of solar-water panels on all new buildings and the proposed installation of streetlights on the Jerusalem highway, because they were ugly.[97] SPNI conservation strategies have been criticized for focusing on the attraction that a given site holds for hikers rather than on the biodiversity it supports.[98]

With so many competing constituencies, it is not surprising that some factions became disenchanted with the organization's specific priorities. Left unresolved, disagreements could mushroom into major political controversies. So it was that Tel Aviv's plush Cinerama auditorium became the scene of the most intense internal showdown in SPNI history. The seeds for the controversy had been sown twenty years earlier. In 1978 the Uri Maimon Association, a small network of high-school nature field groups named after a young nature enthusiast from Haifa killed in the Yom Kippur War, joined ranks with the SPNI. They immediately became the organization's elite educational corps. But the "match made in heaven" turned sour. The field groups resented their perceived status as second-class citizens within the organization and demanded more funding. In 1996 the Association left the SPNI en masse, returning to its original independent status.[99] The SPNI responded by creating rival scouting groups and had a thousand kids signed up within six months. Association leaders were dissatisfied with the divorce and sought to seize control from SPNI management.[100]

Unable to contain the acrimony within the family, the SPNI aired its dirty laundry in the newspapers, making SPNI elections a hot item.[101] The SPNI old guard and management claimed that the organization was fighting for its life and soul to stop a hostile takeover by developers. Dror Hoter-Yishai, the founding chair of the Uri Maimon Association and at the time head of the Israel Bar Association, was an unabashedly aggressive developer. Yet the Uri Maimon Association flatly denied Hoter-Yishai's involvement in the elections. Rather, the scouting groups countered that the SPNI had become corrupt, stagnant, and out of touch with its original mission.[102]

The controversy proved to be a remarkable excuse for a membership drive. Amotz Zahavi came out of "retirement" to lead the campaign, which brought thousands back into the fold to give a vote of confidence to SPNI management. On January 13, 1998, chartered buses from kibbutzim in the

Galilee and from the Arava poured into the heart of southern Tel Aviv as twelve thousand SPNI members rejected the challengers.[103] But the great victory celebration by the SPNI management rang hollow; there are no real winners in civil wars. The campaign revealed an organization that was increasingly divided into separate fiefdoms, with little collective sense of a common mission. Most of the scouting groups' substantive complaints about excessive salaries, commercialization, nondemocratic organizational culture, and a diminished relevance to youth were not refuted. Rather, there were promises that change was on the way.

AN ORGANIZATIONAL OVERHAUL

Aware of past ossification, as director, Gidalizon sought to change the personnel strategy of promotion from within and was inclined to seek new blood. In 1998 he began employing staffers with environmental, planning, and even legal backgrounds. Emily Silverman became director of the increasingly activist Tel Aviv branch of the SPNI. She was an example of the organization's new professionals, hired because of her experience as a consultant to grass-roots activists. Under the new management philosophy, she downplays past tensions created by centralized control:[104]

> Today, if a branch thinks an issue is a priority, it enjoys autonomy. For example, the Jerusalem branch wants to focus on the Jerusalem Forest. The Nature Protection Department sees the Judean hills as a bigger priority, because there's more to protect there. So present thinking says: "Fine, go ahead and do it yourself." The problem is when there is a disagreement between the headquarters and the local branch over the issue itself.[105]

This was the case in the mid-1990s, when SPNI national leadership decided to support Minister of the Environment Yossi Sarid's decision to expand Beer Sheva's sanitary waste facility at Dudaim. The trash from the central Tel Aviv region was to be carted to the southern site. Other locations proposed were deemed worse for a variety of reasons, from issues of transportation and hydrology to national heritage. In contrast, local SPNI branch activists were furious at their leadership for not backing the protests against the decision of the Ministry of the Environment to turn Beer Sheva into a national garbage dump.

The clash of interests may lead to positive results, however. During the 1990s local activists in Jerusalem influenced the headquarters' strategy to prevent development in the Arazim Valley, which lies at the city's western entrance. They argued that the issue needed to be framed in the context of parks for Jerusalem's public, rather than merely nature protection. The subsequent campaign generated considerable interest in the press, bumper stickers, and even a ceremony involving a visiting chief of a Native American tribe from Canada.[106]

The real challenge has always been getting the branches excited about any environmental issue. During the past twenty-five years, twenty-two branches have cropped up across the country.[107] The larger ones eventually came to embrace activism. For instance, the Haifa branch was a leading player in coalitions that began fighting for improved air quality during the late 1980s and, more recently, for coastal preservation. The indomitable Tzipi Ron for years ran a militant Jerusalem branch that curbed development, saving historic sites such as the house that once hosted Theodor Herzl.

Most SPNI branches, however, were begun as independent initiatives by locals with their own agenda. Run by volunteers, often retired, they have their own way of doing things. Hiking, not rabble-rousing, is what they enjoy.

By the end of the 1990s, Director Eitan Gidalizon became a strong advocate for more aggressive activism within the organization. During a four-year period of large budget cuts, he increased funding for the activist Nature Preservation Department threefold: from 630,000 to two million shekels.[108] A hard core of several hundred SPNI members was formed to serve as "organizational shock troops" in the field. It was this cohort who could be called upon to demonstrate on short notice. For instance, in the 1998 campaign to pass a bottle bill in the Parliament, dozens of them sent hundreds of empty beverage containers in the mail to key Knesset members to make sure that the issue was not forgotten.[109]

During that year, a new SPNI initiative to foster student activism on campus gained momentum. Eran Benyamini, an ebullient Tel Aviv University undergraduate, had little time to study as he established "Green Course" chapters on every campus in the country; these became increasingly independent of their SPNI patrons.

This new stage reflected the realization that action would not come from the field schools, which had increasingly become commercial entities. In its promotional literature, the SPNI describes them as outposts for conservation work. In fact, field-school involvement and initiation of campaigns became increasingly arbitrary and inconsequential. The change also reflects the makeup of field-school personnel: Despite high turnover, the women soldiers continue to be exceptionally gifted guides and teachers.[110]

They cannot take a leading role in advocacy work against government policies, however.

When the Israel State Comptroller's 1997 report reviewing field schools lambasted the Ministry of Education for offering the SPNI preferential treatment in receiving government contracts, it did not suggest that the Ministry was buying an inferior product.[111] Still, the report revealed a sagging interest among the Israeli public in the "field school package."[112] The report offered several reasons for the drop. Most important was competition from newer and often less expensive travel companies and kibbutzim, many of whom, ironically, hire former SPNI guides. As a result, many field schools could not cover their expenses.[113]

In response, management scrambled to cut costs and make the field-school operation more efficient. Ultimately it was Director Eitan Gidalizon who made the tough decisions (and took the resulting political heat) for consolidating the field schools and culling the educational staff. The more fundamental question, however, received less attention: The field-school network plays a major role in distracting the organization from its original and officially highest mission—nature protection. As an objective observer, Israel's comptroller called field schools the organization's central activity.[114]

With the SPNI's agenda still largely one of conservation, other environmental organizations sprang up, in particular during the 1990s. Wary of competition, SPNI staff were not always thrilled with the new kids in town. During the second half of the 1990s, however, partly owing to the collegial orientation of Leshem and Gidalizon, the SPNI came to embrace publicly the pluralism that emerged among Israel's environmental groups. Naomi Tsur, a convivial British immigrant, a far cry from the stereotypical "Sabra with a backpack" SPNI activist, created a new model for the organization's chapters. Tsur galvanized a motley collection of local groups into a powerful coalition, working together as "Sustainable Jerusalem."

Even if today it personifies Israel's green establishment, the SPNI experience remains a highly relevant adventure for many Israelis. Membership levels, one indicator, remain remarkably high. In the mid-1990s, when numbers dropped 20 percent below 1986 levels, the SPNI responded by offering a basic membership for 20 NIS (six dollars). It allowed for basic affiliation without corresponding services. The move seems to have returned membership to previous levels, which sits steady at thirty-four thousand families.[115] Ultimately membership has more political than economic significance.

Although membership was once a major component of revenues, during the 1990s dues provided only 1 percent of the SPNI budget.[116]

Despite an increase in support from international Jewish foundations, money remains a perennial problem at the SPNI. It is, of course, much easier to smugly criticize an organization's source of support than it is to find alternative funding for it. And given the expectations of the increasing numbers of those hoping to build careers at the SPNI, as well as the growing costs in areas ranging from field-school maintenance to sound systems at demonstrations, new moneymaking schemes need to be hatched. One of the early innovations involved the Society's serving as an agency for international travel. This arrangement was reasonably profitable and offered a bonus to the SPNI guides who led the trips. Some of the old guard, predictably, could not reconcile themselves to the organization's transformation into something like a business.[117]

International travel packages were just the tip of the iceberg, however, and people questioned what seemed to be a growing commercialization. During the early 1990s it seemed that SPNI sponsorship was available to almost anyone, for a price. Some deals were decidedly ill-advised:

- The cellular Pelephone and later Cellcom companies enjoyed the Society's endorsement despite growing concern about the technology's possible association with brain cancer and growing public discomfort with the proliferation of relay antennas.
- During Passover vacation in 1993 the trails of the Galilee were littered with bottled-water containers that bore the SPNI logo.
- Recently, with much fanfare, the Electric Company cosponsored bird conservation initiatives with the SPNI, although people concerned about air quality have historically considered Israel's Electric Company to be environmental enemy number one.

Distasteful "greenwash" perhaps, but these efforts may be critical for reducing SPNI dependence on government funding, which hovers around 30 percent of the overall budget.

When the issue of government funding arises, Director Eitan Gidalizon invariably would tell about his meeting with then Minister of Finance, Avraham "Beige" Shochat. In 1993 the Supreme Court upheld a ruling that required the organization to pay four million dollars to Johnny Cohen, a young American tourist who had become paralyzed in 1981 through negligence on the part of an SPNI guide during a hike.[118] It was a very harsh verdict. Bank credit to make the huge payments was available only if the government of Israel would cosign the loans.

The portly Shochat greeted the SPNI delegates at his home in Arad, clad in undershirt and shorts. On his table were a series of full-page color advertisements that the SPNI had sponsored, lambasting the Trans-Israel Highway, one of the Rabin government's flagship infrastructure projects. "Don't you think it's a little cheeky," chastened the Minister, "coming to us for help when you're printing this sort of stuff?" Gidalizon calmly told him that one issue had nothing to do with the other and proceeded with the presentation.[119]

The dependence of the SPNI on governmental funding may not affect organizational priorities and positions, but it does make the Society vulnerable to changes in government proclivities. When Netanyahu's new Likud administration, led by Minister of Education Yitzhak Levi, cut back on SPNI funding and school visits to field schools during the late 1990s, the organization was financially devastated, and hundreds of employees had to be released.

It is not uncommon, or necessarily unhealthy, for organizations to resolve cognitive dissonance by coming to favor what began as a necessity. Financial interactions (and personal friendships) during the 1950s cemented close bonds between the SPNI and government managers. Today an "insider's" orientation is a central component of SPNI strategy to influence planning decisions. This reached a peak during Yossi Sarid's tenure as Minister of the Environment in the mid-1990s. Mickey Lipshitz, the SPNI's Director of Nature Protection Department, took a job as deputy director at the Ministry of the Environment. At the same time, Chairman Yoav Sagi headed the government's River Reclamation Committee and was drafted to serve as the environmental consultant on the peace-negotiating team with Jordan.

Since its inception, the SPNI has faced a continuous onslaught of challenges surrounding preservation issues. Like the scrambling hero in an arcade video game, no sooner does the organization fend off one monster than it is beset by three more. It is difficult to identify the specific ramifications of a given effort. Sometimes the ripple effect of a given campaign clearly runs beyond the specific resource in question. For example, starting in 1978, a ten-year fight to reduce the landscape damage caused by the Karmiel-Tefen road did far more than just prevent an unsightly scar along the karstic hillsides of the southern Galilee.[120] It also revolutionized the environmental sensitivities of Maatz, Israel's highway construction agency.[121] An in-depth look at three of the leading SPNI campaigns in the past decade reflects the organization's capabilities and dilemmas and illustrates the fine line it walks as an NGO watchdog and governmental partner.

THE VOICE OF AMERICA CAMPAIGN

In 1984, with his reelection assured, President Ronald Reagan spared no expense in his relentless battle to topple the Soviet Union's "evil empire." In addition to military hardware, reinforcement was needed in the propaganda war. The Voice of America (VOA), which sent its message of hope and democracy beyond the Iron Curtain, was called on to augment its broadcasting capacity. It proposed the largest radio transmitter in the world: sixteen 500-kilowatt transmitters, two ground stations to interface with satellites, an extensive control center, electromagnetic monitoring equipment, and forty-seven 180-foot antennas.[122] A location was sought that was well within broadcasting range of the Soviet Union. After Greece and Turkey—not wishing to aggravate relations with the Communist world—demurred, the Israeli option took the lead.[123] A site in the central Arava desert, twenty kilometers south of the Dead Sea and adjacent to the agricultural village of Moshav Idan, seemed the ideal location. There didn't seem to be much there.

Yet though there may not be many people near Idan, there are quite a few birds. As a land bridge joining three continents, the Arava Valley is a key junction for five hundred million birds that migrate from Europe and Asia to Africa and back every autumn and spring. Of the 280 species that have been identified flying over the Arava, none is more impressive than the raptors. In 1988 ornithologists counted 1.2 million of these imposing birds of prey—the largest such migration ever recorded.[124] Yet their breathtaking flight failed to appear on the radar screens of American or Israeli decision makers.

The project brought with it an estimated investment of three hundred million U.S. dollars just for construction by local companies. Its economic benefits, however, were ultimately of secondary importance. After enjoying twenty years of remarkable magnanimity in the form of military and civilian aid, it seemed the least Israel could do for its most generous benefactor. In 1985 Prime Minister Shimon Peres notified the Americans of Israel's general interest in the project. Yoav Sagi immediately fired a letter off to the Prime Minister on behalf of the SPNI, requesting that the environmental impacts of the project be considered before a decision was made.[125] But the wheels had already begun to turn. By 1986 Israel's

National Planning and Building Council called for the drafting of a formal Master Plan for the station, as well as an accompanying environmental impact statement. In June 1987 Minister of Communications Amnon Rubenstein signed a contract with the VOA that was later approved by the U.S. Congress.

Sixteen million dollars were appropriated and transferred to Israel for Negev development. (The SPNI would not be bought off, and refused a five-million-dollar grant to renovate the adjacent Hatzevah Field School.) No sooner had the ink dried on the contract than Israel's Ministry of Communications set up a permanent steering committee called Tomer, whose job was to shepherd the project through the local planning and building bureaucracy. It would also try to make peace with potential opponents.

Parallel teams from the SPNI and the Nature Reserves Authority were chosen to prepare the bird surveys that were to be part of the impact statement. Developers openly acknowledged ulterior motives behind the selection: Once the leading governmental and nongovernmental conservation agencies signed off on the project, its passage was assured. There was internal opposition within the groups to this cooptation.[126] In retrospect, it seems ironic that slipping through an incomplete environmental impact statement proved to be the VOA's fatal tactical error.

Bilha Givon is one of the few women who managed to survive for any length of time in the predominantly male SPNI hierarchy. A no-nonsense, heavy-smoking, former biology teacher from a Beer Sheva suburb, she cultivates a tough image and obviously thrives on a good fight. It was her successful campaign that saved Israel's last remaining "Great Sand Dune" in Ashdod; planning committees were convinced of the merits of her position and canceled a proposed neighborhood of apartment complexes to be built there.[127] When she heard about the Voice of America proposal, she decided that she had found her next campaign: "I asked, 'What's the Society doing about it?' People said, 'It's already at the National Planning Council. Peres promised the Americans, we were involved with the bird survey, so it's really going to be hard to object to the plan.'"[128]

If the SPNI was lethargic, developers were not. A Master Plan for the VOA station and accompanying environmental impact statement were completed by April 1989. When the National Planning Council met in July, it decided to pass the plans on to the Southern District Planning Committee for comments before making a final decision.

Despite their isolated desert location, word of clusters of cancer cases associated with high exposure to electromagnetic radiation began to reach the farmers living near the proposed site. No one denied that the largest radio transmitter in the world would generate prodigious amounts of low-level, nonionizing radiation.

The local Regional Council filed a Supreme Court action to disqualify the National Council's decision on procedural grounds. They argued that the VOA transmitter was not a national issue and should be reviewed only at a local or district planning committee (where the influence of the affected residents was presumably greater).[129]

Faced with a specific proposal and the clear opposition of the local residents, the SPNI was forced not only to take a stance but to decide what level of attention it would give the transmitter. During that same summer (1989) the Israeli Air Force requested new training grounds to replace those in Training Area 90, which would be compromised by the VOA facility. The best substitute was identified to the south in the Nahal Nekorot area. Lying near the Ramon Crater, it is an idyllic desert spot with unique geological formations and is a popular destination for hikers.

Givon brought photographers to the existing training grounds on Saturday when military maneuvers were suspended, to show them the likely fate of the Nekorot site. She assisted local residents in setting up an action committee. And she continued to lobby her boss, Yoav Sagi, to take a stronger stand.[130] Sagi relented, launching one of the most sophisticated environmental campaigns in Israeli history.

The international nature of the project and the sensitivity of bilateral United States-Israeli relations was only part of what made it an unusually complicated case. Electromagnetic radiation remains a poorly understood phenomenon, with enormous uncertainty with regard to its effect on health. To discuss the issue intelligently requires knowledge of physics, aeronautical engineering, epidemiology, and ornithology. Outside of birds, the SPNI had no experience in these areas.

Volunteer experts joined the cause, and the Nature Protection Department began to move up the learning curve. By 1990 the SPNI had formulated its opinion—a strong "No." Substantively the campaign argued three points against the station: loss of pristine open spaces owing to expanded military training; human health damage from radiation; and damage to the millions of migrating birds.[131] Of course it was the birds that caught the attention of the press. And coverage was not always sympathetic.

"We haven't lost hope," mocked the contentious journalist (and later politician) Yosef "Tommy" Lapid. "Maybe it will yet become clear that the electromagnetic field from the transmitters will force the birds to shift a little to the right, and that surely is reason enough for the station not to be built, and for tens of millions of dollars not to be invested in the Arava and for hundreds of workers not to be employed and not bring in the most sophisticated transmitters in the world and not to fulfill our commitment to the Americans. It's all up to the pelicans—humanity's final hope."[132]

Tactically the campaign undertook four simultaneous courses of action: political pressure organized by Givon with the local residents; a direct appeal to cancel funding in the United States, coordinated by Yoav Sagi with the help of twelve American environmental organizations; intense lobbying of ministers and Knesset members, as well as the National Planning Council; and another legal action to enjoin the project based on flaws in the environmental impact statement.

On February 18, 1990, the Supreme Court rejected the arguments of the Arava residents in the original suit and ruled that the unique nature of the project justified the National Planning Council's involvement.[133] When the plan finally came up for the vote in June, Prime Minister Shamir was apparently unaware of the degree of environmental opposition. With a typically poor turnout at the National Planning Council, the majority rejected the plan. Citing the lack of a quorum, Shamir demanded a revote. No chances were taken during the second round, and even peripheral members of the thirty-two-member Council were pressed into service. His office pressured the environmental ministry's representative on the National Planning Council, Valerie Brachiya, to support the project against her best professional judgment. She requested that these circumstances be recorded in the protocol during the vote, but this time the proposal to build the transmitter passed.[134]

The SPNI did not despair. Politically the Israel campaign had run its course, but another legal action was promptly filed to disqualify the Board's decision based on the inadequacy of the new impact statement. At the same time progress on the American front appeared promising. With the Iron Curtain fading into historical irrelevance, the project increasingly seemed to be a boondoggle that would benefit only a few economic special interests. Chief among these was the project's chief backer: the high-profile publishing-magnate scion (and 1996 Republican contender for the presidency) Malcolm "Steve" Forbes. Forbes offered a new rationale for the station, even closer to Israel's strategic interests: broadcasting an anti-Islamic-fundamentalist message to the Arab world.

Yoav Sagi set out for the United States, where a coalition of American groups arranged a high-powered Washington trip. Sagi joined leading U.S. environmentalists before a U.S. Congressional committee, where he politely told members that Forbes was not telling them the truth when he assured them that the transmitter had met all Israeli environmental requirements. (Ever the NGO activist, Sagi decorated the staid committee room with posters of Negev landscapes and the Arava.[135]) By the end of the visit he had acquired the signatures of eighteen members of Congress opposing the project. He spoke about the issue on CNN and enlisted an American law firm to prepare stateside litigation. International pressure was elicited in the form of 110 conservation organizations from around the world who joined in the protest. But it failed to move decision makers at Israel's Ministry of the Interior and the Prime Minister's office. With the access road completed and with increasing pressure from the first Bush administration to begin work, it looked as if the transmitter and the expansion of the military training zones was ineluctable. But help came from a surprising source.

Israel's Supreme Court typically does not cancel development projects. It can delay them, however, by correcting procedural flaws. A second impact statement had been prepared regarding the lands that were to replace the existing Air Force training grounds. It was a sloppy job, and the SPNI and the Arava residents became aware of it in time to act. On May 20, 1991, in one of her last decisions as a Supreme Court judge, Justice Shoshana Netanyahu ruled in favor of the petitioners, requiring changes in the impact statement.[136] This gave the campaign the additional time it needed. A new bird survey was conducted, this time independently by Ben-Gurion University ecology professor Berry Pinshow and his associates. The survey included 180 days of radar surveillance by the Swiss Ornithological Institute and extensive ground observations.[137]

When Yitzhak Rabin was elected Prime Minister in July of 1992, the political equation changed. During his first visit to President Bush, at Kennebunkport, Maine, the VOA issue came up, and Rabin promised that Israel would keep its part of the bargain.[138] Upon his return, he began to pressure for a decision.

When the long-awaited report from Professor Pinshow arrived in August 1992, it confirmed that the migrating birds flew at a mean altitude well above two hundred meters and therefore estimated that no more than thirty might be injured on an average night during the migration season. Use of appropriate lighting would reduce this number by as much as 90 percent.[139] VOA supporters were delighted with the results and rushed them to the media.[140] Interviews with Steve Forbes, a man used to getting what he wanted, showed a confidence that the project would commence promptly.[141]

The report was a setback, but the SPNI was still undaunted. It redoubled the media campaign and demonstrations, stringing grilled chickens in front of government offices to dramatize the cruel fate of birds that crossed the transmitter lines.[142] The SPNI took the offensive against Tomer (the steering committee in the Ministry of Communications), charging the council with corruption and falsification of documents.[143]

Although the prime minister may not have been moved by the protestations or a petition signed by more than half the members of the Knesset opposing the project, these developments clearly had an effect on a new player in the debate—Yossi Sarid, the recently appointed Minister of the Environment. Unlike his predecessor, Ora Namir, who had a tendency to see the world (and the VOA) in terms of reduced unemployment,[144] Sarid was a great nature lover. He vociferously opposed the project and called for its cancellation, creating the sense that the issue was very much alive. When word of American hesitation started to filter back to Israel, the Minister of Trade and Industry, Micha Charish, demanded an immediate cabinet meeting to expedite the project.[145] On December 31, 1992, the Cabinet overruled the opposition of the Minister of the Environment and officially confirmed its support for the project. But it was too late.

A joint U.S. Senate and Congressional committee had already called for the freezing of funds for the VOA project.[146] Even before the elections, vice presidential candidate Al Gore, who opposed the project in the Senate, wrote to the SPNI assuring them that the VOA project would be canceled if he and Bill Clinton were elected.[147] He was as good as his word. In April 1993, newly elected President Clinton sent the SPNI official notification of the Arava project's cancellation and its replacement by a more modest transmitter in Kuwait.[148] In January 1994, ten years after Reagan's initial overture, the Israeli government agreed to the project's cancellation.

Some critics say that the VOA case should actually be categorized as an SPNI defeat, or that if kudos is to be handed out, it is the American environmental lobby that deserves to be lauded.[149] But this is a twisted interpretation of the facts. Without the SPNI involvement, local activism would not have been as potent. The delay caused by the SPNI litigation was crucial, as was its work in convincing the majority of Knesset members and later Yossi Sarid, that the project was ill conceived. And, of course, the American environmental groups would never have heard of the issue without the SPNI's initiative. The institutional stamina alone is notable, and it can be said with some certainty that the station would be in operation today had the SPNI not opposed it. To be sure, they were lucky. But there are times when delay works in environmentalists' favor.

The point has clear tactical implications as well. The SPNI may formally have raised all the arguments, but allowing the press to depict birds as the centerpiece of SPNI concerns was unwise. While other salient, less ecocentric issues could have led the fight, bird preservation was scientifically indefensible and left the organization open to ridicule and caricaturizations. The SPNI's deep involvement in the original bird surveys for the impact statement may have started the bias. It also served to slow the organization's reaching an independent position about the case. One can also argue that this level of involvement crossed a line. Among the SPNI's long-held policy positions is its opposition to the developers' dominant role in impact-statement preparation. But why should it be assumed that advocacy groups are any less biased?

As in so many controversies, an indignant public, along with vociferous rank-and-file SPNI workers, had a key role in pushing the organizational leadership to take an aggressive stance. This goes back to the most problematic dynamic of the campaign—its slow start. It took five years for an SPNI position to crystallize. Valuable time was lost. In other similar cases, hesitation would be fatal.

THE RIVER JORDAN'S LAST ROAR

The campaign to stop construction of a hydroelectric station on the Jordan River is an unhappy example of an SPNI campaign that made its move too late. Most of the river has been so diverted and tapped by agricultural development projects as to make it unrecognizable in comparison with its former state. Yet in the northern Galilee, a fourteen-kilometer segment north of Kinneret Lake remained intact. Its steep gradient created the only formidable white-water rapids in the country.[150] For years this area lay below hostile Syrian snipers in the Golan Heights. After 1967 it was considered too wild for most hikers and was physically inaccessible. At the end of the 1980s, however, small, entrepreneurial rafting businesses caught on. As word got out, Israelis began to taste the excitement and the raw beauty of the untamed river.

Kibbutz Kfar ha-Nasi lies a kilometer away from this river segment, which became known in Hebrew as the "mountainous Jordan." The kibbutz was also aware of the river's potential as an electricity generator. It developed a plan to divert a third of the river's water into a reservoir that sat fifty-four meters above sea level. Taking advantage of a thirty-four-meter drop, the water would be channeled through pipes to turbines. The facility would actually only generate about fifteen megawatts, one twentieth of

1 percent of Israel's annual power production. But this translated into 1.5 million dollars of net profit a year for the kibbutz, a handsome return for a modest four-million dollar investment.[151] With solid financial incentives, the kibbutz guided the project through the labyrinth of government bureaucracy.

Kfar ha-Nasi representatives first approached the Nature Reserves Authority about the project in 1985. Tentative agreement was given after the kibbutz promised to spruce up the area around the reservoir, for tourists. After three meetings in January 1988, and agreement of the Water Commissioner, the National Planning Council changed two conflicting national Master Plans, paving the way for approval by the Regional Commission.

An environmental impact statement, prepared by Technion Professor Yorik Avnimelech, did not foresee major ecological damage to the river but expressed concerns about safety. After adding two pages of revisions based on the suggestions of the environmental impact statement, the Regional Planning Committee submitted the program for public comments and objections.

The SPNI and the newly formed Ministry of the Environment both filed formal legal objections. The SPNI argued that the change in the water level would damage the biological systems along the river. In dry years the kibbutz would find a way to pump more than the plan allowed, leading to irreversible damage downstream. The issue of the Kinneret's water quality was also raised. Moreover, the landscape itself within two kilometers of the plant would be forever changed, and not for the better. Alongside the ecological complaints were broader issues. The kibbutz, a private corporation, would benefit, whereas the public would lose yet another piece of its natural heritage.[152] And there was the underlying ethical question arising from the fact that this was the last untouched stretch of the Jordan.[153]

But the Regional Planning Committee was not sympathetic. In February 1990 it rejected the SPNI's concerns as unfounded, ruling that rather than damaging the Jordan River, the development would constitute an enormous improvement for hikers and tourists over existing conditions.[154] The proposal was sent to Jerusalem for the signature of the Minister of the Interior. After additional modifications, the approved plan was published on December 6, 1990.

Three months later, the bulldozers began the groundwork for the hydroelectric plant.[155] It was only then that environmentalists woke up. In its initial campaign, the SPNI brought twenty thousand people to the site, reportedly the largest rally of any kind in Israel that year.[156] It made no dent, however, in public complacency, and for most Israelis the Jordan River hydroelectric plant remained a nonissue until June 1991, when the plant became the hottest news item in the country. In the first act of environmental civil disobedience in Israel's history, a group of activists from the area padlocked themselves to the bulldozers and construction equipment, stopping work at the site. Then they threw the keys into the river.

The environmentalists, mostly radicalized former SPNI activists such as Danny Rabinowitz, were arrested and taken to jail in Rosh Pina. The early morning move was timed to allow television cameras to bring the sympathetic videotapes back to Jerusalem to be featured as the top story on the evening news. The coverage was dramatic and the entire country seemed to share the protesters' indignation.[157]

Officially the SPNI distanced itself from the lawbreakers. Although it opposed the facility, the organization maintained its law-abiding image. The organization had been divided over the appropriate tactics, and it reached a compromise: No top-level SPNI personnel would take part in the demonstration, but the Society would help organize it and provide publicity.[158]

No one anticipated the reverberations of the protest. The Jordan River controversy became a permanent feature of the evening news, complete with a logo and daily update. Politicians from all sides got involved. The Knesset's Interior and Environment Committee took a field trip to the site, where it called the project a national disgrace. Avram Burg, a Labor politician and later Speaker of the Knesset, went so far as to propose a law that would supersede the planning commissions' approval and cancel the project by fiat. An SPNI petition drive reflected the public's mood. From street corners to book fairs around the country, Israelis called for cancellation.

In contrast to the Voice of America campaign, where litigation was essential, in the case of the Jordan River campaign it would prove detrimental. The SPNI and other environmental groups had chosen not to file legal actions, primarily because the chances of winning seemed extremely remote. The kibbutz had gone strictly by the book.

But suddenly, the Council for Quality Government jumped into the fray. The Council usually serves as a watchdog for good government, filing high-profile suits against corruption and political excesses. It had never been involved in environmental issues before and never would be again. Yoav Sagi beseeched the Council's well-known attorney, Aviad Shraga, not to file the petition, telling him, "If you lose on procedural grounds, people won't understand. When the Supreme Court doesn't stop a project, [people] just assume that it's fine."[159] But his protestations went unheeded.

The suit was more ideological than legal, except for a dubious procedural argument claiming that preliminary decisions were made before the impact statement was prepared. In a ten-page decision, Justice Theodore Or showed little sympathy in rejecting the petition.[160] In addition to losing, the Council was ordered to pay the relatively high sum of twelve thousand shekels to the defendants. The environmental price was much higher.

Once the court became involved, the issue became *sub judice,* and the press backed off. Politicians, with their keen sense of the spotlight, also lost interest. Why be accused of circumventing the court? And the public's attention span, already stretched beyond its normal capacity, dissipated. The ruling came less than three weeks after the court began deliberations, but that was enough. The Jordan River reverted to its previous "nonissue" status. By the winter of 1992 the station was in full operation, and the SPNI's symbolic protests, staged at the formal dedication in May, only highlighted the failure of the campaign.[161]

The bitterness of an unnecessary loss does not quickly go away, and there have been many retrospective evaluations of what went wrong in the Jordan River campaign. All point to the early "complicity" of the Nature Reserves Authority with developers as a major hurdle in the early stages. But in other campaigns that did not stop the SPNI's more principled position from prevailing. Uri Marinov, Director General of the Ministry of the Environment, argued that once the plant was formally approved, the entire campaign was ill advised. In a 1992 Tel Aviv symposium dedicated to evaluating the Jordan experience, he reproached the SPNI for launching an unwinnable battle that served to publicize environmental vulnerability and weaken the movement as a whole. Developers should never have been given the impression that they could so easily ignore environmental concerns.

Political scientist Avner De-Shalit sees a connection between the outcome and the philosophical basis of the SPNI's campaign. Unable to hang the decision on a public-health issue, it never clearly enunciated an anthropocentric position that framed the problem in human terms.[162] To be sure, these damages were presented from the perspective of people, but De-Shalit doesn't believe the message got through. Danny Rabinowitz views the loss as part of a sluggish strategy and an overall lack of tactical innovation that deprived the issue of the passion it deserved: "If Azariah Alon had begun a hunger strike in front of the Knesset, they never would have built the power station."[163]

None of these views is wholly satisfactory. For example, in the past, the SPNI never really needed human health damages to win campaigns. Fear about the impact of "lost battles" on Green deterrence is a formula for paralysis, reflecting a governmental rather than a nongovernmental perspective. Planning commissions are often rigged against ecological interests. If the SPNI accepted unfavorable rulings as final and called off the troops every time its objections were dismissed, it would have missed many of its most important triumphs. And while officially the SPNI did not partake in civil disobedience, it is unlikely that the illegal protest would have happened without its support. Subsequently, the Society knew how to capitalize on the momentum it created.

The greatest irony in comparing these two campaigns of the early 1990s is that, in many ways, once it got rolling, the Jordan River campaign was more successful than that of the Voice of America. The rallies were bigger, the press coverage was more dramatic, and the swelling of public support and involvement of politicians were certainly more extensive.

The great "what if" of the case involves the ill-considered extraneous lawsuit: Would the outcome have been different if the political campaign had run its course? In short, one could argue that in this case the SPNI was simply unlucky. But when a campaign begins after the bulldozers have already removed most of the vegetation at the project site, there is no margin of error for bad luck.

THE TRANS-ISRAEL HIGHWAY

Another example of a major SPNI campaign, originating in the 1990s, has yet to be fully played out, but many lessons are already apparent. It is not a complacency born of past "come from behind" victories that explains the chronic delays in challenging environmentally unfriendly projects. Rather, the SPNI's penchant to declare war at a late stage of the game can be traced to its involvement in the planning stages. It is impossible to be entirely successful at the bifurcated game of participating in government committees while suing, picketing, and trying to undermine the very project that involvement implicitly accepts. This explains both the SPNI's vacillation and its late entry into the Trans-Israel Highway controversy.

Elisha Efrat, now a professor emeritus of geography, remembers penciling "Highway Number 6" into the National Master Plan for Roads as an afterthought when he worked for the Ministry of the Interior during the 1960s. It was to be a modest north-south route, to the east of the crowded coastal region.[164] When traffic conditions worsened in the bottleneck region of greater Tel Aviv, in the early 1990s, the "afterthought" was revived, but in a much more grandiose form.

The new vision was an eight-lane turnpike stretching from Beer Sheva to Metulah, splitting into two Galilee highways.[165] The 304-kilometer road would eventually gobble up sixteen thousand dunams of land on its route. It was not just the immediate disturbance to the forests, countryside, farming villages, and Arab towns along the way that concerned environmentalists, but the broader implications: The superhighway would spawn development in some of the last remaining green areas of central and northern Israel.[166] The billions of shekels that went into the highway would probably come at the expense of the country's long-neglected railroads.[167] Israeli dependence on cars would be further reinforced at a time when concentrations of air pollutants from automobiles already exceeded ambient standards.

On June 16, 1992, the National Planning Commission called for preparation of a detailed master plan for much of the Highway 6 route. While not opposing the idea, Yoav Sagi, sitting on the Council as a representative of Life and Environment, Israel's umbrella group for environmental organizations, requested a broader discussion of the highway's environmental impacts and called for a transportation master plan. Sagi's concerns were duly noted but did little to slow the planning process. A government corporation was created to move the project forward, with former military chief of staff Moshe Levy as its chair. Work on the detailed plans began, as did a series of environmental impact statements on different highway segments.

Sagi conducted his own assessment of the highway through the DESHE "landscape preservation" think tank he had established at SPNI. In 1992 the transportation team of DESHE conducted an intense review of the Highway 6 project. The team's perspective was dominated by Ilan Salomon, an articulate and charismatic professor from Hebrew University. Environmentalists became concerned that Salomon might be co-opting the SPNI position rather than the other way around. Professor Salomon had always been pessimistic about the potential of rail transportation in Israel and was an open proponent of the highway and expanded road capacity in general.[168] He even sat on the Highway 6 board of directors.[169] A year later the DESHE transportation team released its Highway 6 position paper.[170] The opening paragraph read:

> The team recognizes the need to develop the transportation system and is under the impression that Highway 6 has the ability to do this. At the same time the staff is convinced that determining the characteristics of the highway and the method of implementation requires comprehensive, systemic assessment and comparison of alternatives.[171]

It was not just other environmental groups that were surprised by the submissive tone of the SPNI position. When Mickey Lipshitz, then Director of Nature Protection at the SPNI, met with the Board of Directors at Adam Teva V'din (the Israel Union for Environmental Defense), in 1993, he quietly urged them to consider the highway independently of the SPNI and hoped that they would reach different conclusions. (Adam Teva

V'din eventually filed two unsuccessful Supreme Court petitions against the highway. The first, in 1994, challenged the lack of a comprehensive environmental impact statement for the highway.[172])

Frustrated at SPNI indifference, some SPNI staff members even offered to resign and help with alternative campaigns against the highway. This backlash among members had an effect. Within two years the Nature Protection Department declared its antihighway campaign to be its top organizational priority.[173]

The slow start would once again hurt the effectiveness of SPNI efforts. The highway's statutory approval was over and done with by the time the organization began to beat its war drums. SPNI thinking may also have affected policies at the Ministry of the Environment. Yossi Sarid, then Minister, consulted frequently with the SPNI about planning matters, and his opinion about the highway also waffled. He was an early supporter of the highway, but in 1996, after the Rabin assassination, he passionately led an unsuccessful fight in the cabinet to cancel the project. When asked about his own vacillations, Sarid insisted that it had nothing to do with having a new prime minister who was less committed to the project. "Here I don't think it was my best performance. It took me too much time to realize that the road was a disaster. Not that it would have mattered, because had I made up my mind earlier there was still no majority against the project in the cabinet. Why was I late? Many people misled me. I mistakenly believed some advisers that supported the road."[174]

The SPNI continued its efforts, unfazed by the road's apparent progress. It staged bikers' protest hikes along the route and stopped construction equipment.[175] It commissioned an economic analysis that cast doubt on the highway's viability. It invited foreign experts opposed to the road to join in a round of lobbying targeting decision makers.[176] It continued work in the Knesset, leading to a proposed law to freeze the project (the law was scuttled at the last minute by the ruling Likud coalition head, Meir Shitrit). When the bulldozers began running, the SPNI funded a consortium of Green groups to set up a teepee at the construction site. The teepee became the launching site for nonviolent direct actions by the Green Course students to stop the work. This time the young radical approach won out, and SPNI Director Eitan Gidalizon, along with other Green leaders, was arrested for disturbing the peace.[177] Cumulatively it amounted to a substantial effort, although it seemed misplaced; the right thing at the wrong time—five years late. By the twentieth century's end, Highway Six construction raced ahead, leaving a black partition across the land—a dispiriting testimonial to the collective failure of Israel's environmental movement.

There were many other environmental campaigns during this period, with a mixed scorecard. For example, SPNI efforts failed to prevent the Knesset from extending to the Dead Sea Works a concession that granted the industrial complex de facto environmental immunity.[178] On the other hand, manure runoff from dairies in the Golan Heights, reaching Lake Kinneret, was reduced after an SPNI campaign sparked investment in treatment infrastructure.[179] Beach cleanups succeeded in bringing thousands of young people to the seashore, but did virtually nothing to reduce Israel's ugly littering habit.[180] A solid-waste reform initiative was not well publicized.[181] Enormous efforts and legal action led to the removal of the illegal "Eddy's Beach" from the Eilat coastline, only to have it replaced, albeit by a more benign substitute. The SPNI's participation in planning subcommittees led to improvements in national master plans such as those for quarrying and forestry.[182] The most important achievement of the 1990s probably was the successful campaign to slow the profligate development of Israel's coasts, which the SPNI entered as part of a broad coalition.

Tactically the biggest change at the SPNI was a retreat from large rallies and a movement toward legal actions. In 1997 suits were filed to stop Mediterranean coastal projects as well as residential developments in and around Jerusalem. The SPNI's disagreement with the Ministry of the Environment over resort development adjacent to newly flooded portions of the Huleh swamp, for instance, reached the Nazareth District Court.[183] The first thing Mickey Lipshitz did after assuming the position of the SPNI Director in the summer of 2001 was to file suit against the Water Commissioner for pumping water out of Lake Kinneret below the hydrologically safe "red line."[184]

The changes in tactics are, in a certain sense, a sign of the times. Emily Silverman argues, "Let's say we bring a thousand people down to the Dead Sea. Will this really help? Mass demonstrations are not always the right recipe."[185] The actual efficacy of rallies remains an open question. While one Trans-Israel Highway demonstration attracted close to ten thousand people, attendance at most of the SPNI-sponsored protests against Highway 6 was pitiful. Clever gimmicks such as painting roads on the green bodies of high school students achieved the same degree of media attention with a fraction of the legwork. SPNI activism increasingly relies on behind-the-scenes and formal committee work.[186]

THE ROAD AHEAD

The SPNI's rapid ascent to an exalted position in Israeli society left it vulnerable. Exciting initiatives were adopted that with time became an economic burden. The very size of the organization ensured that it could not retain a single operational focus and that divergent and competing interests would emerge. Quality control became harder.

Amotz Zahavi warns against the tendency to glorify the past, saying that "the first 80 percent is always the easy part." Today's SPNI has far more employees and a far larger budget to focus on nature preservation than ever before, with two million shekels earmarked for activism.[187] Yet although the Society may be more professional than ever, it is also more ponderous. Its very size may have slowed its ability to adapt to the new rules of the game in the rough and tumble of current environmental conflict. During the 1960s and 1970s developers were unprepared for Green campaigns, and government decision makers largely identified with the SPNI message. In today's conflicts, on the other hand, private developers and industry bankroll a stable of "environmental experts," motivated by windfall profits. Money talks louder than ever in the game of government lobbying. A new level of professional sophistication is required that runs counter to an organizational culture where workers would rather be out hiking. And often there are just too many holes in the dike to plug.

In a retrospective survey it is always easier and perhaps more interesting to highlight failures. The many cases where natural resources were quietly preserved did not often make for good press. Sometimes the price of winning is discretion. All this makes the list of SPNI successes more impressive. Mistakes were made, but the overall record is dominated by environmental victories. The map of Israel is literally dotted with lovely corners and even a few species that would have long ago disappeared without the SPNI's uncompromising voice.

The Society has never played entirely by the rules. The positive results it gets by such tactics sometimes surprise even its own veterans. It is possible that it can continue to play its complex game of parallel identities. It may succeed in running an inspirational educational empire and still play the environmental pit bull; it may be able to sit on State committees and enjoy substantial support from government ministries while holding their friends' feet to the fire; it may be possible to be both gigantic and nimble. But the organization may also have to make some hard choices.

Nature protection is a tough business. Veteran activists quip that "there are no real victories—only stays of execution." For almost half a century the SPNI has been under fire in this rewarding but frustrating line of work. Many national treasures survive as a result. Whether they constitute "temporary stays of execution" or not depends on many factors, one of which is the continued evolution and influence of the Society for the Protection of Nature in Israel.

6 A General Launches a War for Wildlife

It was a brisk winter's morning on the old Eilat–Beer Sheva highway in February 1959. Avinoam "Finky" Finkleman was doing his daily run, driving the truck from Kibbutz Yotvata's dairy to the north of Israel. As he approached the Nahal Chayon stream bed, a large animal meandered across the road. From a distance Finkleman thought it might be a monkey. As he got closer, it appeared to be a very large dog. The animal began sprinting alongside the truck at the astonishing speed of eighty kilometers per hour. After driving for about two kilometers, Finkleman braked to a stop, and the animal climbed atop a nearby hillside to look the driver over. Finkleman admired his speedy escort, identifying it as a large spotted cat. "Must be a leopard," he thought.[1]

Every kibbutz has at least one nature fanatic, and Giora Ilani was the resident expert at Yotvata in those days. He seated Finkleman in front of a series of illustrations and had him go through the zoological equivalent of a police lineup. The identification was definitive: What Finkleman had seen was no leopard. Its head was too short, its forehead too high, its body too sinewy and "greyhoundlike," and its tail too bushy. The animal was a cheetah. The location also supported the classification. The highway cuts across a wide-open plain that is a favorite grazing area for gazelles.[2] Leopards and most other large cats are not fast enough to catch them, but cheetahs are. That cheetah was the last wild cheetah ever seen in Israel.

During the nineteenth century, zoologists visiting Israel reported that cheetahs were more common than the leopards that prowled throughout the wooded and rocky desert regions.[3] Like many African species that migrated north, they thrived on the 128 species of mammals and the disinterest of a dispersed human population that rarely ventured into the wild. But all that changed with the advent of accurate firearms and the population explosion that Zionism spawned. By midcentury, only 400 dorcas gazelles remained of the herds that had once covered the plains of the Negev.[4] The ibex was so rare that the only specimen available for years was a taxidermy artifact.[5] As the immutable laws of all food chains decree, the predators were soon to follow.

The body count is long and discouraging. In 1966 a dead ostrich was washed onto the edge of the Dead Sea during a flash flood. Although in recent times several Israeli farms have been raising the big bird commercially, it is not the indigenous subspecies; none of those remain. When the first Jewish settlers arrived in Palestine at the end of the nineteenth century, its skies were filled with thousands of imposing vultures and birds of prey that cleaned the land of debris. Pesticides have ravaged almost all thirty-nine of these raptor species,[6] and only a handful of the majestic lappet-faced vultures can be found in Israel today. They are confined to pens as scientists try to coax a few fertile eggs from them. A 1996 study summarized the damage to the breeding avian population at fourteen extinctions, with fifty-eight species presently threatened.[7] When the Huleh Valley was drained, several fish species disappeared forever. In this way an eclectic collection of reptiles, deer, bats, bear, birds, and other long-term residents of the land of Israel has quietly vanished.

Israel's twenty-six hundred plant species (including 130 that are endemic only to Israel) and almost 700 vertebrates (including 454 bird species) reflect a unique biological juncture where Africa meets Europe and Asia.[8] By the 1960s, trends suggested that precious little would be left of such biodiversity. The mighty thrust of Zionist progress was too great for the vulnerable creatures in the land. When the Knesset established the Nature Reserves Authority (NRA) to serve as an independent agency, its members were uncharacteristically pessimistic. With the pace and pattern of Israel's development, many simply felt that "it was too late."[9] But it was not.

The Zionist vision from the Diaspora had no clear concept of the Jewish relationship to the other creatures that called Israel a homeland. Herzl's conception of Palestine's biodiversity was completely theoretical and not particularly friendly. In his manifesto, *The Jewish State*, he called for the clearing of wild beasts in the new country by "driving the animals together, and throwing a melinite bomb into their midst."[10]

The early settlers were less hostile. Among them were botanists and zoologists who catalogued the flora and fauna of their new land with a passion that was almost unparalleled. And the people of the Yishuv and in Israel quickly came to know and cherish their nonhuman neighbors.

Tapping this affection, from 1963 the NRA began to spread a checkerboard network of lovely and diverse sanctuaries that now covers thousands of square kilometers. The National Master Plan for Parks and Nature Reserves approved by the government in 1981 envisions the eventual establishment of additional reserves and set aside a full 25 percent of Israel's land. (To date, 159 reserves on 575 million acres of land are formally protected, and an additional 373 reserves on 1.3 million acres are planned.[11]) Real estate alone, however, does not convey the full picture.

The wildflowers that had begun to thin in the wake of Israelis' passion for springtime blossoms have made an astonishing comeback. Viper bites doubled in the 1960s after the Egyptian mongoose populations fell victim to massive government-run poisonings, but today the ecological balance has been restored: The mongoose is back, and snake bites are down. In a land where neither Muslims nor Jews eat pork, there are more wild boars running around than anyone knows what to do with. Hawks, vultures, fallow deer, wild asses, and the elegant white oryx have all been successfully reintroduced to the wild. For thirty-five years the NRA has made a compelling case that human intervention with the natural world can be beneficial and that trend need not be destiny.

Nature preservation in Israel constitutes a success story that should be proudly told, but it is not without its caveats. Reserves often are tiny and frequently must be shared with the military. Rangers cannot control the growing hordes of motorbikes and all-terrain vehicles that invade the reserves almost at will.[12] Supporting habitats, such as wetlands and sand dunes, dwindle into insignificance, and some indigenous animal species, such as the Arava gazelle, may no longer survive in the wild. The expanding network of highways strands animals in a matrix of "island" habitats, unless they take their chances crossing the road.[13] As these and other hazards grow, the budget for nature preservation remains woefully inadequate. The resulting personnel shortages make it difficult to ensure the integrity of the reserve system. And the growth of the human population continues to turn up the pressure.

These and other challenges test the political prowess of conservation leadership and the professional acumen of agency staff as never before. They have large shoes to fill. During the 1960s and 1970s, Avraham Yoffe, a charismatic general, shaped the Authority in his own image: clever, passionate, defiant, and pragmatic. For eighteen years Yoffe was the dominant personality on Israel's environmental scene and one of the most effective (and colorful) government leaders in modern environmental history. He also managed to assemble an impressive supporting cast. His legacy is in no way ensured, however. The past may not hold all the answers, but it certainly offers inspiration.

FIFTEEN YEARS AN ORPHAN

The British Mandate was not oblivious to the issue of nature preservation during its thirty-year tenure. The British hunting laws and the Forest Ordinance, although narrow in scope, offered a modicum of protection. Yet frequently the gap between policy declarations and implementation was enormous.[14] By 1948 the depletion of wildlife stocks had reached dangerous levels, and a variety of mammals stood on the verge of extinction. Professor Heinrich Mendelssohn is largely credited with persuading Dr. Freund, a veterinary official in the Ministry of Agriculture,[15] to ban all hunting during the first year of the State as part of the general effort to return the land to a calm equilibrium.[16]

Once the ban was lifted, it became clear that the primary problem lay with the somewhat undisciplined army, which had most of the weapons in the country. The Joint Nature Protection Committee of the Zoology and Biology Societies (which evolved into the SPNI) wrote an impassioned plea to the generals of the Israel Defense Forces calling on them to stop the unrestrained hunting of gazelles by soldiers. The appeal succeeded in producing an IDF general order in 1951 that strictly prohibited all hunting of gazelles. Pressure on the small remaining herds once again subsided.[17] But no sooner had wildlife been relieved of the scourge of military hunters than it was faced with a more perilous threat: poison.

During the Mandate, although rabies was not uncommon, it was hardly considered the paramount public health problem. In the early 1950s, however, the disease had the young country in a panic. A baby boy's blood-stained clothes were found near Kiryat Shmoneh, and it was assumed that he had been attacked by a wolf. Said Warwah, a Christian Arab from Nazareth and first-rate hunter, was called in to track it down. He shot two mad dogs and discovered remains of the child in the stomach of one of them. In those days it was possible to keep such incidents out of the press, but the government decided to take action.[18]

The Veterinary Service of the Ministry of Agriculture assumed that the primary carriers of rabies were jackals. A massive campaign was launched to eradicate the animals. "I'd sit in back of the government veterinarian's pickup truck with a box of strychnine and a pile of chickens," former ranger Alon Galili sheepishly recalls in describing his young "cowboy" days. "We'd stuff the strychnine in the chicken's mouth and just throw it into the open spaces. We would throw out a thousand chickens a day. Whatever ate it died. It didn't take long for the area to become littered with dead animals. The thing is, there really wasn't much more rabies than there is today in Israel."[19] During the 1956 Sinai campaign, the French sent their Israeli allies crates of canned horsemeat to airlift to soldiers in the desert. The military rabbis banned the nonkosher provisions. The meat eventually ended up at the Ministry of Agriculture, providing cheaper bait than chicken.[20]

Despite the massive carnage, the incidence of rabies increased, reaching a peak in 1955, when 180 dogs (but only 11 jackals) were found to be rabid. In 1956 the numbers were unchanged (177 dogs and 12 jackals). Stray dogs were clearly the primary vector of the disease, but the policy was continued. Eventually the poison began to produce results, and 1960 witnessed the last human fatality from rabies for thirty years.[21] Ecologically it left a terrible scourge. Predator populations, from badger and mongoose to vulture and eagle, were decimated. In a 1971 article, Professor Mendelssohn opined that, although hunting had wiped out only one species (the otter) since the creation of the State, poison threatened scores of others.[22] Otters are alive and kicking today, so reports of their demise were premature.[23] But, for the first time in thousands of years, the lonely howl of the jackal marking his territory did not pierce the night in the hills of Judea and the Galilee.

As hunting and fishing were under the purview of the Ministry of Agriculture, it was, ironically, also deemed the appropriate body to take on issues of nature preservation. (The Ministry's conflicts of interest were not limited to the contradictory areas of poisoning and protecting wildlife. For years the Ministry oversaw crop yields and pesticide regulation as well as water quality and supply.) The lack of a clear institutional will for protecting habitat and wildlife was manifested in the limited funding and manpower made available for these purposes.

The Department of Beekeeping was the best bureaucratic solution the Ministry of Agriculture could come up with. The spiritual significance of milk and honey notwithstanding, the Department did not provide full-time work for its head, David Ardi. So it was that during the early 1950s, issues involving hunting and wildlife protection in Israel came under the purview of beekeepers.[24] Two years later, in 1956, a national hunting inspector was appointed to join the department; Uri Tzon was the first to occupy this post. It was hardly an effective bureaucracy, however: Tzon had to share a jeep with a gardening specialist.

When Peretz Naftali became Minister of Agriculture, things began to change. Naftali was a German immigrant who enjoyed the company of

Professor Mendelssohn. The two would shmooze in their native tongue for hours.[25] Mendelssohn prevailed on him to sponsor a toughened version of the British Hunting Ordinance. Working with the government veterinarians, Mendelssohn helped write a remarkably stiff statute,[26] and in 1955 the Wild Animals Protection Law passed the Knesset.[27] The law's stringent provisions are still in force today, with only minor amendments.

Mendelssohn's law took a novel approach that has since been adopted in other countries around the world. Rather than making lists of animals that are off-limits to hunters, the law bans hunting in general but allows for licenses to permit the hunting of specific species that can sustain losses.[28] All nondomesticated animals are protected from hunters, from February until September. During hunting season, the Minister stipulates the limitations on the type and number of game. The law bans certain forms of hunting entirely, including the use of poisons, drugs, traps, nets, glue, or explosives. Pursuit in a motor vehicle is also forbidden. The maximum penalty for violation is one year's imprisonment and a fine. Faced with new statutory responsibilities, the Minister of Agriculture established an independent hunting department.[29]

The new regulations enabled concerned citizens to get involved. In 1957 Dr. Reuven Ortal was a typical fifteen-year-old, growing up outdoors at Kibbutz Maoz Haim. One day he and a friend were wandering past Kochav ha-Yarden, the Crusader castle overlooking the Jordan Valley. Suddenly they spotted a hunter, shooting at birds indiscriminately. Feigning interest in the prey, Ortal engaged the hunter in a discussion and eventually got him to show him his hunting license. Once he saw the name, Ortal revealed his real motive and charged that the conditions of the license were being violated. The hunter put a bullet in the chamber and told him that he and his friend had one minute to clear out of his view. Ortal made a beeline for safety but immediately reported the incident to the agricultural ministry. The evidence he presented in the judge's chambers (as was required for minors) led to the hunter's criminal conviction, cancellation of his hunting license, confiscation of his gun, and a sentence of a five-hundred-lira fine or forty days' imprisonment.[30] Perhaps Ortal was not entirely typical. Seven years later he was hired as an inspector by the brand-new Nature Reserves Authority and today is its most experienced employee.

In those days, SPNI Director Amotz Zahavi enjoyed excellent relations with the Deputy Director General at the Ministry of Agriculture, Asael Ben-David. Zahavi saw the Ministry, with its existing authorities in water and land resources, forestry, grazing, and hunting, as the appropriate place to base governmental conservation activities.[31] In practice, Ben-David was the "strongman" in the Ministry. In 1958 Zahavi convinced him to change the name of the Hunting Department to the Department for the Protection of Nature. Naturally Ben-David turned to Zahavi to find an appropriate person to run it. Uzi Paz, who had some experience in the area (as an SPNI inspector in Eilat), was chosen for the task. Paz (see Figure 12) was a devoted nature lover throughout his childhood and also as a member of Kibbutz Sasa. He would remain associated with the Israeli government's preservation efforts for almost forty years.

Under Paz's management, the Department would slowly increase its budget and grow to include a handful of inspectors. (It finally received funding in 1962 to pay for a departmental vehicle.[32]) The small team set to work, doing what it could to curb the anarchy that characterized hunting in Israel. For example, Paz banned hunting altogether in two areas—Ramat Isachar and the pools near the coast at Ma'agan Michael, protecting gazelles and birds, respectively. This set the groundwork for what would later become nature reserves.[33]

The general idea of *reservatim* was in the air, and a number of proposals were advanced.[34] The vision of Zahavi and the SPNI was by far the most ambitious. As part of a scientific symposium in 1953, Zahavi helped Mendelssohn compile a map of those areas in Israel that should be declared nature reserves.[35] A government agency would be required to oversee them, of course. In 1958 the SPNI passed the proposal on to the Ministry of Agriculture.[36] Early optimism, however, proved deceptive.

Although there was considerable support among governmental ministries for a nature reserve statute, the government was unstable and collapsed twice, forcing new Knesset elections. Each time, the legislative process went back to square one.[37] An even more serious threat to the proposal came in the form of a small branch in the Prime Minster's Office called the Department for Improving the Country's Landscape (or "Landscape Improvement Department" for short).

As director of the Prime Minister's office, Teddy Kollek (later Jerusalem's renowned mayor) was responsible for promoting tourism. In 1956 he set up a committee to give the subject a boost. It was a high-powered forum. Joining Kollek as cochair was former military chief of staff Yigael Yadin and other leading archaeologists. Their orientation was completely different from that of the SPNI. They envisioned a network of parks and tourist attractions and established a permanent department to further their cause.[38] Although Yadin formally chaired the Landscape Improvement Department, it was run out of the Prime Minister's office by Yan Yanai, a former general and head of the Communications Branch of the army. Even without a proper framework of legislation, Yanai established thirteen national parks. Among these were the ancient Roman ruins at Caesarea, the burial caves at Beit Shearim, and archaeological sites in Ashkelon, Beit Shean, and Avdat.[39] The national parks were based upon restored archaeological sites and frequently included swimming facilities.

The nature protection camp's suspicion of Yan Yanai's Landscape Improvement Department was grounded not so much in territorial politics as in ideology. They wanted reserves, not tourist attractions. The lines were drawn for one of the most interesting parliamentary struggles in Israel's environmental history.

TOWARD A NATURE RESERVE AUTHORITY

In 1962 Moshe Dayan was appointed Minister of Agriculture. On the whole, his attitude toward the intrigues and nuances surrounding nature preservation policy can best be characterized as indifference. "When we first approached Dayan, he didn't know what we wanted from him," recalls Uzi Paz. "But he was practical enough to appoint a committee. He assigned the job to the head of the Israel Lands Authority, Yosef Weitz. Weitz, in turn, appointed Nachman Alexandron to head the forum."[40]

The Alexandron Committee began to sift through seventy different proposals, trying to reach a consensus among its very diverse members. But Yan Yanai and his Landscape Improvement Department would not wait for them to reach conclusions. Frustrated by six years of improvisational initiatives without real statutory authority, Yanai may have sensed the growing influence of Paz and the preservationists on the Alexandron Committee. Preempting the Alexandron process, Yanai lobbied for passage of a National Parks Law that had been collecting dust since its preparation in 1956 at the Ministry of the Interior.[41] With the continued hesitation of the Minister of the Interior, Moshe Haim Shapiro, Yanai took his case straight to his boss, the Prime Minister. Ben-Gurion decided to submit the proposed law on behalf of the entire government.

The bill created a single National Parks and Nature Reserves Authority. It established two ruling bodies—a broad Council to recommend to the Minister of the Interior which lands should become parks and a more narrow overseeing Authority. Both were composed primarily of government representatives. The brief, three-paragraph explanation of the law clarified its orientation. It was not nature preservation.[42]

It was at this time that JNF strongman Yosef Weitz became concerned that his laissez-faire approach to the new bureaucracy might have been mistaken. He launched a last-minute bid to bring the entire subject under the auspices of the JNF.[43] Had he taken this view from the outset, he could probably have steered Dayan and his cronies in the government in this direction. But he was too late. Knesset members, who had grown increasingly resentful of his dictatorial status in the JNF, were not interested in granting him even greater authority.[44]

Three times during December 1962 the Knesset would deliberate the bill. The speeches constitute the most profound discussion about the human relationship with the land of Israel in the country's legislative history. Their topics, ranging from Jerusalem building codes to the National Water Carrier's impact on aquatic habitats, offer a rare snapshot of the extent and form of ecological awareness during that period.

THE KNESSET'S GREENEST DAY

From his opening statement on December 3, 1962, on the Knesset rostrum it is clear that Ben-Gurion's vision of the new Authority was closer to that of the nature preservationists than to those who supported the establishment of national parks. The Prime Minister took the unusual step of quoting a previous speech, by Knesset member Yizhar Smilansky, who was also a well-known literary figure, published under the pseudonym S. Yizhar. He had already emerged as the greatest parliamentary proponent of nature preservation and was openly troubled with the proposed law's orientation. Ben-Gurion apparently sympathized and turned Smilansky's speech into Israel's quintessential plea for nature preservation:

> It is impossible for man to remain without vistas that have not been mended by his own hand. It is impossible to exist in a place where everything is organized and planned unto the last detail, until all remnants of the original image, the natural and organic signs of the earth's creation, are erased. It is a necessity for man to have a place to go to shake himself off and refresh himself from the city, from the built, from the enclosed, from the delivered and to absorb the refreshing contact with the primal, with the open, with the "before the coming of mankind"—if there ever was such a time. A land without wildflowers through which winds can blow is a place of suffocation. A land where winds cannot blow without obstruction will be a hotel, not a homeland. . . .[45]

Eleven Knesset members (MKs), including most of the women in that body, waxed eloquent about the importance of conservation. Leftist Mapai party representative Rachel Zabari led the calls for much broader membership in a Nature Preservation Council, to include scientists, landscape architects, and artists.[46] Coached by her friend and kibbutz neighbor Azariah Alon, MK Ruth Haktin praised the SPNI's past role in the area and pressed its platform. Along with many other speakers, she strongly opposed the imposition of entrance fees to parks and reserves. Seven MKs criticized as "undemocratic" a provision that made arbitration mandatory for landowners whose real estate might be expropriated to create parks.

The underlying philosophical and political power struggle between the champions of nature preservation and those of Yan Yanai's developed National Parks occasionally crept through the otherwise apolitical remarks. Yizhar Smilansky, the nature lobby's closest parliamentary ally, directly challenged the anthropocentric orientation of the law. Yanai had also lined up his supporters. MK Gidon Ben-Yisrael extolled the Landscape Improvement Department's past experience and contributions, countering that it should be granted a central role. All speakers felt the law was late in coming and lamented favorite pristine corners of their beloved land, laid waste by the fourteen years of rapid development that followed independence.

Ben-Gurion's[47] closing comments, apparently off the cuff, offer one of the most forthright expositions of the founding father's ideas about preserving the country's national and natural heritage:

> I think that all the speakers, and I place myself amongst them, are imbued with the beauty of nature, of trees, of the sea, of the magnificent mountains, [and] see in them grandeur and loveliness. But I also see this in man's creation. Not just the creation of nature and not just the literary and artistic creations of man, but also economic and technological creations. . . .[48]

Ben-Gurion concluded the debate on an upbeat note: "The debate itself may not have changed our country's landscape, but it may have slightly altered the Knesset's landscape."[49] The bill passed, with a consensus that a joint committee (Interior and Education) would have to make massive revisions before its final second and third readings.[50] The serious politicking could begin.

SMILANSKY'S SOLOMONIC COMPROMISE

With the proposed law passed, the Alexandron Committee appointed by Dayan to consider the issue was in danger of becoming irrelevant. After its slow start, it was forced to reach conclusions swiftly. Discussions focused on the scope and boundaries of the possible nature reserve system. The committee eventually recommended that ninety-three sites be set aside, covering 120,000 dunams from Beer Sheva to the Galilee.[51] No less important was the demand submitted to the Water Commissioner to guarantee "water to the landscape of Israel." This would ensure a continuing supply to areas such as the Huleh, as well as the Ahmud and Kziv streams.

Zahavi and Paz were fearful that under the proposed bureaucratic framework, nature preservation would be co-opted by Yadin and Yanai's vision of grassy archaeological parks and swimming areas with fences and admission fees. The two Knesset committees served as arbiters between the competing bureaucratic perspectives of Yan Yanai from the Prime Minister's office and Uzi Paz from the Ministry of Agriculture. Yanai enjoyed the advantage of being able to highlight the popular new park system in demonstrating the advantages of his approach.[52] When feelings ran high in committee, Paz became concerned that the Knesset would send the bill back to the overseeing ministers to work things out.

As a tactical response, the nature lobby called for a formal distinction between "parks" and "reserves." Under the two-tier system that MK Smilansky and other parliamentary allies pushed through the joint committee, the Nature Reserves Authority and the National Parks Authority would be run entirely independently. Nature reserves were to be left alone, whereas national parks were to be developed. God's commission to Adam about the Garden of Eden (Genesis 2:15) is sometimes invoked to clarify the difference between the two entities: "to work" (parks) and "to protect" (reserves). The distinction creates confusion, because the term "National Parks" in countries like the United States has come to mean something much closer to Israel's "Nature Reserves."

The other issue that troubled Paz and Zahavi was the absence of any provision that extended nature protection beyond the limited boundaries of the reserves themselves. From an ecological perspective, Israel is a tiny country, and most of the nature reserves envisioned by the Alexandron Committee were miniature in their dimensions. Plants and animals needed broader protection. Yet the concept of "protected natural assets" did not appear in the proposed law and was mentioned only in passing during the Knesset debate, by Rachel Zabari.[53]

Despite the wall-to-wall support for the concept, discussions over details in the joint committee bogged down. In the summer of 1963, Israel's Knesset decided to disband once again for early elections. The adjournment threatened to erase the considerable parliamentary progress that had been made. At the last minute, the law was rushed through committee, and an entire section about natural protected assets was inconspicuously inserted. In August 1963, just prior to the recess, the Knesset passed the law.

In the general haste surrounding the vote, the text came out ragged. The legislative experts at the Ministry of Justice forgot to include a prohibition on sales of "protected natural assets" that was to have been part of the package.[54] But the revised law, with its provision of parallel bureaucracies for parks and reserves, passed. (The system survived until 1998, when the two authorities were finally merged.) For nature preservation in Israel, the Knesset vote was a triumph.

AVRAHAM YOFFE ENTERS THE STAGE

Although it fell under the general purview of the Ministry of Agriculture, the Nature Reserves Authority was an independent government corporation, ruled by an eleven-person board and advised by a scientific committee. The law stipulated that Authority members represent a broad range of ministerial, scientific, and public interests. The group's first meeting, in January 1964, included most of the key players in the political maneuvering and manipulations that surrounded the legislation: Yadin, Mendelssohn, Smilansky, Weitz, Alexandron, and Paz. To preside over this unruly cast, Moshe Dayan appointed a new face: an active IDF general, Avraham Yoffe.

It was also natural for Dayan to appoint the head of his nature preservation department, Paz, to serve as Director of the new NRA (or, as it quickly became known within nature circles, "the Authority"). Yet once the flush of victory wore off, Paz was left with his existing staff of hunting inspectors. During its first year there were never more than sixteen workers in the entire agency, although conditions were somewhat better than before the establishment of the Authority. For instance, at his previous job Paz had no access to a private car, but as director of the Authority he was important enough to receive a vehicle. However, he still lacked sufficient status to merit a telephone in his home for emergencies.[55]

In January 1964 the Authority set up shop in an old German house under the literal shadow of the central office of the Ministry of Agriculture in the Kiriyah, a government/army complex in downtown Tel Aviv. Fearful that their rivals would try to send the law back to the Knesset and cancel the two-tier system, Paz and his staff worked around the clock to create user-friendly reserves that might be showcased in defense of the independent Reserves Authority.[56] (Ironically, the flurry of trails, parking lots, and signs placed in a few small reserves such as the Tel Dan streams and the Tanur waterfalls actually blurred the distinction between parks and reserves.) It quickly became apparent that the Authority's influence was minimal; the dream of a meaningful, national network of nature reserves appeared increasingly impossible.

Amotz Zahavi and the SPNI leadership still had a paternalistic attitude (bordering on condescension) toward the new Authority. After all, their lobbying had created the institution, and it was being run by one of their own. Still, the leaders of the SPNI recognized their own limitations. If nature enthusiasts today have a quixotic image, during the early 1960s they were considered positively eccentric. Amotz Zahavi came up with the novel idea of enticing General Yoffe (see Figure 13) to take over the job of NRA Director when he retired from his post as Head of the Northern Command. It may have been Zahavi's greatest single inspiration.

Yoffe's biography was indistinguishable from those of other elite "British army alumni" from the Yishuv, the leaders whom David Ben-Gurion ultimately favored to command Israel's military.[57] Although many of these men, such as Yigael Yadin, Ezer Weitzman, and Haim Bar Lev, were ten years his junior, they were Yoffe's peers because of their veteran status; they were his friends and, in many cases, relatives. Yoffe was born in the agricultural village of Yavnael in 1913 to parents who came to Israel during the Second Aliyah, in 1906.[58] He studied at the Yishuv's leading agricultural boarding school, Mikva Yisrael, and distinguished himself as the most mischievous of the many students who would later on assume the leadership of the new State.[59] He was active in a youth movement and helped found Kibbutz Tel Amal (later called Nir David). Yoffe trained under Orde Wingate's[60] night battalions, and when World War II broke out, he formally joined the British Army, in which he served with distinction for six years. The experience left him with a lifelong appreciation for English culture; in his final days he joked about bringing the British back so the country might be run properly.[61]

In 1946 Yoffe returned to his kibbutz but was soon called back to full-time service by the Haganah, the Yishuv's military organization. (At his wife's behest, Yoffe finally moved to Haifa in 1950, yet he always missed the pastoral communal life of the kibbutz.[62]) After serving as a battalion commander in the Golani Infantry Brigade during the War of Independence, he stayed on as a career officer for fifteen more years, until the age of 51. During the 1956 Sinai Campaign, he headed the 9th division, which raced down the Sinai desert to conquer Sharm el-Sheikh at the peninsula's southern tip. Overnight he became a national hero.[63] In 1963, although still on active duty, he received special permission to serve as chairman of the newly formed Nature Reserves Authority Council.

He was not an obvious choice from the point of view of the environmental community. Yoffe was known to be a hunting enthusiast who took advantage of his military privileges to pursue his nefarious hobby.[64]

Ever the good soldier, Uzi Paz was supportive of the idea of assigning Yoffe to head the NRA and volunteered to step down. Abdicating the top position at a new government agency for the good of the Authority constitutes one of the most selfless acts in the history of Israeli public service, much less environmental administration. Paz's wife Batyah was not enthusiastic about the decision and to this day feels that it was never adequately appreciated.[65] But Uzi Paz had no regrets.

Yoffe was hesitant. There were other ideas he had envisioned for a civilian career, including running development projects in Caesarea. Dayan insisted that he run the Authority full-time.[66] Ultimately the lobbying from the conservation community proved irresistible. Yoffe retired from active military service in November 1964, and in May 1965, Haim Givati, the new Minister of Agriculture, appointed him director; Yoffe immediately appointed Paz as his Deputy.[67]

While nationally Avraham Yoffe still enjoys recognition as a military hero, in the environmental community his stature has become mythical. It is virtually impossible to find anyone who worked with him to say anything bad about him. Everyone has a favorite story that highlights his larger-than-life proportions. Yoffe's son Danny claims that people's memories exaggerate his father's actual dimensions. "He really was not that tall. Just wide. He loved food, truly a carnivore. And he ultimately died of diabetes."[68]

His appetite remains the topic of tall tales to this day. Reuven Ortal, a young inspector at the time, remembers a goodwill feast in Yoffe's honor hosted by the mukhtar of the Druze village of Beit Ja'an. So many local guests wanted to meet the army hero that they had to host the banquet in two waves. Yoffe had no trouble eating twice. Wasting no room on vegetables, he devoured all the exotic meat delicacies, especially the tongue and skull.[69] Even in more refined settings, his appetite sometimes got the better of him. Yoffe had a habit of walking into the kitchen of his host, opening the refrigerator, and pulling a leg off of a chicken or helping himself to some other tasty morsel without being invited.[70]

Not only for food did the general have an appetite. His relationship with his second wife Aviva, the elder sister of Leah Rabin, was not always a happy one. (Yoffe's first wife died in a motorcycle accident with Yoffe's brother driving. The family complications did not end there, however. He soon married Aviva, literally the girl next door, who previously had dated his brother, Shaul.) Two daughters were born to the couple, and by all accounts Yoffe was a devoted and beloved father. His exceptional rapport with children was reflected in his rich collection of children's literature. The table at the Yoffe house in Ramat ha-Sharon was open to a parade of guests, and the family was the centerpiece of a social group that hiked together regularly and camped each year by the beach.[71] But Aviva Yoffe's chronic back problems, Avraham's frequent absences, and pervasive marital tensions often left a certain melancholy in the Yoffe home.

Taking the job as the head of the Authority changed him as much as it did the Authority. While Yoffe had always felt close to nature, his son Danny insists that with the appointment he was born again. Yoffe gave away all his hunting rifles and never shot at another animal, even during the "hunting" season, when it would have been permissible.

A SAFE PLACE FOR NATURE

When the NRA was created, Amotz Zahavi perceived it as an implementing body, with the SPNI providing direction and overall ideology. Because they were already filling government enforcement and inspection functions, it seemed natural for SPNI workers to continue working in this capacity. So Zahavi proposed a hybrid unit to be called "Shachal," an acronym for Society/Authority Cooperation.

The experiment lasted a year. Once he took over, Yoffe's approach quickly became, "The Society is all well and good, but I'm in charge here."[72] Many workers, with a foot in both worlds, found the experience particularly wrenching.[73] From the Society's perspective it was always clear that Shachal was a transitional stage. Although the new and understaffed agency needed the reinforcements, Zahavi told his SPNI recruits, "The second they feel strong, they'll throw us out." It was hard for Zahavi to imagine that it would happen so quickly. The split presented the SPNI with an identity crisis. Until then they had had a clear hands-on activist mission to go into the field and protect nature. Suddenly the organization had created a competitor that had more resources and authority and could therefore do the job more effectively. The story of the child who outgrows the father may be common but still involves painful transition.

In retrospect, though, environmentalists have few regrets that Yoffe ran the Authority "his way." The Authority's immediate task was to get as much land under its protection as possible. Without Yoffe's unique combination of obstinacy, connections, and charm, only a fraction of today's reserves would have been saved.

The law requires a tedious process of consultation prior to an area's actual declaration as a protected reserve: The Minister of the Interior can make a declaration only after consulting with the Minister of Agriculture (today the Minister of the Environment). He must receive feedback from all affected local authorities and, in the event that the reserve might affect a holy or ancient site, the Ministers of Religion and Education. The site has to be approved in all relevant zoning schemes, requiring the cooperation of the slow-moving local planning committees. The law also holds that if a reserve is proposed on a site that has significance for national security or that might potentially be used for military training, the Minister of the Interior must comply with any directives from the Minister of Defense.[74]

The bureaucratic hurdles were deliberately designed to ensure a balance of interests, and they successfully clipped the Authority's wings. Despite the enthusiasm and momentum after the unanimous Knesset vote, during the Authority's first year, only three sites were declared nature reserves.[75] Yoffe, after taking over, not only increased the number of sites but their size as well. For example, in the largest undisturbed area of the north, Mt. Meron, the Alexandron Committee recommended three separate reserves that collectively did not exceed twenty-four thousand dunams.[76] On December 9, 1965, Yoffe pushed through Ariyeh Sharon's integrated ninety-six-thousand-dunam reserve and then began work on expanding it.[77] To this day, the Meron Reserve is the jewel in the NRA crown. By the end of his tenure, close to one hundred reserves had been declared.[78]

Strategically Yoffe was faced with a dilemma. On the one hand, he could create as broad a network as possible, meaning that most nature reserves would be small, capturing pockets of particular natural and scenic value. Alternatively he could concentrate on the larger reserves.[79] A decade before the so-called "SLOSS" debate raged among American conservation biologists over the ecological advantages of a "single large" reserve (SL) versus "several small" reserves (SS), Yoffe faced one of the seminal questions of nature preservation. Typically, he decided to reject what he perceived as a "false dilemma" and pursue both routes simultaneously. In doing so, he knowingly subjected his staff and its successors to the painstaking bargaining and minutiae that accompanied the declaration of each reserve. It also meant spreading thin his modest staff of rangers. At the same time, however, he would preserve a much richer variety of resources. The results can be seen in some basic statistics: Of existing and planned reserves in Israel, 63 percent are less than one square kilometer in size and 25 percent are between one and ten square kilometers.[80] At the end of the century, thirty-five years after the process began, some 40 percent of the land designated to be reserves still has not been formally protected.

While Yoffe took the lead at the national level, the local political wheeling and dealing behind the declarations were not trivial. Most of the lands were located in rural areas, so it was important to gain the support of the surrounding communities. Dan Perry, an SPNI recruit, had been a general secretary and business manager of a kibbutz. Yoffe assumed that he could relate to farmers and sent him out to make peace with these communities. Perry describes the climate of the 1960s and early 1970s:

> In general the country was very naïve in those days, and the agricultural settlements saw nature preservation as something positive. Every kibbutz had some members who were known as enthusiasts, and they were natural allies. The country was less greedy during this period; there was a willingness to sacrifice for the public good.[81]

Not all communities were completely altruistic, however, and deals had to be cut. For instance, Kibbutz Dan gave up land, in return for which it received a promise to allow development near the Tel Dan reserves. Other kibbutzim got grazing rights.[82] The SPNI was critical of some of the concessions, but without the agreement of the local planning commissions, controlled by representatives of the same kibbutzim, declarations could be stalled indefinitely.

While he could bank on the goodwill and public support that the SPNI had built up over the years, it was Yoffe's personal connections that were of paramount importance. This was particularly so in the NRA's many dealings with the military. Nervous underlings would often caution Yoffe that he was being too ambitious and that his grandiose proposals would backfire. Anticipating the competing bureaucracies' tendency to nibble away at his initial request, however, his typical reply was, "If you've got mice in the pantry, you have to start with a *very* big cake."[83] Usually his maximalist strategy proved right.[84] In one crucial area, however, he may have miscalculated.

Yoffe haggled over the creation of reserves only in the northern half of Israel. Regarding the southern Negev region, he backed a broader vision, proposed by Azariah Alon, that turned the usual presumption on its head: The entire Negev would be declared a reserve, and any settlement there would require the permission of the Authority.[85] Here Yoffe encountered opposition. Haim Kuberski, the powerful Director General at the Ministry of the Interior, generally sympathetic to environmental interests, felt that it was too much. Nothing came of the plan. In a sense, Yoffe went for broke and ended up empty-handed.[86] Only after his retirement did the NRA change its orientation, but by then much of the Sinai peninsula had been returned, and the military was much stingier. It is likely that a more piece-meal approach during the boom years of Yoffe's regime would have generated more than the 30 percent of Negev lands that were ultimately designated to become reserves.[87]

FLOWER POWER

With a system of postage-stamp-sized reserves, it became crucial to expand protection beyond the boundaries of the reserves themselves. The "protected natural assets" statutory provisions offered the NRA just such an extraterritorial vehicle. Ultimately, hundreds of ferns, fish, trees, shrubs, and well-known mammals were listed for protection regardless of their location.[88] The list of proscriptions even includes the damaging or taking of four different types of geological formations. The rules were first tested on Israel's wildflowers; the wildflower protection campaign of the 1960s showed the potency of government and public-interest groups working in concert.[89]

Springtime in Israel carries resplendent blossoms, and meadows are transformed into multicolored quilts. Since the advent of Zionist settlement, wildflowers have inspired a spontaneous race to the countryside. Israelis delighted in bringing home the colorful anemone, the edible lupine, and the rare iris by the sackload. Picking wildflowers was in fact part of nature education. As Israel's population grew, however, the childhood pastime became destructive. Entrepreneurs sold commercial quantities of posies by the roadside. With no time to germinate, the blossoms began to disappear.

Moshe Dayan had finally gotten the nature nuts out of his hair and was unwilling to entangle himself in any more legislative squabbling on environmental issues. And so Uzi Paz quietly prepared a draft, and in 1964 laureate Yizhar Smilansky shepherded it through the Knesset as an amendment to the National Parks and Nature Reserves Law. The amendment prohibited sales of protected natural assets without a license (granted by the Director of the NRA[90]). The campaign was ready to begin. Immediately thereafter a list of protected wildflowers was submitted to the newly formed National Parks and Nature Reserves Council for approval.

Passing the law was easy; the primary challenge was educational. The first step was to simplify the list of protected flowers. For instance, by including all types of irises, many varieties that were in no way endangered rode on the coattails of the threatened species. In the same way, blanket protection was afforded to all twenty endemic fern species, even though some were quite common. Extremely rare flowers were left off the list entirely. The presumption was that people were unlikely to encounter them, so there was no point in their having to memorize them.

Aesthetics was also considered. Only flowers that were attractive made the list. Initially there were two flower designations: protected and defended. The public was allowed to pick ten flowers each from the former list, which included cyclamens, narcissus, buttercups, gladiolus, and the conspicuous red anemones.[91] But with time this only confused the public and the inspectors in the field, and so picking flowers from this group was banned as well. In all, seventy wildflowers were "defended."[92]

Consultation with public-relations experts did not produce any novel ideas. The initial slogan for the campaign, "Don't Pick, Don't Uproot, Don't Sell, and Don't Buy," came off heavy-handed. Eventually it was changed to "Go Out to the Landscape, but Don't Pick," a rhyming jingle in Hebrew, which survives to this day. So does the campaign poster of wildflowers on a black backdrop, based on paintings by Heather Wood, a British artist who had impressed Yoffe during his travels.[93]

Yoffe raised the equivalent of forty thousand dollars for the campaign. Every national newspaper published the new regulations and sported a wildflower-of-the-week column in their expanded Friday editions. National lottery tickets featured wildflowers on them. The NRA sent the wildflower poster to tens of thousands of public institutions. But the marketing strategy targeted children, especially in the compulsory kindergartens (children around five years old). They would bring the message of self-restraint home to their parents. Elementary and preschool teachers were the frontline troops, and their effectiveness exceeded everyone's expectations. And to beef up enforcement presence, the Authority appointed hundreds of "volunteer inspectors."[94]

Except for a few cynics, who saw the whole campaign as a front for the T'nuva agricultural cooperative of commercial flower-growing interests, it was an extremely popular campaign.[95] Many factors can be put forward to explain its phenomenal success, relative to other public appeals: the idealistic spirit of Israel during the 1960s; the population's homogeneity; the availability of an alternative, inexpensive flower supply; and the lack of inconvenience (as opposed to efforts to increase public-transport ridership or antilitter campaigns). None of the explanations is particularly satisfying. The simplest may be the best: Israelis love their wildflowers and came to understand that without collective self-discipline they would disappear. The NRA and the SPNI had a simple message, and they stuck to it, together.

Flowers tend to be more popular with the public than are other types of plants. The energy and creativity surrounding their protection also served as an exception that proved a rule. Ironically it was the zoologist, Mendelssohn, who in the early 1970s admonished both the NRA and the SPNI for neglecting the preservation of relatively arcane flora species in favor of less endangered but more attractive wildlife.[96]

RANGERS AND NATURALISTS

Not only the wildflowers but also the staff of the NRA grew under Yoffe's leadership. In his first three years at the helm, the number of inspectors in the field increased from fifteen to thirty-three. After ten years the number reached 156.[97] Although it may have been an institutional flop, the Shachal unit introduced to the Authority a generation of young SPNI personnel who chose to stay with Yoffe in government service. This group included such figures as Adir Shapira, Dan Perry, and Alon Galili. They brought with them not only a passion for the subject matter, but knowledge and field experience.

Their presence also created a clear dichotomy among the NRA workers. Despite their lack of formal training, the SPNI alumni were perceived as the experts. Primarily from kibbutzim, they wore their sandals year-round and shared a youth movement educational orientation. On the other side were the more macho military types that Yoffe had brought with him from the army. Many of these "rangers" were attracted to nature because of hunting. For instance, Uri Horowitz, one of the early NRA inspectors, had been convicted of hunting violations.[98] Yoffe also brought on board Arab and Druze inspectors who were known to be skilled hunters. The rangers departed from conventional international stereotypes. Hunting enthusiasts are typically found in forestry rather than nature preservation agencies.

Although the new agency probably benefited from the different skill sets and capabilities each group brought with it, there was tension between the two camps. It was not only their attitudes toward nature that differed, but their deportment. The coarseness of the military types, who freely burped and cursed, made the more refined and younger SPNI inspectors somewhat uncomfortable.[99] Yoav Sagi was one of the few Shachal rangers who chose not to leave the SPNI, maintaining a formal "dual affiliation" as chief ranger and field-school director in the Meron region until the end of the 1960s. He believes the essential difference in the organizations was the approach to decision making. Raised in a kibbutzlike organizational culture, SPNI rangers worked around the clock but expected their views to be influential in decision making. Yoffe, on the other hand, was used to giving orders. Eventually, however, he came to rely on the naturalist contingent, which were more committed to the new agency and which ultimately replaced him at the Authority's helm.

It did not take long for a synthesis of the two styles to emerge. The macho "sheriff" function was manifested in the tough oversight of the two thousand hunting licenses the NRA granted during the 1970s (and of the many others who hunted anyway).[100] It also was reflected in the alacrity with which rangers themselves picked up rifles. In 1966 the NRA forged a formal agreement with the Veterinary Service. Rangers agreed to shoot any stray dogs they encountered, and the veterinarians agreed to suspend all poisoning activities.[101] During the 1960s and 1970s, seven thousand to ten thousand dogs were shot each year. Even today the problem remains acute in the West Bank, where several hundred dogs are shot each month.[102] In addition, stray cats in natural settings were gunned down in order to preserve the genetic purity of the indigenous wildcats. Recently, at the request of animal rights groups, the NRA instructed its field personnel to try to call stray dogs before shooting them, but the 1966 agreement remains in force.[103] Sometimes a rabies outbreak, such as a 1997 scare in the Arava, requires broader intervention; in this case NRA rangers shot dozens of potentially rabid foxes inside the perimeters of area kibbutzim.[104]

None of the rangers had advanced zoological or botanical training. Yoffe was aware of this and, in 1966, brought D'vora Ben Shaul (see Figure 14) onto the NRA staff. Ben Shaul did not fit into any particular mold. Raised on a farm in east Texas, she believes that her passion for animals was forced on her by a pet dog who would chase down a variety of rabbits, partridges, and squirrels and then tote them home to her. Having acquried doctoral degrees in biology and theology, she spent much of her time in the 1960s working as an unpaid curator at the Jerusalem Biblical Zoo. When she came to complain to Yoffe about the pervasiveness of poisonings and their impact, he ended up hiring her as a wildlife supervisor.[105]

She was not welcomed into the Authority's pervasive macho culture. (Years later, the inspectors told her that they actually staged a protest when they heard they might have a female boss. Yoffe threw them out, telling them he had "enough balls" already and wanted some brains.) It was not just her hands-on experience with animals that eventually gained the confidence of her coworkers: Ben Shaul was from Texas and could drink many of the inspectors under the table.

In 1970 the NRA founded a Poison Department to deal more systematically with the issue of pesticide abuse. Ben Shaul was pressed into service to begin the long task of reining in the agrochemical industry and its allies at the Ministry of Agriculture's Plant Protection Department.[106]

Coalitions were formed with the Ministry of Health, but their influence was not extraordinary. In 1973, for example, there were two thousand documented cases of protected animals being poisoned from agricultural sources;[107] birds were especially affected. By 1972 Professor Mendelssohn had enough evidence to indict the exaggerated (and illegal) application of the rodenticide thallium sulfate as the primary reason for the dwindling raptor population.[108] The sins of the past could last as long as the chlorine-carbon bonds in a persistent pesticide.

The next scientist that Yoffe recruited was also a woman and no less a maverick. Aviva Rabinovich also grew up on a farm, but in Israel near Rehovoth. As a child, her constant forays into the fields to bring back plants and animals baffled her immigrant parents. At age seventeen she ran away to join the Palmach. During her five years of service in the noted Har El division, she was involved in many of the most decisive battles of Israel's War of Independence. She thinks that in later years veterans of the unit, among them Arik Sharon, Rafael Eitan, and Yizhak Rabin, listened to her ecological ravings only because they remember her as the only woman wounded during the War while charging an enemy position.[109]

After the war, Rabinovich settled on Kibbutz Kabri, where she still lives. She raised a family and taught high-school biology, chemistry, and physics. By age forty, her command of botany and ecology was so impressive that traditionally inflexible Hebrew University allowed her to skip formal B.A. requirements and complete her graduate studies in a few years. In 1969, at the recommendation of a staffer, Avraham Yoffe asked to meet her. Rabinovich chewed Yoffe out for starting their interview late, and he immediately hired her.[110] It was a good match. Rabinovich remained affiliated with the NRA for twenty-seven years.

Regardless of their differences, the motley crew assembled during the 1960s universally adored Yoffe, perhaps because he never tried to hide his flaws. He was a demanding boss with a temper and had a hard time staying on budget, but he readily delegated authority and backed up his workers. He was also unapologetically autocratic. The eleven-person Authority, to which he formally was subservient, quickly became a rubber stamp. By 1970, there was rarely a quorum at meetings.[111]

NATURAL ENEMIES

There is nothing like an external antagonist to unify a team. The Jewish National Fund played this role perfectly for the NRA staff. Problems began even before the Authority's inception. There was no love lost between the foresters and the SPNI, which was born as a protest statement against JNF reclamation policies. Even a decade later, the JNF was not happy about the creation of a competing agency, and Yoffe's hegemonistic tendencies only heightened tensions. New nature reserves brought the Authority into direct conflict with the JNF, which argued that the reserve system should be limited to the modest vision endorsed by the Alexandron Committee. Yoffe disdained what he felt were petty bureaucrats.

Yoffe also had no love for Yosef Weitz, still the central figure at the JNF during the 1960s. He resented the JNF's enormous budget and the way Weitz could waltz into the offices of his Second Aliyah cronies Prime Ministers Levi Eshkol and even Ben-Gurion and extract huge funds and concessions under the guise of Zionism.[112] (Of course Yoffe did the exact same thing with pals from his own generation.)

There were other areas of friction. NRA field staff complained about the JNF forests' impact on wildlife. Animals had to abandon the forests for lack of food and began to go down to the valleys and bother the farmers. Then they became the Authority's problem.[113] Aviva Rabinovich elevated the acrimony to a new level. Her professional expertise enabled her to explore the relationship between geology and botany, and her surveys found fascinating correlations between rock formations and plant types. Not surprisingly, the JNF's monocultures were an affront to her sense of ecological integrity.

Rabinovich's vituperative attacks on the JNF in any and all public forums often bordered on abuse. Her critique did not always fall on deaf ears, however, and although she remains a particularly reviled figure for most JNF officials, some admit that she influenced their thinking.[114] Indeed, during the 1980s, the JNF placed her on their research committee, and she became an active lecturer in their professional training sessions. Within her own organization, some critics found her continued fixation on forestry (even after the JNF clearly changed their approach) out of line. More important, it led to neglect of other, more important topics, such as the military's redeployment in the Negev during the 1980s after evacuating the Sinai.[115]

Generally the Nature Reserves Authority received good reviews. The truth of the matter was that if Yoffe's personal charm could not win over critics, then the sheer breadth of the NRA's achievements could. In 1971, during the peak of the Yoffe years, Israel's State Comptroller reviewed the Authority's operations. He found plenty to criticize. Safety in the reserves was not addressed properly; jobs were never filled through a formal tender, but on the basis of an interview with Yoffe or his deputy; presigned checks were left in the hands of underlings who had no authority to make purchases; there were no personnel records and no filing system in the central office; worst of all, the Comptroller found that the NRA had exceeded its authority and dabbled in many areas without legal authorization[116]—the scolding, however, got lost in the overall glowing conclusion. "The Authority took care of a large number of reserves, got many ready for mass visitation, and developed educational activities to inculcate nature protection into the public consciousness," the report raved. "In these areas, the Authority attained achievements that are worthy of praise."

THE NEXT GENERATION

In 1978 Yoffe turned sixty-five. He was a member of the Knesset and was ready to step down after fourteen years directing the NRA. There were two clear candidates to replace him: Adir Shapira and Uzi Paz. Paz was much older, with more experience, and better versed in the science of nature protection. He also challenged Yoffe more on professional issues. On the other hand, although Shapira had never been a high-ranking military officer, he had a much more forceful and authoritarian personality. Yoffe chose him over Paz.

Shapira maintained the general momentum that Yoffe had started. Expansion continued. For instance, in April 1979, the nascent Environmental Protection Service empowered the Authority's rangers to regulate oil pollution in the Gulf of Aqaba and the Mediterranean coast.[117] During the late 1970s the Authority reached its peak number of workers—four hundred, most of them in the field.[118]

Under Shapira, the NRA modified its strategy toward the Negev. Shapira commissioned a plan for a Ramon Crater reserve in August 1979; it was approved by the government only in 1982.[119] This was among the first of the large desert reserves. At the same time, the NRA set its sights on the forty-one thousand hectares of the dark granite Eilat mountains that make up the southern tip of the country.[120] Toward the end of the 1980s, when Ariyeh Deri took over as Minister of the Interior, the process moved into higher gear. Deri was the *wunderkind* of the new Orthodox-Sephardi party, Shas. He was a young Yeshiva genius who had hardly served in the army. Unintimidated by the military, he did not hesitate to declare nature reserves in the south. From the perspective of sheer magnitude, Deri should go down as the Greenest interior minister in Israeli history.[121]

Because 38 percent of the lands in reserves overlap with military training grounds, the actual breadth of preservation in the Negev may be overstated. Section 23 of the National Parks and Nature Reserves Law frees security forces from complying with its provisions. Technically the army has free rein to go about its business in the reserves. This may explain why it has never been particularly obstinate in opposing the declaration of reserves, even when the land is already utilized for training. The NRA offers the IDF a letter openly agreeing to continued military activities, and the status quo continues.[122] Preservationists take the long view, assuming that once a reserve is declared, they can expect greater sensitivity from the military. And perhaps one day peace will obviate the need for army maneuvers.

In practice, the Authority tries to reach a consensus with the army on three classifications of land usage in their reserves: nonshooting areas, air force grounds, and tank training grounds. The latter suffer considerable damage, but the massive areas set aside as safety zones around the periphery of firing ranges remain unharmed.

AN AMBIVALENT QUEST FOR GOOD SCIENCE

From the Authority's inception, its field staff contained many talented individuals who quickly attained considerable practical knowledge. Alon Galili still wears his NRA T-shirt, even though he is no longer formally on its staff. Galili's fame/notoriety comes largely from his work as head of the Green Patrol (see Chapter 10). At his home at Sdeh Boqer, it takes little to get him talking about nature preservation. His story is typical of the first generation of workers who joined Yoffe in the 1960s. A kibbutznik from Ein ha-Shofet, he began his conservation career at the SPNI. Galili would compensate for his lack of academic training by sheer proximity. When asked to draft a conservation strategy for wild boars, he literally lived with the pigs for six months—dyeing them, tagging them, marking hooves. The task completed, it was on to porcupines and gazelles.[123]

Yoffe often preferred the common-sense orientation that his field staff brought with them to the more ponderous deliberations of academics. Paz, whose own perspective was colored by his years as an SPNI inspector, was considered the senior scientific voice in the Authority. Yoffe also understood that this was not enough and cited "limited local experience" for his heavy reliance on foreign expertise. His workers remember his frequent trips abroad under the guise of selling Israel bonds. While "in the neighborhood," he would drum up donations and scientific backing for his own preservation projects.[124]

In 1968 Yoffe brought a delegation headed by Sir Peter Scott and other leaders from the International Union for the Conservation of Nature (IUCN) to Israel. Its mission was to offer advice about the management of open spaces; its recommendations provided the conceptual strategy for the

Authority for the next two decades. The year before, he consulted Professor Bob Davis, a visiting limnology expert from South Africa, on the condition of Lake Kinneret. Davis argued that the Sea of Galilee was rapidly eutrophying and that immediate measures were required to stall its demise. Yoffe immediately moved to create a Kinneret Basin Authority.

Part of the attraction of foreign scientists for Yoffe may have been that it freed him from the Israeli academic establishment. Although everyone remembers him as a good listener, he often found local scientists unrealistic. One of the results of Yoffe's ambivalence toward science was the lack of a systematic, empirical approach to internal decision making. For instance, the scientific community could not decide whether reintroduction of gazelles in the Golan Heights was a good idea (owing to high seasonal humidity). In 1968 rangers found an ibex skull in the mountains. It alone was taken as sufficient proof of past wildlife survival to move ahead with a major relocation project.[125]

The generation that came to the Nature Reserves Authority from the SPNI had tremendous practical knowledge about animals. The problem was that they rarely converted it into a scientifically usable form. Anecdotal information survived through a haphazard oral tradition.[126] When he became Director, Adir Shapira decided to change this.

When Uzi Paz decided to take a leave of absence to complete his doctorate, Shapira took advantage of his departure to create a new department with Aviva Rabinovich at its head. Among the first things Rabinovich did was establish scientific criteria for the delineation of reserves, first in the Galilee and then on a national level. Dozens of graduate students were funded to generate the data from the field that would be required for detailed biological mapping. Rabinovich established professional advisory committees to assist the NRA in its work, and an extensive educational program was set up to give rangers advanced training to supplement their skills. Rabinovich established a controlled-grazing initiative based on ecological principles that for the first time allowed herds inside reserves in order to balance the flora of the area.

There were many practical reasons for upgrading the NRA's scientific capabilities. Advocacy was strengthened in all areas of activity if it was backed by even modest expert opinion. Rabinovich became the NRA's scientific "hired gun" in its arguments with Maatz (Israel's highway construction agency), the JNF, the Bedouin, and the Ministry of Agriculture.

Scientific self-reliance intensified the rift between the NRA on the one hand and the academic community on the other, as well as with the SPNI, which often felt shut out.[127] At the same time, the elevated influence of in-house scientists strained relations between the rangers and the scientists. This tension became sharpened when Uri Baidatz succeeded Shapira as NRA Director. In 1988, at the recommendation of one of his former students, Baidatz approached Uriel Safriel, the original SPNI Eilat inspector and by now an eminent Hebrew University professor. Aviva Rabinovich was talking about retiring after almost two decades of commuting to work by bus, and Baidatz thought it was time to bring an ecology expert from academia on board.

Safriel embraced the challenge, but immediately encountered systemic problems. There was little follow-up in monitoring the success of management strategies. Safriel pressed the staff to frame its activities in a scientific context, with a null hypothesis that could be evaluated. This was not the way the pragmatic field staff was accustomed to working, and resistance was great. Inspectors became embittered and muttered about scientists who were only interested in their research findings. After she returned from a scientific mission to Guatemala, Rabinovich was unhappy with her successor's new, academic orientation. She felt that Safriel was out of touch with the reality of the field and opposed his attempt to shift expertise and geographic information system (GIS) databases from the NRA to the universities. To this day she cannot forgive Safriel for dismissing several veteran NRA field scientists.[128] In the ensuing tensions, NRA Director Baidatz backed his Chief Scientist, along with his personnel and policy changes.

When Dan Perry took over as Authority Director from Baidatz, he reverted to the more traditional, pragmatic approach. Perry had come through the system as an SPNI recruit from Shachal. Self-taught, with considerable field experience, Perry was quite knowledgeable. Safriel felt that Perry was not interested in consulting with a scientist at all. It did not take long for Safriel to tender his resignation.[129]

The process of enhanced technical sophistication at the Authority, however, was unstoppable and probably had little to do with any given personality. By the 1990s, science was built into the fabric of the agency's culture. The computerized ecological databases that Aviva Rabinovich had pioneered during the early 1980s now contained hundreds of thousands of observations of natural assets and resources, ranging from the genetic sources of grains and cultivated plants to plants in the wild and fungi. They are supplemented by a GIS, computerized-mapping department. The Authority also initiated long-term research ventures, receiving international funding for monitoring specific species, controlled-grazing programs in protected woodlands, and predator-prey balance.[130] Each NRA region has a biologist who works alongside the rangers. The level of formal training among field staff continuously improves; arbitrary management decisions, with no firm basis in science, are harder to impose on a conscientious and competent staff.

The NRA annual gazelle count is instructive, for instance. Each year the Authority undertakes an inventory of the local stock of gazelles, checking the age and sex distribution of dorcas gazelles and a range of other wildlife. If the survey is done incompletely, results can be misleading. In 1991 Daphna Lavi, the southern region biologist, repeatedly asked that the count be discontinued until a more reliable protocol could be implemented.

When her request was denied, she fired off a scathing report to the NRA's *Yedion* circular: "All told, the count consumed 110 work days," she railed. "It produced very few conclusions, and these are not useful at all. . . . I again recommend presenting the problem of the Negev gazelles to a scientist who will address the subject in an intelligent fashion. . ."[131] In such an open and critical institutional culture, the days when a director could run things on intuition alone seemed as remote as ancient history.

THE WATCHDOG BARKS

Relations between the SPNI and the NRA remained complicated. Even after the wildflower success, the earlier organizational divorce left a bitter aftertaste. Yoffe thought that the SPNI pushed its nose into areas where it had no business. There was also a sense that its unrelenting criticism was tainted with self-interested motives.

On the human level, however, there were close ties. Most of the NRA's professional staff had once been affiliated with the SPNI. Such relationships could lead to awkward situations at times. In fact, the values of staff members were practically identical. Disagreements were primarily over tactics and arose from institutional constraints.

When the two entities managed to work together, nature was the beneficiary. For instance, while the SPNI conducted rallies to stop the proposed power station along the Taninim Stream Reserve in the 1970s, Uzi Paz was lobbying the Knesset on behalf of the NRA. Politically it was a difficult case, because the Hadera mayor, Dov Barizilai, was a right-wing Heirut party (later Likud) member, outside the traditional realm of the nature community's influence. However, Heirut chairman Menahem Begin agreed with Paz that national interests should trump local ones, and he overruled his party member.[132]

But just as often the NRA and SPNI fought like cats and dogs. No issues were more divisive over the course of the two organizations' long relationship than the access to and development of nature reserves. These involve classic dilemmas with which wildlife managers wrestle around the world; the questions are about ethics as much as they are about management.

Reserves are set aside for a variety of purposes. Some are declared because of the existence of a rare species, others for their scenic value. Still others are declared because they are representative of various stages in ecological development. And, of course, there are critical habitats. The functional diversity of the reserves requires flexible management. Officially, the NRA has established three kinds of reserves:

- *Developed reserves,* with basic infrastructure to accommodate the public (e.g., roads, trails, parking lots, and sometimes even picnic tables and rest rooms)
- *Open reserves,* where development is limited to a few trails and the occasional sign
- *Closed reserves,* which contain elements or processes too sensitive to allow visitor exposure

The issue of charging for entry into reserves has been controversial from the minute that debate began on the law. Some see admission fees as inherently undemocratic; others counter that the public will appreciate reserves more if they pay for them.[133] Under Shapira's leadership, the NRA took the latter view. In November 1970, to cover "operating expenses," the Ein Gedi and the Tel Dan reserves began to charge admission fees.[134] Others would follow.

The financial perspective of each group informs their institutional position on the issue. The NRA, always short on funds, is expected to raise much of its own budget, and the entrance fees at the dozen or so "commercial" reserves generate significant revenues. Conversely, among its many functions, the SPNI serves as a de facto consumer advocate for Israel's hiking and nature-loving community. The organization is also in the tour-guiding business and absorbs some of the costs of the entrance fees. The issue of closing reserves is more complex. Although the SPNI understands the need to let natural systems regenerate, it feels entitled to access, on educational grounds, and is quick to cry "elitism."[135]

Far more heated than the debate over entrance fees is the question of the reserves' physical development. Even the seemingly benign marking of hiking trails to show walkers the way engenders criticism. Here, it is the Authority who can cry "elitism." Proponents argue that the trails increase safety and allow the hiker to focus on nature rather than on not getting lost. It is also a question of preservation: Trails minimize human interference with natural processes. The NRA approach holds that people are guests in the reserves. Almost without exception they are not allowed to stay the night. Then there is the issue of handicapped access. Recently, the NRA constructed wheelchair trails in the Tel Dan and Huleh reserves for mobility-impaired visitors.[136] Surely Israel's large elderly and handicapped population deserve to visit a fair share of the country's reserves.[137]

The clear intent of the law is to develop national parks but leave reserves untouched. But by statute the NRA was authorized to undertake development activities, setting up buildings and facilities and managing, arranging, and running them, along with services, for visitors and hikers.[138] From the outset, infrastructure accompanied the declaration of many reserves.[139] The streams of Dan and Banias, the oasis in Ein Gedi, and the waterfall at Tanur all enjoyed these improvements, some of which incensed the SPNI watchdogs.

Amidst all the huffing, little serious effort is made to consider the actual motivation behind and validity of the Authority's development activity. The underlying impulse typically is not recreation (as is the case in initiatives in the national parks), but preservation. By managing the public, nature is better protected.

As NRA Director, Dan Perry was especially reviled by activists when these issues arose, because he was so completely unapologetic. Now retired, Perry has a senior statesman persona in the international conservation community, owing in part to his graying beard as well as his permanent limp from an unfortunate meeting with a land mine in the field. He consults frequently, and he still represents the pragmatic end of the NRA spectrum. Perry claims that compromising to accommodate people has been the key to the Authority's success from the start. Increasingly the world recognizes that reserves need to integrate local populations into their long-term strategies, leading to the concept of biosphere reservations.[140] Jordan's Dana reserve has intrigued visiting Israelis, but Perry claims that in practice, Israel has for years been making similar arrangements in administering reserves. In the case of the Ramon Crater, the NRA concluded that if they did not want jeeps and motorists to cut a thousand trails across the crater, they had better give the public a reasonable road to drive on.[141]

Over time, the NRA's pragmatic approach won over most "purists." The Coral Beach in Eilat was one of the first reserves declared in Israel (November 26, 1964) and was also among the first that charged admission fees. After seeing that the five-hundred-meter-long beach was fenced in, NRA Director Shapira had a kiosk and shower facilities installed. As part of the development package, he even allowed an undersea observatory to be built at the reserve's southern tip. The biologists who worked for him in the Authority were initially furious but later had a change of heart when unprotected areas sustained massive damage.[142]

Even under the watchful eyes of the NRA staff, the reef in the reserve began to suffer, primarily because snorkelers were unintentionally trampling the sensitive corals on their way into the water. Although slow in responding, the NRA ultimately built concrete bridges to enable swimmers to hop over the shallow, most vulnerable section of the reef. The management of the reserve also began to limit the number of visitors in the reserve at any given time, based on estimates of carrying capacity.[143]

DANCING WITH LEOPARDS

Reserves often cannot provide sufficient habitat and refuge for the creatures that inhabit them. Predators in particular need an enormous range to find their meals. The spillover of wildlife causes problems for rural communities around the world—from East African elephants to the coyotes of the American west. In Israel gazelles happily munch on lettuce, and foxes raid chicken coops (and have acquired a taste for watermelon). Plastic irrigation pipes are favorite targets of confused woodpeckers and provide teething rings for young hyenas.

This tension between human settlement and wildlife habitat captured national attention with the NRA's efforts to study and save the leopards in the Judean desert. Although a larger, northern subspecies (*Panthera pardus tulliana*) has apparently disappeared, Israel's southern subspecies of leopard (*Panthera pardus nimr*) remains the natural "high carnivore" of Israel's desert.[144] It is a very large cat with black spots on light brown or white fur, a large solid head, round ears, round eyes, long whiskers, four nipples, and a tail that is more than three-quarters the length of its body.[145] Leopards can be ferocious, and any animal in the desert is a potential item on its menu. The cats live amidst the desert's rocky cliffs, from among which they pounce on their prey, typically hyraxes, ibex, and porcupines.

In 1863 and 1864 the British priest and zoology enthusiast H. B. Tristam traveled across Palestine on the most extensive of his four local safaris.[146] He reported that "the range of the leopard is broader than the cheetah in Israel, but their number is smaller. It is found around the Dead Sea, in the Gilead, in the Bashan, and occasionally in forested areas of the West. A wonderful pair was hunted in the Carmel while we were there."[147]

The many places he found in the region with traditional Arab names that included the word *nimr* (leopard) testified to its versatility. In 1930 the Yishuv's leading zoologist, Israel Aharoni, reported that leopards could be considered extinct in Palestine, yet time and again they would resurface. When a female was shot in Safed in 1942, three of her cubs survived. They were rescued, and one of them, "Teddy," was briefly adopted and had a book written about him. (Unlike Elsa of *Born Free* fame, Teddy lived out his days in a zoo.) Nocturnal sightings of the cat were common. There was even a 1970 case where three workers happened upon a leopard at Kibbutz Ma'ayan Zvi's fishponds near the Tel Aviv–Haifa highway.[148]

The little that is known about Israel's leopard species is the result of research by Giora Ilani, yet some people blame this fascinating man for leading these animals down the road to ruin. Ilani's biography runs parallel to those of his generation of nature professionals. He grew up on Kibbutz Gan Shmuel. Before he enlisted in the army, in 1958 he began working as a volunteer for the SPNI. It was Ilani who actually invented the annual "gazelle count" ritual, long before it was officially adopted by the NRA. During an inventory in 1964 he discovered a new subspecies of indigenous Israeli gazelle (*Gazella gazella acaciae*).[149]

Ilani was even more dedicated to wildlife than his colleagues. Avraham Yoffe once told him, "If you were in charge of this country, then animals would be free and people would be in cages."[150] And in fact Ilani made few friends in the Arava Valley when he shared his belief that people should be banned from living inside the Syro-African rift, leaving it, as it had been historically, for the animals.

Of the twenty streams that drain across the Judean desert to the Dead Sea, only two have water year-round: Nahal Arugot and Nahal David. They define the boundaries of the Ein Gedi oasis. With no direct road to the area until after the 1967 War, Ein Gedi held a special mystique for travelers.

After the creation of the State, Kibbutz Ein Gedi was established south of the oasis. It tapped the streams for its fields, orchards, and highly profitable guest house. Yet it remained a lovely site in the middle of the stark, steep, jejune landscape. Bedouin, with their weakness for hunting, no longer passed through the area. The wildlife population in the area, especially the ibex, rebounded. In the anticipated positive-feedback loop of food chains, the leopards soon discovered the oasis.

In the Ein Gedi region, there had never been a confirmed leopard sighting until Yossi Feldman, who ran the Ein Gedi Field School, reported one in 1969.[151] At the time, Giora Ilani was working as chief zoologist for the NRA. After Feldman's experience, Ilani started to seek them out. It would take him almost five years of following tracks to meet one face to face. The memory of the triumphant moment still moves him: "On October 19, 1974, I managed to photograph one. It was the first time in history that a leopard was photographed in the Middle East. (I remember the date, because the next day my son was born.) After the photograph I cried nonstop, I was so moved and so excited. Avraham Yoffe sent the photograph off to be published in the *New York Times*."[152]

Many would argue that Ilani remembers his son's birthday because of the leopard. Before Ilani's interest in leopards becamse so keen, most of his focus had been on the hyenas that roam Israel's deserts. The experience with the leopard completely changed his life. Not everyone in the Authority was as enthusiastic about his initiative. But Avraham Yoffe, predictably, was captivated. He gave Ilani his full support.[153]

Ilani became extremely resourceful at finding the leopards. It was almost as if he had developed a sixth sense to detect the distinct four-toed paw print, the droppings with the encrusted hair of the last meal, and the smell of urine that marked the leopard's territory. Ilani would offer tips to help find leopards by watching the ibex and hyraxes. They developed a distinct warning call when they sensed the presence of this most deadly predator.[154] By 1974 he had fifty-one confirmed observations of leopards in the area and hundreds of definitive signs.

Ilani pressed on to the next stage. He left food for the animals, stunned them, and then collared them for continuous monitoring. Ilani gave all of them names and befriended them. Almost twenty years later he and his wife still argue over which female was in which wadi. The transmitters enabled Ilani to reach a new level of intimacy with the animals. For instance, he could identify the female "Shlomzion" as a young virgin and record her first sexual encounter—a three-day orgy which involved sixty separate acts of copulation (per day) with the virile "Alexander Yanai"! Ninety-one days later, two cubs were born in the very cave where the revelry took place. Attempts to visit her were foiled by an angry male hovering above the cave. Ilani was never able to ascertain whether the roaring leopard was "Alexander Yanai," the protective father, or "Katushion," a rival suitor.

From the start, Kibbutz Ein Gedi opposed the project. They were neither intrigued with the racy nuances of leopard mating rituals nor thrilled about sharing their oasis with a savage predator. Even Ilani acknowledged that if you startled a leopard at night, it could kill you in self-defense. Such concerns became more palpable in 1975 when leopards began to penetrate the kibbutz looking for food. In the nearby community of Neveh Zohar, a leopard ate fourteen cats. In 1979, "Bavta" ate a couple of sheep in the kibbutz's petting zoo but was shot and wounded before she could return to her cubs. Residents responded to the presence of the leopards by planting poison bait. In Ein Boqeq leopards were killed with strychnine.[155]

The NRA moved to assuage local concerns. Avraham Yoffe felt that the people of Ein Gedi should see their new neighbors as a boon to tourism, which offered a more promising economic future than agriculture.[156] The kibbutz hotel and associated facilities were flourishing, but residents were unhappy about the uninvited guests. They argued that, in fact, the leopards returned from their forays to the kibbutz with diseases that were the real cause of the epidemic of early cub deaths.[157]

In general it is hard to enlist public support for predators. People enjoy watching them in documentaries, but they become edgy when they meet them face to face. During the same period, Giora Ilani served as an expert witness in a 1976 prosecution against an illegal shooting of a hyena. Ilani explained to the judge, in his meticulous fashion, the essential ecological role that the animal plays in clearing the remains of dead animals. The judge retorted that while the letter of the law required him to issue a fine, he was completely sympathetic with the defendant.[158]

In 1982 the NRA installed an electric fence around the kibbutz to keep the leopards out.[159] It reduced the number of infiltrations, but the cats became crafty and learned that the fence short-circuited during rainstorms. Then they would come in and feast on the local pets. The farmers took matters into their own hands, getting out the traps and the strychnine.

The numbers of leopards continued to drop. Females, who were the most adventurous hunters, were disproportionately hunted down. Although they can live past the age of twenty-five, they are fertile only from age three until eleven. The decline in fertile females created a macabre dynamic. The male impulse to perpetuate his own genetic material became a major threat to species survival. As long as they nurse, females are infertile. Male leopards naturally seek to kill the cubs and thereby hasten female ovulation. As the population dwindled, problems of inbreeding and poor genetic diversity began to set in.[160] A 1992 scientific survey estimated that there were between eight and seventeen leopards in the Negev High Mountains, including at least one fertile female.[161] But none remained in Ein Gedi and the Judean desert.

THE LEOPARD'S LEGACY

Ilani's dismissal from the Authority in 1990 enhanced his legendary persona within nature circles and rendered his work a subject of even greater controversy. Ein Gedi residents are the most hostile of Ilani's detractors. Many claim that if he had not stuck his nose in, leopards would have retained their fear of humans and remained inconspicuously in the reserve. Uzi Paz, never happy with the way the leopard project was run (including the feeding of leopards), remains critical. He felt that Ilani had become too emotional, which affected both the scientific quality of his work and the quality of his relations with the residents of Ein Gedi.[162]

Professor Lev Fishelson rejected the criticism out of hand and was extremely favorable about Ilani's actual fieldwork. "Giora may not have been great with public relations, but he was amazing with the leopards. He understood them so well. The problem is you can't catch leopards. You can either kill them or bring them to a zoo."[163]

Others saw the failure as an institutional one for the Authority, which never succeeded in convincing local residents of the leopards' value. The leopards were always perceived as the Authority's problem. If the kibbutz had ever felt that they had a real interest in the reserve, it would have been the best of protectors.[164]

Uriel Safriel, Chief Scientist at the NRA, made the decision to release Ilani. He rejected the view that Ilani's familiarity with the leopards emboldened them to enter the kibbutz. Rather, it was the availability of easy food sources. Dan Perry, NRA Director during this period, shared this view and is surprisingly blasé about the ramifications:

> In the case of the leopards, reality and myth really diverge. Their survival may be an ecological miracle, but ten to twenty predators have virtually no biological significance. If we had put the same efforts in other areas, it would have been more efficacious. It may have appeared important for image and publicity, but really the leopards do not need us. We need to preserve their habitat and not concern ourselves with them.[165]

With ballpark estimates for the minimum viable population size of a species ranging from 50 (in the short run) to 500 (for long-term viability), the handful of Israel's leopards surviving in the wild are so far below sustainable levels as to appear tottering on the brink.[166] Most of Israel's zoological experts actually are cautiously optimistic about the leopards' long-term chances of survival. The question optimists face is: Have we learned anything from Ein Gedi's disastrous encounters during the 1980s with local leopards? There may, in fact, be some room for encouragement. Israel's endangered raptor population suffered owing due to the reduction in available carrion after improvements in veterinary medicine and animal husbandry practices were introduced. To compensate, the NRA established feeding stations that leave carcasses to the birds (as well as opportunistic hyenas, foxes, and wolves). The electric company has even taken pains to reduce the risk of electrocution of vultures.[167] The tourist potential of such active conservation efforts is only now being tapped, yet if the locals are not part of the solution, they will once again become part of the problem. This became painfully evident in the Golan Heights in July 1998, when farmers used a persistent phosphorus poison in traps for wolves that were devouring their cattle. It hit higher up the food chain, decimating the precarious population of vultures that had slowly been making a comeback.[168]

In the meantime leopards continue to pop up unexpectedly. For example, one evening in 1997 Roni King, the NRA biologist from Eilat, responded to an emergency call from the Negev town of Mitzpeh Ramon. A female leopard had been prowling the city streets before collapsing in exhaustion in the courtyard of a local yeshiva. King thinks that even though the cat was young (according to her sharp teeth), she had just grown weary of hunting down food in the desert. After she was fattened up and fortified at the NRA Hai Bar facility, the leopard was released.

Then, on December 21, 2001, shuffling into the cold dark of a desert night after watching a film at the Kibbutz Ein Gedi auditorium, the audience was greeted by a pair of fearsome eyes. A leopard had come back. There were already signs: Five goats had been devoured at the kibbutz petting zoo earlier that week. As it left its prey behind, this particular visitor was assumed to be a male. Yankele Gal-Paz, the kibbutz general secretary, told the press that the returning leopards did not alarm the kibbutz members, who "would do what was needed to protect them—and themselves."[169]

The leopard remains a symbol for Israel's nature lovers. The SPNI puts the animal's image on its promotional materials and its MasterCard. Although the jury is still out on this case, the NRA experience offers a warning about the perils of sentimentality. As Israel becomes more crowded, a laissez-faire approach will only hasten the clash between wildlife and humans.

RECLAIMING DIVERSITY: THE HAI BAR INITIATIVE

Perhaps the NRA's most famous example of proactive management to increase biological diversity is its Hai Bar (literally "Wild Life") reintroduction program. It too is the subject of controversy. In 1960, when beekeeping was still synonymous with nature preservation, Uri Tzon established the Hai Bar Association. Its goal was to return the animals that roamed through the pages of the Bible to their ancient land. Money was raised from donations, primarily through an American-based Holy Land Conservation Fund. The idea appealed to Avraham Yoffe's dramatic sensibilities, and the NRA adopted the project. Indeed, in 1971 Israel's State Comptroller reproached NRA workers for taking advantage of the requirement for annual renewal of hunting permits, to encourage hunters to contribute to the Hai Bar Association.[170]

It took two years to finish fencing off an area of thirteen thousand dunams south of Kibbutz Yotvata in the southern Arava, but the Hai Bar facility was ready in 1968.[171] In 1973 a smaller facility for Mediterranean animals was established in the Carmel Forest. Beginning in 1968 with three pairs of Somali wild asses, the rare creatures slowly began to arrive.

At that time the reintroduction of lost species was a new venture, and international conservation professionals had not yet developed standard procedures. Today, reintroduction is common, and the IUCN publishes formal guidelines.[172] In running the Hai Bar program, Yoffe went mostly on intuition. Ever a maximalist, he sought any animal that he could get his hands on that may have lived in Israel in years gone by.[173]

The return of each species had its own drama. Fallow deer barely made it onto the last El Al plane out of Tehran after the overthrow of the Shah and Iran's severing of diplomatic ties with Israel.[174] Oryxes had been completely destroyed in the wild, and international conservation organizations were hesitant to entrust Israel with even a few. But eight of these majestic white herbivores, whose long horns were mistaken by Crusaders for those of a unicorn, arrived from the San Diego Zoo in 1978.[175] From Ethiopia came ostriches. Addaxes, originally from the Sahara, were procured from a zoo.

The southern facility opened as a safari attraction in 1977 to cover project expenses, but the objective was reintroduction into the wild.[176] The addition of a "predator center" did little to help the project's image. This part of the facility keeps leopards, karakuls, wolves, foxes, and hyenas in small pens and more closely resembles a mediocre zoo than anything else. It may have increased the attractiveness of the facility for tourists, but it is still not enough to make the park break even.

Ironically, the indigenousness of many Hai Bar animals is often called into question. The asslike onagers are a hybrid of an Iranian and a central Asian subspecies, produced in a Dutch zoo.[177] The addax is probably of Saharan[178] or Indian origin,[179] not from Israel. Jordanian zoologists were polite when they were presented with a gift of ostriches during their first visit to Israel. Privately they marveled at how a sophisticated program like Israel's could consider releasing a species into the wild that bore little resemblance to what they believed were the true indigenous subspecies. The unfortunate birds are today stuck in a small pen in a Jordanian reserve.[180]

As Chief Scientist of the Authority, Uriel Safriel realized that for political reasons he could not kill the initiative. Using scientific criteria about carrying capacity, he managed to downsize the expectations of the project, however.[181] David Saltz has spent more than a decade writing feasibility studies, overseeing the reintroduction programs, preparing long-range plans for the Hai Bar project, and carrying out postrelease monitoring. His research indicated that five species were acceptable for reintroduction: Persian fallow deer, Arabian oryx, roe deer, Asiatic wild ass, and ostrich. Except for the ostriches (a Sudanese strand that are as vicious as they are stupid), reintroductions are under way for all, and postrelease monitoring is being carried out.

In retrospect, Saltz believes that Yoffe made two major oversights: in failing to anticipate the enormous costs of running Hai Bar facilities and in failing to ensure the authenticity of the animals' endemic identity.[182] But he points out that reintroduction has become an integral part of conservation biology internationally. The debate therefore is about priorities, not right or wrong.

The Hai Bar initiative's bottom line is reintroduction, and here the verdict seems to be in. By April 1982 the first group of wild asses were assigned color codes, tagged, and released.[183] On May 15 some disappeared near the Jordanian border;[184] it is not clear whether they were eaten by predators or, as Giora Ilani insists, shot by Jordanian soldiers.[185] Those who chose not to emigrate are doing well, with roughly a hundred animals in and around the Ramon Crater.[186]

After the Carmel Hai Bar population of fallow deer reached two hundred in September 1996, the first group received its "visas" to the Nahal Kziv reserve.[187] Slowly, this animal, painfully shy around humans, began to reproduce in the wild and spread into the surrounding JNF forests, even making the occasional foray to local farms.[188] In March 1997, twenty-one oryxes were released in the central Arava. They will probably compete with the local gazelles, but they appear to be surviving in the wild.[189]

There is a legitimate case for attributing ecological significance to the reintroductions that brought back large grazers, an unfilled ecological niche, because Bedouin herds are no longer in the reserves today. There are tactical justifications, too, because the animals increase the perceived value of sensitive areas, making their preservation easier.

The Hai Bar's real benefit, however, is psychological. Conservationists have a sense of always being on the defensive. Reintroduction programs turn the tables. Tourists, of course, love the biblical drama, and even the cynics cannot deny the excitement of chancing upon one of these wonderful animals while hiking in the wild.[190] At a deeper level, there is also a dimension of justice. After so many years of excess, humans should be required to do what they can to restore the ecosystems they have so thoughtlessly obliterated.

"NEW DEAL" FOR NATURE

In 1994 Minister of the Environment Yossi Sarid went to talk with the Prime Minister about the Nature Reserves Authority. "There's no logic in having two separate agencies for parks and for reserves any more," he told him. "No one can tell the difference anyway." "You're right," Yitzhak Rabin replied. "But what Minister would be willing to cede departments under his control?" Indeed, bureaucratic reshuffling has traditionally happened in Israel only as part of coalition agreements and elections. But Sarid did the unthinkable. He went to talk to Minister of Agriculture Yaakov Zur and simply put the case before him. Zur listened and on the spot responded: "You're right."[191]

It would take almost four years and a change in government before these changes received the Knesset's statutory blessing.[192] The amendment finally passed the Knesset in January 1998. By then, the Ministry of Agriculture was no longer the powerful agency it had been in 1962. Agriculture in Israel was in decline, increasingly irrelevant to the country's high-tech economy. So after thirty-five years, the Nature Reserves and the National Parks systems were merged, just as Yan Yanai first proposed, only under the auspices of a Minister of the Environment. It is not yet clear whether the new institutional supervision bodes well for nature. Dan Perry believes that historically there were clear advantages to being associated with the Ministry of Agriculture in spite of the undeniable conflicts of interest, from chemical poisonings to water allocation contests.

Although it would be unreasonable to expect today's heirs to the NRA legacy to be as ambitious as Yoffe, many observers have sensed a gradual enervation in the vigilance and scope of the NRA's agenda. Perhaps this change is just a sign that new institutions like the Environmental Ministry are now on the scene. The Authority may not need to go as far afield as it used to, doing battle against pesticide abuse and oil spills. Still, there is a growing perception that to maintain an image of moderation, directors subconsciously (or sometimes openly) build concessions into their opening bargaining positions.

This is altogether different from honest mistakes, which are inevitable. For instance, when Israel was leaving the Sinai during the early 1980s, the Authority's management focused its attention on the military redeployment going on in the Negev. Precious little effort went into confining the network of Jewish "mitzpim," or small settlements, that were scattered across hillsides in the Galilee.[193] The sprawl has been extremely problematic for wildlife.[194]

In other crucial battles, however, rather than standing up for nature, the Authority has maintained a reticence. For example, Highway 6 will have a devastating effect on Israel's wildlife, yet the NRA has done little to bolster opposition efforts. It is hard to imagine Avraham Yoffe sitting on the sidelines in a matter of such importance. In the Jordan River case, the NRA's conciliatory position undermined the efforts of the SPNI and other environmentalists who fought for preservation. A gap between the field staff, for whom it is relatively easy to hold uncompromising opinions, and central management in Jerusalem, who face political constraints, is unavoidable. Yet recently it seems to have grown.

Part of the problem with the Authority is the dilemma created by its quasi-independent status. The Ministry of Finance has come to expect the NRA to generate much of its own budget, even though it is a government entity. This creates an excruciating dilemma for management, which desperately needs revenues to continue the Authority's many activities but which must often pay a price. Yet, in 2001, when a new Minister of the Environment, Tzachi ha-Negbi, attempted to funnel the Authority's budget through his Ministry, nature advocates, including Uzi Paz, called foul and attacked the Minister as undermining the agency's traditional independence.[195]

Ultimately the ministerial home of the Authority is less important than the quality of its leadership. After a lackluster performance by Shaika Erez, a former general, Aaron Vardi was appointed as the first Director of the combined Nature and National Parks Authority. He too was a senior military man. A personable kibbutznik, Vardi had been an effective Director General at the Ministry of the Environment. After the 1996 elections, Rafael Eitan, the new Minister, immediately replaced him, only to remember that he actually knew and liked Vardi from the army. So he gave him the tedious, eight-month-long task of merging two beauracracies that had competed for decades.

To be sure, by the 1990s, the institutional divisions between the two agencies seemed wasteful and anachronistic. As more nature reserves charged entrance fees and national parks came to hold increasingly large tracts of unmanaged lands around their historical attractions,[196] it was often unclear whether there were any salient differences. Creating a shared identity and common sense of purpose, however, was hardly an overnight phenomenon. The combined forces of six hundred workers (350 and 250 respectively) offered a formidable team to preserve the country's natural and historic heritage.[197] The magnitude of the challenges probably requires a much larger army.

A BASIS FOR HOPE

Nature preservation in Israel is a success story. Bushes and indigenous trees have begun to flourish, relieved of the pressures of overgrazing by goats. The wildflowers are back as never before. The gazelle count now reaches into the thousands.[198] There are even increases in numbers among Israel's thirty-four bat species after so many years of their being hunted down and gassed in their caves, unjustifiably maligned as enemies of agriculture.[199]

One could also paint a completely different picture. Only a handful of Arava gazelles survive south of Yotvata, and their numbers are dwindling.[200] Birds of prey, for example the griffon vulture, rarely nest in Israel. Sand foxes have almost disappeared outside of the Mitzpeh Ramon Crater.[201] Professors Yoram Yom-Tov and Heinrich Mendelssohn of Tel Aviv University, probably the country's most eminent zoologists, estimate that even after three decades of NRA's efforts, 40 percent of vertebrates have either suffered extinction or a substantial drop in numbers during the twentieth century.[202]

Nature preservation also requires constant vigilance: For instance, in 1993, reports surfaced of massive wildflower picking during the springtime by new Russian immigrants. The Ministry of Absorption hastily issued updated versions of the standard wildflower message, in foreign languages.[203] It is the new threats, coming from unanticipated directions, however, that are the scariest.

Motorization is one problem that does not receive sufficient attention.[204] Habitats are crisscrossed by divided highways, which lock animals into small "fragmented" areas. Alon Galili has been raising this issue for almost twenty years: "Reptiles can't get across the roads. Rodents can't get over. Maybe a cat can under some circumstances. But on the whole, highways close animals in genetically."[205] Presumably, one solution is to build a system of tunnels under the roadway. Except for gazelles, animals generally like tunnels. They could be coaxed through, tempting them with water and food at the ends. It could make an exciting educational program. But such an initiative requires a commitment.[206]

Even more pernicious may be the vehicles that invade the reserves. All-terrain vehicles, jeeps, and motorbikes bring noise and leave unsightly tire marks everywhere. In the sensitive desert terrain, the signs can last for eons. It is not just a question of aesthetics. The ten thousand 4 × 4 vehicles in use today increasingly ravage the sensitive biota in the desert wadis that are the food source for many creatures.[207]

Another new scourge for wildlife has emerged in the form of Thai farmworkers, who have a penchant for trapping animals of all types to supplement their meals. What began as a joke about the Thai workers' willingness to eat anything that moves is suddenly not funny at all. Professor Yoram Yom-Tov reported to a Knesset committee that Thai traps had killed 90 percent of the Golan Heights' gazelle population, which is now on the verge of disappearing.[208]

In 1992 the NRA commissioned an independent review of the condition of the southern reserves.[209] The picture painted by Menahem Abadi was pathetic. Abadi described personnel constraints that made attempts to police the illegal vehicles laughable. Pirate contractors pilfered rocks, sand, and trees. Concomitantly, the pressure from military maneuvers grew more acute as West Bank and Golan firing ranges moved south in anticipation of geopolitical contraction.[210] Then there were the problems of water and electricity lines, mining concessions, and seismic tests, all of which required roads and infrastructure. And human settlements continued to expand.

Abadi's report saw no alternative for the NRA but to abandon 60 percent of the Negev reserve areas and regroup. Feedback by area rangers suggests that the report was both too harsh in its generalizations and too defeatist in its recommendations.[211] Yet the problems raised are very real and—without major adjustments—will only get worse. Personnel justifiably gripe that the direct investment in nature reserves in countries such as England far exceeds that in Israel, even though the other countries have less to preserve.[212] Many of the northern reserves, such as the lovely Habonim Coastal Park, are too small to be able to hire a resident ranger. These small gems of nature can rapidly turn into garbage dumps, owing to seemingly benign neglect: The NRA continues to plod ahead, preparing the groundwork for the declaration of an additional hundred reserves. However, if the Authority is unable to manage its existing lands successfully, it seems a futile exercise.

International coordination may offer the greatest opportunity for expanding wildlife habitat. Moreover, preservation itself can facilitate international cooperation. Indeed, nature helped resolve a deadlock that threatened to destroy the peace negotiations between Israel and the Palestinians. The bone of contention was whether an Israeli interim withdrawal from the West Bank should be 10 or 13 percent.[213] The compromise position set forward in the Wye River Memorandum in November 1998 held that a disputed 3 percent of West Bank territories be transferred to the Palestinians but remain as a nature reserve.

Wildlife management will never really succeed without a regional strategy. For years Amotz Zahavi has spoken of a binational biosphere park connecting Jordan's Dana and Israel's Shizaf reserves. The Sinai remains largely uninhabited, and leopards may thrive there. Peace negotiations with Jordan have included talk of peace parks, especially what would be billed as "the lowest park on earth," in the Dead Sea region. There are also partners with whom to work.[214] On the Jordanian side, the Royal Society for the Conservation of Nature is particularly competent. As a membership organization and manager of Jordan's nature reserves system, it is a combination of SPNI and NRA—but smaller than either.[215] Their educational efforts are both diligent and innovative, but they too have a huge task before them. The gaps between the countries are more than political.

Reuven Hefner is the NRA expert on wolves today, much as Giora Ilani used to be for leopards. By collaring them with transmitters he has discovered that Israel's desert wolves are nomads and can move as much as forty kilometers in one night in search of food. Many of his wolves cross the border into Jordan and can be followed only by satellite tracking. (As almost all the Arava sand dunes in Israel have been mined for construction, Jordan will continue to have more to offer them than Israel does.) For the wolves, however, it is often a one-way trip. Jordanian colleagues returned three transmitter collars that were taken off wolves shot after crossing the border.[216] Hefner reports that although predators avoid humans in Jordan, hyenas, once in Israel, are not at all afraid of vehicles, and wolves somehow have also learned that they are safe within the Green Line.

If the wolves and the hyenas can sense a difference, then Israel must be doing something right. One feels it amidst the quiet of a reserve when a particularly beautiful flower catches the eye or a herd of ibex scrambles over the rocks. Before long twenty million people will be squeezed in between the Jordan River and the Mediterranean Sea. Whether this insatiable species of ours will find a way to accommodate the other 2600 types of plants and 500 types of animals that also call Israel home remains an open question. If there is any basis for hoping that the answer will be in the affirmative, Israel's Nature and National Parks Authority deserves the credit. It got the country off to a very good start.

7 The Quantity and Quality of Israel's Water Resources

No natural resource was as important to Zionism as water. In that sense the new Jewish State almost instinctively adopted a traditional Jewish inclination. Like the Eskimos' reputedly rich vocabulary for snow, the Hebrew language has separate words for the first and last rainfall, dew, different levels of floods, and half a dozen types of drought. The word "water" itself appears 580 times in the Old Testament.[1] The Hebrew patriarchs concerned themselves with digging and protecting wells. Water is a prerequisite for a variety of ritual purifications. There is no more common metaphor in the religious liturgy.

History holds particular importance for understanding the present condition of this environmental medium. Israel's Coastal Aquifer, the country's largest single source of fresh water, lies roughly thirty meters below an unsaturated zone of sandy soils. In many areas it takes roughly a year for pollutants to seep down a distance of one meter towards the underground reservoir. Today's contamination can quite literally be traced to activities that took place during the 1960s.

Within the twentieth-century context of Jewish nationalism, water was the key to creating a vibrant agrarian economy and a fulfilled rural citizenry. It held the power to translate the pioneers' lush European aesthetic into a greener Middle Eastern landscape. Water resource development was both a symbol of technology's unlimited potential and the prosperity that the Jewish revival could bring the land. The Jewish farmer was an ideal, and the water that irrigated his land became an integral part of the national identity. Not simply a commodity, it belonged to the realm of ideology.[2] Hydrological considerations influenced foreign and defense policies.

Once Israel was free of the confining restraints of British rule, water became the new frontier. And there was much to be done.

David Ben-Gurion himself set the tone when soon after the State's establishment he bemoaned:

> Water and power, these are the two main things lacking in our country. . . . The groundwaters, springs, rivers, and brooks of our country are limited and scanty. Even these have not been fully exploited; the water of the Jordan flows down to the Dead Sea, and the Yarkon water falls into the Mediterranean: a considerable proportion of the water of Lake Kinneret evaporates and even the rains, plentiful in the north and minimal in the south, flow wasted, in large measures to the Mediterranean or the Dead Sea, without fully benefiting the thirsty soil.[3]

Water development had political backing at the highest levels. During Israel's first twenty years, Pinhas Sapir (the seemingly omnipotent Minister of Finance) and Levi Eshkol (the third Prime Minister) were undoubtedly among the country's five most influential politicians. A high point of their pre-State résumé was establishing and managing Mekorot, the Yishuv's water cooperative.[4] In a rare act of cooperation, in 1937 the major development agencies in the Yishuv founded a single company to establish, operate, and administer hydrological projects for irrigation and household needs. Under the able management of Levi Skolnick (later Eshkol), the company cornered the market on water supply and development. Mekorot also assumed ongoing operational responsibility for many of the wells and pipes that delivered water to the Yishuv's cities and farms. When Eshkol went on to head the Jewish Agency's settlement department, he passed on the Mekorot portfolio to Sapir.

Neither ever lost his enthusiasm for the subject of water, and both made it a priority budget item when they became leaders in government. This passion can be seen as almost a Shakespearean tragic flaw. On the one hand, it fueled innovative water development projects on a scale hitherto unknown in the Middle East. On the other hand, it created an unrealistic appetite and blinded decision makers to the long-term effect of stress on a fragile resource. The argument could be made that they almost loved Israel's water resources to death.

Like most frontier histories, Israel's water experience should have passed through two stages.[5] During the first phase water was available, but not at the desired places or times or in the required quantities. During this developmental stage, engineering obstacles constituted the primary policy challenges. A second stage, characterized by resource constraints, might have followed immediately thereafter. With a growing population and no readily available supplementary sources, the time had come to prioritize consumption needs, enhance conservation, improve efficiency of delivery, and focus on preserving water quality. The fundamentally ideological approach to water, however, prevented a successful transition to the more mature, sustainable stage. The unrealistic optimism and myopia of Israel's political leaders until the present was to some extent a function of the enormous success of the access stage. No one was more responsible for these grand engineering achievements than Simcha Blass.

A RELUCTANT WATER CHIEF

After a few years of bumming around kibbutzim during the early 1930s, Simcha Blass opened a Tel Aviv office as an engineering consultant. The private sector suited his fast pace, disdain for mediocrity, and somewhat irascible disposition. There he planned and supervised implementation of water development projects for the settlements of the Yishuv. From its inception in 1937, Mekorot hired him on a retainer to serve as its Chief Engineer. Blass designed most of its major initiatives, beginning with the Kishon River irrigation project.[6]

As soon as the war was over, Blass was ready to pick up where he had left off as a consultant. Haim Halpren, the Director of the nascent Ministry of Agriculture, had other ideas. He offered Blass a job running the Water Department. As the Mandate's Water Directorate had been based in the agriculture department, functionally it meant putting Blass in charge of the country's water resources. Blass was friendly with many of the ministers in the new cabinet, who well knew his outspoken and impatient demeanor. They told him that he might last a few weeks as a civil servant. Blass was inclined to agree but had heard a rumor of who the alternative candidate for the position was. Blass felt him to be so incompetent he accepted the offer and stayed at the job for four years.[7]

The country was short on everything. The first thing Blass did was to order pipes, which were unavailable locally. The Ministry of Finance and its foreign-currency-conscious clerks felt that any more than thirty thousand tons of pipe a year would be wasted. Blass ordered ninety thousand, but these soon ran out.[8] In 1949 Israeli soldiers stormed southward, reached the Red Sea, and conquered Eilat without a shot. Then they discovered that there was nothing there for them to drink. The army began to desalinate seawater, using a crude, high-energy process that cost six cents a cup, or 240 dollars a cubic meter—more than a thousand times the cost of well water.

Blass felt certain that water could be found in large quantities below the desert floor. He sent four drilling teams down to the surrounding Arava plains to find it. By the time they finished, he was 400 percent over budget, and the drillers had gone unpaid for several weeks. Yet in the fall of 1949 they struck water at Beer Ora, fifteen kilometers north of Eilat; a brackish but potable water source was tapped.[9] And so it went. With much guesswork and improvisation, a water system was created.

Despite the hassles, these were exciting times for water engineers. After being limited to a small percentage of Palestine's lands, they suddenly had the water resources of an entire country at their disposal. More than 75 percent of the freshwater supply came from three sources: Lake Kinneret, which receives the waters of the Jordan River watershed; the Coastal Aquifer, stretching down the Mediterranean coast from Haifa through the Gaza Strip; and the Yarkon-Taninim, or Mountain Aquifer, which runs parallel to the Coastal Aquifer to the east.[10] This much was clear in 1948. What Israel's first water managers did not know was how much water these three reservoirs contained.

The few experts with hydrological expertise sensed that the 248 million cubic meters of water utilized was only a fraction of the recharge potential available, although they had no idea of what that fraction was. Characteristically they were overly optimistic and guessed that with full development, supply might one day reach 3.5 billion cubic meters of water a year.[11] Israel's replenishable volume is now thought to be roughly half that amount.[12] Hazarding long-term estimates was something of an indulgence. Within eighteen months of independence the country's population had grown by 50 percent.[13] The new nation needed more water, and drinking water was the least of it: The Israeli farmer was impatient to stake his claim.

During the 1950s Israel's agricultural development was astonishing, showing over 500 percent growth in yields. Such growth was based solely on groundwater.[14] With its high water table, the Coastal Aquifer was the easiest resource to tap. Dozens of wells were drilled.[15] The environmental problems caused by this burst of productivity were already clear, but as Blass would later write, the exigencies of the period overrode even prudent professionals' sense of caution:

> Lack of food during the first years of the state caused speedy development of water sources that could be achieved through shallow wells. In this frantic effort, exploitation of the Western part of the coast was particularly prominent. It was the necessity of the time,

> though we knew that, in the near future, overpumping was liable to draw sea water into the fresh water surface and cause salination.[16]

Overpumping leads to pollution, not only because of the vacuum (slowly filled by seawater) created when the water table drops, but also because the natural flow of water to the sea, flushing salts and minerals out of the aquifer, is interrupted. When Tel Aviv's wells became too salty for drinking in the mid-1950s, it was a harbinger of things to come.[17]

WATER PLANNING: WHO'S IN CHARGE?

There was a need for a coordinated national plan for water resource development, supported by a legislative framework that would enable the professional water managers to work effectively. Anarchy and duplication of efforts were undermining the single most critical resource of the new nation. But who would plan it? Bureaucratically, the cast of vested interests guaranteed a turf war.

For instance, the British Mandate had operated a hydrological service, whose job it was to measure and estimate water resources. It was run by a hydrogeologist, Dr. Martin Goldsmith, a British Jew. Nevertheless, the Yishuv's Jewish Agency did not trust him, so they set up their own hydrology department. After the war, the Jewish Agency staffers had no intention of closing up shop. Blass had a grudging respect for Goldsmith's professional skills and kept him on as head of the Hydrological Service, which was to serve Blass's department at the Ministry of Agriculture. To this day, Israel's Hydrological Service generates most of the data and technical information (qualitative and quantitative) about Israel's ground- and surfacewater resources.

Then there was Mekorot. When the State was created, controlling shares of the water utility were passed on to it and the Jewish Agency.[18] Naturally, Mekorot wished to continue in this capacity nationally and to expand its control. In 1959 the passage of Israel's Water Law catapulted Mekorot to official status as the national water utility, by appointment of the Minster of Agriculture (with approval by the Knesset and the Government).[19]

Yet another player in the water business was the Ministry of Health. In 1949 the Israeli Hebrew Language Academy created a new word for sanitation: *tavruah* (until then, the Yishuv had just used a Hebraized form—*hegeniah*—of the English term "hygiene"[20]). Aaron Amrami was one of the few Israelis who had formally studied sanitary engineering at the graduate level; his mentor was Professor Walter Strauss, a German-trained hygienist. The Minister of Health was happy to put Amrami to work. He took the other Israeli with professional training in the field, Hillel Shuval, as his deputy. Shuval (see Figure 17) had just immigrated from the United States, arriving in July 1948 at the height of the War of Independence. He had served in an American engineering unit during World War II and received a sanitary engineering degree under the G.I. Bill. The nascent army asked Shuval what he could do for Israel's war effort, and he was put in charge of chlorinating drinking water in the new Engineering Corps.[21]

For the first few years, Amrami and Shuval were literally the Ministry of Health's Sanitation Department. They faced a daunting laundry list of responsibilities: sewage and water supply planning; oversight of drinking water and wastewater quality; monitoring of industrial effluent discharges; supervision of municipal solid-waste disposal; city cleanliness services; monitoring and grading of food production and services; assessment of milk quality; insect, rodent, and pest control; regulation of swimming pools and beaches; sanitary oversight of schools and summer camps; sanitation in ports; quarantine services for imports; health conditions in immigrant camps and villages; and air pollution control. With no one to take on the regulation of radiation, it was soon added to the list as well.[22] During the 1950s the Sanitation Department was the only Israeli institution seriously concerned with water quality. Although it consistently sent its staff for advanced environmental training in the United States, the seven permanent sanitary engineers and technicians were simply spread too thin to stem the growing tide of contamination.[23]

In addition to the planning departments of the Health Ministry and Mekorot, the Jewish Agency had its own planning department. And of course, as Director of the Ministry of Agriculture's Water Department, Blass saw himself primarily as a planner. It was a mess. At the same time, Blass began to find the constraints imposed on Israeli civil servants unbearable. More than his Spartan public-sector salary (which forced him to sell his ten-room house and move into a four-room flat[24]), he resented the never-ending haggling for reimbursements and the awkward personnel procedures that were impediments to attracting qualified professionals.[25] This frustration with the government bureaucracy, rather than any environmental logic per se, led in 1952 to the creation of Tahal, an acronym that stood for "Water Planning for Israel."

Rather than just empowering one of the existing entities to centralize planning functions, this new corporation was to advise the government about water planning.[26] Blass happily took over as its director. As part of the restructuring, Blass's Water Department at the Ministry of Agriculture became the Water Administration, run by development guru Pinhas Sapir (also the chairman of Mekorot and Tahal).[27] Later this Administration became Israel's Water Commission.

Levi Eshkol, who held about four jobs at the time, including Minister of Agriculture and Director of the Jewish Agency's Settlement Department, did not like the arrangement. He saw the new agency as a Mekorot competitor. But after four years of suffocation, Blass refused to work out of a large bureaucracy. So Eshkol agreed to the creation of the new nongovernmental institution and then appointed himself to be both the governmental and the Jewish Agency representative on Tahal's Board of Directors.[28] The JNF board representative did not bother to show up for the meetings, so, with Eshkol covering for him, Blass had all the latitude he needed. He inherited the staff members of the competing departments[29] and rolled up his sleeves.

A DESERT FANTASY

Within four years Blass engineered the two mammoth public-works projects that have defined Israel's water supply strategy until today: the Yarkon-Negev pipeline and, a decade later, the National Water Carrier. Israel's natural water distribution is not suited to its geographic and economic circumstances. Although 78 percent of the rain falls north of Tel Aviv, most of the lands that need the water are in the south.[30] Moreover, fields, groves, and lawns are thirstiest in the summertime, precisely when the rain does not fall. These dynamics drove the form and content of the two infrastructure projects. In the hydrological equivalent of the prevailing socialist ethic, Israel's water system took from the rich and gave to the poor. Blass had been waiting thirteen years for the opportunity.

In 1939 British land restrictions had essentially shut down expansion of Jewish settlement in the north of Israel. Although land transfers and agricultural settlement were permissible in the south, British "flexibility" appeared disingenuous, because there was no water there. That spring Blass, then a private consultant, bumped into Dr. Arthur Rupin, the economist who chaired the Jewish Agency. Rupin said to him, "Mr. Blass, maybe you can propose to me a fantasy that would irrigate the Negev." Blass jumped at the business opportunity; he collected maps, chartered a rickety two-seater for aerial observations (losing his lunch during the bumpy flight), and then hunkered down to work. A few months later he submitted a three-stage program: Pump water from the wells closest to the Negev; take water from the Yarkon headwaters at Rosh ha-Ayin; and carry water from the Jordan in the north. To get the fantasy moving, Blass proposed starting the first stage immediately. For the bargain price of seven hundred thousand pounds sterling, he could provide half a million cubic meters of water a year to three settlements.

The project was tabled until after World War II. Its resurrection is part of Zionist legend. During the summer of 1946 the British imposed an eighty-hour curfew on Tel Aviv and arrested scores of Zionist activists, an action which became known in the Yishuv as the "Black Sabbath." Levi Skolnick (Eshkol), who by then was head of the Jewish Agency Settlement Department, sought a creative form of revenge. The night after Yom Kippur, he staged a lightning campaign, creating eleven new Negev settlements on JNF lands. Blass was drafted to design the water delivery system. All the Mekorot team had to work with were tiny, recycled six-inch pipes that had been used in London during World War II to help firefighters counter the bombing attacks during the blitz. Now the pipes were to wind past Beer Sheva and irrigate the desert. Pinhas Sapir himself came down to command the operation, barking out orders in his imperious fashion. Measurements were taken a few steps ahead of the diggers, and safety procedures were completely ignored.[31]

The project was completed before either the Mandatory government or the area's Bedouin could interfere. Only weeks before the outbreak of the War of Independence, a pair of pipelines connected the northern Negev desert to the center of the country. Together the pipes could carry only a million cubic meters of water a year, but stage one of Arthur Rupin's desert fantasy was complete (after the war the pipes were replaced by a system whose capacity was fifteen times greater).[32]

Once at Tahal, Blass and his new staff focused on stage two: the Yarkon-Negev pipeline, the first major water development project in the new State. The very speed with which the project was carried out ensured snags and, of course, improvisation. Menahem Kantor, who later became Israel's Water Commissioner, was on the Tahal team. He recalls the paucity of hydrological data but also a willingness to modify the project when it became evident that mistakes would be made.[33]

The Yarkon-Negev project was completed in July 1955. It delivered one hundred million cubic meters of water to Negev settlements.[34] The line was based on a sixty-six-inch diameter pipe, the largest of its kind in the world (see Figure 18). The pumping stations were all built underground for security reasons. Blass insisted that they be run by diesel engines, because he was convinced that in the next war the Arabs would succeed in bombing Israel's electric power stations.[35]

When it was completed in July 1955, Israel's politicians were happy to trumpet the Yarkon-Negev project as the pinnacle of Zionist ingenuity—opening up "barren" regions for settlement and making the desert bloom. But it also epitomized the environmental price tag of Zionist achievement—transforming the mighty Yarkon River into a smelly trickle of sewage and setting the stage for subsequent environmental disasters. The river's demise did not go unnoticed but was considered a necessary sacrifice for progress. That was before the days of environmental impact statements. Tahal did not think about the mess it left behind, and government leaders were not inclined to worry about it too much either.

The Ministry of Health's Sanitation Department, however, was concerned. Its field staff had checked oxygen levels across the Yarkon River and predicted an anaerobic stench in the heart of greater Tel Aviv.[36] The Reading power station lay at the outflow of the river to the sea, so the Ministry lobbied for the Israel Electric Company to release the station's cooling waters a few kilometers up the river. The Tel Aviv section of the Yarkon would enjoy a cleaner flow, even if it was seawater. There was also concern about the potential transfer of bilharzia (as had occurred in African water projects) as the water was moved from the Yarkon to the Negev; the Sanitation Department imposed snail control measures in response.[37]

The water pipeline produced at least one important institutional benefit: The Yarkon River problems forced Tel Aviv and its suburbs (Petah Tikva, Ramat Gan, Givataim, and B'nei Brak) to think about a regional solution to their waste discharges. The outcome was the establishment of a Dan "Union of Cities."[38] This consortium of city governments teamed up to build a temporary sewage outflow, eight hundred meters into the sea (until then, the effluents were released at the edge of the Tel Aviv coastline, with predictably fetid results). By 1962 the Union of Cities completed a pipeline to a regional sewage aeration facility on the sand dunes south of the city, at a cost of ten million dollars.[39] The level of treatment was not high, and the odors from the settling ponds were sufficiently foul to provoke legal action.[40] Moreover, the overflow system was a major source of marine pollution. Yet it was a vast improvement over dumping raw sewage onto Tel Aviv beaches. More important, in retrospect, it was the first stage of what would eventually become the Dan Sewage Project, the largest advanced treatment facility in the Middle East.[41]

TVA ON THE JORDAN

The final and most ambitious stage was the construction of the National Water Carrier. The list of people claiming patent rights to the project is long.[42] In his 1902 novel, *Altneuland,* Theodor Herzl envisioned the Jordan River tributaries providing water for the drier half of the country. (He also envisioned water from the Mediterranean Sea driving electrical turbines as it dropped down through a canal system to the Dead Sea.) In his 1944 book, *Palestine, Land of Promise,* Walter Clay Lowdermilk, the American soil scientist, devoted an entire chapter to a proposal he called a "Jordan Valley Authority."[43] Lowdermilk based his model on the Tennessee Valley Authority (TVA), an American public-works agency that had successfully tapped rivers for irrigation and hydroelectricity.

Emanuel Neuman, an American Zionist leader, was inspired by Lowdermilk's idea and wanted to convince British authorities that it was not an unrealistic dream. With Chaim Weizmann's blessing, Neuman approached James Benjamin Hayes, an internationally renowned water engineer from the TVA, and commissioned a more detailed blueprint for a Jordan River system.[44] Hayes's 1949 report, *TVA on the Jordan,* projected a total water supply of 2.5 billion cubic meters by diverting Lebanon's Litani River, along with the Jordan tributaries and available groundwater sources.[45] This would be sufficient to produce electricity, allow for the irrigation of 2.8 million dunams, and provide for the domestic needs of close to four million people.[46]

It was Simcha Blass and the Tahal engineering staff, however, who would turn the vision into reality. In 1950, while still at the Ministry of Agriculture, Blass established a committee to oversee the planning of the National Water System. Composed of eighteen experts, this committee provided for scientific peer review, a rare and refreshing act of prudence for a young country in a hurry, with little time for formalities. Over six years the committee would hold seventy-five meetings, and its opinion was elicited at every important juncture.[47]

"We asked a very simple question: 'What information do you have on the quality of the water in the Jordan River and Lake Kinneret?'" recalls Hillel Shuval, the young sanitary engineer at the Ministry of Health who attended these meetings with his boss, Aaron Amrami. "The answer was that the only tests that had been done were for salinity, to see what crops could be irrigated. But it was clear that the National Carrier would provide drinking water as well, and they admitted that they had never conducted bacteriological or chemical testing."[48]

Blass was happy to leave water quality problems in the hands of the Ministry of Health. He had more pressing engineering problems of his own to contend with. For example, the Hayes plan would require a system of twenty-nine reservoirs to hold water until the summer months as well as to provide storage for drought years. It was not clear whether the porous local soils were sufficiently impermeable to adopt this American approach. Between 1953 and 1955, nine pilot reservoirs were built, at a variety of locations. The results were conclusive: A solution other than artificial reservoirs for storage would have to be found.[49]

Then there were financial problems. Blass demanded fifteen million dollars a year for eight years for the project—an astronomical sum for the young State, considering that its total foreign currency earnings for 1950 were eighteen million dollars.[50] (The final price tag when the National Carrier was completed in 1964 reached 420 million lirot in 1964, or roughly 175 million dollars.[51]) Yet Ben-Gurion's vision of a green Negev would not let him rest, and money was found.

The real challenges were technical in nature. Israel had never attempted an engineering job on this scale. Conceptually the plan was relatively simple. The collective flow of the three main tributaries of the Jordan River was 520 million cubic meters of water a year. These tributaries met on the hills and plains of the Galilee, north of the Huleh swamp, before the river made its three-hundred-meter dive down to Lake Kinneret. The original plan took advantage of this height differential and diverted the waters for the National Carrier north of the lake, letting gravity lead them downhill to the south. On the way, the Netufah Valley would serve as a billion-cubic-meter reservoir, about a quarter the size of the Kinneret. Depending on the specific topographic conditions, the National Carrier was to utilize three different mechanisms to deliver the water across the country: canals, tunnels, and pipelines. Assuming that tunnels would constitute the bottleneck in the program, in 1952 Blass began excavations of Israel's first, the Ibon. The plan for a National Water Carrier itself would not be approved until four years later.

WATER WARS: ROUND 1

The project's most conspicuous obstacle was geopolitical. Of the three main tributaries of the Jordan, only the Dan lay squarely inside Israeli territory. The Hatzbani came from Lebanon, and the Banias was still in Syrian hands. More important, diverting or even tapping the Jordan River would change the flow and water quality of the river. Along with the Yarmuk, the Jordan River was Jordan's primary source of surface water. The Hashemite Kingdom's consternation should not have surprised anyone.

In 1953 Blass pushed to begin the work to divert the Jordan River anyway. The area was technically under Israeli jurisdiction but was part of a demilitarized zone between Israel and Syria. Beyond concern about international pressure, there was fear for the safety of the construction crews. At the time, the general in charge of Israel's northern command was Moshe Dayan. He decided to gamble that Syrian ruler Adib al-Shishakli's own domestic problems and unrest would serve to neutralize him. Dayan ventured that not a single bullet would be shot in response. He was right. On September 2, 1953, Israel sent two bulldozers to begin the diversion at the Jordan River's B'not Yaakov bridge, ten kilometers north of the Kinneret. While the Syrians quickly massed troops across the river from the construction site, they held their fire. Israel expanded its workforce.[52]

Syria had other means of voicing its displeasure. It complained to the commander of the United Nations Peace-Keeping Force, Danish general Wagen Benika. Benika demanded an explanation from the Israelis. The Syrians told him that the water was to be channeled to the Kinneret for hydroelectric power (which was part of the Hayes plan). Syria then lodged a formal complaint with the Security Council. And so it was that in October 1953, the United Nations got its first lesson in the politics of Middle Eastern water resources.

It was Lebanon's and Pakistan's turn to join the Security Council rotation in addition to the five permanent members. Although they did not have veto power, their hostile presence weakened Israel's position. Under pressure, the peerless Anglo-Israeli diplomat Abba Eban agreed to halt work along the river until the hearings commenced. During the interim two weeks, Israel did its homework. Blass flew across the United States collecting the equivalent of affidavits from three of America's top water management experts, who confirmed that not only did Israeli utilization of the Jordan not contradict a regional development scenario—it actually expedited it.[53]

At the United Nations session itself, Syria's representative, Dr. Zein al-Din, made a strong two-hour speech in front of a packed auditorium, accusing Israel of violating the cease-fire, stealing his country's water, and eventually causing the starvation of fifteen thousand citizens. Abba Eban responded with one of his typically eloquent presentations. Eban countered that the demilitarized zone was under Israel's sovereignty and that, according to the preexisting Sykes-Picot agreement between France and England, the British Mandate received all rights to the Jordan River. Israel, he argued, inherited these rights. Eban charmed the diplomats, but he could not stop the Soviet representative from vetoing an American resolution that would have allowed Israel to continue its diversion work.[54]

To try to defuse the volatile situation, American President Eisenhower launched an independent diplomatic initiative. Eric Johnston arrived in the Middle East in 1953 as the president's personal emissary and "roving ambassador" to the Middle East to work out a compromise. Johnston impressed the Israelis. A tall, trim, youthful sixty-year-old businessman, his experience in mergers and acquisitions in the corporate world had honed his negotiating skills. Most important, he was not part of the State Department's foreign service establishment, whose sympathies were felt to lie with Arab interests.[55]

During five trips from 1953 to 1955 he hammered out the "Johnston Plan." Lake Kinneret was to be the main reservoir for waters of the Yarmuk and Jordan Rivers. A dam would be constructed on the Yarmuk, and Israel would transfer water to Jordan via a canal. Under the plan, Israel would receive 40 percent of the available water allocation. (Israel's opening position demanded 60 percent, and the Arabs agreed to 20 percent.) Jordan and Syria were allocated 45 percent and 15 percent respectively. More important, Johnston's package explicitly recognized the legitimacy of Israeli transfer of water to its southern regions.[56] The United States sweetened the deal by agreeing to bankroll many of the spin-off development projects, which it assumed would provide employment for Palestinian refugees.

Just as the agreement was about to be finalized, the Arab League got cold feet. Rather than embarrass President Eisenhower openly, their October 1955 debate on the proposal was "postponed." Then they directed their energies towards blaming Israel for collapse of the negotiations.[57] While Johnston did not achieve a de jure arrangement, his plan became the de facto baseline for water allocation in the region. Its ground rules also gave Israel the green light it needed to begin the National Water Carrier.

STAGE THREE: THE NATIONAL WATER CARRIER

Given the new political constraints, Israel decided to shift from the Jordan approach and instead base its National Water Carrier around Lake Kinneret, which would function as the central reservoir for the system. (The Netufah Valley site had turned out to have too high a percolation rate to serve as the central reservoir.[58]) Ecologically, this was fortuitous. The freshwater in the Jordan River naturally has only 5 percent of the chloride concentrations found in the Kinneret. Tapping this primary source of dilution before it reached the Kinneret would have doomed the aquatic ecosystems in Israel's only freshwater lake.[59]

Yet from an engineering perspective, abandoning the B'not Yaakov bridge created a serious problem. Lake Kinneret lies more than two hundred meters below sea level.[60] Water would have to be pumped to a height of 151 meters above sea level before making its trip south, requiring copious amounts of energy. As a result, today the National Water Carrier consumes one hundred megawatts of electricity, or 2 percent of the electricity produced in Israel.[61]

From the moment that the government approved the final plans for the National Carrier in 1956, it took eight years for Mekorot to complete the project. For almost a decade it swallowed 80 percent of the Israeli investment in water infrastructure.[62] In June 1964, the gigantic, thundering pumps began to heave prodigious amounts of water up the Galilee hillside to begin a trip that remains unchanged to this day. Upon reaching its northern peak, the water flows into a scaled-down reservoir at Beit Netufah for purification treatment. After being properly sanitized, it makes an eighty-six-kilometer journey through enormous 108-inch, Ashkelon-made steel and concrete pipes. The route is interrupted by three additional tunnels, blasted through mountains and rocky terrain. The longest of these, the Menashe tunnel, stretches for 6.5 kilometers, beginning near the Megiddo-Yokneam road. At Rosh ha-Ayin, east of Tel Aviv, the water links up with the Yarkon-Negev system.[63]

Construction was not entirely free of international complications. As work on the National Carrier reached an advanced stage, Syria lodged another complaint with the Security Council. This time the United Nations chose not to intervene, and in December 1963 the Arab League decided to take matters into its own hands. It called for diverting the Hatzbani and Banias tributaries of the Jordan from their natural flow into Israel. (This would have left the National Carrier high and dry.) In February 1964, Syrian and Lebanese construction crews began to build canals to this end. Israel declared it a provocation and shelled the bulldozers. That summer the crews returned to the task, and this time Israel's Air Force took out the equipment. With only 1 percent of the work complete, Syria backed off.[64] The National Water Carrier was finally free to finish its "sacred" mission, which had started with Herzl's romantic vision.

It was a Herculean task. Many contemporary decision makers doubt whether the societal commitment exists today for such an enormous investment in national infrastructure and water resource management.[65] Ultimately the National Water Carrier irreversibly changed the terms of reference of Israel's water policy. Of the 850 million cubic meters of water that reach Lake Kinneret, 30 percent evaporates. That leaves roughly five hundred million cubic meters that can be tapped and delivered literally anywhere in the country. Typically 95 percent of this amount is pumped into the system.[66] When it was completed in 1964, the Carrier provided two-thirds of the country's water. Of this, 80 percent went for irrigation. With the change in Israel's demographics, however, today half goes to the drinking water supply, and by 2010 this may be as much as 80 percent.[67]

The National Water Carrier also solidified Mekorot's role as the only serious player in the field of water supply and development (by 1965, more than 60 percent of Israeli households were receiving their water directly from Mekorot[68]). It is no wonder that there was a major battle over who would get to build the enormous system. Blass, who had spent six years planning the project and who had already overseen the first tunnel's excavation, was convinced that Tahal was best suited for the task. Levi Eshkol, Minister of Finance at the time, was concerned about Blass's tendency to go over budget. Eshkol was also worried that building such an enormous project would shift Tahal's role from one of planning to implementation, undermining institutional distinctions that were already muddled. Pinhas Sapir, still chairman of Mekorot, insisted that it run the project.

As usual Sapir got his way. Blass quit in a furor and never again returned to government service.[69] It is ironic that, in looking back on the National Water Carrier, all the old-time water managers give Blass full credit for making the project happen. History, however, is written by politicians, not engineers. The National Carrier's pumping station and museum is called the Sapir Center. From there the water flows to the Eshkol Reservoir at the Netufah Valley.

THE NATIONAL CARRIER: THE ENVIRONMENTAL UPSHOT

There were serious water quality problems associated with the National Carrier, most of which were solved by Mekorot's engineers. By far the most significant was the salinity of the water, caused by the salty springs that drained into the Kinneret. In 1964 salinity levels in the lake were measured at 390 milligrams per liter of chlorine (chloride ions).[70] (Although Israel's drinking water standard allows chlorine to reach concentrations of 600 milligrams per liter, it recommends 250 milligrams per liter to reduce the risk of hypertension and improve the taste.) In response, Tahal planned a Saline Carrier that collected the flow of the surrounding saline springs and released them into the Jordan River, south of the lake. Chlorine levels in the Kinneret were immediately cut in half and now measure roughly 200 milligrams per liter, a far more acceptable level.[71]

The Saline Carrier was among the many areas of disagreement with Blass, who insisted that the saline streams' diversion precede all other work (he was overruled, and as a result from 1964 to 1967 Tel Aviv's drinking water was very salty[72]). Even after the diversion, the high chloride content of the Kinneret posed a problem for irrigation. Mekorot responded with a system that diluted the water in transit by injecting clean, low-chloride water into the flow from the lake.[73]

In addition, from its survey during the 1950s, the Ministry of Health was already aware of high levels of coliform bacteria in the Jordan. This was primarily attributed to the runoff from dairies and fishponds (manure is a relatively popular fish food). Chlorination is the standard treatment, but it poses other risks. Chloroform and other trihalomethanes are created in water when organic sediments mix with chlorine gas. These compounds are some of the unfortunate by-products of disinfection and are associated with cancer of the liver.[74] (Given the levels of suspended solids in the Kinneret, it was no surprise when trihalomethanes were discovered in Israeli drinking water.[75]) To make matters worse, the Kinneret has a relatively high bromide level (two parts per million), which, when chlorinated, oxidizes to become active bromine. The bromination of these organic materials may pose an even more significant carcinogenic risk.[76]

Even if the cancer risk levels are disputed, the sediments alone created a serious turbidity problem. During the winter months, the Jordan River can turn a dark cocoa-brown from runoff. In those days, the Ministry of Health measured turbidity according to Jackson turbidity units (JTUs), with the benchmark (as opposed to legal) standard set at 10 JTUs. Today, the allowable level has dropped 1000 percent, corresponding to 1 JTU. After rainstorms, however, frequently even the 10-JTU standard could not be met, and turbidity levels went into the thousands, leaving the water unfit for human consumption.[77] This produced the naturally corrosive conditions and murky waters found downstream in the Kinneret.[78]

These factors determined Israel's treatment strategy for the waters delivered to the south. The Ministry of Health ordered that the Carrier include treatment by coagulation and sedimentation along with chlorination to reduce the pollutant concentrations. No one had any experience in a project this size, so Ben Nessen, the Chief Water Quality Engineer from New York City, was brought over to Israel to help prepare the treatment plan.[79] Nessen's disinfection strategy has changed little in thirty years.[80]

Israel took a chemical rather than a biological approach to drinking-water treatment. For instance, Mekorot engineers for years were nervous about introducing fish into the National Carrier reservoirs as part of the water's biological treatment. They still had not gotten over the surprise of 1959, when consumers in the south found tiny fish flowing from their taps (the fish had swum out of a reservoir built as a section of the Yarkon-Negev line[81]). By the end of the 1970s, however, evidence of the beneficial properties of fish was compelling. Fish were introduced into the Eshkol Reservoir to eat algae, snails, and other small aquatic creatures, which they devour voraciously.[82]

In evaluating the impacts of Israel's national water system, there is ultimately no single environmental bottom line. It certainly hurt the Dead Sea. Without the Kinneret spillover, the saltwater lake quickly began to shrink, producing an unsightly naked shoreline and reducing the attractiveness of the recreational experience.[83] Lake Kinneret itself shows mixed trends. Although salinity dropped, other parameters, such as nutrients, rose. To ensure water quality, the carrier withdraws water at a relatively shallow intake that is low in nutrients. This increases the Kinneret's overall concentration of ammonia and sulfides and over the long term may exacerbate the eutrophication or aging processes.[84] Nitrogen, rather than phosphorus, was identified as the nutrient most linked to this phenomenon.[85]

The Carrier made the entire country dependent upon Lake Kinneret's water quality, elevating its preservation to a national priority. During the winter months, newspapers post the level of the lake in their front sections alongside weather forecasts, where it is followed with a devotion otherwise reserved for the standings of the National Soccer League. While far from being springwater, the quality of Lake Kinneret water is reasonable.[86]

The alternatives may have been worse. Blass's impatience was not based solely on the need to expand agriculture but also on the overpumping and rapid salination of the Coastal Aquifer. Groundwater is less subject to remedial measures than the Kinneret. "The powers-that-be thought that I was only trying to scare them," he later wrote. "The fact is that by the time the Carrier went on-line, there was already a deficit of 500 million cubic meters along the coastline."[87]

NEW WATER LAW, NEW WATER COMMISSIONER, OLD WATER POLICY

Once the National Carrier was underway, it was clear that Israel had already tapped the majority of its replenishable water. It was time for the country to institute a policy that focused on conservation and preservation rather than simply allocation and expansion. Such a strategy had to be anchored in law, with a strong arbiter empowered to sort out competing needs. As early as November 5, 1952, Blass presented thirteen principles for a Water Law to a committee of experts and government ministers. But he was more interested in developing water sources than in writing laws regulating their use.[88]

A subcommittee was appointed, chaired by Haim Halpren, who had moved from being Director of the Ministry to Director of the Agricultural Bank of Israel. At the committee's first meeting, one of its five members, Pinhas Sapir (who was both the Director of Water and the Director of the Ministry of Finance as well as the Chair of Mekorot and Tahal), announced that he was against a water law in principle. At the time Blass was busy planning the National Water Carrier, and it is little wonder that it took seven years for the committee to complete the task.

When it was finally enacted, the Water Law was hailed as an innovative and comprehensive statute.[89] Among its key provisions were the elimination of private water rights, along with a vague guarantee of the public's right to receive and use water as long as it did not deplete a source or cause salination. A Water Commissioner—appointed by the government, based on the Minister of Agriculture's recommendation—oversees the allocation of water. The law created a Water Council to advise the Minister of Agriculture, a Water Court to offer judicial review to the Commissioner's decisions, and a system for setting water prices. Tahal retained its planning function, and by government agreement, Mekorot remained the primary water utility.[90] Yet it fell short of what might be called an environmental statute. The word "pollution" did not appear in any of its 150 sections.[91]

It fell to Moshe Dayan, who then served as Minister of Agriculture, to appoint the first Water Commissioner. (The relationship between the Commissioner and the Minister of Agriculture can be compared to that between the Israeli Army's Chief of Staff and the Defense Minister.) Here again, Dayan made the right call, setting an important precedent by appointing a professional rather than a politician for the job. Even at age thirty-seven, Menahem Kantor was probably one of the most experienced water engineers in the country. Kantor moved to Israel in 1922, when he was one year old. Fourteen years later he joined the Haganah and soon thereafter found work at Mekorot as one of its first five Haifa-based employees. When Mekorot's planning department moved to Tahal, Kantor went to work under Blass as head of the Hydrology Department. Among his many tasks was assessing just how much water was available in the country.[92]

Retired today at Kibbutz Ma'agan Michael, Kantor describes the water bureaucracy during his tenure as Commissioner as one big happy family.

He has only good things to say of his compatriots, including his old boss, the controversial Simcha Blass. Kantor's quiet, authoritative demeanor was quite a contrast to Blass's emotional style, and he managed to quell some of the turbulence that his former boss enjoyed stirring up. As Water Commissioner, he instituted weekly meetings with the directors of Tahal and Mekorot. Even academia was co-opted. In those days the Technion had an even tighter monopoly on technical training in the field than it does today, and Kantor quickly acted to bring its faculty on board as consultants and thereby ensure a consensus.[93]

The record is not quite so cozy. For instance, in 1967 a committee was established to recommend ways to reduce friction among the different water institutions.[94] As might be expected, the tensions continue to this day.[95] Kantor was respected and consistently came out a survivor in the internecine quarrels. He remained at the job for twenty years, more than twice as long as any subsequent commissioner, and then he went on to run Tahal. His longevity was not hurt by a consistently paternalistic approach towards agricultural interests. The Minister of Agriculture remained the government reference for water until 1996 and was unquestionably the Commissioner's boss. The agricultural sector consistently received the vast majority of overall water allocations, enjoying a much cheaper price than industrial and domestic users, far below the actual costs of extraction. The record shows precipitous deterioration in water quality during Kantor's administration, but water supply ambitions still dominated Israel's hydrological agenda.

For a time, desalination was thought to be the key to a continued strategy of resource expansion. Although it typically connotes high technology, the process is hardly new. Aristotle wrote that saltwater became sweet when it turned into vapor and was condensed. During the siege of Alexandria, Julius Caesar used stills to desalinate seawater for his soldiers.[96] It was only natural, then, for Israel's history-conscious leaders to attempt to give this ancient idea a modern application.

If there was an environmental area in which Prime Minister Ben-Gurion displayed his visionary qualities, it was that of desalination:

> The purification of seawater by an inexpensive process is not only vital for Israel—it is a necessity for the world. Hundreds of millions of the inhabitants of the great continent in which we live suffer from lack of food, but as yet only a small part of the earth's surface is tilled. . . . If Israel succeeds in desalting the water of the sea, it will bring great benefits to the entire human race, and the task is not beyond the power of Israeli science. . . . The irrigation of the desert with purified seawater

> will appear a dream to many, but less than any other country should Israel be afraid of "dreams" which are capable of transforming the natural order by the power of vision, science, and pioneering capacity. All that has been accomplished in this country is the result of "dreams " that have come true by virtue of vision, science, and pioneering capacity.[97]

Spurred by this ardor at the highest level, in 1965 Tahal proposed a fifteen-year, one-hundred-million-dollar massive desalination venture. It was officially adopted by the government, but by then, Ben-Gurion was no longer running the show. This time the Cabinet was not inclined to sign a blank check. Israel tried to interest the U.S. government in joint funding for the project, but the Americans balked at the price, considering the project technologically premature and economically unfeasible.

At the same time, the agricultural lobby was decidedly unenthusiastic. Even as late as the 1990s, in the circular logic that ruled the agricultural lobby, the idea prevailed that once water was desalinated, farmers would be forced to pay the full treatment price.[98] And so the pursuit of desalination technologies was abandoned. Without an engineering fix, Tahal had no alternative water strategy. Instead Tahal lamented inadequate infrastructure and began to assume the prophetic role of "predicting water doomsday."[99]

THE MIXED BLESSINGS OF WASTEWATER REUSE

Once desalination was deemed prohibitively expensive, water policy makers sought other ways to squeeze more water out of existing resources. Cloud seeding, for example, is still practiced, but it produces only a modest increase in rainfall.[100] (The process uses silver iodide, which causes ice particles to form in the clouds. Mekorot estimates that these efforts lead to an additional eighty million cubic meters of rain each year within Israel.[101]) Sewage water was much more promising.

With only a small percentage of freshwater resources being utilized during the 1950s, it is not surprising that Blass took little interest in effluents as a source of irrigation while he was Director of Water. Yet, in contrast to desalination, irrigation with sewage effluents had two powerful proponents lobbying for it as early as the 1950s. One was Aaron Amrami at the Ministry of Health. His motivation was sanitation. Reuse offered an incentive for cities to collect sewage, treat it, and sell it to agriculture. The other interest group was the farmers themselves. On their own initiative, several kibbutzim established small-scale wastewater irrigation projects and solicited the treated wastes of nearby cities and towns.[102]

The quantities that required treatment were rising fast. During the Mandate, most of the Jewish population had access to running water and flush toilets. For disposal, however, homes relied primarily on septic tanks. Independence did little to change this. Of the thirty new towns that Israel built during its first twenty years of statehood, only one had a waste treatment facility.[103] Certain sectors, such as the kibbutzim and the Arab villages, had neither septic tanks nor central sewage. In the former case this was an ideological decision. Abraham Herzfield, perhaps the central figure in both the Jewish settlement movement and kibbutz budgeting during the formative years after Israeli independence, believed flush toilets to be a luxury and violently opposed them. A rumor made the rounds of the sanitation community that when the first kibbutz installed a proper toilet, he burst in with an axe and destroyed it.[104] Herzfield could not stop progress, though, and by the 1960s most kibbutzim permitted their members the frivolity of a flush toilet. Progress in the Arab sector was somewhat slower.

It was the cities, however, that held the key to wastewater reuse. Because of the crowding and heavy, impermeable soils in many neighborhoods, septic tanks and pits began to clog and spill over. The stench, health risk, and general nastiness became unacceptable. With public outrage mounting, the Ministry of Housing began to make central sewage systems a priority.[105] During the country's first ten years, Israeli cities and towns began to link homes and businesses systematically to central sewage systems. The trouble was that nothing was waiting for the wastes at the end of the pipe. Sewage was dumped, with little or no treatment, into the closest stream, wadi, or body of water.

Once concentrated, the untreated or partially treated wastes created an enormous nuisance. For instance, during the 1960s, 90 percent of mosquitoes were attributed to untreated sewage outflows, which provided an ideal breeding ground for pests.[106] Groundwater also suffered, although it would take some time to learn how badly. It was not just overpumping and salinity that forced Tel Aviv to close its wells during the early 1950s. The city's ubiquitous septic tanks leaked into the wellheads. Then, in 1956, hundreds of residents in Kiryat Bialik and Kiryat Motzkin, north of Haifa, came down with gastrointestinal diseases.

The initial explanation for the epidemic was food poisoning. But the Ministry of Health sanitary engineers quickly noticed that only specific communities were affected and that the breakout occurred simultaneously in several places. Drinking-water maps revealed that the affected individuals all drank water from a well in nearby Afek; it did not take long to find a sewage source one meter from the well.[107] Based on this experience Shuval, who took over as Chief Sanitary Engineer in 1958, began to press for preventative chlorination. This provided a reduction in short-term risks but addressed the symptom rather than the cause.

The most pervasive problem associated with the neglect of sewage treatment was the contamination of rivers (see Figure 19). By 1967 practically all of the streams south of the Galilee were utilized as sewage conduits.[108] A 1970 description of the central region's Alexander Stream is typical:

> The appearance of mass quantities of dead fish on the banks of the stream during the month of August 1970 was a depressing sight. The day before the appearance of the fish kill, a powerful stench of sewage was detected in the area of the train bridge. At the same time, the level of water in the river was high as a result of a clogging of the exit to the sea. With the opening of the stoppage the next day, the stream level dropped and the dead fish appeared on the banks. After a few days, things returned to their regular state. Testimonials and photographs point to a typical fish kill, caused by an extreme drop in the level of dissolved oxygen in the water. There was no proof seen nor provided beyond that of Netanya's sewage that was discharged at the time.[109]

It was during this period that the tightly knit agricultural community came to accept sewage as a valuable nutrient resource. Wastewater reuse meant foregoing control over the timing of fertilizer applications; crops might suffer some nutrient deficiency during the winter and then excess nutrients (and weeds) later in the season.[110] Still, farmers were impressed to learn that agricultural yields from sewage-irrigated plots were significantly higher than from comparable plots with normal water, even when an equal amount of fertilizer was added (in some cases, yields were two to three times greater).[111] With sewage disposal problems mounting and an agricultural sector willing to be part of the solution, reuse seemed to offer the proverbial two birds with one stone.

In 1956 Tahal drafted a national Master Plan for Israel Irrigation Development. It projected 150 million cubic meters of wastewater for agricultural usage.[112] For once, Blass's predictions turned out to be pessimistic. Today Israel recycles twice that amount of water. In 1962, fifty projects connecting Israeli farms to municipal sewage treatment centers were up and running. By 1972 the number had climbed to 120, using 20 percent of all urban sewage.[113] Today, Israel's 66 percent reuse rate is spectacular alongside that of other countries; the United States, for example, recycles only 2.4 percent of its sewage.[114] There were certain disadvantages from the farmer's perspective, however. For instance, the small number of permissible crops that could receive the effluent limited the combinations of rotations. And then there was the smell! But once Israel bought into sewage as an irrigation source, it never looked back.

In the short term at least, wastewater recycling was more cost-effective than desalination. Yet here too the government was unwilling or unable to generate capital locally to launch a major reuse initiative. It therefore turned to the World Bank. In 1972 the Bank provided a thirty-million-dollar loan to establish the National Sewage Project. Run as a revolving loan program, it identified seventy-eight subprojects involving seventy-five local authorities. By 1981 the long list of projects was completed. Although not a panacea, this loan gave Israel's environmental infrastructure a substantial boost.

The jewel in the public-works crown was the Dan Regional Wastewater Reclamation Project, which received 40 percent of the overall National Sewage Project budget. By the late 1960s, Tel Aviv's problematic aerobic settling ponds had been redesigned and the offending smells dissipated.[115] The initial cooperation showed that economies of scale held clear advantages and that a regional approach was more efficient. A half dozen more coastal cities wanted to join the consortium, bringing the total amount of sewage in the greater Tel Aviv area that had to be collected and treated to ninety million cubic meters a year. The system urgently needed an upgrade. On the collection side, the National Sewage Project funded five hundred kilometers of sewers and force mains, eighty pumping stations, and forty treatment works.[116]

For a change, the investment in sewage *treatment* was as serious as the investment in the *plumbing*. The Dan Union of Cities enlarged the aeration ponds (approximately two thousand dunams) on the Rishon L'Tzion sand dunes. It installed additional secondary and advanced treatment technologies, based on activated sludge and denitrification. As the final stage of treatment, the wastes were recharged into the aquifer (through flooding and drying). After a prolonged period of dilution underground, the water is generally clean enough to drink.[117] For aesthetic reasons, however, when the third line to the Negev became operational in 1989, all the eighty million cubic meters of water a year it sent south were designated for agricultural use. The total price tag for the initiative was only 115 million dollars,[118] but it increased the national water supply by 6 percent.[119]

While sewage's popularity grew as a source of water, the Ministry of Health moved to confront a potentially lethal public-health problem. It was the pathogens and bacteria in the wastes, rather than the high salinity in the effluents, that bothered them most.[120] In 1953 the Ministry recommended some of the first wastewater irrigation standards in the world, disqualifying raw sewage as an irrigation source. Yet even primary and secondary treatment of wastes does not always rid sewage of pathogens and bacteria. Thus, the Ministry limited the crops that could be grown with treated sewage to cotton, fodder, and produce that is not consumed raw.[121] If these proscriptions were followed, then using effluents was deemed safe. For instance, a study of eighty-one kibbutzim during the 1970s found little significant difference between the health of communities that used sewage effluents in irrigation and those that did not.[122]

The problem was that the Ministry of Health's irrigation recommendations were often ignored. Gastrointestinal illness was part of life in Israel until the 1970s. For instance, during the 1960s, 6 percent of hospitalizations and 8 percent of outpatient visits[123] were related to digestive-tract problems; fruits and vegetables were among the main causes. Perhaps the loudest wake-up call to the effluent-irrigation problem was the 1970 outbreak of cholera in Jerusalem. Some 250 laboratory-confirmed cases were reported between August and October of that year. Because of the higher incidence among Jerusalem's Arab population, it was deduced that the disease was caused by vegetables sold in East Jerusalem markets. Typically they were irrigated and contaminated by raw sewage water.[124]

The disease perpetuated itself as patients' excrement made its way through the city's sewage system. (The Environmental Health Laboratory at the Hadassah Medical School found 18 percent of sewage samples for the city to be positively contaminated with the *Vibrio cholerae* bacteria.) Even though treatment could not neutralize the pathogen, sewage water continued to be utilized by many Arab small farmers, who were dependent on it for irrigation, in particular during the summer months. The Ministry of Health scrambled to close restaurants linked to cholera and to give locals and tourists the painful cholera vaccination. But the infected sewage continued to flow into the Dead Sea through the Kidron Stream or to the Mediterranean via the Sorek.[125] In November there was an outbreak in the Gaza Strip with an additional three hundred cases; half of the victims were children.[126] The Knesset responded by amending the Public Health Ordinance. It empowered the Ministry of Health to promulgate legally binding standards for sewage treatment prior to wastewater irrigation.[127] The Ministry did so—eleven years *after* the cholera outbreak.[128]

SALINITY: THE POLLUTANT OF CHOICE

One consequence of the Six-Day War of 1967 was Israeli control over all of the Jordan River's tributaries. There had been at least a dozen water-related cease-fire violations in the Jordan River basin since 1951.[129] But, for the time being, the issue of the Kinneret's water supply seemed resolved—quite favorably, from Israel's perspective. (Transboundary water quality problems would not be recognized until thirty years later.[130]) The National Water Carrier was up and running. Water was even discovered in considerable quantities in the Arava, and subsequently a string of settlements was established.[131] With water quantity falling into place, sustainable allocation and water quality should have at long last become the Water Commissioner's paramount issue. Yet this was not to be. Spurred on by the new resources liberated during the Six-Day War, Israel's political leaders were not yet ready to abandon a paradigm that had so successfully lubricated past development. Pollution would have to wait for government attention—but of course it did not.

Israel's water is naturally alkaline and hard, quickly coating the country's electric teakettles with the chalky white residues of calcium carbonate. Salts are often added by industries and laundries to soften the water. Intensive human activities quickly upset the natural water chemistry, however. The Water Commission was well aware of a steady deterioration in water quality. Yet initially its experts did not fully grasp the scope of the problem. The only pollutant that garnered serious attention was salt. Salinity takes many forms, but is usually regulated according to chlorine concentration (salt is composed of sodium and chlorine atoms). It can also be roughly assessed by measuring electrical conductivity because the atoms are ionized. At high enough levels, salinity is associated with hypertension and elevated risk of heart disease.[132] Its negative impacts on crop yields has been apparent since the dawn of agriculture. Of course, its effect on taste is often what people notice first.

The fixation on salt makes sense, given the perspective of water managers who were obsessed with expanding access to water resources during Israel's early years. Salinity is a parameter directly linked to issues of groundwater mining and overpumping. Even today, former Commissioner Kantor speaks of no other environmental parameter.[133] This narrow water quality orientation is reflected in the original text of the 1959 Water Law, purportedly the most modern water statute in the world at the time. Although the law did not include the terms "pollution," "contaminant," or "water quality," it did prohibit activities causing salination.

During the 1950s and 1960s, scientists did not seem to fully grasp the contribution of surface pollutants. For instance, Israel's romance with wastewater reuse exacerbated the salinity levels in groundwater. Wastewater is saltier than freshwater, because conventional treatment does not really remove what is added during its original use, such as the leftover water softeners from industry. The salting of kosher meats makes slaughterhouses the single greatest contributor to Israel's effluent salinity.[134] Furthermore, effluents sit in storage ponds until farmers need them in the summer. The resulting evaporation raises salt concentrations even further.[135]

Dan Zaslavsky, a professor of engineering at the Technion, served as Israel's Water Commissioner during the 1990s. He describes the perils of the prevailing management strategy of the period:

> Intuitively, water managers thought that the amount of water that you could pump should be equal to recharge. But they didn't consider two things: First, you are adding salts all the time from above, and you have to rinse them out. Second, you may be getting salts from lower geological strata. I still have to explain to hydrologists that if you don't allow water to flow to the sea, salts will concentrate indefinitely. The major problem for water managers in Israel and in fact in most arid and semiarid countries is to change the balance of solutes: to import less into the source and export more.[136]

It is little wonder that Zaslavsky has emerged as Israel's most outspoken promoter of tough, drinking-water-level treatment standards for sewage or discontinuation of wastewater reuse in Israel altogether. Yet he is quick to defend the conscientiousness of Israel's first generation of hydrologists, whose determination made the desert green.

The practical way to prevent overpumping was to set an optimal level of usage, beyond which no pumping of water would be allowed. Unfortunately, this required more hydrogeological knowledge than was available. Kantor explains the somewhat arbitrary origins of Israel's most famous environmental standard:

> I thought we were dealing with 3.5 billion cubic meters of water per year, and today I know that I'm a lot poorer. But I also knew from the first minute in the job that I had to establish a "red line" beneath which we wouldn't pump. This required a national consensus. Not

> among citizens, but in academia. Almost all the Israeli academics were working for me already. We'd supplement them every five years by hosting a forum of international experts on hydrology. We'd send out materials, and they would prepare for the meeting. And that's how we established the red line in the early 1960s. It was in the concluding session of one of these meetings.[137]

The designation of red lines in Israel's three major sources of water was ultimately a somewhat arbitrary "guesstimate." Once they were set, however, the Kinneret's and the aquifers' red lines assumed enormous significance and quickly entered the national consciousness. Weather broadcasts refer to them, informing the public about the practical importance of rainstorms. As Water Commissioner, Kantor took them very seriously.

But not everyone was as conscientious as Menahem Kantor. As Israel's Water Commissioner, Dan Zaslavsky took bold action to reduce allocation and bring water levels up above the red line. Zaslavsky remains furious at the lenient attitude towards red lines and the chronic overpumping that prevailed among previous Water Commissioners during the 1970s and 1980s, when agriculture received whatever it asked for, regardless of the available reserves.[138] Meir Ben-Meir, a lifelong advocate for agricultural interests, served as Water Commissioner twice, during the early 1980s and at the end of the 1990s. His cavalier attitude towards the red lines and allocation policies, which seemed to reflect a belief that overpumping could continue with hydrological impunity, made him a reviled figure among many in the environmental community.

At the end of 1990, State Comptroller Miriam Ben Porat prepared a scathing review of water allocation policy at the Water Commission. The collective deficit of the aquifers and the Kinneret had reached 1.6 billion cubic meters—an entire year's worth of water. The drop in water table levels caused by the overpumping allowed the seawater to encroach 1.5 kilometers inland.[139] Sixteen percent of the wells in Israel's largest Coastal Aquifer were deemed unusable, because of chlorine concentrations too high even for agriculture.[140] By then it was clear that pollutants other than salinity might be just as deleterious to Israel's water resources.

BLUE BABIES AND NITRATES

On February 15, 1953, Dr. S. Wago, a pediatrician from the Zrifin hospital, admitted a three-week-old infant from Ramla and immediately diagnosed her condition as acute methemoglobinemia. In layman's terms, the baby was suffering from blue-baby syndrome. Methemoglobinemia occurs following exposure to excessive nitrate levels. Infants (and some stomach cancer victims) have bacteria in their stomach that reduce nitrates to nitrites. When nitrified water is ingested, the hemoglobin in the blood shows a distinct preference for the nitrite ion and binds with it to the exclusion of oxygen. When oxygen levels in the blood drop too low, asphyxiation sets in.[141]

In this case, the infant was lucky and was only in an intermediate stage of distress. After a shot of methylene blue, her color returned to its natural state. A day later the baby could be released.[142] Two years later, another Ramla girl was brought to the emergency room with the same symptoms. The treatment produced similar results, but she returned a day later in a critical blue condition even after the family had begun to drink water from another faucet in the neighborhood. Samples showed nitrate levels at her family's home reaching sixty-seven milligrams per liter (the recommended drinking water standard was forty-five milligrams per liter, although Israeli regulations allowed levels to go twice that high). In her report Dr. Wago hypothesized that the reason that more infants were not affected by the high nitrate levels was because most babies nursed. Even if methemoglobinemia never reached epidemic proportions, there was room for concern about the carcinogenicity of nitrosamines, a by-product of nitrates after they are metabolized.[143]

The Ministry of Health was consulted on this and three subsequent cases of the disease.[144] The Sanitation Department sensed that nitrates posed a more severe public-health problem than salinity, even if the chemical was of little interest to the official water agencies. On this issue, they found an ally in the form of a Mrs. Esther Foa, who worked at the Hydrological Service. Although she never published in professional journals, she was deeply concerned about nitrates and began to collect copious quantities of data about the steady increase in concentrations, which she passed on to the Sanitary Department.[145] At the turn of the century, nitrates were almost nonexistent in Israeli wells.[146] By 1970 Foa reported three hundred wells within the Coastal Aquifer that had reached the recommended forty-five-milligrams-per-liter ceiling, with several showing nitrate concentrations as high as 100 milligrams per liter.[147]

Stopping nitrate pollution is a particularly daunting regulatory challenge, because its sources are so diffused. Nitrogen is a basic nutrient without which plants cannot survive, and synthetic nitrogen-based fertilizers (e.g., ammonia) have been the crucial factor for the last century in the increased productivity of farmers around the world. (Cow manure is such a good fertilizer because it contains fifteen times the nitrate concentrations of human wastes.) When fertilizer is applied at a rate faster than the plants can absorb, however, the chemical leaches into the groundwater. In nonagricultural areas, the problem was particularly acute in wells located near septic tanks.

Eventually the planners at Tahal began to wake up to the problem. In 1972 Dr. Chen Soliternick prepared the first systematic evaluation of the problem. Soliternick estimated that about forty-eight thousand tons of nitrates reached the soil, beyond the plants' ability to absorb them. Most of these nitrates could be traced to agriculture. Of the six identifiable sources, over 50 percent of the loadings came from manure and fertilizer applications.[148] Soliternick also identified wastewater irrigation as a major nitrate source.[149] Data indicated that it would not take long before the water from entire sections of the Coastal Aquifer would be rendered unfit for human consumption. Of even greater concern was the discovery of increasing nitrate concentrations in the Mountain Aquifer. Although its nitrate levels were still within the drinking-water standards, this much deeper source of water previously had been considered to be in pristine condition.[150]

Regardless of the pathway, most of the nitrate contamination could be traced to Israeli agriculture. But Kantor and subsequent water commissioners had a paternalistic attitude towards farmers. They never got tough on the issue. For example, Section 20(D)(2) of the Water Law amendments of 1971 calls for the issuing of regulations to control agricultural cultivation and fertilizer usage. The Water Commissioner never drafted any such regulation for the Minister of Agriculture to sign. Nor did a Water Commissioner ever promulgate any directives to protect water from the growing menu of pesticides, even though the Water Law empowers him to set such regulations as well. Ultimately the Water Commission never owned water quality as an issue and never felt compelled to go beyond the bacteriology- and pathogen-based standards of the Ministry of Health.[151]

Even at the very end of the twentieth century, Water Commissioner Meir Ben-Meir preferred to accept nitrification of Israel's aquifers as inevitable and focus instead on well water treatment.[152] The selective electrodialysis provided by Israeli companies (using EDA, a Mitsubishi subsidiary's technology) can reduce nitrate concentrations by some 50 percent for a fee of roughly one shekel per cubic meter.[153] The price and concentrations will rise as the aquifer's nitrate levels continue to mount. Combating pollution at its source required sophisticated enforcement efforts and political resolve, rendering it an unpopular policy option.

For forty years Israel's water policy towards agriculture was consistent: Copious quantities were provided at a low, highly subsidized price.

Farmers understandably took the path of least resistance. The lack of any clear regulatory signal spawned inefficiencies. Water-intensive crops such as cotton were introduced. No one considered whether it made sense for Israel to export its scarce supplies of water (in the form of produce) or whether there was an associated environmental price tag.

Kantor today becomes angry when it is suggested that water policies were influenced by an agricultural lobby. "It is just silly and ignorant to claim that there was an agricultural lobby," he says. Lapsing into the Zionist catechism, he explains that "in those days we knew that the water was available and that if we wanted to make the country blossom we had to develop it. You know, not long ago if you drove fifteen kilometers south of Tel Aviv everything was brown. Government bureaucrats did not change this. Farmers did. And I don't know a single farmer who got rich from his work." Kantor is appalled at present policies that have reneged on the national commitment to the Israeli farmer. He calls present agricultural policies "a national larceny."[154] But a firmer regulatory approach to agriculture, designed to preserve water quality and impose an ethos of hydrological frugality and conservation, might better serve the long-term interests of Israel's farmers.[155]

It is important to emphasize that Israeli agriculture was never the one-dimensional dinosaur that some environmentalists like to vilify. Beyond farmers' achievements in wastewater reuse, Israel is deservedly proud of being among the few food exporters in the Middle East, where hundreds of billions of dollars go to imported food to meet a growing population's basic needs.[156] And, of course, its paramount innovation has been drip irrigation.

ISRAEL'S IRRIGATION MIRACLE

Once again Simcha Blass emerges as an unheralded hero. In terms of global significance, the tiny, individual drops of water that Blass learned to release through his irrigation pipes dwarf the rushing streams of water flowing in his massive engineering projects. Like many inventions, it was something of a fluke. According to his memoirs, sometime during the 1930s Blass visited Abraham Lubzovsky, a Second Aliyah pioneer near his Karkur home. Lubzovsky proudly showed Blass an enormous tree that had been watered by a leaky pipe that had left droplets of water on the seemingly dry surrounding soil. In Blass's inimitable words: "Water droplets raising a giant tree hit me like a mosquito in the mind of Titus the Evil."[157]

It would take Blass another twenty years to find the time to perfect the technology. By the end of the 1950s, however, low-cost plastic piping enabled him to develop a system that used a fraction of the water with much greater efficacy than conventional sprinklers did.[158] Rather than flooding the plant's root zone, water (and fertilizer) is spoon-fed to trees and plants drop by drop through narrow black pipes whose drippers regulate the amount of water released. Computer systems eventually optimized the rate and timing of applications.[159]

Blass set out, peddling his innovation to Israel's economic establishment. Although they heard him out, the contraption was politely dismissed as harebrained by the major economic corporations of the period, including several kibbutzim who later came to regret their position.[160] Eventually Blass settled on Kibbutz Haterzim, a young settlement near Beer Sheva, as a partner. On August 8, 1965, they created the Netafim Company, and production began.[161] Thirty-three years later, for Israel's fiftieth anniversary, a team of experts picked Blass's drippers as the most important Israeli invention since the founding of the State.

As Netafim grew to become the undisputed world leader in drip irrigation, racking up over two hundred million dollars in annual sales by the end of the 1990s, Israel's reputation as the state-of-the-art leader in water engineering reached new highs.[162] For instance, a Netafim-supplied ten-thousand-acre cotton plantation in Arizona was the biggest drip irrigation site in the world.[163] (Although Netafim representatives feigned embarrassment, it probably did not hurt their business when California law enforcers attributed much of the earnings in the state's lucrative marijuana crop to savings from the efficient Israeli technology.[164]) Drip irrigation solved any number of technical problems. Evaporation is greatly reduced relative to sprinkler or flood systems. Even steep terrain and shallow soils, as well as coarse sands and clays—which had always posed a problem for traditional irrigation techniques—responded favorably.[165]

Much of the water delivered to Israeli farms is recycled sewage, and drippers hold the additional environmental advantage of not producing aerosols, which can form from conventional sprinklers and drift 750 meters downwind. When recycled sewage was delivered to vegetables via drip irrigation systems, they were free of infectious bacteria.[166] With the steady increase in wastewater reuse, Israeli farmers reduced water demand per yield to new lows.[167] From the perspective of conservation, Israeli farmers were justifiably admired around the world as a paragon of thrifty water management.[168]

INDUSTRIAL POLLUTION

Despite the country's sentimental attachment to agriculture, it was industry that quickly came to serve as the backbone of Israel's thriving economy. For the first ten years of the State, factories enjoyed carte blanche when it came to the environment. Eventually, however, the pollution became too offensive to ignore. With the Water Commission's blessing, in 1971 new draconian antipollution provisions to the Water Law were enacted.[169]

Under the amended section, the definition of water pollution was broader than comparable definitions in other environmental laws passed at the same time, such as the U.S. Clean Water Act. The amendments gave the Water Commissioner almost dictatorial powers to defend water resources.[170] The Commissioner is authorized to institute a system of environmental permits, force polluters to clean up environmental damage, and even order without warning the cessation of water supply to a pollution source. The trouble was that this formidable environmental arsenal was never really used. Industry's insouciance was therefore not surprising.

Despite Kantor's convenient recollections of his "standing up to industry," general government obsequiousness to, or even collusion with, industrial interests has been glaringly apparent to this day. Water Commissioners made only token efforts to hold industry's feet to the fire. Between 1974 and 1976 the Commissioner issued twenty-eight orders against polluters, requiring the upgrading of effluent quality. Compliance was extremely spotty.[171]

There were many reasons for the poor record in this realm. Many industries were government-owned and enjoyed functional immunity. The lack of an enforcement team at the Commission that could effectively measure pollution levels did not help either. Frequently, monitoring results were fabricated in reports, with no government follow-up. For instance, as late as 1994, Dr. Mouna Noufi, an analytical chemist working for the public-interest environmental group Adam Teva V'din, measured the actual discharges into the Kishon River from various outfalls of Haifa-area petrochemical plants. The facilities were still operating under the lenient requirements of a 1979 generic permit from the Water Commissioner. Noufi's laboratory results proved all outfalls to be in violation of the effluent standards, with one factory discharging wastes with concentrations a thousand times above the permissible level.[172]

Even without this chemical analysis, biology offered a damning indictment. Dead fish and the sickeningly sweet stench of chemicals in rivers and on beaches could hardly be ignored. When former frogmen from Israel's elite commando unit "Shayetet 13" began to develop cancer, they began to blame their military training during the 1970s and 1980s in and around the Kishon River. A Commission of Inquiry, headed by former Chief Justice Meir Shemgar, was formed in 2000 to consider claims of the Navy's negligence. When Shemgar's commission confirmed the sick veterans' complaint, the Navy temporarily took the unprecedented step of suspending all military diving until it could sample water quality in all training areas, as recommended by the report. The story provided another rare, full-front-page sensational environmental headline that passed from public consciousness as quickly as it had entered.[173]

When the Water Commissioner did move to address a particular facet of the pollution problem, implementation was feeble. Industry did not seem to take his rules seriously. The ban on hard detergents that the Minister of Agriculture promulgated in 1974 is illustrative. In the mid-1960s the United States and England banned laundry detergents that were less than 80 percent biodegradable. In these "hard" detergents, conventional sodium (as in soap) was replaced with organic sulfur, because its solubility was not limited by the level of water acidity. When rinsed into a body of water, however, the resulting suds interfere with aerobic processes and lead to fish kills. The Water Regulations (Prohibition of Hard Detergents), promulgated in 1974, forbade the import or manufacture of hard detergents without a permit from the Water Commissioner.[174] Over two years later, a study showed that most laundry detergents in use were still categorized as hard, in flagrant violation of the regulations.[175]

As Israel came to assume a position of world leadership in the area of wastewater reuse, industrial noncompliance became particularly problematic. As mentioned, some negative hydrological impacts from sewage recycling are inevitable, because effluent waters are much saltier than freshwater and tend to exacerbate groundwater salination. Contamination by industrial toxins, however, is avoidable. Most domestic sewage systems received wastes from the city's industrial zones as well as from residential homes. When pretreatment at the factories was poor, conventional sewage purification facilities could not remove the toxic contaminants, such as heavy metals and organic solvents. As farmers watered their fields with treated sewage, they had no idea that they were systematically spreading industrial chemicals across the soil. With the exception of chromium, concentrations of heavy metals generally remain low in Israel's sewage.

Nonetheless, 5 to 10 percent increases in cadmium at the plow level have been measured.[176]

As early as 1974, Tahal hydrologists were predicting contamination of the Coastal Aquifer by toxic chemicals and metals in sewage effluents reclaimed for irrigation. The models suggested that the northern part of the aquifer would be contaminated by the year 2000.[177] It did not take that long. In the late 1980s, research conducted by Dr. Leah Muszkot indicated that industrial chemicals had already reached rural wells.[178]

Muszkot was an unlikely ecosleuth. An analytical chemist working out of the Volcani Institute, her research took place within the agricultural establishment, funded and supported by the Ministry of Agriculture. Muszkot was careful not to sensationalize the results of her research at conferences or with the press.[179] She was unwilling to show environmental organizations maps indicating the precise location of well sites. But she did publish the results of samples taken from wells at thirty-meter depths in areas with twenty years of wastewater irrigation.

Her mass spectrometer identified a veritable toxic cocktail of chemicals.[180] Israel did not even have drinking-water standards for chemicals that she reported on, such as benzene and toluene and the many organic compounds used in the plastics industry. For years it had been easy to discount concerns about industrial contamination as doomsday exaggerations and hide behind the pervasive uncertainty in the hydrology models. The presence of industrial, carcinogenic chemicals in groundwater underlying nonindustrial regions such as the Sharon suggest that not only Navy commandos will pay the price for pervasive environmental casualness among Israeli manufacturers and those whose job it was to monitor them.

MUNICIPAL RESPONSIBILITY: THE MISSING LINK

From an institutional perspective, these disparate water quality issues share a common denominator. According to the traditional English administrative orientation that survived in Israeli laws, municipal governments were the key institutional address for combating public-health insults and nuisances. In fact, however, they were often the problem. For example, Israeli cities were responsible for delivering water to the tap but typically lacked the technical staff to ensure optimal chlorination levels. In addition, city governments profited from the water they sold to residents, undermining any incentive to encourage conservation. Although the Licensing of Business Law empowered them to set environmental conditions for factories and commercial enterprises, protecting their local property tax base constituted a more compelling interest.

Drinking-water quality is another example of the bureaucratic quandary created by a multiplicity of central and local agencies. Technically the Minister of Health was empowered to set drinking-water standards but failed to do so for twenty-five years. Instead, between 1958 and 1970, a nonbinding instruction sheet adopted by the Israel Standards Institute was used by regional doctors to determine the potability of the country's drinking water.[181] In practice Mekorot pumped and delivered 60 percent of the water to municipalities, but the cities provided the delivery, including chlorination.

Mekorot was a public utility that needed to be regulated itself; it was not designed to be an overseeing agency. But with its meager resources, the Ministry of Health's monitoring capacity was limited. The Ministry could take only about 6500 samples of water each year during the 1950s, testing for a very narrow number of drinking-water parameters.[182] By the mid-1960s, this number had quadrupled, if Mekorot's own testing was considered, but the financial constraints also meant that only bacterial concentrations were checked regularly, and small population centers went virtually untested.[183] For the most part, water quality seemed acceptable.

The serious water problems were to be found in areas where there was not only no information, but also no running water. By 1957, ten Arab villages in the Galilee had been connected to Mekorot's water system, and their drinking water quality was comparable to the Jewish sector's. But that was only 14 percent of the Israeli Arab sector.[184] A special committee was formed to address the issue, but it would take years before the gap began to close.

As chemical analysis became more precise, and environmental epidemiology advanced, the number of potential drinking-water contaminants expanded. Testing was expensive. The Ministry of Health's own 1974 standards required only that drinking water sources be given a full chemical checkup once a decade![185] (Bacteriological testing was much more frequent, but cities with as many as twenty thousand residents were checked only on a monthly basis.) The Ministry of Health had little choice but to leave the drinking-water issue at the doorstep of local governments, even though it knew that they lacked the budget, interest, or technical capabilities to treat drinking water properly.

Pollution from sewage, arguably the most severe environmental problem of the period, is also the best example of municipal failure. Although the Water Commissioner was responsible for water resources in general, sewage was framed as a local issue. Under the British Mandate's Municipal Sewage Ordinance and the 1962 Local Authorities (Sewage) Law that replaced it, city governments are responsible for sewage.[186] Under the 1962 statute, they are granted special authority to levy two types of sewage taxes to pay for infrastructure. The surcharges have been hailed as an expression of the "polluter pays" principle.[187] But until the 1990s, Israeli mayors made a mockery of the concept, diverting sewage charges to more politically popular budget items, such as festivals and football stadiums.[188]

The Ministry of the Interior, the agency responsible for enforcing the sewage law, was also responsible for the solvency of local government. Perennially short on resources, the Ministry usually found it easier to ignore long-term pollution problems and to focus, for instance, on finding the funds to cover a city's education budget deficit. Moreover, the Ministry of Health was already involved in the issue of sewage water monitoring and standard setting. The Minister of the Interior and the Water Commissioner could place their limited resources in other areas and pretend that sewage treatment was somebody else's problem.

Symbolic efforts were made in response to the capacity gap. For example, Israeli kitchen sinks almost universally lack garbage disposals, in order to reduce the organic loadings on the overloaded municipal treatment centers. The inadequacy of investment in sewage infrastructure, however, has prevailed until the present.

THE KINNERET SECRETARIAT: A RARE POINT OF LIGHT

The Kinneret Secretariat, in retrospect, is the closest thing to a water-quality success story from this period. Yet it too reflects the Water Commission's dubious commitment to water quality during the 1960s and 1970s. As the Kinneret's role as a national reservoir became clear, considerable energy went into limnological research in order to understand and protect it better. By 1968 over five hundred articles, books, and reports about the lake had been published, mostly during the preceding ten years.[189] The picture that emerged was not encouraging.

When the JNF drained the Huleh region, it removed an essential "nutrient sink" that absorbed much of the nitrogen and phosphorus from the surface runoff into the Kinneret. In addition, the rich organic peat, newly exposed and nitrified, was washed down into the reservoir. The resulting 50 pecent increase in nutrient loadings threatened to induce eutrophication.[190] Straightening segments of the Jordan River also increased the amount of sediments flushed directly into the lake. In 1964, Mekorot, responsible for operating the National Water Carrier, recognized the danger and approached both the Minister of Agriculture and the Water Commissioner with a request to create a Kinneret Watershed Authority to protect water quality. Kantor agreed in theory and set up a committee to discuss the possibility but then stalled.[191]

It was a peculiar dynamic. Generally, a national Water Commissioner, charged with protecting Israel's water resources, should serve as the main promoter of water quality. The regulated utility is expected to take a more evasive role. In the Kinneret it was just the opposite. Perhaps the best thing about putting Mekorot in charge of operating the National Water Carrier was that it forced its managers to take a long-term look at water quality.

When Bob Davis, a South African limnologist, joined his son in Israel in 1971, Haim Gofer, the new Kinneret Committee head, asked him to prepare a comprehensive assessment of the lake.[192] Davis's report suggested that the Kinneret was in an advanced trophic state. He predicted that if immediate measures were not taken to reduce its nutrient loads, the lake would die.[193] The trouble was that the engineers at the Water Commission had never really taken into account the Kinneret's role as an aquatic habitat but rather saw it only as a reservoir.[194] When the Water Commissioner did not respond to Davis's recommendations, it was leaked to the press, creating quite a stir. (Kantor later claimed he never saw the report.) Yigael Alon met with Davis and created a special committee of Ministerial Directors General to field a Kinneret strategy. The Committee banned all construction within fifty meters of the lake and also called for a new regional plan to protect the Kinneret. The Water Commissioner fired Gofer, apparently for insubordination.[195]

In all fairness, it should be said that the Kinneret Secretariat, declared in 1971 and formally authorized as a Regional Drainage Authority, made remarkable progress. Drainage authorities are created primarily to prevent flooding and to continue the job of land reclamation. They are not considered to be environmental pit bulls. But, after thirty years, the Kinneret Secretariat has shown what a committed drainage agency can do if it has high-quality data, a cogent watershed orientation, and political support. The Secretariat's activities included developing and managing the shoreline, supervising fishing, and monitoring bacteria and pesticide residues, as well as launching environmental education initiatives. Perhaps most important has been its policy of no effluent discharge, which led to upgraded sewage treatment and to improved solid waste management in sanitary landfills outside the watershed.[196]

Environmental indicators, for a change, reflect meaningful progress. During the past twenty years eutrophication has not advanced, and indicators such as primary production and chlorophyll did not rise between 1972 and 1993 (at the same time there has been an increase in total nitrogen concentrations and algal biomass in the lake).[197] This remains a far cry from a fairytale ending. Today the Kinneret is still polluted by a variety of point sources, including drainage pipes and outfalls from industrial areas such as the Zemah complex.[198] Overpumping during the drought years of 2000 and 2001 shrank the lake's banks as never before. Yet, had the Coastal Aquifer been protected as energetically as the Kinneret was by the Secretariat, Israel's natural resource portfolio would be much richer today.

THE END OF THE OVERDRAFT ERA?

Miriam Ben-Porat, Israel's State Comptroller, was used to breaking barriers. She was the first woman to represent the country in the State Attorney's Office. Then she was appointed as the first female justice in the Israel Supreme Court. When the mandatory retirement age forced her to step down at age seventy, she was considered to be at the peak of her professional powers. It made sense to make her the first female State Comptroller. As always, she was fearless, and the public loved her for it.

Each spring the State Comptroller releases a voluminous report covering various areas of inefficiency, corruption, and mismanagement across government agencies. By the end of the 1980s the water situation was sufficiently bad to warrant a special report. Ben-Porat did not mince words:

> Since the middle of the '60s, the water reserves of Israel have deteriorated, and today, November 1990, in the three main reservoirs of the State—the coastal and mountain reservoir and the Kinneret—the amount of the water deficit reaches 1.6 billion cubic meters, close to the annual amount utilized by the country. As a result of overpumping, the levels in these reservoirs dropped, the reserves depleted completely, and a severe drop in the quality of water ensued. Water allocation, particularly to agriculture, in a quantity that exceeds the water that is replenished from rains on an average basis, is what caused the overpumping and the liquidation of operational reserves.[199]

It was the top story in the news, made particularly relevant by weather reports that indicated a third consecutive winter with low rainfall. The Comptroller's report did not happen by itself. Rather it was the culmination of a long campaign. As early as the mid-1970s, the Water Commission's own environmental reports from over 1200 wells openly documented the same unsustainable trends.[200] A decade later, Israel's scientific community finally spoke out about the public-policy implications.

Hillel Shuval, by now a senior professor at Hebrew University, had been declaring a "crisis" situation for Israel's water resources in academic forums for years.[201] As time went on, the chorus of disenchantment grew louder. In 1986 a group of engineers and scientists decided to go public with the message in the form of a Committee of Scientists for Water Affairs. Except for Darcy's law, there is little that is simple about hydrology. As a result of Israel's diverse geological circumstances, its groundwater science is riddled with nuances, uncertainties, and competing models. To the layperson, technical discussions about water can quickly turn arcane and mystifying. So the scientists kept the message for Israel's media simple: Overpumping to sate agriculture's unquenchable thirst was destroying the quality of Israel's water reserves.

The media blitz led to meetings with the Minister of Agriculture and even the Prime Minister but did not translate into any policy changes. On the contrary, between 1987 and 1989, the Water Commissioner allocated from 8 to 14 percent more water than the amount recommended by Tahal and his own Hydrological Service.[202]

Despite the technological transformation of drip irrigation, domestic policies encouraged wasteful habits. With such low water prices, farmers were not always assiduous about fixing leaks. It was not uncommon for a farmer to turn sprinklers on fallow fields, rather than risk a reduction in the next year's water quota for not utilizing present allocations. Environmentally, farmers were even less conscientious. Reduction in the application of fertilizers and pesticides was not given a prominent place on anyone's agenda.

Like the captain of the *Titanic*, agricultural leaders ignored warning signs and sailed on defiantly toward disaster, showing no intention of voluntarily reducing consumption and resenting every drop that reached the Dead Sea unnecessarily.[203] They trumpeted the plight of the farmer, complained that water subsidies in America were four times higher than in Israel, and assumed that the engineers at Mekorot would continue to find a way to provide huge quantities of high-quality water at artificially low prices. Most of all, they ignored unsustainable trends, refusing to recognize the incompatibility of present practices with demographic growth in the country and the region. The international community of experts, once generally sycophantic, became critical.[204]

By the 1990s, agriculture's status was also completely different from when Simcha Blass made irrigation the paramount national infrastructure priority. Irrigated agriculture's contribution to Israel's gross national product had dropped from 30 percent in the 1950's to 3 percent.[205] Of the many Zionist axioms, agriculture's paramount role in national well-being was among the first to lose its luster. The image of the kibbutz suffered from thirteen years of Likud rule and disingenuous smears, branding the collectives elitist and even parasitic. Even so, objectively, the richest kibbutzim no longer made their money farming. In many instances, the tractors and fields seemed a sentimental gesture to old-timers—a living museum to an ideology whose time had passed.

In a phenomenon that continues to this day, the difficult agricultural work in Isrel was done increasingly by Palestinian day laborers who were bused in from the West Bank and Gaza. (Later, when political tensions made the workforce unreliable, they were replaced by a combination of Thai, Chinese, and Filipino laborers.) Some dispassionate water experts even openly advocated the end of agriculture in Israel. Although still iconoclastic (and economically foolhardy), the view was no longer sacrilegious.

Meir Ben Meir, the two-time water commissioner and life-long advocate for Israel's agricultural sector, linked the issue to the basic right to an occupation, a right recently conferred quasi-constitutional status. More compelling were arguments in favor of continued agricultural water subsidies as an indirect way of preserving open spaces and landscape heritage in the face of the relentless "Los Angelization" of Israel's heartland. Yet, the traditional ideology, glorifying the spiritually edifying experience of working the soil, became increasingly irrelevant as a societal force. Only 2 percent of Israel's population worked as farmers, and the contribution from agricultural commodities to gross domestic product continued to drop, settling at 5 percent by the end of the century.[206] *Economically*, it was difficult to justify the allocation of 70 percent of water to agriculture. *Environmentally*, it was impossible. Water policy reform offered a rare zone of agreement between these two frequently incompatible disciplines. And as the 1990s approached, grim water-quality indicators vindicated the urgency of calls for change.

THE WATER CYCLE

Change appeared to come in the form of Dan Zaslavsky, a bespectacled professor of agricultural engineering at the Technion given to wearing suspenders and exploring alternative energy sources. Zaslavsky was born into a family of engineers. Following the 1948 war, he started out his adult life on a kibbutz, but it proved too confining for a man of his considerable talents. After completing a Ph.D. in soil physics at Iowa State University in 1960, he returned to Israel and thrived in academia and as a consulting engineer. Zaslavsky's research and personal interests generally went beyond the theoretical, and during the 1980s he briefly served as Chief Scientist at the Ministry of Energy. Yet until 1990, Zaslavsky was just another of the many illustrious Haifa-based academics.[207] All that changed when he was unexpectedly appointed to be Israel's Water Commissioner.

Rafael "Raful" Eitan became Minister of Agriculture after the 1988 elections and brought a reputation for honesty to the job. Soon recognizing that Israel's water resources were being cheated, Eitan wanted to fire the Water Commissioner, Zemach Yishai, immediately. But Yishai was active in Liberal Party politics and was close friends with Likud minister Moshe Nisim, who backed him. With the Labor Party tacitly supporting Yishai's proagricultural positions, Eitan was initially unable to muster the majority required in the Cabinet to throw him out. When the State Comptroller dropped her bomb about Israel's water management, he had the excuse he needed to replace Yishai with an expert.

Zaslavsky never learned from whom the Minister of Agriculture got his name. Originally Eitan offered Zaslavsky chairmanship of the Mekorot water utility, a position which did not interest him. So Eitan told him, "You know what, you'll be de facto Water Commissioner, and I'll work with you until Zemach Yishai leaves."[208] In fact, the Water Law gave the Minister of Agriculture most of the key powers to make policy decisions. So until Yishai stepped down, Zaslavsky worked as "adviser to the Minister," focusing on long-term planning after years of apparent neglect.[209] Once appointed Commissioner, he focused on the short term. The situation was considered critical: With the Coastal Aquifer and the Kinneret rapidly approaching the red lines, the threat of a drinking water shortage loomed for the first time in the country's history. There was no time for changes in price structures, deregulation schemes, or new laws.[210] Eitan heard Zaslavsky's remedy—drastic cuts in allocations—and decided to push the bitter pill on his constituents.

The State Comptroller's report had raised the profile of the water shortage, and the drought weighed heavily on the public consciousness. After the Iraqi Scuds stopped falling in the spring of 1991, Zaslavsky started to make his changes. He tried to use economic tools, cutting subsidies and raising water prices. But the Knesset grandstanded the changes, and his boss, the Minister, was powerless to help. Then Zaslavsky tried to cut allocations to the cities and to dry up grassy parklands he considered wasteful given the need for austerity. The nation's mayors called his regulations illegal, and the law was on their side. So Zaslavsky appealed directly to the public. During the two years of drought, there was a voluntary drop of 10 percent in overall urban usage—not withstanding the increase in immigration and economic growth during the period. Zaslavsky recalls the remarkable swelling of civic spirit and how citizens would proudly tell him how they tiled over gardens for the national good. Only the mayors were recalcitrant.[211]

Seventy percent of Israel's water was still going to the farmers, though, and here Zaslavsky was resolute. On the whole, farmers were extremely resentful of even minor cuts. One Friday morning, Zaslavsky was invited to speak in Tel Aviv, at a meeting chaired by former Water Commissioner Meir Ben Meir, who then headed the Agricultural Research Center. He never got a chance to talk. One after another, speakers launched into tirades. Part of the problem may have been Zaslavsky's presentation. It took little time for farmers to recognize that he was smarter than anyone else in the room and knew a lot more about water; some agricultural advocates found this hard to take. Years later, one agricultural lobbyist would quip that three years of drought were not as destructive to Israel's agriculture as two years of Dan Zaslavsky.

The cuts Zaslavsky spoke of were tantamount to a hydrological earthquake. Ultimately the issue reached the government. But the Minister of Agriculture backed his man, and Zaslavsky got his way. Yitzhak Shamir, who was the Prime Minister, presided over a very heated debate:

> I didn't need to be an expert in water. I had enough experts who were quite dedicated and serious about the subject. We heard Zaslavsky, who certainly had his own theories on the subject that seemed quite original. Then we heard from others who felt differently than him. It was hard to say who was correct. I didn't feel that I had to sit as a judge and decide between two views. In this case there was a Minister in whom I had confidence.[212]

With the government's blessing, Zaslavsky did what he had to do and issued reforms. After the Water Commissioner cut back, God did His part to get the nation's water balance sheet back in shape. The summer of 1991 was moderate, and the winter of 1991–1992 was astonishingly wet. Zvi Ortenberg, Chairman of the Lake Kinneret Administration, reported that a total of 1.2 billion cubic meters of water poured into the lake from December 1991 until May 1992. It rose over four meters, from an all-time low of 213 meters below sea level to an astonishing 208.9 meters below sea level, with a threat of flooding.[213] The aquifers also bounced back much faster than the most optimistic hydrological projections.[214]

But in the 1992 summer elections, the Labor Party reclaimed power. Yaakov Zur was appointed Minister of Agriculture. Zur was a kibbutznik with strong ties to the agricultural establishment. Getting rid of Zaslavsky was a priority, and he wasted no time.[215]

Zaslavsky never received his due for trying to change the course of Israel's water policy, and some farmers openly celebrated his unceremonious dismissal. As is the case with Water Commissioners, the vast majority of Israelis have never heard of him. Gidon Zur (no relation to the Minister) was appointed in his stead. He was a career civil servant, and, while not quite the fighter that Zaslavsky was, he harbored environmental sensitivities. As rainy years improved the water balance, however, water allocations slowly but surely began to resemble the agricultural "largesse" of the 1980s. Meir Ben-Meir, appointed for his second term in 1996, was unabashedly open in the priority and subsidies he brought the agricultural sector and dismissive of the "doomsday" criticism of the hydrological community. The Water Commission celebrated its fortieth year with a strategy that did not look altogether different from what had characterized the Menahem Kantor administration.

Given the climatic cycle of drought, and the steady growth in demand, Israel's water policy remained a far cry from sustainable.[216] Enthusiastic pumping and the generous ceding of water rights to Jordan during the wet years of the 1990s left little reserves for the inevitable dry span. Whether a result of global warming or not, rainfall during the winters of 1999, 2000, and 2001 fell far short of average, bringing Israel's water resources to their lowest recorded levels, far below the red lines.[217] The public was aghast at the frequent photos in the press of anomalous islands that now stuck out in the middle of the lake and the parched and exposed shoreline of the Kinneret on land where clear water had once glistened and children waded.

When Shimon Tal, the competent and congenial chief engineer from Mekorot, was appointed in Ben Meir's stead in 2000, he slashed agricultural allocations again. He could not, however, bring the Knesset to approve the massive cuts in urban usage (including a ban on watering lawns) and the conservation policies that were needed to meet the growing hydrological challenge. A special Knesset Commission of Inquiry began to investigate the issue during the summer of 2001, and the parade of witnesses—including former water commissioners, scientists, and politicians—was quick to point blame and bring historic citations to the legislators, documenting just when "I told you so."[218] As the politicians considered who the scapegoats should be, real institutional reform once again became a common topic of speculation and discussion.

Tal, the new Water Commissioner, quickly came to support a revolution in management strategy that would cancel the old quota system entirely in favor of a market-based allocation system. In such a system, Water Commission personnel would no longer serve primarily as "allocation referees" but begin to function as water quality regulators. "If someone is willing to pay the costs of desalinated water to wash their car, why should we impose restrictions?" he asked.[219] How Israel's thirsty streams, wetlands, and other aquatic habitats might find the funds to compete in such a new system was not yet resolved. Nor was the ultimate effect on Israel's greenhouse gas emissions of this belated embracing of energy-intensive desalination. And of course, it wasn't clear that the beleaguered agricultural community was in fact ready to accept economic subsidies in areas beyond their water bill.

THE LAST DROP

The story of Israel's water resources reflects Zionism's finest hours and its most glaring deficiencies. A delivery infrastructure created by zealous politicians, bureaucrats, and engineers jump-started an economy and led to unprecedented achievements in semiarid cultivation. The development of water resources solidified a healthy national myth that glorified farming as the most honorable of Jewish occupations after a two-millennium exclusion from tilling the land. For most of its history, Israel was a net food exporter. But this hydrological "progress" also left rivers filthy and subjected aquifers to a contaminant bombardment from which they may never really recover.

Successes in protecting the Kinneret watershed or implementing drip or wastewater irrigation technologies confirm that history could have been different. After all, during years when water allocations were adjusted, agriculture did not vanish. And as the issue of competing water rights became a salient controversy in peace negotiations, government leaders were forced to think more realistically about Israel's long-term water management. But it remains unclear whether the policy has truly gone full cycle and whether politicians really want to rewrite the narrative. Even as the data continued to pour in about pollution and contamination, the old Zionist water quantity paradigm cast a powerful shadow. In the wake of the bureaucratic clamor that followed the 1996 elections, the link between the Water Commissioner and the Minister of Agriculture was finally severed. But control of Israel's water resources was not passed over to an environmental ministry, whose mission was preservation, but to a new Ministry of Infrastructure, responsible for development.

8 Israel's Urban Environment, 1948–1988

The Politics of Neglect

As part of its obligations to the United Nations after the 1972 Stockholm Conference on the Human Environment, Israel's Parliament held an open plenary session about the country's environmental situation. The summer recess of 1973 was rapidly approaching. It was the end of a golden era for Israel. The trauma of the Yom Kippur War was only a few months away but completely unimaginable. Most of the nation was still intoxicated by the 1967 six-day military triumph and the stunning territorial acquisitions it delivered. Israel's economy continued to soar. Morale was high. Yet at least one government leader was deeply concerned.

Yigael Alon was one of the most eminent heroes of Israel's War of Independence. When Prime Minister Levi Eshkol died of a heart attack in 1969, Alon and Moshe Dayan were the front runners to replace him. (Even though only 3 percent of voters surveyed supported Golda Meir relative to Dayan and Alon, to avoid a power struggle she ultimately became Mapai's consensus compromise candidate for prime minister.[1]) But Yigael Alon remained a formidable figure in the government and was generally considered among the country's most erudite Sabra leaders. Having atended Oxford and Harvard, Alon managed to stay abreast of global trends, despite his immersion in parochial Israeli politics. This included an abiding interest in the environment. His early attitude toward development was representative of the 1950s and 1960s: Build everywhere and preferably horizontally.[2] By 1973, though, he was deeply concerned about pollution and had taken the time to learn about the subject.

As Deputy Prime Minister and Minister of Education, Alon had almost single-handedly pushed the government to confront the pollution problem. It was only natural that he offer the official government view of the country's environmental situation. His presentation to the Knesset was remarkable for its thoroughness and thoughtfulness. The laundry list of environmental hazards presented in his speech reflected the scope and severity of Israel's environmental problems at the time:

- During the two previous decades, Israel's population grew five times and its GNP increased by 11 percent per year, which was the largest relative increase in the world. The population density rose from 43 people per kilometer to 152 people per kilometer.
- Most of the nonperennial rivers in Israel, in particular those near the coast, were polluted by sewage.
- Large quantities of wastes were flowing each year to the Kinneret, including industrial effluents and agricultural runoff from the Galilee's settlements.
- Forty percent of the world's oil was transported across the Mediterranean despite its small size, leading to massive destruction.
- During the preceding decade, chemical production grew by 160 per cent and production of detergents by 100 percent. Israel showed the third highest increase in coal and fuel use in the world.
- During the same period, agricultural land use increased 5 percent while use of nitrate fertilizers grew 40 percent.
- Annual energy consumption was growing at a rate of 10 to 12 percent.
- Solid waste was growing rapidly—in the central Dan region alone, the average quantities increased between 5 and 6 percent annually. Only 15 percent of the trash was disposed of properly from a sanitary and environmental standpoint.[3]

The only thing that Alon skipped over in his remarks to the Parliament was air pollution. In a 1971 national conference, "Humans in a Hostile Environment," twenty-one separate papers were presented on the subject. Conferees reported dismal news: Higher concentrations of ambient sulfur dioxide were already recorded in Israel than in the United States or England, and lead was rapidly approaching the levels found in these industrialized nations;[4] estimates at the time suggested that emissions were increasing at a rate of 7 to 12 percent a year, depending on the pollutant and area of the country.[5]

Israel's environmental crisis had arrived. The trouble was that only a handful of people knew it. The small but dedicated group of civil servants who were trying to do something about it were, like most Israeli pioneers, an unconventional lot. Their efforts were mostly frustrated, at times pitiful, and on occasion inspiring. Frequently they lacked even the most basic data for decision making. Their tiny budgets reflected their government's priorities and the strength of the current against which they invariably swam. Yet without their contribution, Israel's environment would have suffered far greater injury.

PLANNING FOR A NEW NATION

Israel's environmental woes were not due to lack of planning. Rather, they may have been the result of an excessive adherence to a well-conceived master plan that had run its course. The father of the plan was a charismatic architect named Ariyeh Sharon[6] (no relation to Prime Minister Ariel "Arik" Sharon). In August 1948 Sharon began overseeing a team of 150 planning professionals that produced a comprehensive twenty-year blueprint for the new State. It was a gifted group who were trained in Europe but whose wings had been clipped by British snobbery and bias during the Mandate period. Once the State was established, however, these professionals came into their own.[7]

Sharon began his work at the Labor and Construction Ministry and later moved to the all-powerful Prime Minister's Office. He completed the project in two years, and it was published in 1951.[8] During the course of the plan's preparation, Israel's population had doubled, with ad hoc immigrant camps heavily concentrated in the coastal region. Sharon was well aware that the exigencies brought on by this astonishing growth posed a threat to the viability of a long-term master plan.[9] He offered his vision as an antidote to the chaos and improvisation that characterized Israel's first two years. For the next twenty years it was followed fastidiously, even though it did not enjoy binding statutory authority.

The plan was based on a twenty-year projection of a population of 2.65 million people (a tripling of the 1948 population), a prediction that proved prescient. Despite an ideological commitment to agriculture, the plan was pragmatic and assumed that the vast majority of Israelis (77 percent) would remain in cities. Its goal was to settle them in some four hundred new settlements and towns, far away from the coastal strip between Tel Aviv and Haifa, which was home to 82 percent of the local population in 1948.

The next two decades witnessed a remarkable explosion of economic growth, as the Sharon map became a reality. With a clear macrostrategy delineated for physical development, the enormous energy and resolve of the new State could be focused on thousands of simultaneous microchallenges.[10] The results left many landmarks that today appear almost as ancestral parts of Israeli geography: the port in Ashdod, the National Water Carrier, and Beer Sheva, as the capital of the Negev. Israel's borders were dotted with new development towns and agricultural settlements, filled for the most part by immigrants who had little say about where they landed. The early planners' and political leaders' agoraphobic obsession to fill up the country's many empty corners (lest the world take the land away from them) found an outlet in the plan.[11]

Although the transformation fell somewhat short of Sharon's ambitious objectives, the pre-State trends were initially reversed. The percentage of the population in the Tel Aviv district dropped from 43 to 34 percent, while the Negev increased from 0.8 percent to 6.9 percent.[12] The trouble was that the planning establishment quickly became locked into Sharon's notion of dispersed and scattered settlements, which rapid development soon made obsolete. When the plan was updated during the 1960s to meet the needs of a nation with four million people, there were no conceptual adjustments. By then, what had been an underpopulated country in 1948 was already beginning to fill up. The unattractive development towns could not satisfy their inhabitants, and Tel Aviv became the uncontested destination for upwardly mobile Israelis. Planners, however, were not inclined to make Tel Aviv a modern, high-rise metropolitan area, thus setting the stage for today's Dan Region megalopolis that some people call Nashdod (that is, containing the sprawl that runs from Netanya to Ashdod).

In retrospect, the environmental balance sheet Ariyeh Sharon and his team of planners left behind is a mixed one. Their vision opened up peripheral and marginal areas for settlement. By dispersing the country's population, they temporarily eased pressure on the crowded coastal region. In making agricultural production a paramount priority, they actually preserved enormous quantities of open spaces.[13] In addition, the Sharon plan put the first nature reserves on the map and squelched the potential for profligate real estate speculation.[14] But it also set in motion planning paradigms that became problematic as the population doubled again and again. The dissonance between, on the one hand, citizens' desire to live in big cities and, on the other hand, national policies that pushed them to settle in the periphery was ultimately resolved democratically. The people won. But a calculated neglect of urban infrastructures in the three large cities, while their populations spiraled upward, served as a backdrop for many of the nuisances and hazards that came to mar the quality of city life for Israel's urban majority.

ISRAEL'S LITTLE-KNOWN FLING WITH COMPOSTING

As it left the starting blocks in its historic sprint for development, the young State found that it stumbled on a number of environmental hurdles. Solid waste was among the first urban problems encountered. For Tel Aviv residents the problem was particularly acute. When the debris finally settled from the War of Independence, they felt that their garbage could no longer be ignored. The city's municipal dump, located near Mikveh Yisrael School, produced an odor that nauseated its neighbors. One of the new housing projects was built one hundred meters from the site, and residents could not tolerate the flies (because of the warm climate, it was estimated at the time that twenty-six generations of flies could breed during a single year in Israel[15]). Residents filed suit against the city for creating a nuisance. The problem was not unique to Tel Aviv. After ten years of national independence, only three cities (Jerusalem, Bat Yam, and Haifa) bothered to cover their garbage dumps with soil or ash in a sanitary fashion.

The Ministry of Health was asked to find an alternative site. The committee run by sanitation experts Aaron Amrani and Hillel Shuval picked a rural location near a deserted Arab village southeast of the city; this area became known as "Hiriyah." The site was selected because it was far from the city, with no neighbors, and seemed to pose little problem hydrologically. (The proliferation of Ramat Gan and the vulnerability of the permeable Coastal Aquifer were still unanticipated.) The city changed not only the routes of its garbage trucks but also its disposal techniques.

Because of the relatively high proportion of vegetables and fruits in the diet, Israel's trash has always had an unusually high organic content. The average moisture in household garbage during the 1960s was 60 percent during the dry season and 70 percent during winter.[16] Given the technologies of the time, incineration was an unattractive treatment option. On the other hand, the Ministries of Health and Agriculture agreed that composting offered a sensible alternative to imported fertilizers. In the 1950s, only 16 percent of Israel's trash was deemed noncompostable. This meant that the country's refuse had the potential for providing 25 percent of its national fertilizer needs.[17] After Amrani took a fact-finding trip to Europe with the head of Tel Aviv's sanitation department, they returned as composting converts.[18]

In response to the Ministry's exhortations, in 1953 Tel Aviv set up a garbage-screening plant on the old Mikveh Yisrael site. It was the first fully mechanized composting facility of its type in Israel and could process a thousand cubic meters of garbage a day. As soon as this facility proved feasible, three more compost facilities opened in Tel Aviv, and similar plants went into operation in Jerusalem and Ramla. Haifa's Dano Biostabilizers Company began to process sixty tons of garbage, converting one-third of the city's trash into compost.[19]

The pièce de résistance was the 1.2-million-dollar, five-hundred-ton-per-day facility at the new Hiriyah garbage dump site. Today Hiriyah is an incongruous "Mount Trashmore," towering above the coastal plain's middle-class neighborhoods and greeting visitors after they land at Ben-Gurion Airport. Although it stopped receiving garbage in 1998, it remains a symbol of Israel's failure to confront its solid-waste problem during the country's first fifty years.[20] Ironically, it started as a promising ecological innovation;[21] when it opened, it may have been the largest composting production plant in the world.[22] It produced a black, odorless humus that was sold for five dollars a cubic meter. The City of Tel Aviv was also able to pay the Dorr-Oliver company a subsidy of seventy-five cents per ton of raw garbage handled, from the savings accrued because of reductions in trash burial.[23]

The waste conversion plants were not without their problems, and many were quickly closed.[24] Nevertheless, many other facilities ran smoothly. By the end of the 1960s, composting reached its zenith. Sixty percent of Israeli trash was buried (half in sanitary landfills and half in open dumps). The remaining 40 percent was composted as an organic fertilizing agent, used primarily by the citrus industry.[25] Yet as the Israeli synthetic fertilizer industry expanded and dumping prices dropped, the facilities began to close, one by one. Squeezed by inadequate budgets, cities often opted for the cheapest disposal method: burial or burning.

Israel's solid-waste woes were not limited to disposal. Cities were spending as much as 20 percent of their budget on cleaning up. And yet a chronic problem of insufficient garbage receptacles in apartment buildings across the country remained unsolved.[26] Conventional trash cans in public areas could hide bombs and constituted a security threat. Litter became an increasing nuisance, in particular in tourist areas that fell between the cracks of municipal jurisdictions. With Israel's generic litter law lacking any specific enforcement mechanism and no serious manpower to clean up public areas outside of cities, garbage became ubiquitous. For instance, in 1969 the shores of Lake Kinneret were so filled with rubbish that the Nature Reserves Authority was forced to step in and organize a cleanup.

Shlomo Bahalul, who supervised the operations, reported conditions so putrid that many of the men on the cleanup crew vomited.[27]

KANOVICH'S LEGISLATIVE FIX

Air pollution is hardly a new issue. Two thousand years ago, talmudic law stipulated that tanneries be sited at a certain minimum distance and downwind from residential areas.[28] Moses Maimonides, the great rabbi and commentator of the Middle Ages, compared the disparity between urban and desert air to turbid versus clear waters.[29] Yet during Israel's first decade, air quality was largely unaddressed. There were very few cars, so the mobile sources that are so problematic today were not a factor. A few factories, such as Haifa's Nesher Cement plant, however, already posed a hazard. Except for the Ministry of Health's Haifa bureau, which established a public committee to take photographs and meteorological measurements, there is little record of formal government air pollution control activities.[30] This changed after Shimon Kanovich drafted and pushed through an Abatement of Nuisances Law, prohibiting unreasonable air pollution (and unreasonable noise and odors).

Kanovich remains something of an enigma, even today. For twenty-five years he was simply a competent German pediatrician, who, like many of his contemporaries, fled the Nazis. His discomfort with the dirt and disorder of the Middle East was typical of immigrants from his cultural milieu. Previously active in the Zionist movement, Kanovich took a nominal interest in Tel Aviv civic affairs.[31] He was in no way associated with the hikers and biologists at the SPNI, nor was he the product of a grassroots environmental group. Kanovich simply seemed to suffer from pollution.

When he was elected as a representative of the centrist Progressive Party to the Knesset, he decided to do something about it. Encouraged by his fellow faction member Minister of Justice Pinhas Rosen, he drew up a simple law that made pollution a crime.[32] The brief Knesset debate during which he presented his law gives a sense of his motives, charm, and naïveté. After detailing the effect of noise and air pollution on the public's mental health, physical well-being, and appetite, Kanovich responded to a heckler and admitted to being a smoker "and enjoying it very much."[33]

Despite his jocular presentation, Kanovich's actual proposal was extremely intolerant. Noise nuisances were defined as any noise that annoyed three people or more.[34] Under the original bill a citizen could go to jail for three years if convicted of a second offense![35] After substantial revisions (the sentence for violations was lowered to six months' imprisonment) the law passed unanimously.[36] Within a year after the law's passage, Kanovich died of a heart attack, but his law would continue to serve as a catalyst for environmentally concerned citizens.

Israel's police force was asked to enforce the statute. Lacking confidence in their technical ability to address air pollution from stationary sources (factories), they went after the country's fleet of diesel vehicles, whose dirty emissions were easy to spot. Within a few days, the police had issued several thousand fines, primarily to truck drivers and buses who were emitting clearly visible dark black smoke. Not only were the courts not ready for the deluge, but the truck drivers revolted. In 1962 they staged a two-day strike in protest, and the Ministry of Transportation took their side. The police were called off the campaign, and a set of regulations was enacted that took the bite out of the law. The regulations gave air pollution violators forty-eight hours from the time of their citation to tune their vehicle up and meet a new sixty-Hartridge-unit standard.[37]

Alexander Donagi at the Ministry of Health was charged with overseeing air pollution control. By the mid-1960s he was extremely discouraged:

> Despite the good intention, the practical implementation turned the law into a sham. Very quickly the vehicle owners learned to adjust their engines for "the test," and when they left the garage, they readjusted the engine to its previous setting, emitting smoke, in order to gain greater power by burning additional fuel.[38]

Periodically a ministerial committee met to discuss the issue of air emissions, but they retained a policy of leniency toward drivers. Technical committees were formed, but they lacked a clear mandate.[39]

Regarding industry, the Ministry of Health was thrust into the role of enforcer. With only one air pollution officer, their capabilities in this regard were limited. Once again, the local sanitary engineers from the cities were called into service.[40] These personnel had no trouble identifying polluting factories. The problem was providing constructive support to help the factories reduce the pollution at the source.

Unfortunately the most basic regulatory tool was missing: quantitative standards to define maximum ambient concentrations of pollutants. Such criteria set an enforceable ceiling above which pollution is unacceptable. Without objective criteria, planners have no basis for imposing restraints on industrial development, and enforcement becomes a subjective (and losing) proposition. Hillel Oppenheimer, a professor at the Technion, suffered greatly from the poor air quality in his city. He finally decided he had had enough. In 1965 he approached Yitzhak Zamir, a young professor of administrative law at Hebrew University, to represent him. Oppenheimer wanted to demand the promulgation of standards to expedite enforcement of the Kanovich law. Zamir offered his services pro bono and filed a petition in Israel's Supreme Court.[41]

The court agreed, ruling that too many years had gone by without a legally binding definition of unreasonable pollution. The Knesset had clearly intended more expeditious implementation from the Ministers of Health and the Interior.[42] In 1971 the Ministers finally signed ambient air quality standards into law.[43] (Years later, after service as Attorney General, Zamir would sit on Israel's Supreme Court and rely on the Oppenheimer precedent to force promulgation of recycling regulations.[44])

If Kanovich had a successor in the Knesset, it was Josef Tamir (see Figure 20), who served four terms there, from 1965 until 1981. As a member of the Liberal Party, he was affiliated with the right-wing Gahal and later the Likud bloc. During his parliamentary career his primary loyalties seemed to be to the environment, dispelling the notion that ecology was primarily a left-wing issue. As its chair, he changed the name of the Knesset's "Interior Committee" to "Interior and Environment," pushed environment-related legislation, held hearings, and generally served as the environmental community's best friend in the legislature.[45] This eccentricity was tolerated by his peers because of his cordial disposition and obvious sincerity.

Tamir's background was in city government, both as an administrator and later as a Tel Aviv City Council member. He was a great believer in strengthening local environmental controls. The environmental inadequacy of municipalities grew more evident with each passing year. It was easy for the central authorities to heap high expectations upon Israeli cities without offering them the professional or financial tools to solve their problems. Things got steadily worse, and by the 1970s air pollution had reached intolerable levels in the Haifa area. Once again, with inadequate personnel, the Ministry of Health opted to pass the buck to the local governments. Minister Victor Shem-Tov's answer on the Knesset rostrum to a 1971 parliamentary query from Tamir is instructive.

Tamir had appealed to the Minister of Health, who was responsible for preventing a repeat of the air pollution episodes that had recently suffocated the Haifa suburb of Kiriyat Benyamin. Shem-Tov answered evasively, "The solution to the problem of air pollution in the State in general, and in Haifa in particular, will be found when the local authorities get organized, in accordance with their responsibility, to implement all required steps to spare their residents from these nuisances and hazards. The Ministry of Health will lend a hand."[46] And so the Kanovich law became a symbol of legislative lip service rather than an actual tool for controlling air and noise pollution.[47] But with time, this would change.

CIRCUMVENTING THE PLANNING LAW: TEL AVIV'S RENEGADE POWER STATION

No law was more important from an environmental perspective than Israel's Planning and Building Law, which passed the Knesset in July 1965 after twelve years of preparation. It was an original Israeli statute that replaced the 1936 Town Planning Ordinance.[48] By Israeli standards, the law's two hundred provisions were positively prolix. The law institutionalized Israel's three-tier planning system: The National Planning Council determines policy and general zoning for sundry activities, from roads to garbage disposal. The six regional planning committees translate the national plans into specific land allocations and oversee the zoning decisions of the sixty-eight local committees in their jurisdictions. The Ministry of the Interior runs the system, and its Planning Administration drafts the specific national and regional master plans.[49]

Although the word "environment" was not specifically mentioned in the statute, certain innovations had direct implications. For a government bureaucracy that harbored a healthy Eastern European suspicion of public involvement, the law was remarkably democratic. It attempted to make the planning process transparent. The law provides that plans submitted to planning committees be published in newspapers (including an Arabic newspaper for projects that might affect Arab communities); the public is empowered to file formal objections to projects that might affect them. And, finally, the Israeli government had to comply with its own zoning constraints in its many construction projects—or so the law implied.

The establishment of Tel Aviv's Reading D power station was undoubtedly the single most controversial air quality issue of Israel's first thirty years. The extent of the public opposition to the project was more surprising than the ultimate outcome of the debate. The case, for the first time, highlighted air pollution and its effect on the urban quality of life. It also showed the vulnerability of the democratic character of the new planning law.

By the 1960s, the surge in electricity consumption (14 percent per year[50]) surpassed even the growth in the country's population and gross national product. In 1962 Israel's Electric Company announced that Tel Aviv's thirty-six-megawatt British plant, named for Lord Reading, would soon be inadequate. Given existing trends, there would be power shortages by 1970.[51] The company began to plan the expansion of the plant in the sand dunes north of the city, converting it into a five-hundred-megawatt facility.[52] Things had changed, though, since the original Reading power station was built in 1936. The greater Tel Aviv–Jaffa region was now home to 391,000 people. Neighborhoods had already crossed the Yarkon River in the north and approached the plant.[53] The Reading Station was perceived as the single greatest air pollution hazard in the city.[54]

Like all of the regional committees, Tel Aviv's Regional Planning and Building Committee was primarily composed of government representatives. Yet they were attuned to public sentiment and were willing to flex the new muscles given them under the Planning and Building Law. Decisions were supposed to be made on the basis of their best professional judgment rather than on political dictates from above. On January 16, 1967, the Electric Company formally submitted its plans, but not before beginning work on the foundations of the new plant.[55] It took less than two months for the committee to reject the building permit request. There would be no approval until the Electric Company could deliver the blessings of the Ministry of Health, verifying the absence of additional public-health risks. Josef Tamir convened hearings on the issue in the Knesset's Public Services Committee and discovered broad-based opposition to the new Reading plant.[56]

The Ministry of Development, responsible for energy infrastructure, did the bidding of the Electric Company. After all, the government owned the utility outright, and the Ministry was responsible for guaranteeing a reliable energy supply. Along with lawyers at the Ministry of Justice, it drew up a one-page law designed to streamline construction and avoid the irritating hurdle posed by the Tel Aviv Regional Committee. The law stated that "the Government was authorized, according to a proposal by the Israel Electric Company, to approve a building plan and the operation of an electric power station in the Tel Aviv area. . ."[57] Once approved, the plant would not be subject to Israel's Planning and Building Law.[58]

The proposed law raised a hue and cry among the public, and two hundred thousand people signed a petition opposing the Tel Aviv site for the plant.[59] The government argued that the facility would be cleaner than the existing Reading A plant (which it would close down) and committed itself to meeting any standards set by the Ministry of Health. In particular, great stock was placed in tall stacks that could send the plume far beyond Greater Tel Aviv.

The Knesset deliberations surrounding the proposal were not only the longest environmental debate in Israeli parliamentary history but also the most heated. No fewer than seventy-one recorded hecklers, from every political party, interrupted speeches to protest the project. There were really only two objections raised by the speakers: The government was selling out the health of Tel Aviv's public for the extra eighteen million lirot it would cost to build the power plant in an alternative southern site, far from residential areas.[60] Furthermore, the government was undermining the democratic character of the planning process that was supposed to "make citizens' rights equal to those of the government in matters concerning planning and building in this land."[61]

The opposition tried to keep the matter apolitical and avoid a factional vote. As maverick parliamentarian Uri Avneri quipped, "This is not a political matter. . . . Does the nose belong to a particular faction? Does the Labor Party have a different kind of nose than the Gahal (Likud) faction?" Menahem Begin interjected that "the nose" was not at issue but "the lungs," and Avneri responded that lungs should be nonpolitical as well. But it was to no avail. Considering that the broad-based national unity government, which was formed in the wake of the Six-Day War, was still in power, it was a surprisingly close vote. (An across-the-board unity government can usually expect to pass legislation at will.) The bill passed, 32–28.[62] It raced through committee, and just six days later, on August 8, 1967, became law. Two years later, Reading D was up and running.[63]

In response to the government's commitment to abide by any conditions it stipulated, the Ministry of Health invited three international experts to visit Israel and offer their opinion. The "Brasser Report" was named for the World Health Organization expert who chaired the team. The recommendations of the group rejected tall stacks as a long-term panacea. Rather, the experts preferred to reduce sulfur dioxide by burning low-sulfur fuel. This grade of fuel was still not available in Israel, so a tall stack and sulfur dioxide removal equipment for the flue gas were recommended in the interim.[64] The government's commitment to Tel Aviv's environmental health, however, was predictably ephemeral. The Ministry of Development countered by inviting a report from an opposing expert committee, headed by a former president of the American Electric Power Company. The authors of the "Sporn Report" were confident that 150-

meter-high stacks would alleviate any problems and recommended that low-sulfur fuel be kept on hand only for emergencies.[65]

Ultimately the Ministries of the Interior and Health set five conditions for plant operation, including a 150-meter smokestack and the closing of the Reading A plant.[66] The case set a lax technological precedent for the future. It would take decades before low-sulfur fuel was burned in Tel Aviv. During the 1970s, "scrubbers" became a proven technology for reducing sulfur dioxide emissions by 99 percent. In 1989 Israel Electric Company officials were happy to wait another decade or two before investing in them.[67]

The impact of the plant on Tel Aviv is a subject of debate. In his memoirs, Josef Tamir calls the Reading smokestack "a symbol of Tel Aviv's disgrace," charging that three tons of soot are released every day.[68] The sulfur dioxide emissions were probably higher, reaching 130 tons per day. Yet at the same time, ambient sulfur dioxide levels in Tel Aviv remained far below national standards.[69] From a narrow, local perspective, the "solution to pollution" *is* dilution: In 1969 Israel had no compunction about taking advantage of the Mediterranean's western sea breezes and exporting its sulfur dioxide emissions eastward into Jordan and the West Bank.

The real fallout from the Reading Power Station was political. Israel's planning process did not seem quite as robust and democratic when it was so easily circumvented. The chimneys of the Tel Aviv Power Plant Law still cast a traumatic shadow. In 1991, for example, environmental leaders would think twice before supporting a proposed law to cancel the Jordan River hydroelectric facility, notwithstanding the plant having received the required approvals of the planning committees (see Chapter 5). Such a law might have produced an immediate environmental victory, but in the long term, site-specific legislation could be a double-edged sword.

The second political lesson was no less discouraging. The Israeli public and a fair number of politicians had finally been roused and had put together a serious fight to stop a polluting facility in the heart of Tel Aviv. An alternative site was available. But their voice was lost amidst the government's insatiable roar for the fastest, least expensive development possible.

The unsung hero in the story was the Tel Aviv Regional Planning Committee, whose members showed unusual ecological sensitivity and fortitude under considerable government pressure. Much more typically, planning committees either ignored environmental concerns entirely or glossed over them.

One of hundreds of examples was the 1973 request by the Frutarom Company for a building permit in Haifa Bay. To this day Frutarom makes

PVC plastic, whose production involves a host of pernicious by-products, including vinyl chloride, a virulent liver carcinogen. The company has always been perceived as environmentally lackadaisical and for years was linked to a number of pollution problems in Haifa Bay.[70] The approval of the company's request was sufficiently outrageous to jar the usually staid Ministry of Health from its typical complacency. Citing mercury levels in nearby fish that were four times the accepted standards, Dr. Eliezer Matan called the decision "scandalous." He charged that any committee that would approve a request from a polluter known to be dangerous, in contravention of three professional opinions, was "beneath contempt."[71]

Through most of the 1950s and 1960s, Israel's public health was probably not inordinately affected by high pollution concentrations. Although there were modest increases in cancer incidence during the 1960s, and women had relatively high levels of lung cancer,[72] these could be attributed to high smoking rates. Still, the first, isolated signs of environmental health impacts on the population were starting to surface. There was much coughing in Haifa, dysentery was common nationwide, cities were noisy and dirty. Much like Casey Jones's proverbial train, they signaled trouble ahead. Government agencies were well aware of the long-term implications of pollution. Yet planning committees, the Water Commissioners, local governments, and even the Ministry of Health's sanitation department were helpless to slow the pace of the degradation.

AN ENVIRONMENTAL AGENCY SLIPS THROUGH

Despite her self-deprecating descriptions, Prime Minister Golda Meir was much more than just another chain-smoking grandmother. Her wide range of interests, however, did not include the natural world and ecology. (The only real reference her autobiography makes to the natural world involves childhood fears that associated the wetlands around her home in Pinsk with the coming of Cossacks.[73]) When Josef Tamir came to speak with her about pollution and the importance of establishing a government ministry to address it, she listened politely. Then she dismissively directed him to speak with the Director General of her office.

In March 1973, Meir flew off to America to meet with President Nixon. It provided a rare window of environmental opportunity. During the Prime Minister's stay in Washington, Deputy Prime Minister and Minister of Education Yigael Alon chaired cabinet meetings. This was during a time of rising global environmental consciousness. A year earlier, Israel had sent a high-level, twelve-person delegation to Stockholm for the

United Nations' historic Conference on the Human Environment.[74] The June 1972 meeting of 113 nations served as a turning point for international environmental cooperation and led to the founding of the United Nations Environment Programme (UNEP). Although Foreign Minister Abba Eban received the headlines in the Israeli press, some of the participants mostly remember him popping into plenary sessions to hear a few minutes of discussion before running back to his hotel to continue dictating a book to his secretary. But when he made his presentation, Israel's most facile English orator did not disappoint.[75] In his speech to the delegates Eban eloquently voiced Israel's concern that humanity now existed "on water and soil polluted by poisons." Even though the Conference preceded the Arab oil boycott by sixteen months, he presciently emphasized dwindling petroleum resources.[76] Stockholm also served as an international ecological debut for Uri Marinov, a young environmentalist. The delegation relied on him to represent Israel in the conference's smaller, more technical discussions.

Marinov was by then a rising star who had hitched his professional fate to environmentalism. Like many idealistic young Israelis, for his military service he joined the Nahal—the Pioneer Fighting Youth Corps, which mixed military service with the establishment of kibbutzim in sensitive border regions. After high school, Marinov and his friends founded Kibbutz Nahal Oz near the Gaza Strip, where he remained as a member for six years.[77]

In 1956 a U.S. news crew from CBS television came to Israel. Reporter Edward R. Murrow was doing a portrait of Israel, while his colleague Howard K. Smith was profiling Egypt. Morrow was captivated by his young host and ultimately joined him at the top of the kibbutz watchtower for a prolonged interview. A national American television audience watched the apparently guileless Marinov steal the show, talking about his dreams of living in peace with his neighbors. It looked very good alongside young Egyptians who called for "throwing Israel into the sea." A star was born. The United Jewish Appeal drafted Marinov for a U.S. speaking tour, and he appeared in follow-up television interviews after the Sinai Campaign. Later he would spend several years in the States and complete a degree in veterinary medicine and master's training at Iowa State University, with the idea of returning to the kibbutz.[78]

But it was hard to keep the ambitious Marinov down on the farm. Upon returning from his studies he took a position as a physiology researcher at the Hadassah Medical School in Jerusalem. It did not take him long to realize that he might never amount to much as a scientist.

Marinov began to look elsewhere, taking a job as head of the life sciences branch of the government's National Council for Research and Development, which preceded the Ministry of Science. The environment was still not on his agenda.

When Hillel Shuval (by then a Hebrew University environmental science professor) headed a water research committee for the council, Marinov began to sense an opportunity. Marinov expanded the branch's environmental research agenda and started an air quality committee.[79] In 1970 he began circulating a newsletter, *The Biosphera*. For twenty-seven years it would be Israel's only continuous publication about environmental affairs.

In 1971 Yigael Alon began to explore a government response to Israel's pollution problem. Marinov had a head start on the issue. So Alon brought Marinov's operation together with representatives of the government, the JNF, the NRA, and the SPNI to create a National Biospheric and Environmental Quality Committee (called by its Hebrew acronym VIBAS).[80] Marinov was to coordinate ten subcommittees and their international representation.[81] Marinov had his ticket to Stockholm.

The thirty-member committee was a valuable academic forum, and VIBAS published some of the most trenchant, media-specific environmental reports ever issued in Israel. The trouble was the absence of operational focus or function.[82] After receiving a report about the severity of the pollution in Lake Kinneret, Alon created an additional forum composed of Directors General from government ministries.[83] The forum was to meet regularly and coordinate environmental policy. It was headed by Tsvi Turlow, the brilliant but impulsive Director General of the Ministry of Justice. Its members were very busy people whose primary loyalties lay with their respective ministries' other priorities; it was not an effective framework for implementing environmental policy.

Inspired by the spirit of the historic 1972 UN environmental conference, from which he had recently returned, Marinov lobbied to promote an ambitious regulatory vision. But he found considerable resistance to a full-blown environmental ministry. Ministries with vested interests in related areas, such as the Ministry of the Interior, openly expressed territorial anxiety.[84] Surprisingly, SPNI General Secretary Azariah Alon also opposed full ministerial status. He feared that a Ministry of the Environment would quickly become marginalized. Instead, he proposed an Environmental Commission.[85] Not discouraged, Marinov drafted and circulated a proposal for an Environmental Protection Authority that would primarily serve in an advisory capacity.[86]

Yigael Alon was supportive. When Maurice Strong, the Director of the newly formed United Nations Environment Programme, visited Israel, Alon met with him and assured him that Israel was going to take the environment seriously; he lobbied Strong to make contamination of the Mediterranean an international priority.[87] Soon thereafter, Alon asked Josef Tamir what he thought of the Environmental Protection Authority idea and who he thought should head it. Alon had narrowed his list of candidates down to two: Avraham Yoffe and Uri Marinov. Tamir felt that General Yoffe was not cut out for the kind of bureaucratic minutiae associated with modern environmental regulation and supported the young veterinarian.[88]

And so it came to be that on March 20, 1973, while Golda Meir was away charming her American hosts, Yigael Alon snuck Marinov's proposal past her and got Cabinet approval for an Environmental Protection Service.[89] Despite a few snickers from peers at his enthusiasm, Tamir had little problem in galvanizing support for the Service in the Knesset even before Meir returned from the United States.[90]

Alon called Marinov into his office to inform him of his future. Marinov knew that the SPNI had made inquiries about the new bureaucracy with the idea of sending one of its cronies to run it. Typically blunt, Alon told him that he had looked all over for a better candidate but could not find anyone appropriate.[91] With three years of experience, Marinov may well have been the most qualified person in the country for the job. He was certainly the most energetic. It would take all of Marinov's considerable stamina, cunning, and obstinacy to complete the tedious, two-decade-long process of building an environmental bureaucracy.

THE UPSTARTS

The Environmental Protection Service (EPS) got off to a slow start. Based as a small department in the Prime Minister's Office, it had an amorphous mandate to coordinate government activity between the different ministries and to act as an adviser to the government. But Israel's bureaucracy was a scattered, cacophonous orchestra that saw no need for a conductor. If this were not bad enough, the 1973 Yom Kippur War neutralized the country for six months. With the budget cuts required after the war, the projected staff of eighteen was cut in half.[92] Later, Marinov could smile recalling how he wangled an extra employee out of the Ministry of Finance by not counting himself, already a paid civil servant.[93] But the new agency's survival was hardly assured.

The Prime Minister's Office was distracted by the political vicissitudes and controversies surrounding the less than triumphant military results. Its top managers seemed to have forgotten about their new department, and in April 1974 Marinov began to take advantage of the situation. Rather than publish formal job descriptions, he looked around for the kind of people he wanted in the EPS. If someone qualified turned up, especially with environmental training from abroad, he built a job around the person. This ad hoc approach created a multidisciplinary style and an international orientation that became EPS trademarks.[94]

Another defining characteristic at the EPS was its declared emphasis on prevention. Environmental planning was considered the most promising way to avoid future follies. The strategy was to penetrate the planning system from within, rather than to assume an external adversarial posture.[95] So in addition to hiring scientific experts, Marinov began seeking out advisers to the planning committees, often sneaking them in as consultants when he could not offer them formal civil service status.

The original EPS workers fondly remember the esprit de corps at the new agency. They did not seem to mind the indignities that were epitomized by the floor drain in Marinov's office, which he covered with a rug.[96] "All we had going for us was our brains and our common desire to clean up the environment, working together," recalls Richard Laster, the EPS's first attorney. They managed to have fun. Most of the original staff retain pleasant memories of piling on minibuses for field trips to see the ecological reality beyond the walls of their Jerusalem offices (see Figure 21). Despite the protestations of his office manager, Marinov purposely would set a route that brought the staff home late, having found that meaningful discussions never really started until darkness settled in.[97]

As he became more influenced by the culture of 1970s environmentalism, Marinov struck an unlikely figure as a government manager. He preferred to take the steps rather than the elevator, to save energy. Even when his status later entitled him to a large government Volvo, he bewildered his driver by insisting on an economy car. Most of all, Marinov read up and became knowledgeable about a range of issues. He was even invited to serve as an international consultant on environmental planning issues. This enhanced his status among his workers as not just a boss but a zealous environmental leader.

Morale was high, but there was also a feeling of frustration. The formal mandate granted the Environmental Protection Service by the government was in fact extremely narrow. At the same time, no agency was more aware of the severity of Israel's environmental problems or more ambitious about solving them. Soon after the Service set up shop, Marinov called a meeting of all ministries involved in environmental activities. They listened to his suggestions, but not one representative showed up for the next meeting he scheduled.[98] The Ministry of Health was already involved in issues of air, water quality, and sanitation. Justifiably, its staff members were not enthusiastic about the perceived duplication when their own ranks were so thin. They did little to make it easy for the newcomers.

Marinov's original strategy for the Service was a highly idealistic one. Because the environment is affected by the activities of almost any government ministry, the EPS was to assist each agency in integrating its own appropriate environmental policy. Israel's highly territorial government culture, however, was not accommodating and made a mockery out of the EPS's well-meaning efforts. Marinov's frustration is reflected in his 1975 letter to the EPS newsletter, *The Biosphera*: "As an advisory body, the Service finds itself advising ministries that aren't interested in getting advice."[99]

One of the few allies that Marinov could find in the government was Haim Kuberski, who had been Director General of the Ministry of the Interior for some time. Rising through the ranks of the National Religious political apparatus and the Ministry of Education, he became the power behind party chairman Dr. Yosef Burg. Kuberski had attended the Stockholm conference and had an appreciation for ecological issues. Afterward he continued to run the Director General's committee on the environment. When Marinov suggested that the EPS leave the Prime Minister's office and move to Interior, Kuberski was receptive. In the fall of 1975 the government approved the move.[100]

Once the talented but motley crew arrived on the third floor at the Ministry of the Interior headquarters, it was clear that they had little in common with their hosts. In those days, the Ministry of the Interior was a haven for politicos and cronies from the National Religious Party. Its staff held a well-deserved reputation for having the most unimaginative clerks in Israel's civil service. Their work ethic was not stalwart. The EPS professionals were almost uniformly secular, openly enthusiastic, each with two or three university degrees, and indifferent to government office closing times. The Service shared a floor with the projection room of Israel's censorship bureau. In late afternoons, the religious elders came to view, maybe enjoy, and then slash the most prurient sections of the 1970s' raciest films. They were surprised to find the EPS staff still working.

The cultural mismatch was epitomized in the decision by the EPS staffers to paint the hallway once they moved into their new home. They received free advice (although they paid for the paint) from the Tamboor Paint Company and insisted on doing the work themselves. The Ministry of the Interior workers would walk through and shake their heads at the staff's peculiar behavior, so unbecoming of government bureaucrats, not to mention the melange of colors they were using. Attorney Laster saw symbolism in the act. "Painting the hallway, like the idea for traveling the country in a minibus, was to stay the power of bureaucratic arteriosclerosis. It enabled us to keep a young image and fight off the bureaucratic aging process."[101]

The contrasting mentalities between the environmentalists and their hosts at the Ministry of the Interior never really grew closer during the thirteen years that the EPS stayed in the painted hallway. Looking back, Marinov remembers the late Director General Kuberski with great fondness, giving him credit for many of the EPS's achievements.[102] If he deserves such credit, it did not come easily. All of the EPS workers can remember the constant shouting between the two as the upstart crusader pushed his establishment boss another step further. "Any environmental document has to sit in Kuberski's office for a good few weeks to 'get used to the temperature on the first floor' before he'll take a look at it," was a common sentiment of workers. But Kuberski cared about Israel's environment and gave the Service the foot in the door it so desperately sought in the realm of physical planning.

ENVIRONMENTAL PLANNING

Marinov joined the National Planning Council, which set the strategies for so much of the country's physical and economic development. Although technically he held only advisory status, he immediately set out to utilize the forum to promote environmental interests. Eventually he found funds to pay environmental advisers to counsel the six Regional Planning Committees. Although the EPS had talked a great deal about environmental impact statements and even proposed a pilot framework, none had actually ever been prepared.

When the "post-Reading" generation of power plants came up for discussions, it was decided to locate the next one in the Hadera coastal region. Marinov had his test case. The EPS took a moderate position, which was to be the trademark of Marinov's environmental management strategy. Sensing that direct confrontation would produce few environmental victories, for tactical reasons his EPS rarely opposed development outright.[103] Marinov's tendency to accommodate industry and his willingness to compromise up front were the subject of criticism both throughout his public service career in the 1970s and 1980s and especially during the 1990s as a somewhat nondiscriminating consultant for the private sector. Defenders of this pragmatism cite the overwhelming absence of environmental consciousness or precedent in Israel at the time. With no real statutory authorities, the young environmentalists could rely only on moral suasion and Marinov's own powers of persuasion.[104] Accordingly, the EPS did not object to the establishment of a coal-fired plant in Israel, arguing that it could be built with sound environmental controls.[105] At the same time, Marinov demanded that an environmental impact statement (EIS) accompany the Israel Electric Company's proposal.

The decision to require an EIS in this case proved to be a crucial precedent. Marinov fought to bring the deliberations from the Regional Planning Committee to the National Council, where he had more allies and leverage. (The previous power station built in Ashdod had been approved by a regional committee, on the basis of a decision that was only two pages long, containing only half a page of environmental provisions.) Predictably the Electric Company lobbied against the venue, arguing that the National Council was not a professional body, having members who represented youth and women's organizations.[106] Yet Kuberski, who chaired the Council, backed the environmental demands, and the utility acquiesced. In May 1976 the impact statement was submitted to the EPS for approval. It was based on a questionnaire drafted by the EPS staff, and it showed that the site proposed by the Electric Company was actually less appropriate than the alternative site next to the Hadera Stream, which was favored by ecologists.[107]

Six years later, the Minister of the Interior signed formal regulations into force that required environmental impact statements for a variety of large projects. The regulations also empowered planning committees or ministry representatives who sat on them to request an EIS for a project.[108] In theory, if for any reason impact statements deviated from the guidelines set by the Ministry, the planning authorities were duty-bound to suspend discussions concerning project approval. By 1985 the EPS planning department could boast forty separate impact statements that it was either reviewing or whose preparation it was overseeing.[109] By 1997 the regulations had given rise to 236 such statements.[110]

The Israeli environmental impact statement was by no means state-of-the-art. It did not require alternative analysis (including a no-action option), a crucial tool for evaluating the actual benefits of a project. It ignored indirect or social impacts. It left the responsibility for drafting the document in the developers' hands, occasioning an optimistic bias in Israeli environmental evaluations. Most of all, many problematic types of projects, such as superhighways, were not on the mandatory-EIS list. Indeed, only a small fraction of the seven thousand plans going through Israel's planning system each year are evaluated via impact statements.[111] Yet the framework was a marked improvement, and in many important cases impact statements succeeded in getting planning agencies (and developers) to stop, think, and sometimes modify building plans.

A SOLID-WASTE MASTER PLAN DIES FROM DELAY

Because of the National Planning Council's ostensible openness to environmental ideas, the EPS focused much of its energies on effecting change through national outline schemes and planning policies.[112] This hardly guaranteed success. For instance, solid waste and recycling had been picked by Marinov as a flagship issue, perhaps because these areas seemed a niche that the environmental service could easily occupy. A steering committee chaired by the Chief Engineer of the Ministry of Health, who was responsible for overseeing solid waste, oversaw the preparation of National Master Plan (TAMA) No. 16 for Disposal of Garbage. The forum was dominated by Green members. The EPS expert, Dr. Mordechai Lapidot, joined the other Interior representatives as well as the SPNI's Azariah Alon on the committee.[113] The master plan drafted reflected their environmental thinking.

Between 1973 and 1976 the steering committee gathered information. Their research showed that each Israeli produced a kilogram of trash every day. Israel's garbage continued to be very "wet" even during the summer months. Roughly 50 percent of the waste stream was in this biodegradable group; paper made up less than 20 percent of the trash, synthetic plastics 13 percent, and metals and glass together no more than 6 percent.[114]

The waste management strategy selected by the committee involved consolidating the hundreds of existing dumps into two national landfill sites and thirty smaller disposal areas. These would be fed by a series of transfer stations in local municipalities. Burial was in no way envisioned as the preferred treatment strategy. Rather, the plan was based on integrated waste management, which included recycling and incinerating materials after they had been sorted by sanitation workers.[115]

Drafting a master plan and implementing it were two totally different matters. Only in January 1986, ten years later, did the National Planning and Building Council approve Master Plan No. 16. For it to become legally binding, the approval of the Cabinet was required. This stage dragged on until 1989. The sixteen years that elapsed from the time information gathering was begun to the day that the Cabinet approved the plan undermined the relevance of the strategy.[116] For example, when the original waste inventory took place, disposable diapers were unavailable. During the years that followed, the recycling and reuse sector had gone from bad to worse. With the exception of paper (about a quarter of which was recycled at the American-Israeli Hadera Paper Factories), everything was buried. During the 1970s, practically all beverage containers had been returnable, with consumers rewarded with the return of their deposit fee. By the 1980s, plastic ruled, and only a small fraction of wine and beer bottles were accepted for reuse or recycling. Israel's solid-waste habits and the actual burial sites in no way corresponded to the plan that eventually was approved by the government.

With no legal basis for pressuring municipalities to take an environmental high road, these all found the cheapest possible way to bury their trash. (Only the small Haifa suburb of Tivon had a policy of separation and recycling, and that was due to an idiosyncratic mayor.) The result was a nationwide pathology of foul smells, groundwater pollution, and seasonal fires (spontaneous summer combustion as well as premeditated blazes). The conflagrations gave rise to horrible odors and emissions of deadly dioxins.[117] The Hiriyah garbage dump was designated only as an emergency site under the Master Plan, but by the 1970s it assumed the role of a dubious Tel Aviv landmark. The birds attracted to the garbage threatened the safety of flights landing at Ben-Gurion Airport.[118] The mountain of trash threatened to topple onto the adjacent highways, which it eventually did during the winter rains of 1997.[119] Recycling and composting had missed the boat.

Disposal of hazardous waste was another part of the national strategy that got lost.[120] When the steering committee began its work, there was no facility anywhere in the country to receive toxic residues. It was assumed that these would be either buried in regular landfills or discharged into the sewage system. The results of thirty years of unaccounted dumping are still largely unknown, and construction crews increasingly stumble onto unpleasant toxic surprises buried in the ground.[121] Orphan sites in which hazardous chemicals were buried during those years periodically make an appearance. One prominent example of the phenomenon is a site in Beer Sheva north of the Machteshim chemical factory where the Dead Sea Bromine factory buried its wastes before it moved to the Ramat Hovav industrial complex.[122] For more than a decade, no one has been willing to pay for the cleanup of the abandoned acid and waste pits. It remains cordoned off, an eyesore that blocks development of an important area of the future metropolis and endangers train and sewage routes.[123,124] The impact on the groundwater will probably not be benign.

Initially it seemed encouraging that a response to an impending hazardous-waste crisis would not have to wait for the formal approval resolution of the entire solid-waste strategy. In November 1979 the Ramat Hovav hazardous-waste disposal facility opened its doors to receive the country's toxic residuals. The site, located twelve kilometers south of Beer Sheva, was selected because of its isolation, favorable wind direction, and apparent freedom from hydrological vulnerability.[125] It was part of an industrial complex to which most of Beer Sheva's most-polluting industries were moved. The concept, promoted by the government planners, was to establish Ramat Hovav as a national treatment center for neutralizing and recycling chemicals prior to burial. The complex was run by a private corporation, and there were no clear laws defining acceptable treatment processes. It quickly became a dumping ground for a perilous cocktail of poisons.

The place was run negligently. Wastes often did not undergo pretreatment before arriving. Storage was improper, and toxic residues frequently went unlabeled. There was no serious monitoring of chemical composition. Barrels grew rusty, and reactive materials were often stored together near cyanide deposits. It was a time bomb waiting to go off, and it did. On April 2, 1982, the site caught fire, and a toxic plume wafted over the city of Beer Sheva. The Ministry of Health immediately sought and received a court order to close the Ramat Hovav facility.[126]

An interministerial committee went back to the drawing board. Running a hazardous facility properly is not a profitable enterprise unless high prices are charged for disposal. It was therefore decided to vest the operation of the facility in the hands of a government corporation. To this end, the Mivneh Tasiyah (Industrial Building) Company was established in 1984 as a government subsidiary[127] and charged with constructing a neutralization facility and running the plant. The government was less expeditious about a six-million-dollar investment in a hazardous-waste treatment plant.[128] Finally in 1986 work began on such a facility, using French technology.

In 1987 twelve thousand tons of toxic materials arrived at the newly upgraded facility for disposal. By 1988 the amount had reached forty-four thousand tons, but it was thought to be only half the actual amount of the hazardous wastes generated in Israel.[129] The problem was partly legal. Until 1990 there was no law requiring that hazardous wastes be sent to Ramat Hovav at all.[130] After such a regulation was passed, management again proved shoddy. A 1991 government committee branded Ramat Hovav's operation as criminally negligent.[131] Throughout the entire Ramat Hovav debacle, the Environmental Protection Service tried to play the role of watchdog. But its bark was feeble indeed as Israel stumbled along a never-ending series of toxic blunders.[132]

FROM PLANNING TO ACTION

Without a national network of local staff members, the Environmental Protection Service, like everyone else, turned to the municipalities. To its credit, some financial assistance was part of the EPS package. In 1976 the EPS announced its intention to start fifteen environmental protection units in cities with populations greater than eighty thousand people.[133] After an initial year of full subsidy, half of the local environmental officials' salaries were paid by the municipal government and half by the Ministry of the Interior.[134] Persuading towns to join the system was not always easy.[135] Amram Pruginin, an ebullient environmental geographer, was brought on board to oversee the initiative. As the Ministry of the Interior controls the municipal budgets in Israel, larger cities proved receptive.

The local environmental coordinators were supposed to be the EPS's "soldiers in the field." They would answer the complaints, measure the noise, run the educational programs in the schools, and, of course, represent the environmental interest before local planning committees. In the hands of the right person, a local unit's activities translated into impressive environmental progress.

For others, environmental regulation proved a bit too ambitious. Many units could not field staff with the required technical qualifications. In addition, municipal employees knew who their boss was and avoided taking on the mayor about contentious issues surrounding pollution. Over time, the units became an entrenched part of the city hall bureaucracy. Exigency soon became ideology. Even after the Ministry of the Environment was created and its independent regional offices were available to replace the local units, the central government continued to subsidize them.

The rationale was simple. Most environmental nuisances were local in nature. Throughout his long career as a public servant, Marinov would advocate the view that a central government had no business overseeing emissions from a neighborhood bakery or noise from a particularly raucous discotheque. Just as city government provides education, welfare, or even health care, it was the appropriate level of government to provide environmental services. Citizens should not have to travel to distant towns to solve their immediate environmental problems, the argument went.[136]

While many minor nuisances received attention, the meager professional qualifications of available local staff hindered many of the units' ability to tackle the most troublesome issues. Often, a pollution source was located beyond the geographic boundaries of an affected municipality, or the technical issues involved exceeded local enforcement capabilities. It was clear that the units were not a panacea. In 1978 Marinov called enforcement "the weak link in the wide chain of environmental management."[137] The local soldiers were no substitute for a centralized, independent, regulatory presence.

Two developments during the 1980s allowed the EPS to depart from its institutional personality of adviser and assume the role of "regulator." The first was the establishment of a marine pollution prevention unit. The second was a series of emissions controls imposed on major air polluters in the Haifa region. Although neither effort solved all of the related problems, they confirmed Marinov's analysis and showed what could happen when the enforcement link was tightened.

PROTECTING THE MEDITERRANEAN

By the end of the 1970s both of Israel's seas were in trouble. When UNEP's new Regional Seas Program picked the Mediterranean as its first target for international cooperation in 1974, it was fortuitous. From an *environmental* perspective, it is hard to think of a more deserving candidate. Moreover, *politically* the Mediterranean had the additional advantage of bridging the concerns of first-world European countries with those of developing nations in North Africa, who held equal stakes in an imperiled resource. Israel, of course, was right in the middle.

A small, enclosed body of water, the Mediterranean Sea requires close to one hundred years for its waters to be fully circulated and replenished.[138] With 460 million tons of crude oil crossing the Sea each year and 80 percent of the sewage from coastal regions flowing into it with little or no treatment, conditions were appalling.[139] Release of ballast waters posed a chronic problem. (Ballasts are used for washing the cargo tanks of oil tankers and providing additional weight to empty vessels in transit so that they can sit at an optimal depth in the water.) It was not just fish that suffered from Mediterranean oil pollution. During the 1960s and 1970s the stains on Israeli beaches from the tarry petroleum residues reached disgusting levels. After swimming, bathers had to scour the soles of their feet in kerosene to remove the gooey black gobs of petroleum wastes.

When the United Nations convened a conference in Barcelona in 1975, all of the Mediterranean nations sent delegates, with the exception of reclusive Albania. The resulting convention and its five operational protocols created a Mediterranean Action Program[140] and one of the first coordinated regional environmental regulatory efforts.[141] During a period of unprecedented diplomatic isolation, Israel was delighted that its participation was not only tolerated but welcome. The Barcelona convention protocols called on the parties to enact stringent laws banning oil releases and limiting dumping and land-based sewage discharges; the Israeli government was amenable.[142]

Passing statutes in and of itself was not enough. Like Kanovich's Abatement of Nuisances Law, however, the UN-mandated laws remained a symbol of well-meaning legislation that floundered, without a commitment to implementation. As is so often the case, the key to implementation was money. Marinov teamed up with Adir Shapira, who had replaced General Yoffe at the helm of the Nature Reserves Authority, to raise the funds. It made sense to begin with the Gulf of Aqaba at Eilat, where the aquatic ecosystem was especially vulnerable and damage to the coral reef was already evident.

During the 1970s, when the Suez Canal was closed, Iranian crude oil heading for Europe was delivered to Eilat and piped north to Israel's Mediterranean ports.[143] The Eilat-Ashkelon Oil Pipeline Company (or "Katza," according to the Hebrew acronym) continued to operate even after the fall of the Shah and the loss of the Iranian connection. (Israel is the only country to import oil via the Gulf—the city of Aqaba receives its Iraqi fuel by truck.) The Katza oil jetties are located directly north of the Coral Beach Nature Reserve and loom as a potential menace to the sensitive aquatic systems. They receive the 2.5 million tons of oil from the Sinai's Abu Rodeis oil fields that Egypt sells Israel each year under the 1978 Camp David agreements.[144] It probably was not a pure sense of civic duty, but rather some subtle regulatory arm twisting, that motivated Katza to assent to Marinov and Shapira's request to bankroll a marine protection unit to be run by NRA rangers. Work began, but the team lacked full legal authority.

When Israel decided to ratify "Marpol," an international marine treaty, as well as the Barcelona oil pollution protocol in 1979,[145] it required a major revision in the 1936 Oil in Navigable Waters Ordinance.[146] Again, the big question was money. Funding mechanisms were left out of the original draft of the amendments. The all-powerful Ministry of Finance was always stingy when it came to environmental matters. It was also zealous about controlling tax revenues, so the very idea of a specially designated fund for marine protection would have been laughed out of their dim corridors. But to the delight of environmentalists, when a new oil pollution prevention law passed in 1980, it contained an independent Marine Pollution Prevention Fund, paid for by shipping fees and fines.[147] Seemingly miraculously the provision had slipped past the watchful eye of the government economists. Actually this development was the result of well-timed legislative manipulations.[148]

When Yuval Cohen returned to Israel after completing his doctorate in marine biology abroad, he came to Marinov to consult with him about employment opportunities. Marinov recognized Cohen's energy and managerial potential and suggested that he oversee the EPS's marine protection efforts. Dissatisfied with the latitude the Nature Reserves Authority gave its rangers, in July 1983 Cohen brought the operation "in house" and opened a Marine Pollution Prevention Department.[149] Suddenly the EPS had a clear regulatory mandate and the personnel to carry it out.

The results were dramatic. Under the 1983 regulations, the Port Authority collected surcharges from all ships calling at Israeli ports and oil terminals and sent them to the Marine Pollution Prevention Fund.[150] In its first year the fees generated $420,000 for the Fund, an astronomical sum by the very modest standards of the EPS.[151] Revenues steadily grew as the Fund doubled in size.[152] The money made it possible to put Jerusalem attorney Yehudah Raveh on retainer to represent the EPS in actions against marine polluters. Raveh already provided prosecution services for the Nature Reserve Authority, empowered by a special authorization from the Attorney General. Private firms presumably were more responsive to government clients than the overworked lawyers in the State Attorney's office.

By 1987 the Fund was generating $650,000 a year. The money paid for oil spill control equipment, jeeps, and eventually a small patrol boat. The fund also bankrolled the salaries of six inspectors. It was money well spent.[153] By 1987 the team conducted six hundred annual inspections of vessels, most of which showed compliance with international limitations on discharge of oil sources.[154] Ship captains grew accustomed to the plucky Israeli inspectors poking about and reviewing "oil records" to see whether or not ballast or oily bilge waters had been released before the ships reached the port's reception facilities (by 1988, Israel retrieved 25,000 tons of oil annually from "deballasting" facilities at its three ports).[155] Violations were categorized as strict liability. It did not matter whether captains were aware of oil leaks; if they were caught releasing oil, they were guilty. The law also did not distinguish between large and small discharges. Yacht captains were prosecuted for tossing French-fry oil overboard in the same way that an oil tanker captain would be for unleashing an ecological catastrophe.[156] The enforcement system was self-supporting. According to the law, fines were fed back into the Marine Pollution Prevention Fund.

Ultimately it is not the number of inspections that serves as the bottom line for assessing ecological progress but the actual quality of the marine environment. At least one parameter showed stunning success. The Oceanographic and Limnological Research Institute monitored the concentrations of tar on Israel's Mediterranean beaches between 1975 and 1987. They found that levels plummeted, from 3.6 kilograms per front meter to 20 grams—nearly a two-hundredfold drop![157] The decrease was partially due to the reduced oil transport to Europe following the revolution in Iran.[158] No less important, though, was the international environmental regime and Israel's own uncompromising enforcement posture. During a period when pollution parameters all showed dramatic increases, the tar level was an important anomaly. Sound environmental policies made a difference.

PLAYING HARDBALL WITH AIR POLLUTERS

The other area where the EPS began to flex its newfound muscles was in controlling air pollution from factories. Just as the Service went after the "big boys" in getting environmental impact statements started, it began ambitiously in the air pollution business as well.[159] Marinov and his legal advisor, Ruth Rotenberg, targeted Haifa's Electric Power Station, Oil Refineries (see Figure 23), and Nesher Cement plant. It was no simple task.

The Kanovich law empowered the Ministers of Health and the Interior to jointly issue emissions permits, which came to be called "personal decrees."[160] Despite the law's clear directive, for twenty years neither ministry had much interest in the issue of air pollution. Fewer than a handful of decrees had ever been issued. The first was a 1972 order to Haifa's Nesher Cement factory requiring installation of electrostatic precipitators.[161] This step was taken only after the State Comptroller's report characterized the dust particles from the factory as the "principal polluter in the Nesher locality."[162] When that order didn't help, Haifa-based Technion professor Antonio Feranio went to the Supreme Court and argued that residents in the Nesher community had been struggling for fifteen years to get the plant to clean up its act.[163] The Court's favorable decision left the ministers no choice.

As the EPS settled into its new identity at the Ministry of the Interior, it latched onto any authorities its minister had. Suddenly air pollution control was no longer an orphan. The trouble was that both potential parents seemed to wake up to their responsibilities simultaneously, and a custody battle ensued. The Ministry of Health continued to resent the competition, even though their Sanitation Department was primarily focused on sewage and drinking water quality.

In 1980, for example, despite the existing auto emissions standards under the Kanovich law, the Minister of Health promulgated his own separate regulations for vehicular pollution under the Public Health Ordinance based on less-than-precise visual assessments made by inspectors.[164] Only Ministry of Health personnel were authorized to enforce the new "Ringleman" emissions standard.[165] Although the EPS and the Sanitation Department staff quibbled over which vehicle emissions criterion was more appropriate, they both seemed to miss the point. The black smoke they were trying to measure came from diesel-powered trucks and buses—only 10 percent of the country's fleet. The steady buildup in emissions of nitrogen oxides and hydrocarbons from gasoline-burning automobiles was harder to see and smell. Unfortunately, controlling these gases was primarily the province of the Ministry of Transportation, which was for the most part apathetic to the issue.

In 1981 Marinov found a way to bypass the Ministry of Health's professional staff and presented his case directly to the Minister, Eliezer Shustack, who agreed to relinquish his own authority in the area of air and noise pollution—leaving the Kanovich law solely in the hands of the Minister of the Interior. The Knesset approved the change on May 3, 1982.[166] Interior Minister Dr. Yosef Burg was a detached but agreeable overseer. Limited only by Kuberski's naturally conservative disposition, the EPS could finally go after air polluters.

From the time that the Nesher Cement emissions first began to annoy the factory's neighbors in the 1920s, Haifa's air was uncontested as the dirtiest in Israel. By the 1980s the situation had reached intolerable levels. According to Israeli regulations, 99 percent of sulfur dioxide concentrations had to fall below 0.3 parts per million. Under no circumstances could the concentrations exceed 0.6 ppm, at which level health impacts were already acute.[167] (Today's standards are 40 percent more stringent.[168]) In 1982, monitoring stations around the city reported 109 violations of the recommended 0.3 ppm standard for sulfur dioxide. On fifteen occasions that year, Haifa had concentrations at twice that level, posing a serious health hazard, in particular to sensitive populations such as asthmatics.[169]

There were meteorological, industrial, and legal reasons for the city's poor air quality. Meteorologically, the scenic Carmel mountains slope down to form an L-shaped seaport with a bowl-like topography that stifles air exchange and occasionally leads to inversions. The industrial zones of Haifa Bay are home to a large collection of the country's chemical industries. In addition, oil refineries and a large electric power station produced most of the area's sulfur dioxide.

What made the situation so frustrating was that the particularly egregious factories were essentially untouchable, a peculiar legal situation that had its origins in the Mandate period. To ensure the prosperity of the oil refineries, British authorities granted Haifa Bay industrial zone political autonomy as an independent jurisdiction. Plants operating in Haifa's northern suburbs continued to enjoy what became known as extraterritorial status—and they flaunted it. Municipal governments were still considered the key to controlling air pollution (with Ministry of Health oversight) through business licenses. Yet they were powerless to rein in Haifa's petrochemical industry.

The situation was particularly galling for EPS legal adviser Ruth Rotenberg, who had joined EPS in 1978 after working for the legal department at the Ministry of Transportion. Her expertise in maritime law eventually led her to focus on marine pollution issues. Rotenberg, who grew up in Haifa, was keenly aware of both the suffocating conditions in the scenic port city and their origins.[170] She also had the benefit of an unusually competent local environmental unit as a partner. The EPS had encouraged the city of Haifa to bring together the other municipal authorities in the Bay area and establish a regional "Haifa Union of Cities" Environmental Unit. Not only did its Director, Zwy Fuhrer, see himself as a regulator, but his deputy, Dr. Bernanda Flicstein, with a passion for Haifa's environment and a Ph.D. in chemical engineering, had the technical qualifications and the temerity to go head to head with the industry experts.

The first big break for Haifa's air came in a Jerusalem courtroom. Haifa Sulfides had begun constructing a new building without a permit. The Ministry of the Interior, fed up with the extraterritorial insolence, intervened, and the issue reached the Supreme Court. Justice Meir Shemgar ruled that the Planning and Building Law was in force throughout the entire country. To overcome the Mandate exemption, Shemgar called on the Minister of the Interior to cancel the British High Commissioner's order, which he did. Suddenly it appeared that the Oil Refineries might at long last be subject to government controls in other realms. And that is when the EPS and the Haifa regulators set their priorities: the Electric Company, the Oil Refineries, and Nesher.[171]

On May 4, 1984, the Minister of the Interior, Yosef Burg, signed into law a personal decree controlling emissions at Haifa's electric power station. The Electric Company was always more congenial than other Haifa factories on the subject, and a cooperative process began. Although there is nothing in the law that requires consultation, the EPS created an informal notice and comment process. The utility had a meaningful opportunity to offer its input about the technical terms of the decree.[172]

The Haifa Oil Refineries, however, were a much tougher customer. The EPS sought to reduce the sulfur content in the fuel being burned at the sprawling plant. The fuel that was used in the refineries came from the Abu Rodeis fields in Egypt and from Mexico. In terms of sulfur, it was some of the filthiest crude in the world.[173] The refineries argued that they had already processed the sulfur content down to levels of 3.2 percent. Any further reductions would jeopardize the plant's profitability. Four hundred workers would be jobless. The environmentalists from Haifa were even accused of sabotaging the peace process with Egypt, by undermining the petroleum agreement.[174]

In retrospect the attack appears to be cynical "best-defense-is-a-good-offense" industry posturing. (Today sulfur levels at the plant range from 0.5 to 1 percent, and the plant operates with considerable profit.) But at the time Kuberski was afraid that his ecozealots may have gone too far. Negotiations dragged on. The Oil Refineries had no intention of knuckling under like the Electric Company. Just when things were starting to move, Yosef Burg, who had been the Minister of the Interior for more than a decade, announced his resignation in light of the new election results. Rotenberg was desperate. If Burg did not sign the decree before departing, the regulations would once again be postponed indefinitely.

On his last day in his office, she managed to get the most recent draft of a Personal Decree for the Oil Refineries through Marinov and past Kuberski. On September 12, 1984, Burg signed.[175] Flicstein, from the Haifa environmental office, remembers tears streaming down her cheeks when she got the call from Rotenburg informing her that months of efforts had borne fruit and that the controls were finally binding.

Celebration was premature, however. The Oil Refineries were still disinclined to invest in cleaner fuels. With a different government in place, it was a new ballgame. They appealed directly to Prime Minister Shimon Peres, who passed the matter on to his Office Director General, Abraham Tamir. For its part, the Ministry of Energy backed the industry position.[176]A committee was formed to find a basis for compromise. On June 17, 1985, a revised version of the emissions standard was approved that allowed the refineries to continue to burn the high-sulfur fuel during periods when background air pollution levels were low.[177] During more polluted periods, the tougher fuel standards remained in force. Like many compromises, nobody was very excited about it. It did, however, establish the EPS's position as a legitimate regulator of air quality.

Environmentalists were not surprised when the elaborate warning system for signaling the switch to low-sulfur fuels at the Electric Company and the Oil Refineries did not work. It would take another round of personal decrees and court cases during the 1990s to bring Haifa's pollution levels down to permissible levels. Nonetheless, there was enormous importance to the first generation of air quality decrees. No less important were enforcement actions initiated by Rotenberg and the Haifa Union of Cities' team when the Oil Refineries blatantly violated the decree. They had been caught red-handed burning 3.5 percent sulfur fuel during a period when it was expressly forbidden. Flicstein passed on the incriminating evidence to the police.[178] Forty times the refineries had ignored the City's notifications to move to low-sulfur fuel. They also refused to shut the plant down when facilities were not functioning optimally, even though the personal decree required it.[179]

The case produced some of the best lore from Israel's limited air pollution enforcement activities of the 1980s. Haifa's Zvulon Police Station was surprisingly supportive. As the story goes, Sari Mizrachi, a woman sergeant with a specialty in investigating sexual offenses, was given the job of interrogating the chairman of the Oil Refineries (he was Zvi Zamir, a former Director of Israel's intelligence agency, the Mossad, and not very used to being the one having to answer the questions). When Bernanda Flicstein, the Haifa regulator/scientist, tried to tutor Mizrachi so that she could carry out her interrogation duties effectively, the policewoman complained that the chemical formulas were incredibly boring. She did not hesitate to leave Flicstein languishing in her office for hours as she ran out to stage an ambush in her (more interesting) area of expertise.[180]

Eventually, though, Mizrachi caught the environmental spirit. Zwy Fuhrer and Flicstein from the Haifa Environmental Union office went along to the Refineries to spoonfeed her the incriminating materials when she interrogated Gideon Engel, the plant's environmental engineer.[181] The rumor circulated that plant manager and former spymaster Zamir became outraged when it was suggested that, as a criminal suspect, he would have to deposit his passport at the local police station. On December 28,

1987, a plea bargain was reached in which the personal indictments against Oil Refineries managers were erased in return for a guilty plea by the corporation. They admitted to thirteen different criminal counts and paid the maximum fine, which at the time was only sixty thousand shekels.[182] Within the context of Israel's political, economic, and social power structure, the EPS had toppled a giant.

CHEMICAL ANARCHY

The EPS staff members did their best wherever they could. They managed to draft and pass regulations, most notably in the area of noise pollution control.[183] They helped lobby several important environmental laws in areas such as marine pollution control, as well as an amendment that has successfully kept commercial billboards off of Israel's highways.[184] They worked with the local environmental units (which eventually reached twenty in number) and promoted educational activities. In 1988 they managed to get a Yarkon Springs Authority established,[185] the first use of a 1965 law designed to encourage watershed management of rivers.[186] They gave a great deal of advice. They pointed a lot of fingers.

Sadly, the record mostly reflects what they were unable to do.[187] Almost all environmental trends remained negative. Groundwater quality continued to deteriorate dramatically, and three successive Water Commissioners did not seem to care very much. Outside of the Greater Dan region, sewage treatment was not effective, and yet the Ministry of the Interior did little to pressure mayors to invest in appropriate infrastructure. With the exception of lead and sulfur dioxide, air pollution emissions increased exponentially. (Permissible lead concentrations in gasoline were reduced to fifteen grams per liter in 1989, producing an immediate decrease, and the personal decrees managed to keep sulfur dioxide levels steady.) Composting as a profitable waste reduction alternative came to a halt, as did the widespread reuse of beverage containers. With little data available, regulators themselves had trouble following the pace of deterioration.

Israel's misuse of pesticides offered a particularly compelling case of government impotence and the need for fundamental regulatory reform. Pesticide usage was governed under a 1956 statute aptly entitled the Plant Protection Law. It certainly was not designed as an *environmental* protection law. Registration, production, and sales of pesticides remained safely under the patronage of the Minster of Agriculture.[188] The perils of insecticides and herbicides were not unknown to Israel's farming establishment.

Rachel Carson's *Silent Spring* had been translated into Hebrew in 1966.[189] But as cotton became a centerpiece in the economic profile of many kibbutzim, the need for chemical protection for this crop drove public policy. By 1976 a staggering 21,000 tons of pesticides were applied to Israeli farmland,[190] 40 percent of which went to Israel's cotton crop.[191]

The Minister of Agriculture went so far as to set up an interministerial committee to review applications to market new pesticides. Although the committee had no formal legal authority, its ostensible mission was to provide protection against ecologically dangerous chemicals. In fact, review of the committee's protocols during the 1980s reveals that the committee was a rubber stamp for the chemical industry. The Ministry of Health representatives frequently voiced their objections to the registration of high-risk pesticides. But these protestations were routinely ignored by the majority of the committee members, who were appointed by the Minister of Agriculture. They justified their approval on the grounds that the pesticides would help crop yields.[192] By the 1980s, over six hundred different brands of pesticides were on the market in Israel.[193]

The disastrous impact of the chemical proliferation on Israeli wildlife was well known.[194] Ironically, for many years, much less effort went into researching the health impacts of pesticides on humans. For twenty-five years, from the early 1950s until the mid-1970s, there was massive use of chlorinated organic pesticides such as DDT, lindane, and benzene hexachloride (BHC).[195] When the breast milk of Israeli women was finally monitored in the 1970s, scientists were astonished to find BHC concentrations that reached 2500 parts per billion—or eight hundred times the 3-ppb level found in a comparable sample of American women. These chemicals were phased out soon thereafter, but it was too late for many women (breast and other hormonally related cancers were unusually high in Israel during the period).[196] Pesticide production was also hazardous. For a brief period, DBCP, a particularly notorious insecticide, was produced by the Dead Sea Bromine Group. Five workers became sterile as a result.[197]

The chlorinated chemicals were replaced by a different family of pesticides—organophosphates. Organophosphates such as parathion or paraquat are not as persistent and do not accumulate in the food chain, but they have a much higher acute toxicity and pose greater direct dangers to human beings. Generally these poisons work by inhibiting the cholinesterase in the nerve junctions. Acute symptoms involving the central nervous system include nausea, diarrhea, heartbeat irregularities, and, in extreme cases, convulsions and, in women, spontaneous abortions. During the 1980s, emergency rooms in Israel were treating two hundred cases of acute pesticide poisoning each year, primarily among farm workers.[198]

The chronic, long-term health effects observed, however, should have warranted even greater concern. Hadassah Hospital occupational physician Elihu Richter, in a series of studies, showed elevated exposure levels (measured by blood test) among rural kibbutz and moshav populations, apparently brought about by absorption through the skin.[199] Aerial spraying was the most common route of exposure, in particular for farmers who worked near the six hundred thousand dunams of cotton cultivated during this period. "Drift" (toxic mists) became a serious problem, and many agricultural communities experienced high miscarriage rates.[200] Regulations promulgated by the Ministry of Health in 1979 set a fairly lenient, 120-meter spraying limit from residential homes. It survives to this day.[201] (Years later, a three-hundred-meter limit was imposed on aerial spraying near sensitive water bodies.[202] The human environment ostensibly remains less protected than the natural one.) In one Israeli study, it was found that farmers within a two-kilometer distance of aerial sprayers had high visitation rates at kibbutz infirmaries.

In a devastating 1986 report, the Israel State Comptroller uncovered bureaucratic mismanagement and particularly lax oversight in the field. The Ministry of Health had never gotten around to adopting food residue limitations for 57 percent of the two hundred common pesticides in use in Israel. However, it may not have mattered very much, because the Ministry only checked sixty-three fresh fruits and vegetables during the entire year of 1985. The 10 percent of samples that exceeded permissible levels were statistically insignificant. For all intents and purposes, the Israeli food supply was unmonitored. The Comptroller also reported that of the 190 stores that were authorized to sell pesticides, only five were checked to see what chemicals they were actually selling.

There was no oversight of the permits required by law for applying particularly dangerous pesticides such as aldicarb (Temick) and parathion. Most alarming, neither the Ministry of Agriculture, nor of Labor, nor of Health had any personnel whose job it was to monitor the use of pesticides in the field. With anarchy prevailing, it was hard to blame farmers and their suppliers for taking the path of least resistance.

THE SOCIOLOGY OF NEGLECT

Throughout the three decades of the 1950s, 1960s, and 1970s, Israeli authorities seemed to wink at the culture of noncompliance that flourished in many sectors of the economy. The Knesset's environmental laws theoretically adopted a "polluter pays" ethos, and the Environmental Protection Service was among the first to trumpet this slogan.[203] Reality turned the adage on its head: Pollution certainly did pay. In the absence of enforcement and meaningful deterrence, environmental responsibility led to competitive disadvantage.[204] Viewed in economic terms, it was simply bad business for corporate managers to adhere to a law with no teeth, and it was likewise obtuse politically for mayors to give up the political windfalls they could reap for funding more popular public initiatives.

In his 1987 memoirs, Josef Tamir summed up the Israeli government's environmental record to date:

> Founders of the State and all its governments and heads during the first 35 years seem to have been stunned by the glare of the powerful historical events that accompanied redemption and rebirth. The government system was a function of oppressive coalitional exigencies with a lack of any serious consideration of the subjects that needed to be prioritized: the sprawling and threatening urbanization, the industrialization that ignored the use of appropriate technologies to reduce pollution, the astonishing increase in motor vehicles without a corresponding increase in infrastructure or controls for toxic emissions. These wounded the State, the relations between humans and their environment, as well as the population's quality of life.[205]

The statistics reflected the perfunctory government commitment. An extensive economic analysis of government expenditures on the environment between 1971 and 1980 indicated that, when controlling for inflation, there were only "relatively moderate increases, and . . . these are accompanied by signs of stagnation."[206] The only substantial increase was seen in the JNF forestry budget. At the same time, it seemed that the only time Prime Ministers Golda Meir, Yitzhak Rabin, and Menahem Begin remembered the existence of the Environmental Protection Service was when looking for an easy place to cut the budget.[207]

Competing ministries and agencies openly resented the growing ambitions of the Service and its self-righteous carping about environmental responsibility. At the same time, for most of the public, the Environmental Protection Service was essentially an invisible agency; this reflected social attitudes. In the final analysis, it often seemed that the urban environment was an invisible issue. People just did not care very much about it.

Litter was one of many discouraging litmus tests. In 1984 the Knesset passed a new "Protection of Cleanliness" statute.[208] The law itself was less than the EPS had hoped for; it had lobbied unsuccessfully to include a bottle bill (apparently distrustful of sterilization, the Minister of Health killed this section because of concerns about germs spread by reused bottles). Nonetheless, the law instituted several innovations to deal with Israel's seemingly incurable litter habit.[209] There was an unspoken anticipation that Israel might face down its litter problem in the same way it had its wildflower problem twenty years earlier.

But as anyone who walked down an Israeli trash-laden street could see, the public was not interested. The apathy was exemplified in an educational evening the EPS ran to promote the new law among Israel's police force. Hundreds of law enforcement personnel sent in forms expressing their intention to attend. The EPS originally considered hosting the event at an SPNI field school, but settled on renting the Beit ha-Am auditorium in Jerusalem. Nature films were acquired, cakes were ordered, and of course materials about the law printed. Two policemen showed up. One was a retiree, and the other was a musician in the police orchestra.[210]

The EPS later managed to pay for a police officer to join its staff who issued thousands of fines each year, but the penalties did not seem to make a dent in Israeli behavior. A substantial segment of the population thought nothing of tossing their used cigarettes, bottles, condoms, wrappers, and growing menu of plastics into the commons.

Curing Israel's ecological woes required strong educational medicine. Getting the message across seemed a hopeless task. For the most part, the environment was not a newsworthy story for the Israeli press. Only the *Jerusalem Post* and *Ha-Aretz* newspapers had reporters who covered environmental issues, along with their other beats. Israel's one television station was completely uninterested. The public school and university system, still locked into the more traditional biology or chemistry disciplines, was of little help.

It even appeared sometimes that Israelis inculcated environmental irresponsibility to their children from the tenderest of ages. After its initial publication in 1975, Alona Frankel's children's book *Sir ha-Sirim* was the definitive guide for toddlers about toilet training.[211] (The book was good enough to be translated into English as *Once upon a Potty*.) After Naftali, the protagonist, finally succeeds in leaving his diaper behind, his mother flushes the results down the toilet exclaiming, "See you in the sea!"

Marinov wrote to the book's author asking her to modify the concluding section,[212] but "see you in the activated-sludge treatment facility" did not sound as good to preschool readers. The present publication continues to make marine discharges the normative waste disposal system in Israeli children's literature.

By the end of the 1980s, the Israeli public's schizophrenia about the environment baffled educators and activists alike. How could a nation that flocked to SPNI hikes, voluntarily stopped picking wildflowers, and so identified itself with the natural world be so numbed to pervasive environmental deterioration? Marinov used to liken the dynamic to a person standing knee-deep in the morass of a toxic dump, with his eyes transfixed by the birds overhead.

9 A Ministry of the Environment Comes of Age

The environment is a particularly popular arena for policy enthusiasts, because time and again it has shown that intelligent government intervention can produce measurable societal benefits. But the checkered history of Israel's Ministry of the Environment—with corresponding swings in pollution trends and the spirits of the country's environmentalists—suggests that policies are only part of the puzzle. In the somewhat weary debate as to whether individuals or larger social forces shape the course of human events, a decade of environmental inconsistency at the Ministry suggests that individuals matter. Institutional status as well as actual improvement in the country's land, air, and water quality reflect the relative savvy, competence, and commitment of the Environmental Ministry's leadership.

The turnover at the helm of the Ministry has been exceptional. Between 1989 and 1996 the staff had to raise a toast to the success of six different Ministers in almost as many years. Such instability reflects the Ministry's lowly status among politicians and suggests that even in a best-case scenario, environmental gains will be inconsistent. It also makes for a choppy and truncated institutional history. A review of the Ministry's experience during its first decade of work confirms that the personalities of its leaders often had a far greater effect on environmental gains than the political configuration or platform of the ruling coalition. Unfortunately it also suggests that even the best Minister of the Environment faces enormous limitations and that the establishment of a Cabinet-level Ministry is only a crucial first step in Israel's battle against pollution.

ELECTION STALEMATE

A swelling of grassroots activism, culminating in the Earth Day celebrations that swept America in 1970, led to the creation of the U.S. Environmental Protection Agency. President Nixon had little choice but to listen to the pulse of the nation and create a new superagency.[1] In New Zealand, the creation of an independent Ministry of the Environment in 1986 was part of a series of wholesale reforms by the Labour Party, promised in their election campaign two years earlier.[2] The establishment of Israel's Environmental Ministry, however, was a fluke. If the 1988 elections had not ended in a deadlock, the environment might still lack Cabinet-level representation today.

Prior to the elections, the Environmental Protection Service continued to limp along as part of the Ministry of the Interior. Under the administration of successive religious Interior Ministers, most recently from the ultraorthodox Sephardic Shas party, the Service's future became increasingly precarious. Its Director, Uri Marinov, had grown bolder and less diplomatic over the years. He was no longer willing to kowtow to Ariyeh Deri, the upstart party founder who was at first Director General and later Minister of the Interior. When Deri directed him to funnel monies designated for environmental activities to Shas-affiliated religious institutions, Marinov went to the police to complain. Relations that had never been good reached a new nadir. Deri disconnected Marinov's phone and telex machine. Marinov decided to leave government service if things did not change quickly.[3]

Deliverance came from the unanticipated electoral stalemate. As the results of the elections came in, it became clear that neither the Labor nor the Likud party would be able to form a majority government. Once again, Israel's two large political blocs would have to find a way to compromise and share power in a "national unity government." The principle that drove the negotiations was an equal number of Ministerial posts for Likud and Labor leaders, to ensure a balanced Cabinet. But after all the minority parties had been rationed their portfolios, there was an odd number of Ministries left.

The odd man out was Ronni Miloh, a very bright young politician who often played the role of aggressive hatchet man for the (right-wing) Likud Party. An attorney who had spent all his life in the sophisticated Tel Aviv milieu, Miloh was perfectly comfortable among leftists and in 1994 was elected mayor of Israel's most liberal city. He had been a Likud activist in university politics and for many years remained steadfastly loyal to the camp affiliated with Prime Minister Yitzhak Shamir. By 1988 he had waited long enough to enter the highest echelons of the Likud power structure. The question was not whether he would be appointed a Minister but what Ministry would be created for him?

Marinov had thought about this from the time election results came in. Together with his legal advisor, Ruth Rotenberg, he prepared an entire plan for an Environmental Ministry, including drafts of all the requisite government decisions for establishing it. Aware of the weakness of politicians for homage, Marinov even found room for spacious Ministerial chambers, giving up his own office in the existing Environmental Protection Service. As negotiations for establishing the government dragged on, Marinov had time to refine the proposal and consider who would be its best advocate. He gambled on Miloh, who had been sympathetic to environmental issues in the past.[4] It was a prescient choice.

A smiling Miloh appeared on Israeli television on the morning of December 25, 1988. He described the dreidel spin of one of Israel's craziest all-night coalitional marathons. At one o'clock in the morning he was going to be Minister of Transportation; then later in the negotiations he was moved to Tourism; then that job was snatched by someone else.[5] At six in the morning, he was to be Minister without portfolio. Rumor had it that the possibility of creating a Ministry of Sports was suggested as a possible area of jurisdiction. It was then that Miloh pulled out Marinov's plan and said, "I want to be Minister of the Environment." Yitzhak Shamir had already been lobbied intensively on the issue by Josef Tamir, who now served as head of Life and Environment—Israel's umbrella group for environmental organizations. He was very supportive of the idea.[6] It was the easiest solution, and everyone was very tired. Under these inauspicious and somewhat random circumstances, the Ministry of the Environment was born. It was long overdue. Some 125 nations had already created independent environmental bureaucracies.[7]

Miloh's brief tenure at the Ministry (see Figure 24) is remembered favorably for several reasons, one of which was the high-minded professional nature of his selection of personnel. His first act as Minister was to retain Marinov's services as Director General. Other key appointments, like that of Professor Yoram Avnimelech as Ministry Chief Scientist, involved experts who were at odds with Likud positions. The choices raised eyebrows in his own party. Miloh recalls that when he told Yitzhak Shamir that he intended to appoint Marinov as his Director General, Shamir questioned the wisdom of the move, as he had heard that Marinov held leftist views. Miloh told Shamir that whatever Marinov's politics were, he was the best environmental professional around. When Shamir asked whether he did not fear the response of the Likud Central Committee, Miloh told him, "If I succeed at this job, everyone will clap for me. If I fail, they won't forget it."[8] In retrospect, most people in Israel believe that Miloh did succeed. The degree of his success, however, particularly in terms of the institutional mandate he gained for his Ministry, is the subject of debate.

LABOR PAINS

The fanfare surrounding the creation of a new government agency seemed to be an exciting turning point for the environment, both among the public and within the government.[9] In fact, the new Ministry floundered initially. It managed to win only minimal authorities because of a series of fractious political clashes. By March 1989 no governmental agency had voluntary ceded its authorities to the new Ministry, and Miloh was threatening a coalitional crisis. The government appointed two professors of public administration, Abraham Atzmon and Yehezkel Dror, as arbitrators to recommend the scope of the powers that should be given to a Ministry of the Environment.

It took only a few weeks for them to offer a comprehensive and thoughtful ruling. Their promptness was all the more impressive because they worked without the cooperation of either the Minister of Agriculture or the Minister of Health, who refused to appear before them.[10] Both Ministries stood to lose from the creation of a new Environmental Ministry. That, however, was not the real problem: Each was run by a Minister from the Labor Party who wanted to do as little as possible to help Ronni Miloh, at the time a particularly unpopular figure among Labor circles.

The professors prefaced their recommendations with a call for restraint, that is, to grant the Ministry a critical mass of authorities, but no more. After paying this lip service they went on to propose a powerhouse Ministry. It included regulatory control in the areas of water, air, and hazardous and solid wastes and overseeing a range of agencies from the Nature Reserves Authority to Israel's Meteorological Service.[11] This turned out to be a futile exercise. Rather than adopt the far-reaching recommendations, Prime Minister Shamir opted to file them away and let each Ministry work the matter out directly with Miloh and his staff.

And so a tedious chain of negotiations between the Ministry of the Environment and its colleagues ensued. It is not by chance that Shamir remembers resistance at the Ministry of the Interior. Ariyeh Deri was hardly enthusiastic about bolstering the status of Uri Marinov, who took the lead in most of these negotiations. But in the end, Deri was the first to sign an agreement. On April 2, 1989, the government approved the transfer of the Environmental Protection Service staff, authorities, and budget (including support for the municipal environmental units) to the new Ministry.[12] This offered a cohesive corps of workers who could hit the ground running in such areas as marine protection, air pollution monitoring and regulation, planning, and litter control. The Ministry also inherited the EPS offices in the Interior Building. After they had been spruced up a bit, Miloh really did move into Marinov's old offices, according to the script.

The Ministry of Agriculture was next in line.[13] It was not enthusiastic. Prime Minister Shamir transferred the National Parks Authority from his Office to the new Ministry as a gift, but the Nature Reserve system stayed in the Ministry of Agriculture.[14] More important, it maintained its control over Israel's powerful Water Commission. Even though the Environmental Ministry would open an agroecology department, it was essentially shut out of any real influence over pesticide registration and oversight policies. The agreement with the Ministry of Health was the longest in coming and was only approved a year later by the government, on January 21, 1990. Here, the Ministry of the Environment came out ahead. It became the lead agency overseeing the Licensing of Businesses Law, except for products that were designed to "go in people's mouths" (such as food or pharmaceuticals).[15] Those remained under the Ministry of Health's purview. The Ministry of Health also divested itself of responsibility for hazardous materials regulation, pest control, air pollution, and other nuisances.[16] As part of the deal, Marinov and the Environmental Ministry inherited much of the same Ministry of Health technical staff with whom the EPS had bickered so consistently in past years, as well as the experienced professionalism of Dr. Shmuel Brenner, a senior government scientist, and his Tel Aviv University–based analytical laboratories.

When the dust settled, the Ministry of the Environment certainly had considerable power on paper. Formally it held authorities under thirteen different statutes. Yet a closer look showed that it was relegated to supporting-actor status on key environmental issues such as air emissions from vehicles, drinking-water and sewage treatment, pesticide registration and usage, nature preservation policy, radiation, and, of course, physical planning.

From Miloh's perspective, more important than the statutory limitations were the financial constraints. The money allocated to the Ministry left him absolutely no room for maneuvering or initiative. Josef Tamir fired off a furious editorial, where he branded the twelve million shekels (six million dollars) allocated to the Ministry "embarrassing." It represented 0.018 percent of the 66-billion-shekel overall government budget.[17] Miloh went to speak about the matter to Shimon Peres, who served at the time as Minister of Finance, and his assistant, Yossi Beilin, who was a strong proponent of the Ministry. For an hour and a half, he and Marinov detailed the objectives and strategic plan of the Ministry. At the end of the presentation Miloh concluded, "Mr. Minister, we are going to need to double our budget if we are to accomplish any of this." Peres listened attentively throughout. Then he looked at Miloh and said curtly, "From my perspective, the budget can be zero." Miloh walked out with Beilin chasing after him, assuring him that everything would be fine.[18]

MILOH'S FIRST FIGHT

Miloh was impatient to make a statement about his new job and put his Ministry on the map. Once the Interior Ministry transferred the personnel and authorities from the Environmental Protection Service, Miloh was ready to hit the ground running. By May he had issued a personal decree against the problematic Castel quarry on the outskirts of Jerusalem.[19] He promulgated regulations that made the 1988 law controlling land-based dumping into the sea operational.[20] Other regulations finally made it onto the books to require garbage dumps to take active measures to prevent fires, such as covering trash with a fifteen-centimeter layer of dirt each day.[21] Ironically, if not inappropriately, it was Kanovich's air pollution statute that provided the legal foot in the door to regulate solid-waste facilities. Miloh's active participation in raising money for the trees of the scorched Carmel Forest in a twenty-four-hour telethon on Israel's only television station raised three million shekels, as well as the profile of the Ministry.[22]

But most of Miloh's time went into institution building. For example, he insisted that the Ministry go beyond an advisory role in planning at a regional level and establish formal district offices in each of the six regions of the country.[23] He also tried to push the Ministry's enforcement capability. Miloh had a sober perception of the feudal nature of Israeli governmental culture. It required each Minister to field a private militia to protect his fiefdom. The income-tax people had their commandos. The ultraorthodox had a Sabbath police (run by Druze), which they operated from the Ministry of Labor. Miloh had something similar in mind for polluters. He went to the Ministry of Finance to negotiate a budget for some reasonable office space and proposed the comfortable Migdal ha-Ir high-rise in the heart of downtown Jerusalem. Predictably, the Ministry of Finance clerks balked at the exorbitant price. Miloh agreed to forgo the offices temporarily in exchange for funds to create an environmental patrol, jeeps and all.[24] The five-man patrol was set up initially under the auspices of the National Parks Authority.[25]

The battle over Haifa's air quality, however, was to be the defining contest of Miloh's tenure as Minister of the Environment. When he asked his new staff what area was ripe for bold action, they did not hesitate: Go after air pollution, they said. They had a heap of unfinished business from the Environmental Protection Service. On April 30, 1989, only three weeks after Ariyeh Deri made Miloh responsible for the country's air quality, the highest sulfur dioxide levels in Israeli history were recorded in Haifa.

Weather patterns usually made air pollution worse during the spring and fall seasons along the coast, but the concentration recorded in Haifa, 2688 micrograms per cubic meter, was genuinely dangerous. It was more than three times the allowable level at the time and five times today's standard.[26] Miloh jumped at the chance to tackle a high-profile problem. But if he was looking for a painless initiation to environmental politics, he had picked the wrong issue. The subsequent effort to improve the City's air quality was indicative of both an Environmental Ministry's potential and the intensity of the political brawl required to green the Israeli government's public policies.

The problem was the 3 percent sulfur fuel used most of the time by Haifa's power plant and oil refineries. When the winds died down and there was no air dispersion, even 1 percent sulfur fuel was not always enough to keep sulfur dioxide levels within acceptable levels.[27] The Israel Electric Company board responded by falling back on dilution—raising the pollution plume high above the land so that it would disperse before reaching the exposed population below.[28] But because of Haifa's unique topographic layout, the fifteen million dollars spent to raise smokestacks to a height of three hundred meters could not solve the problem.

To give Haifa's air the intensive care it demanded, Miloh first needed to revise Israel's national ambient air pollution standards. The 1971 standards posed an obstacle to progress. First of all, the list was very narrow and did not include such basic pollutants as ozone, the leading indicator of photochemical smog.[29] The World Health Organization recomended a new list of air quality criteria that was almost three times as long.

A national committee of health experts had been comparing it with the allowable air quality levels in Israel and found several Israeli standards to be too lenient.[30] Ostensibly there was no reason that the WHO recommendation could not be signed into law.

The new, expanded list of pollutants was prepared for Miloh's signature, with new standards set as absolute ceilings on air pollution levels—unlike the old limits, which allowed for violations 1 percent of the time (a standard that made little sense from either a public health or an ethical perspective). It seemed like a simple solution. On May 7, 1989, the Minister was only too happy to issue the new standards as the basis for his crackdown on polluters. But the powers-that-were at the Haifa Oil Refineries and the Israel Electric Company had other ideas, and they were used to getting their way.

Under a 1985 Directive from the Attorney General, Israeli government Ministers must consult with all other Ministries that might be affected by their regulations before secondary legislation can come into force.[31] In 1989 Moshe Shachal, a long-time Labor politician, was the Minister of Energy. His job put him in the position of patron for the Oil Refineries and the Electric Company, both government corporations. They told him that the standards would be too tough for them to meet and would cost the government millions. The fact that Shachal lived in Haifa did not seem to faze him, perhaps because his villa was on the other side of the mountain from the industrial facilities. The Energy Minister formally opposed the regulations and prevented their publication.

Miloh did not hesitate to go after Haifa's megapolluters, but these efforts were also frustrated. On October 18, 1989, Miloh signed amendments to the existing personal decrees (stack emission limits) against both the Electric Company and the Oil Refineries.[32] They were to go into force in three months' time. The new directives forced the Oil Refineries to meet a specific sulfur dioxide emission standard at all times, never spewing more than 1.3 tons per hour into the atmosphere. If background air pollution levels went up, the refineries would have to halve their emission levels, to 0.6 tons per hour. Yet once again, the Energy Minister Shachal intervened and brought the issue to the Cabinet.

To avoid a coalitional crisis, a committee of four Ministers was appointed to consider the matter. As the precarious unity government required symmetry in all matters, the committee included two Likud and two Labor Ministers. The political balance once again called into question the assumption that the environment was always a left-wing or liberal issue. Within the committee, the right-wing politicians represented Green interests, and the leftists backed the industrial polluters. They quickly reached a standoff.[33] So Miloh and Shachal were charged with finding an arbitrator to resolve the technical disagreements. They settled on Professor Haim Harari, a brilliant physicist and President of the Weizmann Institute.

Harari may have been a physics genius, but he was not an experienced environmental regulator. When his report was submitted in February 1990, it failed to make anyone happy.[34] Harari reduced the sulfur dioxide criterion from 780 to 500 micrograms per cubic meter but left it a "statistical standard" (allowing for a doubling of the standard 0.25 percent of the time). Sulfur levels in fuels were to be cut immediately to 2.5 percent, and during pollution episodes, 0.5 percent sulfur fuel was to be used.

Miloh took a "damn the torpedoes" approach and left his original personal decrees in place. They were to come into force on March 18, 1990. It did him little good. Both companies immediately brought the issue before the Supreme Court, calling for cancellation of the emissions limits as arbitrary and capricious.[35] The Court granted a temporary injunction until it could rule on the matter.

By the time of the Court's judgment, however, Miloh had one foot out the door. With new prestigious Cabinet portfolios opening up, Miloh quickly forgot about how essential an Environmental Ministry was. Miloh was offered the Minister of Police position, and did not think twice before accepting it. Later, he would tell stories about a visit to England in his new capacity. His Ministerial colleagues in London heard of his decision with incredulity. In England the Interior and Environment Ministry is a far finer feather in a political cap than Police Chief. But the political calculus in Israel was different. Looking back, Miloh claims that he reached the decision because after two years of internecine squabbling to keep his Ministry running, he was just psychologically worn out.[36]

In 1990 an attempt by Labor Party chief Shimon Peres to reshuffle coalitional loyalties and unseat Prime Minister Shamir failed. Labor was kicked out of the coalition, and a Likud-led coalition survived with a narrow parliamentary majority. The government was left with a small, poorly funded Environmental Ministry that no one had really wanted in the first place. By default, Prime Minister Yitzhak Shamir assumed the portfolio and held it until Yitzhak Rabin's Labor victory in July 1992. Fulfilling a debt to the National Religious Party, he eventually appointed Yigael Bibi, a good-natured former mayor from Tiberias, as Deputy Minister. Bibi had excellent intentions but lacked initiative and, unfortunately, did not sit around the Cabinet table (being only a *Deputy* Minister). He left Marinov as Director General to keep the place running. In political terms, the Ministry of the Environment had become an orphan.

A MINISTRY WITHOUT A MINISTER

The epilogue to the Haifa case offers a good indication of the Environmental Ministry's second-class status from 1990 to 1992. Only after a legal action was brought by a new public-interest environmental law group—*Adam Teva V'din,* the Israel Union for Environmental Defense—against the Prime Minister[37] did the government decide to adopt Harari's recommendations. The press declared it an environmental victory. At least publicly, environmentalists were less than jubilant, gloomily calling Haifa's air a symptom of a Ministry without a Minister.[38] It would take another Supreme Court petition,[39] but in April 1992, three years after Miloh had originally signed them into law, the amendments to the personal decrees were published.[40] It was worth the wait. Harari's compromise formula proved efficacious. Haifa's air quality quickly improved, and levels never again came near the perilous concentrations that characterized the late 1980s.[41]

The Ministry, however, was not in good shape. In Jerusalem alone, its workers were scattered across three different offices in as many neighborhoods. The budget had grown but was still meager. The general scarcity was especially acute in the operations of the six regional offices. In a typical example, one staffer was responsible for overseeing industrial pollution for the entire northern quarter of Israel. The Ministry could not provide him with a vehicle. Gasoline funding in the budget could barely pay for the trip from his home in the Golan Heights to the Nazareth office. Luckily, his kibbutz believed in his work and bankrolled any travel to the field he had to do.

More embarrassing was the general lack of clout, in particular when it came to enforcing environmental standards. The sewage discharges into the Kinneret Lake during 1991 and 1992 were illustrative. After the summer of 1990, following three consecutive years of low rainfall, the scenic boardwalk of the City of Tiberias sat high above the lake's low waterline. Limnologists at the Kinneret laboratories were stunned in January 1991 when they spotted a serious leak in a sewage pipeline from the tourist establishments; it was jutting right above the waters.[42] Raw wastes were pouring into the national drinking-water reservoir.

The scientists spoke to Tiberias Mayor Josef Peretz, who claimed that he lacked funds for the substantial repairs required. By April 1991,

Figure 1. Rescuers seeking Australian athletes in the Yarkon River after the collapse of the bridge in the Maccabiah's opening ceremonies, 1997. Image appears courtesy of the Government Press Office.

Figure 2. Bedouin herdsmen, circa 1940. Image appears courtesy of the Government Press Office.

Figure 3. The Reading Power Plant alongside Tel Aviv's Yarkon River, under construction, 1937. Image appears courtesy of JNF Archives.

Figure 4. First spraying against malaria in the Sharon Valley, circa 1940. Image appears courtesy of JNF Archives.

Figure 5. Menahem Ussishkin hosting Professor Albert Einstein in Palestine at JNF nursery, 1940. Image appears courtesy of JNF Archives.

Figure 6. "Father of the Trees" Yosef Weitz inspects JNF forest, 1953. Image appears courtesy of JNF Archives.

Figure 7. Ussishkin and Weitz at JNF tree-planting ceremony at JNF nursery, circa 1935. Image appears courtesy of JNF Archives.

Figure 8. Amotz Zahavi, the SPNI's indefatigable founding director and later zoology professor (right), chatting with Israel's President Yizhak Navon. Image appears courtesy of SPNI Archives.

Figure 9. Azariah Alon, the SPNI's "trumpet" and most eloquent spokesman for nature in Israel. Image appears courtesy of SPNI Archives.

Figure 10. Yoav Sagi, first Director and later Chair of the SPNI's Nature Protection Department. Image appears courtesy of SPNI Archives.

Figure 11. Yossi Leshem, SPNI leader and bird advocate extraordinaire, circa 1997. Image appears courtesy of Yossi Leshem.

Figure 12. Dr. Uzi Paz, who lobbied to get the Natures Reserve Authority through the Knesset and became NRA's first director. Image appears courtesy of Uzi Paz.

Figure 13. General Avram Yoffe, who waged war for Israel's nature reserves and won. Image appears courtesy of David Rubin.

Figure 14. Dr. D'vora Ben Shaul, who came all the way from Texas to be the NRA's first woman employee and first scientist (1999). Image appears courtesy of Jocelyn Sheffer.

Figure 15. Dan Perry, NRA director and lifelong nature advocate. Image appears courtesy of Dan Perry.

Figure 16. Hyrax, indigenous desert species. Image appears courtesy of the Government Press Office.

Figure 17. Professor Hillel Shuval, member of the Ministry of Health's original "sanitation" team and water quality advocate. Image appears courtesy of Professor Shuval.

Figure 18. Laying pipes for the Yarkon-Negev water carrier, 1954. Image appears courtesy of the Zionist Archives.

Figure 19. A polluted stream leading to the Palmahim beach, circa 1974. Image appears courtesy of Government Press Office.

Figure 20. Josef Tamir presiding over the first meeting of the Knesset's Interior and Environment Committee, 1972. Image appears courtesy of Josef Tamir.

Figure 21. At Environmental Day in 1972, hosted by Israel's President Katzir; Green Parliament member Josef Tamir speaking. Image appears courtesy of Josef Tamir.

Figure 22. Israel's Environmental Protection Service on a field trip to Mitzpeh Ramon, 1982. The director, Uri Marinov, is at far right. Image appears courtesy of Liora Belkin.

Figure 23. The Haifa Oil Refineries, site of the first major government action. Image occurs courtesy of the Government Press Office.

Figure 24. Ronni Miloh, Israel's first Minister of the Environment. Image appears courtesy of the Government Press Office.

Figure 25. Yossi Sarid, unlikely Green icon as Environmental Minister. Image appears courtesy of Sarid and his Meretz Party.

Figure 26. Minister of the Environment Yossi Sarid (left) lobbying Prime Minister Yitzhak Rabin in the field, 1995. Image appears courtesy of the Government Press Office.

Figure 27. The Hiriyah garbage dump—not only a source of water pollution, odors, and periodic fires, but also a real danger to incoming planes because of the concentrations of birds it attracts, 1980. Image appears courtesy of the Government Press Office.

Figure 28. Nehama Ronen, who ran the Ministry of the Environment as Director-General, 1996–1999, declaring her candidacy for the Knesset in a new "green" party. Image appears courtesy of Nehama Ronen.

Figure 29. Firefighting efforts in JNF forest, 1974. Image appears courtesy of JNF Archives.

Figure 30. Prime Minister Levi Eshkol visiting Beit Ja'an, circa 1966. Image appears courtesy of the Government Press Office.

Figure 31. Bedouin shepherds outside Beer Sheva's bus station seeking pastures. Image appears courtesy of the Government Press Office.

Figure 32. Herschell and Shirley Benyamin, EcoNet's tireless crusaders, 1998. Image appears courtesy of the Benyamins.

Figure 33. Reuven Yosef, saving Eilat's bird habitat, with Levna sparrowhawks, 1998. Image appears courtesy of Reuven Yosef.

Figure 34. A rally by activists protesting against inappropriate treatment of medical wastes in Modi'in's municipal landfill, 1999. Image appears courtesy of the Government Press Office.

Figure 35. Elli Varberg, head of the marine pollution prevention station at Eilat, inspecting equipment after Aqaba cleanup, 1995. Image appears courtesy of Varberg.

four months later, they were out of patience and reported the problem to the Ministry of the Environment. The Ministry demanded swift action but for an additional eight months had no choice but to swallow the Mayor's excuses and apply moral suasion. Finally, on December 2, 1991, Uri Marinov forwarded a formal warning to Peretz, notifying him that the City's evasion constituted criminal activity under the Water Law.[43] Israel's Water Law is a criminal statute but had rarely been used as such. Yet, if knowingly contaminating the national drinking-water reservoir was not a reason to prosecute a polluter, it is hard to think of any case that would be. When the Ministry's formal threat and call for action did not produce results, its legal staff called for the indictment of the Mayor and city officials. The Ministry staffers even wrote out the legal documents and put together the list of witnesses for prosecutors. But Esther Gofer, the Northern Region District Attorney, refused to file it.[44]

Eventually, word of the pollution got out. Adam Teva V'din sent letters to the Mayor and government authorities demanding action. When copies reached the press, it reported the case.[45] Things suddenly began to move, but only after the City received one million shekels to undertake the work. The Mayor was eventually indicted,[46] but soon after construction of the pipes commenced, the District Attorney canceled the suit without even bothering to consult the Ministry of the Environment.[47] For two years, the Ministry of the Environment and its Deputy Minister had sat by impotently as Israel's largest single source of drinking water was being poisoned. The Mayor's flippant assumption that the environment was not really a serious matter was confirmed by the District Attorney's complicity. It was demoralizing.

The case was not exceptional. It took just as long to convince Israel's Attorney General to prosecute Rafi Hochman, the Mayor of Eilat (the resort city on the Red Sea). The farmers in the area were willing to use the city's wastes for irrigation but could not afford the cost of the treated water. For its part, the city was unwilling to pay the expense, estimated at somewhat less than one-hundred thousand dollars a year. Hochman discharged his city's sewage into the Red Sea, upstream from his hotel district, without a permit. Then he openly taunted the Ministry of the Environment in the press to do something about it. The Eilat sewage suit survived a bit longer than the Tiberias legal action,[48] but the judge was openly hostile to the prosecution. After work began on a pipeline to take the treated wastes to the fields of the neighboring kibbutzim, the indictment was buried by the legal authorities.

THE RUSSIANS ARE COMING

Perhaps the greatest environmental setback of the period came in the form of "emergency legislation" passed by the Knesset in July 1990 that essentially eviscerated Israel's planning and building system. The Soviet Union, in its final days, at long last opened its gates, and hundreds of thousands of immigrants flocked to Israel.[49] The Israeli government, fearing an acute housing shortage, decided to streamline the planning process for residential units. Benignly titled the "Planning and Building Procedures Law (Temporary Measures),"[50] the law created ad hoc committees known by their Hebrew acronym, the Valalim.[51] These were to replace temporarily the more ponderous regional and local committees.[52]

Dominated by government development interests such as the Ministry of Housing and the Israel Lands Administration, the Valalim committees were quantitatively and qualitatively overzealous in their work. For instance, the government set an objective of forty-five thousand approved housing units for Jewish year 5751 (1990–1991). As speculators cashed in on the real estate sweepstakes, the Valalim committees obliged, by approving 380,000 units.[53] Technically the emergency committees were supposed to review only those housing plans containing two hundred or more units. In practice, plans ranging from small developments to entire urban, commercial, and transportation infrastructures sped through the committees. Environmental considerations, impact statements, and common sense were conspicuously absent. Until then, the public usually had enjoyed two months to review development proposals and file objections to problematic plans. Under the emergency system, it was limited to twenty days—an impossibly short period to put together a compelling case against a development scheme. A study showed that official review of complex development plans generally took less than a week.[54]

The legacy of environmental mistakes remains traumatic to environmentalists. From 1990 until 1993, when housing starts doubled,[55] all the Ministry of the Environment could do was to catalog them. To name but a few, homes sprang up adjacent to railroad tracks and air strips with no consideration of the noise; neighborhoods were built without sewage systems; and sensitive scenic areas were scarred by the hasty projects. No less important, the law made a mockery of a planning system that for years had been considered a linchpin of environmental policy. It was a free-for-all, and government bureaucracies that had long been stifled by annoying planning procedures were among the most exploitive entrepreneurs.

The law was designed to be a temporary stopgap arrangement, with its provisions in force for only two years. When it expired, a glut of approved housing plans existed. The emergency committees were very popular, however, among developers, inside and outside the government. The Ministry of the Environment staff dutifully tried to mitigate the most egregious plans, submitted reports detailing dozens of abuses, and openly called for the law's natural discontinuation. The Knesset Interior and Environment Committee nodded sympathetically and then proceeded to extend the law's life on four separate occasions.[56]

Faced with the impossible position of having Ministerial responsibility without Ministerial power, Marinov become increasingly truculent. Two decades of fighting for environmental causes with inadequate authority left him with a long list of enemies. They all seemed to surface for a brutal expose in the newspaper *Chadashot*, which painted a harsh picture of his ruthless management style. However, for Marinov there was some comfort in international recognition: In 1991 after several unsuccessful attempts, he was elected to the Directorate of the Mediterranean Action Plan in Cairo.[57] This was particularly gratifying—since declaring Zionism to be racism in 1974, the United Nations had excluded Israelis from any position of real responsibility in an affiliated institution. And of course, there was the global high in June 1992 associated with the Earth Summit in Rio de Janeiro, when Marinov and his deputy, Amram Pruginin, sat in for Prime Minister Shamir, who chose to stay home and campaign for the July elections.

But without a Minister, the Environmental Ministry became a throwback to the days of the Environmental Protection Service. It was mostly the same faces, and they still lacked the power to implement policies that might reduce pollution from pesticides, cars, or industrial waste pipes. In addition, the Ministry seemed incapable of articulating a Green vision that would excite the public. Issues such as public transport, solar energy, economic incentives, eco-labeling, urban aesthetics, and biodiversity were outside the Ministry's operational and conceptual universe.

AN UNPOPULAR BOSS

The Labor Party victory of 1992 put an end to the orphan period at the Ministry of the Environment. However, the staff quickly came to miss the carefree days when they worked without the heavy hand of a Ministerial parent. In this case, the oppressive mother was Ora Namir. Mordechai Namir, her husband, had been the Mapai (Labor Party) mayor of Tel Aviv during the 1960s, giving Namir access to top-level politicians. After his death, she embarked on her own political career and was elected as a Knesset representative in 1973. Namir made a name for herself there as a defender of the disenfranchised, particularly in the area of labor reforms and women's issues.[58] She felt snubbed by her former political patron, Yitzhak Rabin, and in 1992 decided to run against him and Shimon Peres in the Labor primaries, where she came in a distant fourth place. It did not improve her standing with Rabin. When the Prime Minister handed out the Cabinet jobs, he kept the Minister of Labor post she so coveted open in the hope of attracting ultraorthodox parties to the government. Namir had to settle for the Ministry of the Environment. Environmentalists were delighted when such a high-profile politician got the job, but they found that her humanitarian public image was a far cry from her private personality.

The first thing she did was fire Uri Marinov out of "managerial" and not personal or professional considerations.[59] It is not unusual for Ministers to appoint Directors from among their political cronies. Still, there was some surprise at the speed and callousness of Marinov's dismissal. His replacement was Yisrael Peleg, a former head of the Government Press Office.[60] With a doctorate in communications, he had hoped to be appointed head of the National Broadcast Authority, but nevertheless jumped at Namir's offer.[61]

In their first impression of Namir in the Ministry, most environmentalists remember her as being superficial and supercilious. Marinov recalls a post-election meeting. Namir asked what the key environmental problems were. "West Bank sewage" was the immediate response. A few days later Namir returned and charged nastily, "You all misled me. I spoke with a senior military officer, and he claims that the sewage problem has been completely taken care of there." The group was a bit stunned at her vociferous attack. With nothing to lose, Marinov pointed out that the Ministry of the Environment generally believed that putting waste into pipes was not enough—that there should be treatment at the end of the line.[62]

In her early interviews, Namir actually identified sewage as her top priority, but she did not put together a serious Ministerial strategy to confront the issue. She returned from a trip to America enthused about trash incineration and was convinced that the private sector would buy a multimillion-dollar technological solution to Israel's garbage problem. Although a number of important regulations were finally passed during her tenure and she seemed to be working hard, Namir is remembered mostly because she managed to alienate people. The Society for the

Protection of Nature in Israel was furious when she chose "employment" over her professional staff's negative position toward the Voice of America. After Adam Teva V'din filed a Supreme Court action against her Ministry, it found that the once-friendly Namir held grudges. She took umbrage easily.

Her staff suffered most of all. Many of the senior officials at first grumbled at what a bad listener she was. After suffering a few of her legendary temper tantrums, they started to look for alternative employment. Namir seemed to enjoy humiliating her subordinates in public. It went beyond contradicting and insulting them—she went as far as to force them to pick up trash from the floor in front of visitors. When her termagant ways were leaked to the press, her fury reached new heights. Morale in the Environmental Ministry was at an all-time low in October 1993, when Prime Minister Rabin reshuffled his Cabinet.

AN ACCIDENTAL ENVIRONMENTAL HERO

Yossi Sarid was Minister of the Environment for less than three years, but no one has had a greater influence on Israel's environmental self-image. A brilliant and complex figure in Israeli politics for over thirty years, there was nothing that indicated that he would assume the role of environmental icon for the 1990s. Sarid began his political career in 1964, not yet twenty-five years old, as both a protégé and enfant terrible of the ruling Mapai (Labor) Party. The son of two Hebrew teachers, he inherited their facile rhetorical proficiency and staunch commitment to principles.

After immigrating to Israel in 1932, Sarid's father, Yaakov, moved up the ladder in the Mapai and the Israeli educational bureaucracy. Eventually he was appointed Director General of the Ministry of Education and Culture. (In 1965, it was Yaakov Sarid—not Ben-Gurion—who made the fateful decision to ban the Beatles from playing in Israel, lest they corrupt the nation's youth.[63]) Sarid speaks of his late father with reverence, but one can feel that the high expectations placed on this only son produced competing impulses of service and rebellion. For instance, the precocious youngster was snuck into the Labor Zionist youth movement prematurely while only eight. Nine years later, Sarid was thrown out for smoking in high school, taking him off the well-trod Labor Zionist track to rural settlement and the kibbutz.

The army did not know how to take advantage of the gifted but cheeky adolescent. He served for two years, mostly in the artillery. Soon after returning to civilian life, he began studying literature and philosophy at

Hebrew University. Sarid's ability to turn a phrase and his sonorous baritone voice made him radiophonic enough to get a job at Kol Yisrael, then Israel's only radio station. He worked there for four years, until 1964, when he switched to print media. The *Yediot Ahronot* daily newspaper hired him first as a court reporter and then as a satirist. Later that year, looking for a hip but loyal media whiz, the Mapai leadership offered him the position of spokesman. He gained their trust, and in 1965 Prime Minister Eshkol made him a policy adviser. Eight years later, at age thirty-three, he became the youngest member of the Knesset.

From the start he was a player. Sarid claims that in 1974 he convinced Pinhas Sapir, the consummate Labor power broker, to back Yitzhak Rabin for Prime Minister, which led to his narrow victory over Shimon Peres.[64] His political future as leader of the Labor Party seemed assured. But Yossi Sarid was too restless and self-righteous to bide his time. He refused to back down in efforts to expedite the resignation of Golda Meir and Moshe Dayan, whom he blamed for sundry debacles during the Yom Kippur War. He opposed the Labor Party's West Bank settlement policy, arguing that it would be a disaster in the long run. The media loved to quote him, and he always had something clever to say. To many Israelis, he was too smart for his own good. For instance, he took potshots (which he later regretted) at Yitzhak Rabin's drinking problems and was merciless in his attacks (which he did not regret) on religious right-wing chauvinism and the secret religious underground dedicated to killing Arabs.

Sarid himself realized that he was out of step. After the 1984 election stalemate, Sarid declined to support a national unity government with the Likud and resigned the Labor Party to join Ratz, a leftist splinter party run by human rights crusader Shulamith Aloni. Later Ratz merged with two other leftist parties to become the Meretz Party, receiving almost 10 percent of the votes in the 1992 elections. Prime Minister Rabin could not form a coalition without Meretz. Initially, Sarid was passed over for a Ministerial portfolio. But in October 1993, Rabin realized he would not be able to expand the coalition. He promoted Ora Namir to be Minister of Labor and gave Sarid the booby prize—the Ministry of the Environment. Sarid was knowledgeable about any number of subjects, but the environment was not one of them.

ENVIRONMENTAL EUPHORIA

When Yossi Sarid (see Figure 25) presented his credentials at the Environmental Ministry's headquarters on the third floor of the Interior

Ministry, the atmosphere was reminiscent of the liberated country of the Munchkins in *The Wizard of Oz*. The mood turned positively euphoric once the staffers got to know Sarid. They found that, in contrast to his sarcastic sound-bite image, Sarid was warm, charming, and genuinely sympathetic to environmental interests, (despite his chain smoking). On his first day at the job, he made the rounds of the departments at the Ministry's three Jerusalem offices, introducing himself as the new worker at the Ministry.[65] The staff also learned that Sarid was positively eloquent. At first, however, he mostly listened, absorbing information in an entirely new area. He proved a quick study.

To expand his base of power, Sarid opened his doors to environmental groups, in particular the Society for the Protection of Nature. This annoyed some of the bureaucracy's old-timers, who resented the promotion of SPNI activists to senior managerial positions, without appropriate academic credentials. Uri Marinov still grumbles that under Sarid, government policies and orientation were supplanted by the SPNI agenda.[66] The truth is that the organization's natural vision for Israel's landscape appealed to Sarid more than the technocratic, urban perspective that characterized Marinov's professional experience. Indeed, Marinov would later argue that Israel had to get used to being an urbanized city-state on the Singapore model.[67] Sarid felt that it was not too late to preserve the open spaces remaining in the land of Israel. He was also ambitious enough to pursue an agenda far broader than the government had envisioned for the Ministry of the Environment.

Sarid also infused the Ministry with an ethos of public service. Mickey Lipshitz coordinated nature preservation at the Society for the Protection of Nature until he became a Deputy Director at the Ministry of the Environment. He was excited to find a Minister who had no patience for bureaucracy, who felt that every single public complaint deserved an answer, and whose militancy on environmental issues was typically greater than his own, scolding his deputies that they were "moving too slowly."[68] Workers at all levels in the Ministry recount the overtime they put in out of a sense of devotion to their new boss.

THE MEDIA DARLING

It soon became clear that as chief eco-crusader, Sarid was having the time of his life. He managed to get the press interested in environmental issues that activists had long forgotten. In one of the most bizarre photo opportunities of the period, he took the press with him to inspect the cleanliness in gas station bathrooms, a pet peeve of the Minister's; he even issued a few fines.[69] When there were rumors about high radiation levels in the Small Crater[70] coming from the nearby Dimona nuclear reactor, Sarid took a group of reporters with him to measure radioactivity. They found only normal levels,[71] but later Sarid acknowledged that there had been a leakage of nuclear wastes on August 2, 1992, that was hushed up by authorities, including Minister of the Environment Ora Namir.[72]

One four-page newspaper profile of the Minister opened:

> Since Yossi Sarid has entered the government, suddenly one hears about the Ministry of the Environment. More precisely, one never stops hearing about it. The hourly news and the newspaper headlines report with an impressive frequency about nuclear waste disposal, sewage treatment facilities, and the Hiriyah garbage dump. Environmental problems that in the past were relegated to bottom priority today are squarely on the public agenda, and Sarid sees to it that they stay there.[73]

The environment ultimately proved too narrow to engage all of Sarid's energies and imagination, and the press followed him to foreign lands as well. Disturbed by the images of starving children in Rwanda, Sarid called Prime Minister Rabin and received his permission to lead a humanitarian medical delegation to the refugee camps in Zaire. Sarid also launched an initiative to save Muslims in Bosnia and bring them to Israel. What really interested Sarid, however, was the peace process. Rabin and Peres passed over loyal Laborites to add Sarid to the inside negotiation team that forged agreements with the Palestinians and the Jordanians.

Some environmentalists muttered that Sarid was really only a half-time Environmental Minister.[74] Sarid dismissed the criticism, explaining that his sixteen-hour-a-day routine allowed him time for everything, and pointed out that the foreign travel log of Ministerial junkets has him among the Rabin government's least frequent fliers. His capacity for work was indeed enormous, and early in his term he was hospitalized with cardiac problems. Still, it often seemed that he canceled as many speaking engagements as he actually attended, because of this or that pressing political exigency.

For the first time, Sarid made ecology a mainstream issue. The environment seemed a more common Cabinet subject than security, and there was a seemingly endless supply of anecdotes, mostly starring an exasperated Prime Minister Rabin, who in each would succumb to yet another Sarid-led subterfuge.[75] If nothing else, Sarid can be credited with providing employment for Green journalists. By the end of his tenure, almost every newspaper had a reporter with a part-time environmental beat.

FROM POVERTY TO MIDDLE CLASS

In perhaps his greatest achievement as Minister, Sarid translated his popularity into money. In 1992 the Ministry of the Environment budget was twenty-six million shekels. By 1996 it reached 231 million. The increase was partly due to Yisrael Peleg's long-time political connections in the ruling Labor Party with Beige Shochat, the Minister of Finance. Riding Sarid's momentum, Peleg managed to initiate a four-hundred-million shekel multiyear cost-sharing fund to help Israeli industries pay for pollution control equipment.[76] (Under Sarid, 120 million were immediately utilized.) The quantum leap in funding was most immediately reflected in the spiffy new Ministry offices. No longer the unwanted stepchild at the Ministry of the Interior, the Ministry dedicated its modern complex on the eve of Passover 1995.[77] Appropriately, it was located at "On Wings of an Eagle Street" overlooking the Angel Bakery. Not long before the move, Sarid had signed a personal decree that led to dramatic reductions in emissions at the bakery, a perennial Jerusalem air polluter.

Part of the reason for Sarid's budgetary success was his unique relationship with the Prime Minister (see Figure 26). The two men were not close politically or personally prior to Sarid's tenure. Indeed, Sarid had chastised Yitzhak Rabin when, as Minister of Defense, he called for "breaking Arab bones" in response to the Intifada.[78] But something clicked between the two soon after Sarid joined the Cabinet. Rabin never became an environmentalist, but he did give Sarid and his Ministry unusual latitude.

Sarid was never bogged down in the day-to-day minutiae of running a government Ministry. Yisrael Peleg was content to fill this role, happily operating under Sarid's shadow and overseeing budgetary details. It was Peleg who pushed for a comprehensive administrative restructuring. He claimed that despite the professed commitment to decentralized management,[79] when Marinov was Director General, in addition to his six regional coordinators, twenty-nine people were directly responsible to him. Peleg found this to be a managerial disaster and moved to appoint Deputy Directors to control different sectors. Peleg's most important contribution, however, was initiating the Year of the Environment in 1994.[80]

AN ENVIRONMENTAL YEAR

Yisrael Peleg never had any pretenses about his capabilities. He was not an environmental expert. Nor was he a charismatic figurehead—there were plenty of both at his Ministry. Peleg, however, did have some understanding about communications. In September 1992, soon after Ora Namir's appointment, he pushed her to have the coming Jewish calendar year declared as the Year of the Environment. Every year the Ministerial Committee for Symbols and Ceremonies picks a national topic for general promotion and public edification. Some subjects are more exciting than others (the Year of Democracy was a bit flat; the Year of Jerusalem, during its 3000th birthday, somewhat more flashy). With Peleg working eighteen hours a day as producer, and Sarid playing the starring role, Israel's Year of the Environment burst onto the public consciousness.

In a typical Peleg touch, it began on September 6, 1993, with a symbolic visit by Sarid and Ezer Weitzman, Israel's President, to the Hiriyah garbage dump with cameras conveying everything but the stench into Israel's living rooms.[81] Even Prime Minister Rabin was compelled to speak at the ceremony that evening at the President's house, pledging government commitment to solving environmental problems.[82]

Sarid's paean to environmental harmony transfixed the audience there:

> I see myself going into our national children's bedroom. They all sleep safely, our children, breathing easy; a good smell wafts in the air, for the atmosphere is intoxicatingly clear, the heavens are clean, the stars are out, and there is silence all around. It is possible to hear the waters churning in their channels, all the purified streams flow to the sea, and the sea is no longer dirty.[83]

When Prime Minister Rabin got up for his perfunctory address, he rolled his eyes and muttered that "Yossi was a hard act to follow."

Peleg brought in marketing concepts that were foreign to the applied scientists at the Ministry. For instance he insisted on using "focus groups" to consider possible logos to accompany the year, choosing a globe with a heart on it and the slogan "To the Environment with Love." Peleg tried to make environmental jargon more appealing, getting Israel's Academy for the Hebrew Language to adopt his new euphemism *matminot* (literally, places of digging) for "garbage dumps" in place of an older term that was an acronym for "sites for waste disposal."

The year was packed with environmental events: concerts for the environment; environmental film festivals; twenty-five specially produced educational television programs; the new ecologo on all government mail; an environmental curriculum for every grade level; soldiers marching in the name of environmental awareness;[84] a campaign to draft 250,000 volunteer litter-inspecting "cleanliness trustees" (the 100,000 who ultimately signed up were impressive enough); and the annual Independence Day torch-lighting ceremony featuring unsung environmental heroes.[85] The Ministry of the Environment was the hottest bureaucracy in town.

The country got an intense dose of ecology and seemed interested. Yet environmentalists are paid to worry. When they were not being overwhelmed by invitations to give lecture appearances, long-time activists invariably turned to each other and asked: "But what will happen next year—when the environmental fad passes?"[86]

A GREENER MINISTRY

Sarid's record as an ecological public relations genius is not questioned. Whether or not he put policies in place that materially improved Israel's environment is a tougher question. Sometimes the Ministry's enhanced status alone was enough to produce positive results. Under the Sarid administration, the Ministry of Justice was less likely to dismiss pleas for prosecution of egregious polluters than it had been before.[87] Before his tenure, environmental workers were never sure whether there would be political support for them if they stuck their necks out or tackled a powerful vested interest. Sarid justifiably prides himself in infusing some fighting spirit into the Ministry staff. When staffers would come to the Minister with explanations for a polluter's poor performance, he would berate them saying, "What's with this empathy? He can do a lot more! The fact is he managed to worry about electricity and hook up the water. Why shouldn't he worry about the environment too?"[88]

Emboldened by the Minister's support and the Ministry's status, environmentalists became more aggressive. In areas where the Ministry enjoyed regulatory experience, such as industrial air pollution, Sarid took implementation up another level. He issued personal decrees to all the operating quarries in the north of the country.[89] When the Haifa Oil Refineries violated the personal decree, Sarid signed an order that would close them within two weeks. The Refineries stopped fooling around and switched to 1 percent sulfur fuel year-round.[90]

On some policy issues, Sarid quite simply was Greener than his predecessors had been. He unabashedly admitted to adopting the perspective of nongovernmental environmental groups and praised them as his teachers. It was more than just fatuous flattery. Sarid's handling of the methyl bromide controversy is just one of many changes in Ministry policy arising from his instinct to "do the right thing" environmentally.

Methyl bromide is probably the most effective soil fumigant yet to be invented. After it is injected into the soil, practically nothing moves. The chemical quickly became an essential control for pests in crops such as strawberries or Galia melons. It is also incredibly lethal. Israeli farm workers have died when they failed to follow prescribed precautions and were directly exposed to the chemical.[91]

International attention began to focus on the pesticide in the 1980s, when scientists recognized its role in destroying the stratospheric ozone layer. The chemical percolates up through the soil and wafts toward the stratosphere, where it readily bonds with the reactive ozone (O_3) molecules. Although only about 25 percent of the methyl bromide in the stratosphere is produced by man (the oceans are the largest generator), it does not take much to upset the natural balance. Methyl bromide, it turns out, is thirty to sixty times more effective at destroying stratospheric ozone than more notorious halocarbon compounds such as Freon-based aerosols. The United Nations brought together a forum in 1992 that included dozens of the world's most influential atmospheric scientists, who reached a consensus. Anthropogenic emissions of methyl bromide used for fumigation applications could have accounted for one-twentieth to one-tenth of the current observed global ozone loss of 4 to 6 percent and could grow to about one-sixth of the predicted ozone loss by the year 2000 if methyl bromide emissions continued to increase.[92] Suddenly, stratospheric ozone depletion became a domestic Israeli issue. The Dead Sea holds the world's richest reserves of bromine,[93] making Israel the world's second-largest producer of the chemical.

Rarely are environmental issues front-page stories in Israel. When Greenpeace did a back-of-the-envelope calculation suggesting that Israel was responsible for 3 percent of the planet's ozone hole, however, it made the front page of *Maariv*.[94] The Dead Sea Bromide Corporation was not only the producer of a third of the world's methyl bromide. Its company representatives had served as Israel's official delegates at international conferences on ozone protection conducted under the Montreal Protocols of the Vienna Convention.[95]

The treaty led to the international phaseout of CFCs (chlorofluorocarbons, such as Freons) and was considered one of the outstanding triumphs of international environmental cooperation. Now the parties were focusing on methyl bromide as the ozone layer's second most potent enemy. As a developed nation, Israel's ratification responsibilities under the treaty involved payment to a multilateral fund to assist Article 5 Parties (developing countries) with projects that phase out ozone-depleting substances. Although Israel wanted to join the ozone-protecting nations, the Ministry of the Environment could not come up with the four-hundred-thousand-dollar annual contribution. Uri Marinov demanded that the Ministry of Finance allocate the funds, but as always its clerks were tightfisted. The Dead Sea Bromide Corporation recognized the pivotal impact of the treaty on its product's future and offered to help pick up the tab in exchange for a place on the delegation as advisors.[96]

The deal was cut in June 1992,[97] and industry scientists hastened to attend the next meetings of the parties to the treaty in Copenhagen in November. Dr. Michael Graber, a conscientious meteorologist who had overseen air quality since the early days of the Environmental Protection Service, requested to attend the meeting, but Ora Namir, then serving as Minister, thought it was a waste of time.[98] The stage was set for a public-relations and environmental fiasco.

Once they got there, the Dead Sea team began to rally developing countries to stop the proposed phaseout of methyl bromide, challenging the underlying science.[99] They branded the move as a "Western conspiracy." Attending U.S. government representatives were stunned at what they saw as a crass disinformation campaign by the Israeli delegation.[100] At the end of the convention, delegates decided only to freeze production of methyl bromide at existing levels and to revisit the issue in 1995 at Vienna.[101] The affair put Israel in the role of an environmental villain who put short-term profits ahead of global survival.

Israeli environmentalists backed Greenpeace's demand for a change in Israel's position. For years the government had been long on rhetoric about its commitment to international environmental protection.[102] Now that economic sacrifice was required, it was short on action. The agricultural lobby, however, was unrepentant. Annual bromine production was worth an estimated sixty million dollars, and many Israeli farms were completely dependent on the chemical.[103]

Sarid took a different tack. He wrote Greenpeace that he was deeply concerned with the depletion of the ozone layer and invited it to send its ozone experts to come talk to him.[104] He pledged that he would never again allow industry to represent Israel's national interests. Sarid appointed a committee to study the issue, headed by pesticide expert Professor Yaakov Katan. When the next meeting of treaty nations was held in Vienna, the Ministry's new Director, Aaron Vardi, headed the Israeli delegation. Based on Katan's recommendations, Vardi backed the U.S. position, which called for phaseout.[105] Despite angry protests by farmers, Sarid stood his ground, and the government backed him.[106] The ultimate compromise called for a 70 percent interim reduction by 2003 and phaseout by 2005. (This did not stop Dead Sea Bromide from circumventing the agreement by creating joint ventures with developing countries, China for instance, that were not parties to the agreement.[107])

OPEN SPACES

Methyl bromide was actually a marginal issue for the Ministry. Soon after taking office, Sarid became convinced that preservation of open spaces was Israel's top environmental challenge.[108] The issue was far beyond the Ministry's traditional sphere of influence, and Sarid found he was swimming against a very strong current. He seemed to enjoy the exercise.

Ironically, although farmers may be faulted for their contribution to water pollution problems, they also deserve credit for literally serving as a hedge for open-space preservation during Israel's first forty years. Though cultivated farmlands transformed the landscape, they provided a buffer. This was especially important for maintaining local biodiversity in many of Israel's central and northern postage-stamp-size reserves. The weakening of Israel's agricultural sector during the late 1980s would have unforeseen but extremely grave ecological ramifications.[109] Open spaces began to disappear as farmers sold out. A comparison of aerial photographs in the center of the country showed that between 1987 and 1996, the 250 square kilometers of built-up areas were enlarged by an additional 490 square kilometers. If the trend continued, open spaces would be reduced by at least a half within a generation![110]

The influx of hundreds of thousands of Soviet immigrants could only partly explain the phenomenon. Veteran Israelis were happy to sell them their crowded city apartments to buy up and move to more comfortable suburban developments. Villas had once been considered a luxury for Israelis. Most rural cottages were austere, and detached single-story homes constituted only 10 percent of housing starts. Yet, with prosperity, the amount of residential space per person rose steadily in Israel, and by the late 1980s more than 55 percent of the units built in Israel were private houses.[111] The resulting sprawl brought with it roads and a proliferation of industrial areas and accompanying services.[112]

The change in agricultural status, though, was the primary cause. Until the 1980s, farmland hugged the city limits of many Israeli towns, forcing planners to build up rather than out. Agricultural communities themselves remained small—limited in the moshav sector by the amount of cultivable plots, and in the kibbutz by both the public's hesitancy to embrace its socialist restraints and the selectivity of kibbutz members. Eighty percent of

Israel's sparsely populated lands were under the jurisdiction of Regional Councils that were dominated by rural interests and a Zionist ethos that embraced agricultural interests as a top national priority.[113] Developers knew they were outflanked and focused their energies inside municipal boundaries.

But all this changed virtually overnight in 1990. When the Israel Lands Administration was transferred from the jursidiction of the Ministry of Agriculture to the Building and Housing Ministry, it was more than a bureaucratic shift. Under the pressure of the emergency legislation passed to circumvent the planning committees, all the dams seemed to burst at once. Almost half of the 340,000 new residential units approved for development between 1990 and 1994 were outside city limits.[114] The Agriculture Land Preservation Committee, which had enjoyed virtual veto power to undermine development in rural areas, was quietly denuded.[115] At the same time, a 1990 decision by the Israel Lands Administration Council completely changed farmers' economic incentives. Rather than prohibiting farmers from converting their fields to a residential designation, compensation for doing so was instituted and zoning procedures were simplified.[116] For the many farming communities who were already deeply in debt, whose livelihoods were less and less dependent on agriculture, and who sought to build new neighborhoods for their children, it represented a windfall. For many farmers it was easier just to sell out.

Sarid railed against the disappearing Israeli landscape at every opportunity. He adopted the environmental party line: Only in the desert was there room for suburbia. (Because he was a long-time resident of a northern Tel Aviv high-rise, his message did not smack of the elitist hypocrisy that many environmentalists suffered who themselves enjoyed a rural lifestyle but called on everyone else to live in the city.) Sarid did more than simply raise public awareness about the importance of open spaces. He also pushed resolutions through a Cabinet-level environmental committee that carried the force of government decisions. The craters in the Negev were to be preserved rather than used for mining. The Sharon Park, the last open space between Netanya and Hadera, while still mired in private claims, got a major boost from a Cabinet-level blessing. In a Sarid-brokered compromise, a highway that would have cut into the heart of the park was channeled underground. And the emergency Planning and Building Procedures Law was at long last canceled.

In an astonishing departure from Zionist dogma, Sarid also pushed the government to adopt a moratorium on new settlements inside Israel. As he was fond of saying, "There just isn't any room left." Under a new vertical imperative that many architects with Green credentials called excessive, city planners were suddenly directed by the government to limit construction to high-rises. In an attempt to seize initiative and offer a positive alternative,[117] the Ministry of the Environment's planning department identified a triangle of land in the northern Negev region that was targeted as the next major development zone.[118] Located in an area that was neither hydrologically vulnerable nor a unique landscape, its carrying capacity was estimated at one million people. The Ministry's plan to shift development to the northern Negev region was never actually put into practice,[119] but Sarid got the issue of Israel's open spaces onto the government's agenda, where it needs to remain forever.

SARID'S POLITICAL REALITY CHECK

Despite the fond memories and the general consensus that Yossi Sarid has been Israel's best Minister of the Environment to date, pollution parameters show that his three years in office did not produce a revolution on the ground. Many of his Ministerial failures came down to the old-fashioned issue of political clout. Despite his ability to wield public opinion and his willingness to go head to head with industry on many issues, Sarid would knowingly bite off more than he could chew. His efforts to prevent the passage of the Dead Sea Concessions Law were indicative of the limitations of his Ministry when it clashed with powerful economic interests.

After the British Mandate granted a seventy-five-year concession to the Dead Sea Works in 1930 to exploit the rich mineral resources in the saline lake, the factory steadily swelled until it became an industrial behemoth. Each year it pulls two million tons of potash and 180,000 tons of bromine out of the water, as well as huge amounts of magnesium, potassium, and, of course, salt.[120]

In 1991 the government decided that the factory would be sold to investors as part of the government's general move to privatize. Beyond the annual profit, part of the attraction that the factory offered to buyers was its extraterritorial status. In 1961 the Knesset had passed a law which granted the factory 620,000 dunams of land (3 percent of the entire country) as part of its zone of exploitation.[121] No less important, it seemed to give the factory functional autonomy within this perimeter, even though it included sizable tracts designated as Nature Reserves. For years the Dead Sea Works interpreted the concession as granting it exemption from basic Israeli laws and launched a number of construction projects without bothering to ask for building permits. The southern section of the Dead Sea came to resemble a chemical production zone, having little in common with the sea's international image as a unique medicinal resort at the lowest place on earth.[122]

This industrial immunity was called into question by the Moriah Hotel in 1992. It was the first time that conflicting tourist and mining interests ended up in court. When Israel had decided to support a tourist area along the Dead Sea, it picked the area around the western edge of the Ein Bokek Wadi to build a string of luxury hotels, beaches, and tourist shops. The compound is located due north of the Dead Sea industrial zone. The constant vaporization process required for mining in the sea necessitates the construction of dirt walls. Over time, these walls had to be raised higher and higher as the beds filled with silt. The Moriah Hotel filed suit in the Beer Sheva District Court claiming that the construction hurt its beachfront and had not been approved under the Planning and Building Law. In August 1992 Judge Yitzhak Ban made a surprise ruling, holding that the Dead Sea Concessions Law did not supersede Israel's Planning and Building Law.[123]

The Dead Sea Works management, along with its government patrons in the Ministry of Finance, recognized that this interpretation could have grave repercussions for the company's net value. They had 860 million dollars' worth of new projects in the pipeline, including a joint magnesium extraction venture with the Volkswagen Corporation. The company proposed that the Knesset set the record straight with a special law that confirmed its extraterritorial status, retroactively grandfathered existing structures, and streamlined the approval process for future initiatives. The government was supportive and drafted a law that created a small rubber-stamp committee to oversee the Dead Sea and its new projects.

The Dead Sea Concessions Law generated considerable media attention, and environmentalists tried to frame it as a "Reading Power Law for the 1990s." Did Israel really want to grant a factory special exemption from the law simply because it was profitable, or should such businesses meet the same environmental standards as all other citizens and corporations? In the entire Knesset, only one serious dissenting voice was heard. Political science professor Benny Temkin spoke passionately about the rule of law and democratic norms; he submitted environmental amendments and even considered a filibuster.[124] Temkin's losing effort in the Interior Committee found support from an unlikely source. Even though the bill was a government-proposed statute, the Minister of the Environment broke ranks with the Cabinet to fight it. "I was ready to go a great distance toward the factory, but they wanted me to crawl to them, and the Ministry of the Environment has stopped crawling," Sarid told the press.[125]

In the end, the Dead Sea Concessions Law was one of the environmental movement's many legislative failures during the 1990s. Sarid knew early on that he was beaten but wanted the fight to be as fierce as possible. He personally called environmental leaders, asking that they be more vociferous in their opposition. Sarid quite naturally slipped into his old opposition role, making a stirring extemporaneous speech in the Knesset committee against his own government.

But the Dead Sea Works proved too powerful. The Prime Minister himself backed Victor Medina, the influential chairman of Israel Chemicals, which owned the Dead Sea Works. Medina walked the corridors of the Knesset, quietly convincing key legislators of the law's significance. Dozens of workers were bused into Jerusalem as a backup, lobbying for a law that was crucial to employment in the economically lethargic Negev region. The Dead Sea Works even painted its facilities in fluorescent (almost surreal) colors. But it adamantly refused to provide the Ministry of the Environment with the emissions inventory it demanded. Despite the protests of environmental groups, and a late but passionate campaign by the Society for the Protection of Nature (which took the Knesset's Interior Committee on jeeps to view the damage from the plant firsthand), the law passed handily.[126]

Sarid's feistiness was a genuine inspiration to environmental groups. If the Minister of the Environment could fight this hard for environmental interests, then certainly they could try a little harder. Unfortunately, sometimes even the most valiant environmental efforts are doomed to defeat. Sarid's "last stand" against the Trans-Israel Highway was such a case. In 1995, after the highway seemed a foregone conclusion, Sarid drafted Ministers Yossi Beilin and Yaakov Zur, forcing the Cabinet to revisit the issue. Environmental groups who had fought the highway and lost in the planning commissions and courts were revitalized.[127] Money was raised, rallies staged, and advertisements posted, and a public opinion poll suggested that the Israeli public had actually begun to change its mind. Although the government decided to continue with the highway, the Cabinet vote was surprisingly close, considering the extent of past support for the project. As the elections approached, Sarid even promised to make cancellation of the project a condition of his party for entering a new government.[128]

But Sarid never held those coalitional cards. Although last-minute efforts to stop the project continued,[129] the Trans-Israel Highway represented a glaring example of governmental and public-interest environmental failure.

Thus, even at the peak of the Ministry's "golden days," total defeat in seminal issues like the Dead Sea Concessions and the Trans-Israel Highway revealed the marginal place of the environment in the country's overall priorities. For instance, critics claimed that the Ministry of the Environment invariably caved in when the mere possibility of unemployment was raised.[130] Most environmentalists saw merits in Sarid's willingness to fight the good fight, even when it was clearly going to be a losing battle. Others preferred Vince Lombardi's slogan—"Winning isn't everything; it's the only thing"—and saw such quixotic quests as a waste of time, political capital, and resources.

SYMPTOMS VERSUS STRATEGY

It was not just political strength but also competency that proved an impediment to Ministerial effectiveness. Sarid was a brilliant tactician, but his Ministry often acted without a cogent strategy. There was no effort to quantify environmental risks and to use an analytically rigorous process for setting priorities.[131] Economic instruments were not among the tools in the policy toolbox. "Source reduction," "clean production," and "pollution prevention"—the mantras of environmental Ministries around the world at the time—did not really penetrate the thinking at Israel's Environmental Ministry. Frequently the Ministry appeared impetuous, trying to solve all problems at once, with haphazard follow-through.

Indeed, the very breadth of Sarid's agenda tended to diffuse the Ministry's effort. Sarid pulled the Ministry in unanticipated directions. For example, because no one else in government appeared interested in the area (and Sarid was), he adopted the issue of animal rights. A new law was passed that gave the Minister of the Environment power to intervene when animals were being abused.[132] Sarid took the issue seriously; he prohibited circus performances with animals and banned filming wildlife outside of its normal habitat.[133]

It was an admirable initiative, but at the same time the Ministry was making little progress in other, more traditional environmental issues, such as the handling of hazardous materials, the establishment of a new toxic-waste facility to supplement Ramat Hovav, and the guaranteeing of an efficient, integrated emergency response to accidents involving hazardous substances.[134] The Hazardous Materials Law that passed in the Knesset fell far short of a modern omnibus cradle-to-grave regulatory scheme. The Ministry staff continued to do the best it could with what was essentially a repackaged ordinance from the British Mandate.[135]

In other areas, it was not lack of attention, but narrow vision that limited progress. Solid-waste policy at the Ministry of the Environment was such a case. The amount of trash generated grew at a steady rate of 5 percent per year.[136] Ministry propaganda spoke of a hierarchy of solid-waste management that started with source reduction, recycling, reuse, incineration, and burial as a last resort. While the Ministry helped pass a recycling law,[137] in practice, the record shows that the Ministry of the Environment's efforts went into trash burial and not into reducing the quantities of garbage or to treating it as a resource.

Prior to Sarid's tenure, the Ministry of the Environment's Solid Waste Department decided that its priority should be the immediate closing of the four hundred garbage dumps across the country. Staffers argued that the damage from these sites to groundwater alone dwarfed all other associated environmental hazards. Concentrating all of the nation's garbage into five environmentally responsible sites would allow for more effective regulation. Sarid agreed and got the Cabinet to adopt the position officially.[138]

When environmentalists brought up recycling, composting, and other treatment alternatives, they were dismissed as premature or even sentimental. The Ministry favored a more pragmatic incrementalism. According to the prevailing paradigm, Israel's garbage system was at step one, with illegal dumps scattered across the country. A state-of-the-art, integrated waste management strategy was step three. Before Israel could get there, the country had to move to step two—closing the dumps and burying trash in national centers. Only then did it make sense to pursue other waste management options.[139]

This keen sense of its own limitations also seemed to drive the Ministry's policy in other areas, such as hazardous waste disposal. At the Ministry's inception, Prime Minister Shamir endowed it with authorities to oversee the government corporation that ran the Ramat Hovav disposal site. But the Ministry had never been able to upgrade the facility to reach a reasonable level of safety, much less implement environmentally sound practices.

A certain defeatism was in the air when scientists from Adam Teva V'din and local settlements negotiated with the Ministry over conditions at the country's first hazardous-waste incinerator at Ramat Hovav. The American trial-burn techniques demanded by environmentalists would be expensive. The Ministry wanted to keep the costs of disposal low. The environmental activists countered that high prices should actually be seen as a positive policy outcome, encouraging factories to modify production methods, recycle, and reduce the use of hazardous chemicals. The Ministry's response was that if it could, it would allow free delivery of hazardous chemicals at Ramat Hovav. Well aware that over half of Israel's toxic wastes were not getting to the Ramat Hovav facility anyway, the Ministry felt helpless to root out this environmental lawlessness and preferred to concentrate as much hazardous waste as possible in a single disposal safe.

But this pessimism meant that the Ministry spent its limited resources on the symptoms rather than the causes of Israel's garbage crisis. For example, the Ministry made battery collection one of the Year of the Environment's key community gimmicks. Stores across the country kept boxes in which conscientious citizens could drop off their used batteries. They were then taken for burial at the Ramat Hovav hazardous-waste site. But a "pollution prevention" orientation might have pointed to the simpler route of banning mercury in batteries and providing tax incentives for using rechargeables. Europe had been doing it for years. Similarly, well-publicized beach cleanups seemed to accomplish little, as the litterers were not among the diligent volunteers. Within a few weeks, the debris would be back.

Thus, conceptually, the Environmental Ministry became locked in the vicious throwaway cycle that was the heart of the problem. It seemed to scoff at the laws of thermodynamics and at the ecoslogans such as "everything must go somewhere" that it had always told the public. The fact that Israel buried an astonishing 94 percent of its trash (as opposed to countries such as Switzerland, for which the rate was 17 percent)[140] was somehow considered immutable or an issue to be put off until the future (see Figure 27). Rather than framing recycling and reuse as a moral and civic duty and a tool to teach about scarce resource preservation, the Ministry caved in to narrow economic analysis.

This was a reversion to the "cowboy" or frontier economics of Israel's Ministry of Finance, rather than the spaceship-earth approach to economics that environmentalists around the world worked so hard to promote.[141] With efficiency as the criterion, recycling was not an end in itself but only a tool for increasing landfill longevity. "Our job is to extend the life of garbage dumps as much as possible," wrote Yossi Inbar, head of the Ministry's Solid Waste Branch.[142]

By the fall of 1994, Sarid himself went on the record, calling to stop recycling's momentum:

> Under present conditions, the great mass of garbage won't be recycled. It is certainly a retreat from what we thought in the past. Now that we have clearer concepts about the economic side of waste treatment, it's clear that things have changed. Should we go into mourning because it became clear that trash burial is cheaper? No, the Ministry of Finance should rejoice at this, and all of Israel should rejoice.[143]

The Ministry's position was especially peculiar, because it came at a time when world markets for recycled products had reached an all-time high. The price of plastics, for example, had doubled. Recycling initiatives around the world that for years had operated in the red suddenly became profitable. Indeed, in the absence of local collection, the Ramat Hovav-based Aviv factory imported 960 tons of discarded plastic from Germany for production.[144] The Ministry's 180-degree turnabout was among the several political (rather than environmental) decisions that prompted the resignation of Chief Scientist Yoram Avnimelech.[145]

Sarid's great disengagement on the garbage front disappointed environmentalists, who felt that it undermined years of public education. But it absolutely outraged the residents of Beer Sheva, who were unwilling to pay the price for the Environmental Ministry's optimal scheme for burying the nation's garbage. Beer Sheva had been chosen as the final destination for Tel Aviv's garbage, after an alternative site near the newly declared Beit Jubrin Caves National Park fell through. The vociferous protestations of the nearby city of Kiriyat Gat and of the Society for the Protection of Nature had persuaded the Ministry staff to go further south.

Sarid backed their proposal to convert Beer Sheva's existing municipal Dudaim sanitary landfill into a much larger national site for the garbage of the Tel Aviv region.[146] The Ministry argued that the Dudaim site was not in a sensitive hydrological location and that it was a safe seven kilometers from the heart of the city. Moreover the landfill was already up and running, offering an immediate solution to the Hiriyah mountain, whose bird population increasingly posed a hazard to incoming planes.[147]

A broad coalition of Beer Sheva interest groups joined together to protest what they perceived as a classic environmental injustice. Unimpressed by long-term promises to turn the Dudaim site into a park, they argued that the odors and stigma from the facility would stymie future growth in precisely the area where the city needed to expand. Moreover, Beer Sheva residents resented the fact that the politically powerful Tel Aviv region was exporting its trash and turning the Negev into the national waste bin. Miriam Turkel, the independent City Councillor who led the fight, argued that Israel's trash problem ultimately was not a scientific issue. "It's a social conflict between the poorer and the richer; between Israel's bottom and its top."[148] When Sarid came to visit Beer Sheva, he was met by angry demonstrators who pelted him with tomatoes.

Sarid pushed ahead. The National Planning Council gave its approval, and a tender for managing the site was issued. Enormous efforts were spent in trying to sell the Ministry's position. Initially, the policy sputtered. Legal challenges were filed against the Ministry. With elections in the air, no party wanted to alienate a large voting bloc, and the Knesset supported the local Beer Sheva position. Then Tel Aviv refused to pay the transportation costs to Beer Sheva and expressed its solidarity with its southern brethren.[149] Finally, the losers in the tender sued for an injunction, claiming that their proposal was better and thirty million shekels cheaper.[150] Despite Sarid's pledge, it was not clear when Hiriyah would finally shut down.

Even when progress was made in the solid-waste area, the Ministry did not always move to consolidate it and prevent backsliding. The case of beverage cans is illuminating. Aluminum made from recycled materials is far cheaper and less energy-intensive than that produced from virgin ore.[151] That is why many recycling programs begin with aluminum cans. Environmentalists called for regulations to require aluminum can production, but Sarid countered that the same results could be reached through moral suasion. And initially he delivered. The Caniel Corporation, producer of over two hundred million cans a year, agreed to change its production process (the alloy of aluminum, tin, and zinc that it had been using was not recyclable).

Because of the lower production costs of recycled aluminum, it was assumed that the used cans would generate their own market. With great fanfare and at an expense of three million dollars, the factory made the transition.[152] It was a small but highly visible improvement. Unfortunately, the invisible hand of the market remained hidden. Without the logistics of a deposit and collection system or a bottle bill in place, Caniel's aluminum cans ended up in the trash. Within a year, with the price of aluminum skyrocketing and practically no recycling taking place, the company quietly returned to its old, unrecyclable production materials.[153]

When Sarid left office, he could justifiably boast that dozens of polluting garbage dumps had been shut under his tenure. But no sustainable policy alternative was in place. Three thousand tons of trash still rolled up the eighty-four-meter-high Hiriyah garbage mountain every day. The garbage finally toppled over into the Ayalon River during a rainstorm in 1997.[154]

Medical waste remained completely unregulated.[155] And "recycling" seemed like a word from a foreign language.

OUT OF TIME

In many areas, Sarid simply ran out of time. It was impossible to follow through on the countless hastily launched initiatives. "I need a new environmental impact statement law immediately," he would charge his staff. They would work around the clock to produce an imperfect but improved proposal. Then it would sit around for years while other distractions took precedence, and never even be submitted to the Knesset for a preliminary reading. Car emissions pushed concentrations of oxides of nitrogen to new highs, and there were on average three hundred air quality violations a year in the Tel Aviv area. In April 1996, Sarid promised environmentalists that he would circumvent the Ministry of Transportation and use his authorities under the Kanovich law to sign new auto-emissions standards into force as soon as a version that his staff approved would reach his desk. Regulations were quickly drafted, but he did not get around to signing them. Eventually environmentalists had to sue for modern motor vehicle emissions standards.[156]

Perhaps the most important of Sarid's unfinished business was gaining formal, statutory authorities to tackle the broadened environmental agenda he had already begun pursuing. This has been the most basic obstacle faced by Israel's Environmental Ministry from its inception. New and expanded drinking-water standards were stalled at the Ministry of Health. The Ministry of Transportation did not seem to care about air quality but remained the vested authority for overseeing the inspection of car emissions. The Minister of Agriculture and his Water Commissioner retained the authority to enforce water quality standards. When the Environmental Ministry proposed a map of sensitive karstic and sandy hydrological zones in which no wastewater should be used for irrigation, it was not taken seriously.

No one could castigate polluters in the press like Sarid or make better use of public forums to vilify them. But without full executive powers, his Ministry could only push through a handful of legal actions against water polluters. Sarid was quite aware of the constraints and sought to change the situation, but efforts to convince the Minister of Agriculture to bring the Water Commission to the Ministry of the Environment fell flat. Sarid claims he never had time to realize his vision of an "environmental empire" to be built with authority from the Ministry of the

Interior and the Antiquities Authority, which could stop any construction project.[157]

Finally, Sarid's ingenuous faith in the powers of an Environmental Minister inevitably led to disappointments. In other words, if he had not been so ambitious, he might have looked better. It seemed that the Minister was unable to pass up any issue that involved unfavorable environmental results. Profligate coastal developments, the Trans-Israel Highway, and the Dead Sea Concessions were among the many issues that any other Environmental Minister would have conceded when faced with such poor odds. But many of his assurances—like the one he made to beleaguered neighbors that the gigantic (and noisy) central bus station in Tel Aviv would not be allowed to open—were empty promises.[158]

Yet even when Sarid failed to attain his immediate objective, his efforts usually produced some favorable result. At the other end of Tel Aviv he tried to ban nighttime domestic flights at the Sdeh Dov Airport and even close down the terminal, which is located adjacent to the Ramat Aviv neighborhood where he himself lives. Once again he was ambushed by the Ministry of Transportation. Nevertheless, he imposed a gentleman's agreement on the airlines: Flights leaving after ten o'clock were to be towed out to the runway by a silent electric cart (Sarid was well aware that every night, delayed passengers "made mention of his mother"[159]). Eventually, the parking area for planes was moved west, adjacent to the sea and away from the residential areas—a small but key step toward improved quality of life for thousands of Tel Aviv residents.

In the final analysis, given his point of departure, it is harder to imagine a much better Environmental Minister. Sarid left a stronger, more effective, and more confident Environmental Ministry than he received. Israelis' attitudes to the environment were, often unwittingly, touched by the ardor of his three years in office. So were the many corners of the country that were left healthier and cleaner after Sarid's efforts.

LIFE WITH A MINISTER OF AGRICULTURE AND ENVIRONMENT

Bibi Netanyahu and the Likud's election victory on May 29, 1996, brought the Sarid era to an abrupt end. Environmentalists lost no time in lobbying for an appropriate replacement, but it was hard to find a politician who wanted the job. To keep the Cabinet size down, the Tsomet Party's Rafael "Raful" Eitan agreed to return to the Ministry of Agriculture and serve simultaneously as Environmental Minister. (The position of Minister of Police that he truly coveted was blocked because of an indictment filed against him that was eventually dismissed.[160]) At first it was not even clear whether the Ministry would survive as an independent entity. It did, but Eitan's background as a farmer suggested where his real interests lay.

Like Sarid, Eitan was head of a small ideological party. Both were known for their commitment to principles and were famed as entertaining storytellers. But all similarities stopped there. Eitan was a retired general and had served as the Israel Defense Force chief of staff. Sarid completed military service as a private. Eitan was the consummate hawk and Sarid the original dove. Sarid waxed prolix while Eitan was laconic and blunt. Yet, there was ample reason to believe that Eitan was a strong appointment. As Minister of Agriculture during the early 1990s, he had stood up to agricultural interests and pushed through a more sustainable water policy. Environmentalists hoped that he would continue Sarid's independent tradition of advocacy.

By the time of the elections, Yisrael Peleg had resigned as Director General of the Ministry to make an unsuccessful run in the Labor Party Tel Aviv primaries. Aaron Vardi, a straight-talking retired military officer, replaced him. Once elected, Eitan immediately replaced him with his personal assistant, Nehama Ronen. With Eitan distracted by partisan politics and another Ministerial portfolio, Ronen overnight became the central figure in the Ministry.[161]

Expectations were low, because Ronen had no substantial administrative experience beyond working within Eitan's Tsomet Party apparatus. An attractive blonde who looked younger than her thirty-five years (see Figure 28), she was well aware that her appointment was greeted with skepticism. The story quickly circulated that Eitan (an old paratrooper) had hired Ronen fresh out of the army in 1984 primarily because, as a clerk for the 202nd Paratrooper Battalion, she had jumped out of planes.[162]

After thirteen years of intense political activity, and a strong sixth position on the Tsomet Party list, Ronen was confident she could manage.

> People dealt with me more cautiously, not only because of my sex but because of my age. Most of the people had been working in the field for many years and I didn't come from the environmental world. But today it is easier to see women in key positions, and I think they began to judge me according to the results.[163]

Ronen brought a pragmatic, no-nonsense style to the job, gaining her the grudging personal respect of many within the environmental community.[164]

Given the Ministry of Agriculture's lackadaisical record in the area of water quality, environmentalists were quick to seize on the potential conflict of interests between Eitan's dual Ministerial loyalties. Ironically, Eitan was deprived of the best opportunity for Ministerial synergy when the Water Commission was transferred from the Ministry of Agriculture to the newly created Ministry of Infrastructure. Meir Ben Meir had been reappointed as Water Commissioner and from the start was perceived as uncooperative with the Environmental Ministry's staff. The Commissioner's powerful investigative and enforcement powers were never marshaled for environmental ends.[165]

In her public presentations, Ronen tried to paint the fact that her boss wore two hats as a source of opportunity. She cited initiatives to solve dairy farm runoff in the Golan Heights and a continued commitment to meet a 2004 phaseout deadline for methyl bromide as examples of environmental dividends her boss had delivered.[166] In retrospect, however, Eitan was rarely inclined to take advantage of his agricultural portfolio to push farmers seriously toward reducing fertilizers and pesticides and toward other sustainable management practices. Pesticide usage in Israel is still among the most intensive in the world; practically no oversight of application takes place in the field, and cancer rates among farmers remain high.[167] Agricultural pollution is an environmental problem that requires rethinking and innovation. Providing farmers with economic incentives helps but may not be enough. International experience suggests that the proverbial carrot may actually be less effective than the stick of regulatory demands to change farm practices.[168] At the same time, the sheer number of agricultural polluters makes command and control regulation a difficult policy to impose on Israel's farmers.[169]

On the whole, Eitan proved capable of leaving his competing interest at the door when he approached his environmental duties. Although he did not exploit his environmental post to further agricultural interests, neither did he create a meaningful new ecological initiative in Israel's rural sector. In the long run, Israel's farmers are certain to be the losers with regard to the lack of a serious policy response: It is their land and water that sustain the greatest damage, as well as their health.

CHANGE AND CONTINUITY

Just as anyone would have been popular in the Ministry after Ora Namir's reign, it was tough coming on the heels of a beloved figure like Yossi Sarid. There was a very conscious decision by Ronen not to compete with his "greener than thou" orientation. The cozy relations that existed between environmental groups and the Ministry provided one such example. Only two months into the job she told the press, "Green organizations will no longer have free run of Israel's Ministry of the Environment." While she wanted good relations with environmentalists, she made it clear that it was unacceptable for them to be as involved in the Ministry's decision making as they had been under the previous administration.[170]

The issue came to a head over the development of a newly reflooded lake in the Huleh region. Much of the agricultural lands redeemed by the JNF after the swamp was drained proved to be useless to farmers (the underlying peat smoldered and collapsed once the protective layer of water was removed). In June 1994, a thousand dunams of lands in the Huleh Valley were once again flooded.[171] Although this was only 1 percent of the greater Huleh area, the effect of the re-created wetlands was astonishing. Within a year, tens of thousands of cranes returned to the new winter resort, apparently en route from Finland.[172] The farmers who had lost fields to the project were promised the rights to establish a guesthouse near the newly formed ecological attraction.

The Society for the Protection of Nature, with its unique historical relationship to the wetlands, opposed any contiguous tourist development. It called for accommodations to be concentrated in existing communities.[173] Ultimately the conflict centered over the appropriate definition for the term "ecotourism."

Although close to ninety years of age, Heinrich Mendelssohn still pulled no punches on ecological issues: "What they are doing is stupid, because all the beautiful birds that people will come to see just won't be there. They will fly off. And then the guests will of course suffer from the mosquitoes so they will have to use insecticides—causing irreparable damage to the site."[174] Sarid had backed this position, but Rafael Eitan had more empathy for the plight of the locals. The Ministry supported a compromise that moved construction away from the water itself but allowed it in the general vicinity. The SPNI denounced the retreat, and the new Ministry management returned fire with fire.

The heart of the disagreement seemed to be the appropriate role of an Environmental Ministry within Israel's executive branch. The Eitan-Ronen administration believed that the Ministry of the Environment agenda should be broader than those of nongovernmental groups and should include general public concerns.[175] In contrast, Sarid took a pluralistic view toward government, arguing that in light of the severity of the situation, environmental agencies should harbor an uncompromising ecological commitment.[176]

Administratively the new leadership also began to chart its own course. It did not seem to mind if the Ministry lost the affection of local governments, and it began to divert funds away from municipal environmental protection units. Environmental enforcement, arguably the Ministry's greatest weakness, presumably was better served by this move, and the number of enforcement agents in the environmental patrol was doubled twice.

Substantively, the Eitan-Ronen administration made some changes in priorities. Cleanliness seemed to have a visceral appeal for the Minister. Considerable funds were allocated for a major antilitter campaign. Some, however, found the poster girl's ubiquitous admonition "Whoever dirties is trash" more annoying than adorable. The campaign certainly caught the attention of the public, but follow-up surveys failed to detect any change in societal attitudes or practices.[177] Ronen explains that the campaign was part of a broader educational orientation at the Ministry: "I believe that this sort of campaign will ultimately lead factory owners to begin cleaning up."[178]

On most substantive issues, however, there was no detectable shift in formal environmental public policy. The Environmental Ministry was still officially opposed to the Trans-Israel Highway, and Ronen even ventured the view that if there had been a concerted effort against the highway from the outset, the project might have been stopped. The Ministry certainly did not become soft on pollution. Once she became acquainted with the situation, Ronen took a hard line on industry, finding funds for more inspectors in the Ministry's field enforcement unit. "I quickly learned that factories won't invest in pollution control technology unless they are forced to do it and see that their friends have been fined for polluting," she explained.[179] But resources were still inadequate to take on the enormous task of turning things around.

The problem was not just quantitative but also qualitative. Tackling serious water and air polluters requires regulators with strong science backgrounds. It is little wonder that the vast majority of the prosecutions initiated by the enforcement branch in the Ministry involve littering, where it takes only a camera to catch a violation.[180] In this sense, the Ministry's inability to turn the Maccabiah disaster (discussed in Chapter 1) into a serious enforcement initiative or a turning point for river restoration was as much an example of inappropriate skill sets as it was flimsy political will.

After the initial purge of SPNI influence, cooperation between the Ministry and environment groups resumed, although it had a more circumspect tone. Even watchdog Adam Teva V'din took time off from its legal petitions against the Ministry to run their fifth Environmental Film Festival together.

Nonetheless, Yossi Sarid's exit left a gaping hole in the Ministry's image and initiative. The lack of a full-time Minister was a blow to the Ministry's prestige. The change in intensity was felt outside the Ministry as well. The environment was no longer an issue that got the press excited. It was also not on the Cabinet's agenda.

The new Ministry of the Environment's administration was by no means a pushover. (It was Ronen and not Sarid who closed down Tel Aviv's Hiriyah garbage dump, and she was crucial in a battle to stop the establishment of a unsightly marina in Haifa.) Rather, the general stock of "Israel Environment, Ltd." had taken a dive.

When Daliah Itzik, a forty-seven-year-old teacher turned Labor politician, was appointed Minister of the Environment by newly elected Prime Minister Ehud Barak in the summer of 1999, she was vociferous in expressing her disappointment. Anticipating the Education portfolio (which ironically went to Yossi Sarid), she pouted about Barak's "betrayal" on the front page of every newspaper.

Itzik eventually offered up the requisite lip service about the importance of her new post[181] and showed considerable savvy with the press. But like the "National Environmental Council" she sponsored (which lacked any discernible authorities), her efforts often seemed largely symbolic. She brought together a Ministerial Committee on the Environment that put on an impressive show in the Knesset with the participation of the Prime Minister but which was quickly forgotten. Itzik gave superb speeches against the Trans-Israel Highway and even brought blankets to protesters camping out by the construction site, but for months, while the bulldozers rolled, she was unable to get the issue on the Cabinet's agenda. Eventually, her bid to cancel the highway was voted down in the Cabinet, seven to five. Itzik also had the good sense to appoint Yizhak Goren, who had successfully served in several top regional and national positions since the Ministry's inception, to the position of Director General. But she often apologized to environmental leaders that she was too busy with other responsibilities in government to meet with them. They could not help remembering that Sarid had always found time for them.

Then, just as quickly as she appeared on the scene, Itzik was gone, leaping at the opportunity to serve in the new Unity Government cabinet in a "real job" as Minister of Trade and Industry.

Tzachi ha-Negbi rose quickly in party ranks as a "second generation" Likud politician, the son of the legendary Jewish underground radio announcer turned right-wing ideologue, Geulah Cohen. His appointment as Environmental Minister in March 2001 was ostensibly a demotion, after his previous post in the Netanyahu government as Minister of Justice. But if he was disappointed with the post, it never showed. His start was encouraging for the Ministry staff that found him to be smart, an excellent listener, and perhaps the first Minister of the Environment who was actually an avid hiker, born into a more environmentally aware generation. More than any other previous minister, he consistently consulted with the full community of Green organizations and sought their input.

Beyond his affable style, his initial decisions were also praised by the environmental community: imposing tough air quality emission permits on the Dan and Egged bus companies, or convincing the Cabinet to adopt a "green government" policy that promotes recycling and other environmental practices. Ha-Negbi hailed in a "recycling revolution" and personally pushed through an ambitious program to increase national recycling rates, which jumped to 15 percent in 2001, to 45 percent by 2005. When he appointed Bina Bar-On, a highly effective Deputy Director, to oversee environmental education, the subject finally began to recieve serious resouces and senior level attention. Still, it is not yet clear whether he will be willing to take a defiant stance when his government moves to adopt anti-environmental policies.[182] And, of course, after his many predecessors' ephemeral tenure the real question seems to be whether he will be at the helm long enough to make a difference."

In retrospect, of the many Ministers who have taken the position, only Yossi Sarid seems to have left a serious imprint on the Ministry of the Environment. But Sarid's tenure seems to have been an anomaly—a brilliant vignette that inspired briefly but did not change the larger, discouraging narrative. It produced the media saturation necessary to let Israelis know that their environmental problems were serious, while conveying a can-do optimism that they could be solved. For a few years, the Ministry of the Environment turned environmentalism into a mainstream concern and gave activists the sense that Israeli public policy might one day move in a sustainable direction. But a new environmental ethic did not have time to coalesce.

EVERYONE'S PROBLEM AND NO ONE'S IN PARTICULAR

Ironically, in many instances environmental progress during the Ministry's first decade had little to do with a given Environmental Minister or his or her policies. Israel's expanding high-tech industry helped. Factories such as Intel's Jerusalem plant brought with them an

American commitment to clean production that went beyond the factory, even including a phaseout of CFCs from the company's dining room refrigerators long before it was required by law.

Sewage treatment, for example, finally began to make the necessary quantum leap forward in cities like Netanya and Kfar Saba. Ministerial pressure and a series of related lawsuits were assisted by the attractive funding schemes offered by a new Sewage Administration established in 1993. But the Administration was run out of the Ministry of the *Interior* by an obdurate but highly effective engineer, Yehudah Bar. By 1997, his agency had doled out more than 1.5 billion shekels in loans and grants for sewage treatment—an unprecedented sum.[183] Bar cajoled, persuaded, and pressured mayors to get serious about addressing this environmental health hazard, and many did. By 1995 he could promise that within three to five years sewage would no longer be flowing into the Yarkon, Ayalon, Poleg, or Alexander Rivers.[184]

Similarly, the most dramatic reduction in air pollution during the 1990s came as a result of new European emissions standards for motor vehicles. The European specifications, adopted almost twenty years after similar American prescriptions were enacted, led to the installation of catalytic converters in European automobiles. After years of lobbying by environmentalists, the Ministry of Transportation finally agreed to import the Europeans' new standards along with their cars. The more stringent carbon monoxide standards were phased in on new vehicles over a three-year period, beginning with the largest engines in 1991.[185] Because lead destroys catalytic converters, gas tanks in the new cars were built to reject the nozzles for leaded gasoline. Slowly but surely, Israel's fleet switched to unleaded fuel. As anticipated, lead levels dropped precipitously in the cities.

If population, affluence, and technology are the ultimate determiners of environmental quality, then it is little wonder that Israel's pollution profile got worse. The Ministry of the Environment was created because these factors had affected the country's environmental indicators so dramatically. When it was on track, the peace process only provided an additional push. As McDonald's and Burger King came to replace falafel stands, the country relished the normalcy of being just another twentieth-century consumer society. Yet the suddenness of the nouveau prosperity was such that Israeli urban culture did not have time to develop the accompanying ecological self-restraint, so much a part of progressive Western nations.

Israel had changed and yet had stayed the same, with the environment losing on both ends. A wealthier nation still loved its land and could viscerally agree with a heartfelt pitch for open spaces. But at the end of the day, it was this very impulse that pushed people to build suburban homes with a half dunam of land, converting orange groves and meadows into concrete and asphalt. Even the best environmental communicator could not convince Israelis to leave the car at home and get back on the bus. Nor did anyone even try to touch the issue of overpopulation or the range of fiscal and immigration policies that encouraged it. Israelis no longer denied that pollution was a problem. Rather, environmental issues belonged to that category of problems that were everybody's in general and no one's in particular.[186]

After more than a decade of work—often very hard work—the Ministry of the Environment had grown. Unfortunately, so had most of Israel's environmental problems. In comparison with most government bureaucracies, the Ministry of the Environment was still a young, committed, and smart place. The public was well aware of its existence and even slowly began to expect environmental services. Certainly they knew where to point a finger when another fire raged at the Ramat Hovav hazardous-waste site[187] or when they read that despite local efforts, the United Nations still called Israel one of the biggest Mediterranean polluters.[188] Sadly, the environment once again lapsed into its old role as a minor issue, to be overseen by Ministers who saw it as a part-time job or as a political insult. In its first twelve years of work, no fewer than eight individuals held the post. This is certainly a reflection of the instability inherent in Israel's governmental system, but also of the alacrity with which politicians fled the position to more attractive Ministerial jobs. The Ministry of the Environment's 230-million-shekel budget for 2002 lagged far behind those of most Western countries on a per capita basis. It was a mere seventh the size of the funding at Israel's Ministry of Religious Affairs.[189]

Israel's environmental experience during the 1990s can alternatively discourage or offer hope. It certainly proved that the mere existence of a Ministry did not guarantee improvement. At the same time, it was clear that the country's pollution problems would not be solved without a powerful and sophisticated Ministry of the Environment.

10 Israel, Arabs, and the Environment

In his last year as a civil engineering student at the Technion in 1978, Basel Ghattas happened to do his final project on the environment. Beyond this, he harbored no discernible Green inclinations. After he graduated, he went home to his village of Rami, where he worked as a construction engineer. On the side he dabbled in local politics. He was upset that, as in the vast majority of Israeli Arab communities, there had never been a central system for disposing sewage in Rami. Yet, unlike most Israeli municipal leaders, politicians in Rami actually wanted to change the situation. For them, collecting the wastes was not enough. They wanted proper treatment at the end of the city's pipe.[1] Ghattas adopted the project as a synthesis of his professional and political interests.

Since 1968 the Rami Local Council had been trying to jump over the long line of bureaucratic hurdles imposed by Israeli law. Things appeared to get off to a quick start. In 1970 the Northern Regional Sewage Committee approved the conceptual master plan that the village had commissioned. It included settling and flocculation technologies at a sewage treatment facility whose purified waters would eventually irrigate olive trees. Yet it was not until 1976 that the Northern Region Planning Committee actually submitted the plan for public scrutiny and objections. In 1977 the National Sewage Project, operating out of the Ministry of Interior, agreed to offer Rami a matching loan for 50 percent of the work. It took two years, but in April 1979 the Rami Local Council provided the guarantee for its half of the funds and was ready to start work. But by then a highway had been approved that swallowed up much of the land designated for the treatment ponds. A smaller site was found, but it required more intensive aeration technologies.[2]

Then, seemingly out of the blue, the Ministry of Health withdrew its support. In October 1979 it announced that it needed to explore the possibility of connecting the village's sewage with the new treatment facility located in the nearby Jewish town of Karmiel. This was ten years after Rami had made the first of many payments to consulting engineers for a sewage collection and discharge master plan. Yet its approved proposal for a local treatment facility was put on hold.

For six more years, debates ensued over the exact path Rami's pipes should take on their way to Karmiel. The agricultural village of Moshav Shazur was situated along the way, and it was not happy with the recommended route. So the path of the pipes was modified.[3] Meanwhile, Ghattas and the village mayor, Elias Kasis, pushed ahead. In 1982 they began collecting taxes from Rami's residents and laid twelve kilometers of pipe to collect and deliver the sewage to the central treatment plant. Then, in September 1985, Karmiel had a change of heart.[4] The consulting engineers now told the town that the plant did not have adequate capacity to receive Rami's wastes, even though it would constitute less than 5 percent of the total sewage reaching the treatment plant. So in March 1987 the Regional Sewage Committee decided to go back to Rami's original 1971 plan. The treatment plant went on line more than twenty years after the blueprints were first designed.[5]

By the time the project was complete, Ghattas had already finished a doctorate in environmental engineering and probably knew more about the bureaucratic and technical challenges associated with sewage treatment for Israel's Arab population than anyone else in the country. He was approached by a health delivery organization, the Galilee Society (the Arab National Society for Health Research and Services), to manage a new project. This was a revolving fund to help Arab municipalities move through the crucial first planning stage on the circuitous road of sewage treatment initiatives.[6] Ghattas's activities helped twenty-six Arab villages prepare master plans and confirmed his experience in Rami: It was easier for Jewish settlements to get a sewage system up and running than it was for Arab villages.

Today, as Director of the Galilee Society, Ghattas remains an unusually good-natured man, his demeanor always amiable. But he is dead serious when assessing blame for the top environmental health problem in Israel's Arab sector:

> I have never seen a case where an Arab village didn't find a way to collect money for its wastes. They may have gone into debt, but in the end they installed the pipes and built the necessary infrastructure. The

problem has always been funding treatment technology. Here, the Arab sector was never a priority for Israel's government.

Israel's Arab population constitutes 17 percent of the country's population. In many ways its environmental experience runs parallel to that of the Jewish majority. The standard of living for Israeli Arabs has improved enormously over the years as reflected in basic environmental indicators: Drinking water has gotten better; air quality has gotten worse; the advent of plastics has created an ugly litter problem. The economy is no longer agrarian; agriculture has become only a marginal factor in an increasingly urbanized consumer culture. Also, any Israeli mayor can sympathize with Rami's bureaucratic nightmare and the headaches of meeting the environmental expectations that the central government has. Yet, in the ecological realm, Israeli Arabs have a story that is different from that of the Jewish majority.

The hundred-year conflict between Zionism and the Arabs of the Middle East has profoundly influenced the Jewish environmental reality. If nothing else, it has left the national budget focused on defense rather than domestic needs. The pervasive tensions spawned by the Arab-Israeli conflict unquestionably affected the treatment of Israel's Arab citizens. Their environmental experience has not been spared the fallout. When the direct and indirect effects of minority status on the physical conditions in the Israeli Arab environment are considered, the picture that emerges suggests that the environment has paid a price for the troubled relations between Arabs and the State of Israel. The relations between mainstream Israeli environmental institutions and Israel's Arab population confirms a prejudice premeditated or unintentional. At the same time, ecology may provide a basis for reconciliation.

MINORITY STATUS

The 1948 war, known by Palestinians as *al-Nakba* (the disaster), left a disoriented and fragmented Israeli Arab population. The Palestinian Arabs' fifty-year battle against Zionism had been resolved—only it had ended in total loss, irreversibly transforming their landscape. Fewer than ten thousand Arabs remained in Jaffa, Haifa, and Ramla-Lod, cities that before the War had collectively been home to over 175,000 people.[8] Three hundred and fifty villages vanished entirely.[9] Only the city of Nazareth had grown, from 15,540 to 16,800, as a result of the absorption of internal refugees. The 150,000 Arabs who remained reluctantly assumed Israeli citizenship. When they began to pick up the pieces, in the 1950s, Israeli Arabs made up roughly 10 percent of both the total Palestinian population and of Israel's citizenship.[10]

Zionism's attitude toward the Arabs of Palestine is the subject of considerable academic scholarship and debate. While the earliest Zionist "thinkers" did not consider too deeply the role that Arabs would play in their Jewish State, most adopted Herzl's optimistic line that things would "work out" amicably.[11] When Arab Palestinians became increasingly intransigent in their rejection of any accommodation, Zionists responded by becoming more militant themselves. The Holocaust only heightened the Yishuv's sense of moral entitlement. After 1949, once this fifty-year conflict was resolved so favorably—from the Jewish perspective—Israel's leaders were sincere in wanting to better the conditions of Arab citizens. There is no commandment that appears in the Bible more often than the injunction to treat equitably the non-Jewish residents (as well as the orphans and widows) "who sojourn among you."[12] This traditional precept, coupled with the general aspiration to be a liberal democracy, constituted the national ideal for the new State of Israel.

At the same time, the first half century of independence was marred by continuous enmity and intermittent violence with Palestinians and Arab nations. The country could not help but view its own Arab citizens as a security threat. The expression "Respect him and suspect him" was almost an official adage. To neutralize the danger posed by a potential fifth column, a strict policy of military rule was initially imposed on Arab communities and eased only over a period of many years. To consolidate Jewish control geographically, many Arab citizens were barred from returning to their original homes. Under the 1950 Absentee Property Act, they were paid compensation that fell far short of the land's actual value. As much as two million dunams, a full 40 percent of private Arab land resources, were confiscated during this period.[13] Today Arabs make up almost a fifth of Israel's population but own only 3.4 percent of the land.[14]

Loss of land was the deathblow to an already beleaguered *fellah* economy that already could not compete with the highly mechanized Jewish agricultural sector. By the 1990s, only 8 percent of Israeli Arabs made a living in agriculture.[15] Despite the painful transition, the Israeli Arab community proved resilient.

In retrospect, it was the 1967 Six-Day War that served as the economic watershed for Israel's Arab citizens. The war came in the wake of the cancellation of military rule in 1966, which freed Israeli Arabs to pursue entrepreneurial ventures. With the new territorial acquisitions, West Bank and Gaza residents replaced Israel's Arabs in many unskilled manual tasks, and Israeli Arabs moved up the economic ladder. Still barred from most "security-sensitive" industries, their prosperity was most conspicuous in the booming construction business. Additionally, a number of successful Arab factories were established during this period, in particular in the areas of textiles and processed food. The Arab sector was not nearly as industrialized as the Jewish sector, but still, by the 1980s some 30 percent of the Arab industrial labor force worked in Arab-owned factories.[16]

A strong commitment to education and a pragmatic approach to their circumstances strengthened the community. Because of their prolific birthrate (which for many years averaged 3.7 percent growth per year) Israeli Arabs have more than kept pace with Jewish immigration. Non-Jewish Israelis presently number over one million. Some academics like to analyze the "quiescence" of Israel's Arab minority,[17] and Palestinian nationalists seek perfidious, conspiratorial explanations. But most Jewish Israelis welcome their "cousins'" natural integration and genuinely appreciate Arab loyalty, in particular during times of strife. The extent of the Arab population's commitment to the country is reflected in polls showing that despite a long laundry list of complaints, only 7.5 percent of Israeli Arabs are interested in moving to a Palestinian State; only 13.5 percent wish to see Israel disappear.[18] During the 1992 elections, a time when ethnic and national origins increasingly influenced Jewish voter affiliation, more Arabs voted for Jewish parties than for Arab ones.[19] The situation during the subsequent decade, admittedly, has beecome more polarized.

Yet because of the salience of the surrounding hostilities, Israel's Arab citizens will never be a conventional minority group. For the foreseeable future, they will have to wrestle with a dual affiliation that, on occasion, leads to profound contradictions and conflicts.[20] When these emotions meet objective economic disadvantage, perceived discrimination, and the simmering conflict with the Palestinian nation, they can be manifested in the sort of violent riots in Arab towns and cities that stunned Israeli society in October 1990.

At the same time, it remains practically impossible to generalize intelligently about such a diverse collection of religions, ideologies, lifestyles, and interests. Certainly, Israeli Arabs who work in the environmental field show none of the meek or embittered spirits that are sometimes used to deride this remarkable community. They most closely fit Dan Rabinowitz's description of the sophisticated modern Arab Israeli professional:

> Their nonassimilating nature is perceived as permanent, acculturation notwithstanding. While consistently acquiring more of the values and the symbols of Jewish Israel and incessantly attempting to gain more

> access to its political arena, they neither want to nor are invited to assimilate.[21]

THE ELUSIVE CONCEPT OF ENVIRONMENTAL JUSTICE

Technically, Israeli Arabs are full-fledged citizens. As an ethnic and religious minority in a Jewish State, however, their status in fact falls short of equality. The various legal manifestations of the preferential treatment for Jewish Israelis have been catalogued elsewhere.[22] Inequality is not always due to direct government discrimination, however. The very existence of a privately funded Jewish Agency, which plays a key role in a variety of social spheres, guarantees unequal results. The Agency has a natural preference for Jewish clients. While there is legitimate debate over the validity of any given social policy, even the most passionate Zionist patriot would agree that Israeli Arabs do not enjoy all the privileges of the Jewish majority. But does this double standard also extend to the environmental realm?

The answer goes to the definition of what is known as "environmental justice."[23] The question is not whether Israeli Arabs suffer from noise, water pollution, and air pollution: By virtue of the fact that they live in Israel, they do. The salient question is: To what extent does this population suffer greater exposures or impaired access to natural resources as a result of its minority status? For example, there is little doubt that when Israel sited Beer Sheva's most-polluting chemical industries, as well as the nation's only hazardous-waste facility, at Ramat Hovav, it did so because it felt that there was nobody there. The scattered Bedouin who camped "across the street" from the hazardous-waste facility were in no position to object. This unfortunate juxtaposition has produced a disproportionately heavy environmental burden for this population and one that falls squarely in the realm of environmental justice (the Bedouin eventually used the exposures as a legal basis for demanding alternative lands[24]). Ramat Hovav represents a case so extreme as to make it almost anomalous. And indeed, there is no shortage of Jewish horror stories involving unreasonable pockets of environmental risk.

Is there systematic environmental injustice towards Israel's Arab residents? If so, identifying it is a particularly vexing conundrum. Just five of many possible caveats reflect the complexity of the issue. First of all, the baseline environmental situation in 1948 was fundamentally asymmetrical. Indoor plumbing and other basic sanitary services were unknown in most Arab villages. The disparities have been reduced dramatically. For instance, by the end of the 1960s, almost every Israeli village had running water. Golda Meir's self-satisfied description of official Israeli efforts to improve the quality of life for its Arab citizens represents more than just an official perception of the situation during the pre-1967 period: "Those Arabs who stayed in Israel . . . had an easier life than those who left. There was hardly an Arab village with electricity or running water in all of Palestine before 1948, and within twenty years there was hardly an Arab village in Israel that wasn't connected to the national electric grid, or a home without running water."[25]

Meir's patronizing tone as well as the implication that the government was doing taxpaying citizens a favor may be annoying to some. Nonetheless, during a period when the young country faced economic hardships and profound security risks, remarkable progress was made. As a result, water consumption in Arab villages tended to rise twice as fast as population growth.[26] The question remains: Was it enough?

The second caveat involves cultural factors that might explain environmental differences—results attributable to community autonomy rather than injustice. Arab Israelis often have different ideas from those of their Jewish counterparts about how to build houses, plant gardens, establish parks, educate, issue licenses, and so forth. Pig farms, not surprisingly, are not a problem in the Jewish sector, but some Christian Arabs own them, and they are a considerable nuisance for many Arab Israelis.[27] In a related matter, having left most of their large pre-1948 population centers, Israeli Arab residents tended to concentrate in small and mid-sized villages.[28] As in any rural sector, objectively there is a different environmental reality—for better or for worse. Such issues lie outside the realm of environmental justice.

The third caveat concerns an inherent tension, heightened by the Arab-Israeli geographic reality, between traditional culture and modern economic activities. The "blue lines" drawn into the local municipal master plans, constraining the expansion of Arab (and Jewish) settlement, exacerbated land shortages within the villages. Fathers traditionally divided their land holdings among their sons. As family size grew and land holdings remained frozen, the size of the subdivision shrank. This led to two environmental dynamics: a shortage of public areas (for example, Israeli Arab schools were frequently built far outside the villages, for lack of available sites) and a tendency to concentrate individual agricultural (and later industrial) activity in increasingly urban family plots. Keeping livestock outside the safe confines of the village is often seen as imprudent. In a related phenomenon, in the Bedouin town of Rahat, there is a high percentage of truck drivers among residents. Attempts to move the potentially explosive gasoline pumps from private homes and concentrate them in a nonresidential location have been singularly unsuccessful.[29] The fuel is perceived as better protected nearby.

A fourth confounding factor is occasional discrimination within Israel's Arab sector itself. For example, members of the Bedouin neighborhood in Shfaram—an Arab city—point to a pattern of harassment that includes a predatory zoning policy imposed on them by the city government. They claim that the policy is motivated by the local Arab leaders' resentment of Bedouin fealty and military service to the State of Israel.[30]

Finally, economic disadvantage does not always lead to environmental disadvantage. Ironically, the lack of heavy industries in the Arab sector (which might be associated with economic inequality) may also mean that environmental exposures to hazardous chemicals is higher among Jewish Israelis. While Israeli Arabs may justifiably resent a per capita car ownership that is only 35 percent of the general population's, certainly it benefits air quality in their communities.[31]

The most basic public health data do not support *environmental* discrimination. Arab cancer rates in Israel between 1960 and 1995, for example, were substantially lower, among both men and women, than those for the parallel Jewish cohorts.[32] Ministry of Health expert Gary Ginsburg confirms that Arab life expectancies are slightly lower than Jewish life expectancies, but this is due to a higher infant mortality rate as a result of poorer socioeconomic conditions rather than exposure to environmental hazards. Indeed, his calculations suggest that in terms of preventable mortality, Sephardic Jews are actually slightly worse off than Israeli Arabs. The closest thing to a pollution-related cancer cluster that his epidemiological evaluation uncovered involved elevated lung cancer rates near the Nesher cement plant in Ramla and near an asbestos factory in Akko, coincidentally two of Israel's four "mixed" Arab/Jewish cities.[33] Nonetheless, there are statistics that suggest that discrimination is manifested in specific environmental media.

The most blatant indicators are in natural resource allocation. Arab farmers, who cultivated over 10 percent of Israel's crop area in 1974–1975, received only 2 percent of the water allocated to agriculture.[34] There has been some improvement; however, by 1996 Jewish per capita use of water was still 120 cubic meters per person per year compared with 42 cubic meters among Israeli Arabs.[35] It is no coincidence that the Jewish Agency owns one-third of Mekorot, Israel's waterworks. Likewise, no Arabs have ever held senior positions in Israel's Water Commission.

Dr. Fayad Sheabar, an MIT-trained public-interest environmental scientist, ranks the top five environmental problems in the Arab sector:

1. An unhealthy drinking water supply
2. Insufficient basic sewage infrastructure and practically no advanced treatment
3. Small industries in urban areas, producing air and noise pollution
4. Proximity of quarries to Arab villages
5. Pesticide abuse

Because of a general paucity of data and empirical research on the subject, such assessments are somewhat impressionistic and less than definitive. Nonetheless, with the exception of pesticide exposures, the severity of the environmental insult to Arab communities appears ostensibly worse for each case than in the Jewish sector. The implication is that there is an ethnic or "environmental justice" association. For instance, bacterial contamination of drinking water is more common in the Arab sector because of the continued presence of septic tanks and the frequently leaky sewage collection systems. Health directives to citizens instructing them to boil their drinking water, especially during the summer months, are far more common in the Israeli Arab sector than in Jewish cities.

Many small cottage industries emerged in the basements, garages, or backyards of Arab cities and villages as a reflection of an industrious but capital-poor population.[36] A number of these enterprises operate without a proper business license. Although most are small, the collective environmental impact of metal and aluminum works, cleaners, and carpenters can be considerable. Establishment of distinct industrial areas seems to be an obvious solution—except that there have been cases where the cure is worse than the disease. For instance, the city of Shfaram created an industrial area inside the city limits. As a result, concrete plants were located four meters from the windows of private homes, producing dust concentrations ten times the legal limit.[37] It is little wonder that the exposed neighbors suffered from chronic respiratory illnesses.

Such dynamics are by no means unique to Israel's Arab sector. For example, during the 1990s, Israel's agricultural moshav settlements began to suffer from an influx of illegal factories and a similar intermingling of industrial and residential uses.[38] The problem in the Jewish sector, however, warranted a high-level commission, investigation, and Cabinet-approved recommendations.[39]

There are less tangible factors used to support Israeli Arab claims of environmental injustice. These include nonquantifiable areas where Arab communities seem to fall short, such as available green spaces or general cleanliness. In "mixed" cities, where Jews and Arabs live together, such as Akko and Ramla, Arab neighborhoods are perceived as grungier.[40] Separating official responsibility from the contribution of the local Arab communities in these instances is practically impossible, and personal predisposition tends to color one's conclusions. Sewage treatment, however, offers a compelling example of differential levels of environmental quality and quality of life.

THE SEWAGE GAP

If there is one environmental medium in which Israeli Arabs have an objective basis for claiming disadvantaged status, it is sewage. The numbers are indisputable. Israel's State Comptroller was not telling the Arab community anything new when she reported with some dismay in 1996 that 80 percent of Arab villages had untreated sewage that was not contained by pipes, flowing in the streets or in open spaces.[41] Until the late 1990s, it was possible literally to cross the two-lane highway separating the high-tech industrial zone of Karmiel and to find fecal wastes flowing through the streets of Der el-Asad—a *fully recognized* town. The stench of sewage flowing in the streets is increasingly rare today, but end-of-the-pipe solutions are still nowhere in sight. The problem is most acute among the many residents of Arab settlements who are classified by Israel's planning authorities as "illegal squatters." The residents of these "unrecognized settlements" find themselves in limbo. On the one hand, like all Israeli citizens, they pay taxes. At the same time they are not entitled to basic municipal services, including sanitation.

Ironically, it was success in achieving relative "equality" in the area of water delivery that set the stage for the failure in sewage disposal. Quite simply, after they piped the water in, Israeli planners did not have time to worry about where the resulting wastes would go. As a result, the vast majority of Israeli Arabs, who live in *recognized* cities, face a discouraging sewage profile. A 1997 survey conducted by the Galilee Society of Municipalities in the "triangle" of Arab settlement found that 75 percent of the communities relied on cesspools and pits as an "end-of-the-house" solution. Cesspools may have been a reasonable solution when there was no running water supply, but once every home had flush toilets and faucets, the quantities of wastes became excessive.[42] Dr. Ghattas for years has challenged the conventional wisdom among Israeli sewage planners that insists on technology-intensive centralized treatment, which is both expensive and inappropriate to the physical reality of the Arab villages.[43] Connecting the houses that crowd along their meandering streets and alleys to a central grid is not a trivial task. It is easier and much cheaper to build a collection network around a matrix of orderly Jewish apartments.

Inasmuch as it is decentralized, cesspool maintenance is irregular, and often the impervious pits fill up and spill over into the streets. Even worse, sometimes the cesspools work, and absorb the sewage. Then waste streams can percolate toward the groundwater or even the unfortunately located and cracked drinking-water pipe. Emptying these waste pits takes place with little oversight and constitutes a major source of stream and groundwater contamination.[44] Moreover, sometimes the ecological damage can be worse when there is central collection. Um el-Faham, the largest Arab city in Israel, with a population of forty thousand, still discharges some 40 percent of its wastes directly into the closest stream.[45] A 1995 Galilee Society survey found that the majority of Arab communities collected their wastes in a network of pipes.[46] (This was a vast improvement from 1988, when only eight of seventy-five recognized Arab villages had reasonable sewage systems in place.[47]) But 60 percent of the cities had no treatment facilities.[48]

Part of the problem can be attributed to an underlying asymmetry in agricultural development. Until 1997, not a single Arab village was using its own effluents after treatment.[49] When kibbutzim were located nearby, they were only too happy to utilize the wastes.[50] As a result, sewage solutions came relatively early to villages like Daburia, Yafia, Kfar Yasif, and Abu Snan.[51] One reason that the Arab city of Sakhnin could finally build the first advanced treatment/reuse plant in the Arab sector was the absence of a nearby kibbutz. Its 150,000-cubic-meter facility went on-line in 1996, and the treated effluents were pumped to Arab olive trees.[52] Yet for the vast majority of villages that could expect no agricultural dividend, advanced treatment and wastewater reuse were admittedly less of a priority than domestic sanitation.

The question of financing remains the most hotly disputed area of controversy. Conventional wisdom among Israel's Jewish population is that Arab municipalities are unwilling to charge municipal property taxes and sewage surtaxes at the same rate as in the Jewish sector. Some justify the disparity because of the Arab sector's lower economic resource base. But many Israelis cannot see the problem rationally and in its totality. Frequently their vision has been blocked by the palatial villas found in many Arab communities. Such Arab homes appear much larger and more impressive than urban Jewish apartments, and their relative prosperity was resented. The sewage gap was attributed, therefore, to an Arab civic culture that presumably was happy to spend money on individual homes but was unwilling to invest in communal infrastructures. This analysis went beyond sewage and included parks, roads, and other social services.

The position is disingenuous for two reasons. First of all, it ignores the systematic discrimination built into the allocations of Jewish nongovernmental development agencies in Israel. During Israel's first thirty years, world Jewry contributed some five billion dollars to the country, 65 percent of which came from American Jews (this does not include compensation payments paid by the German government to Nazi victims). Israeli Arabs derived little direct benefit from these funds, which paid for a considerable portion of the infrastructure as well as the agricultural development in the Jewish sector.[53]

Moreover, practically no Jewish city fully bankrolled its own sewage infrastructure. The National Sewage Project and later the National Sewage Administration combined Ministry of Interior funding with low-interest loans to establish waste treatment facilities. While the law in theory authorized local government to tax its residents, in practice most of the money for activated-sludge and other treatment technologies came from the country's general fund. Israel's Arab communities, which started in a disadvantaged position, deserved the same level of funding.

Arabs suffered most of the resulting public health burden in the form of periodic outbreaks of hepatitis, typhoid, and dysentery.[54] Khatam K'naneh, who in 1988 served as the Ministry of Health's Deputy Regional Physician for the north of Israel, described the current situation:

> Even today, infants die from infectious disease associated with water and environmental hygiene. From an epidemiological perspective, the significance of these isolated instances of death is that they represent the tip of the iceberg that can be seen. For every death caused by intestinal infection, there are far more hospitalizations, and for each hospitalization, there is a far greater number of illnesses in the community for the same reason, even though many are not reported. Statistically, every Arab in Israel is sick twice a year as a result of his water supply and sanitation. In a 1984 survey of Galilee villages, every child under the age of three years was sick five times a year from these diseases.[55]

K'naneh pointed out that even though the death rate among Arab children from sanitation-related disease was eleven times higher than in the Jewish sector and hospitalizations were three times higher, the pollution did not leave the Jewish settlements unscathed. There have been several instances of partially or completely untreated wastes from Arab towns contaminating drinking-water sources that service Jewish communities.[56]

Israeli Arab environmental disadvantage reflects badly on Israeli society as a whole. It especially represents a failure for Israel's environmental movement. During most of the country's history, relations between Israeli Arabs and Israel's mainstream environmental institutions were best characterized by relative degrees of alienation. Recent improvements are a function of change in the orientation of a few of the Jewish organizations coupled with a growing cadre of Israeli Arabs willing and professionally inclined to become active in mainstream Israeli pursuits. Each relationship has its own historic antecedents and controversies, illuminating a different aspect of the Arab Israeli complaint. Yet collectively they show how difficult it is to separate the broader national conflict between Jews and Arabs from Israel's ecological experience.

JEWISH TREES

The Jewish National Fund constitutes the oldest and, in the long run, perhaps the most problematic environmental institution for Israel's Arabs. The inherent tension of a particularistic Jewish player endowed with government functions in a purportedly democratic nation has yet to be resolved. Most Israeli Arabs see the JNF as the perpetrator of systematic exploitation and discrimination against Israel's Arab community and a tool for hegemony in the occupied territories.[57] Although the JNF has a large number of Arab employees, they are limited to positions of manual labor and do not participate in policy making. Fifty years after land acquisition from Arab effendis ceased to be the primary JNF priority, the organization's ongoing concern about Jewish domination of land resources continues to engender animosity.[58]

Many Jewish Israelis agree that forests serve a role that goes far beyond ecology and recreation. For Arabs, the trees are often perceived as a not-so-subtle ploy to constrain the expansion of their villages. Planting is interpreted in a strictly nationalistic light—a declaration of Jewish sovereignty. The JNF's controversial involvement in West Bank afforestation efforts and land acquisition initiatives in East Jerusalem served to reinforce those who see the JNF in a purely political rather than ecological context. Israeli Arabs resent the quasigovernmental status accorded JNF lands and see its discriminatory policies as a festering sore on the face of Israel's democracy. These are lands owned by Jews, for the Jews—and are off limits to many of Israel's citizens.[59] A small, less-than-flattering literature has sprung up in which a few historians and sociologists consider the political implications of the JNF's work.[60]

There is an ugly ecological side to the political acrimony. Israel's trees have paid a price for Arab disenfranchisement. Even before the 1980s, when the Intifada marked JNF forests as a priority target for Palestinian violence, politically motivated arson was a constant threat to the country's woodlands, in particular during the dry summer months. Forest fires in Israel are political events that dominate headlines.[61] Indeed, many Israeli Arabs perceive arson in the JNF forests as a legitimate form of political expression. An ambush by JNF foresters uncovered an underground network from the Galilee village of Jilabun in 1998. The gang went so far as to print up an elaborate forest-burning battle plan, complete with maps, timetables, and a breakdown of responsibilities.[62] The most famous literary work that takes place in a JNF forest is A. B. Yehoshua's 1968 novel *Opposite the Forest,* in which a Jewish graduate student working as a watchman exhibits growing empathy for the destroyed Arab village over which the forest is planted. Ultimately, he does nothing to prevent its immolation by an arsonist.

In the world of nonfiction, however, Israeli foresters were never sympathetic. They were resolute and resourceful. Along with the fire lines separating the trees into modular compartments, the network of watchtowers established in 1958 has probably proven to be the JNF's most effective response. The forty towers and two hundred vehicles (see Figure 29) used today by the forestry department provide sufficiently early detection to reduce the impact of fires enormously.[63] The arson that blazed throughout the second half of the 1980s, however, changed the dimensions of the problem. In 1976 there were 320 events per year in JNF forests; by 1987 the number tripled to a thousand. Although the source of most fires remains unsolved, at least 25 percent of the conflagrations have been determined to be politically motivated. The actual proportion may exceed 60 percent.[64]

The JNF was quick to respond, beefing up its monitoring capacity to meet the challenge. Despite the dramatic rise in the number of fires, the damage to forests was halved by the end of the 1980s.[65] Today, a remarkable 32 percent of the five hundred annual events are extinguished with only one to five dunams damaged. Almost half of the fires were nipped in the bud completely, with only one dunam of woodland affected. Still, roughly 10 percent of the fires are categorized as forest crown blazes, involving total loss.[66]

Like soldiers fighting against guerrilla warriors, the JNF foresters are limited to offering tactical holding measures. The Israeli public continues to donate generously in highly publicized campaigns to replace burned forest segments.[67] A strategic solution may be linked to a larger, geopolitical reconciliation. Even when a more peaceful climate in the region takes hold, however, the physical climate will always be dry, and forest fires will remain a threat, especially for pines, with their flammable resin. Different tree composition and more broadly grazed buffer zones will also need to be part of a long-term solution.

Most Israeli Arabs would never openly support the acts of arson, but they identify with the motives. The JNF is perceived as an organization that is out to grab every square centimeter of land that Arabs are willing to relinquish. The level of alienation runs high. "The Arab community derived no benefit from these forests," complains one Israeli Arab conservation leader. "The very least they could have done was call the forests by the names of the places they took the land from. Even the JNF can't say that Arabs don't like to go to parks. So why isn't there an Um el-Faham Forest or a Wadi Rabba Forest? That way there might be a chance that Arabs could feel attached to these places."[68] The JNF has begun to explore what a new relationship with Israel's Arab community might look like through some community consultation exercises in and around the region of Sakhnin sponsored by the JNF Community and Forestry Department.[69] But these efforts remain embryonic in character. Until the JNF's agenda is environmental rather than political, such differences may be irreconcilable.

THE SPNI'S ARAB UNIT

The one environmental institution that has managed to penetrate the Israeli Arab sector broadly is the Society for the Protection of Nature in Israel. Two thousand Arab teachers participate in advanced training with the SPNI each year (this constitutes roughly one-fifth of all teachers in the Arab community). There are SPNI branches in such Arab cities as Nazareth and Sakhnin.[70] A professional tour guiding course is taught for Arabs in Arabic. More than fifteen thousand Arab youth typically come out for SPNI's annual rally in the field.[71] Yet even this ostensible success story reveals some of the built-in tensions that contribute to the sense of alienation by Israeli Arab environmentalists.

SPNI's Arab unit was born of a unique partnership between Mahmoud Gazawi and Yossi Leshem. In 1975, after working as a teacher in the southern Bedouin town of Tel Sheva, Gazawi approached the SPNI, seeking work. Azariah Alon, then serving as SPNI General Secretary, was not interested in hiring an Arab. Leshem recalls:

> Azariah figured that it wouldn't work out. SPNI teaches more than a love for nature, but also a connection to the country. He couldn't

> understand how an Arab could teach Jewish youth love for the land of Israel. So Mahmoud came to me one day and said, "Look, I can't seem to get a job at the Society. Azariah always seems to find another excuse why not." So I told him, "Forget Azariah. You'll work in my nature protection department, and I'll find you the money."[72]

Armed with only nominal funding and Yossi Leshem's blessing, Gazawi returned to his village, Kalansawa, as a teacher. On the side, he began an SPNI unit. Quickly it blossomed into a massive success. He initially focused on teacher training because he lacked the resources to work directly with youth. After a few years he had reached most of the interested teachers, and Gazawi decided to appeal directly to pupils. In 1980 a major hiking/educational event for Arab students was planned. When five thousand kids turned up, Gazawi knew he was on to something. Each year the meeting is held in a different place, with the numbers rising steadily.

The SPNI Arab unit clings to a relatively narrow mission, and it stays "on-message." Its focus is nature, and its orientation is education. The unit puts out pamphlets and books, leads hikes, and runs school groups. It does not attempt to address the urban environmental issues that may be most relevant to Israeli Arabs' day-to-day existence. It also steers clear of themes that have any national overtones. The unit keeps its geographical mandate restricted. The border between Israeli Arabs and their Palestinian cousins in the territories is never crossed. Gazawi explains, "Anything connected with water and soil resources there is catastrophic. We don't deal with it. It's just too big, even for us."

A solid man with graying hair, Gazawi attributes his passion for nature preservation to his upbringing.

> I was raised in a rural landscape. My work is especially important to me because of the changes taking place in Arab villages. The natural landscape is disappearing. The children and the general public seem to be moving totally in the direction of Western consumer culture. In the past there was a familiarity with the outdoors. It was part of the lifestyle of the fellah. The entire village identified with the fields, and they would go out to work early in the morning. Today, just try to get people to sleep outdoors! The alienation is drastic. To cook over a fire used to be the greatest pleasure and really a daily event in our childhood. Now young people tell me: "Why cook coffee over the fire when we can use gas?" We used to eat food from the fields. Now we eat Sunfrost.[73]

Gazawi is surprisingly frank about the organizational bias that seems almost as persistent today as it was twenty-five years ago when he first encountered the SPNI. The Arab unit justifiably is a source of pride at the SPNI (and Azariah Alon has long since been won over). Yet Gazawi still perceives Arabs as second-class citizens within the organization. He points out that there still are no Arab Israeli guides in a field school, much less an Arab director of a field school. His unit has to rely on freelance guides rather than full-time SPNI employees.

As a member of the governing board of the SPNI, he is very open about this disenchantment and no longer worries about the impression he makes. There is no shortage of Jewish board members who agree with him. He does see steady progress, though slow. Certainly, his many years of promotional work have paid off. When a survey was conducted among residents of Um el-Faham, it was found that 73 percent of the city's Arabs perceived the SPNI as Israel's leading environmental institution.

THE NATURE RESERVES AUTHORITY

One of the reasons that the SPNI Arab unit has not tried to assume a more aggressive watchdog role is to distance itself from Israel's Nature Reserves Authority (now the Nature and National Parks Authority).[74] One would imagine that Israeli Arabs would have an easy time identifying with such an ostensibly apolitical government agency. But quite the opposite is true. It is hard to grasp the depth of the hostility harbored among Israeli Arabs for the Authority. There may be a variety of reasons, but ultimately two historic areas of conflict with the Authority poison relations to this day: the tension over the Mount Meron reserve and the village of Beit Ja'an, and the activities of the Green Patrol.

Mount Meron was for many years Israel's largest nature reserve and remains the best-preserved natural woodlands in Israel if not the entire region. The unique combination of high altitude, abundant precipitation (nine hundred millimeters per year, including regular winter snows), and historic isolation produces an ecological climax situation.[75]

Druze have been living in the vicinity of the reserve for eight hundred years, mostly in and around the village of Beit Ja'an (see Figure 30).[76] Most Israelis hear of Beit Ja'an only on the evening news, when the village is singled out as the home of yet another fallen Druze soldier. No settlement in Israel, not even among the kibbutzim, has lost as many of its young men, proportionally, as Beit Ja'an—over fifty casualties in a village of 8500 residents. Environmentalists, however, know it as the most troublesome single point of friction between Israel's nature reserves and a civilian population.

The conflict is as old as the Nature Reserve system itself. When the Minister of the Interior gave General Avram Yoffe the megasanctuary he asked for, it included private tracts, tended by small farmers from the village. Many of these lands had been owned by Beit Ja'an residents for generations. The Druze landowners were not offered alternative plots for which they might trade their own lands that suffered the misfortune of ecological uniqueness. Rather, they were allowed to keep them—subject to the draconian limitations deemed necessary for preservation. To be sure, these involved a relatively small section of Beit Ja'an lands located in the most "sensitive" sections of the reserve. Agricultural cultivation was not prohibited altogether, but heavy, earth-disturbing mechanical equipment was banned. Forced to till their families' age-old plots with donkeys rather than modern equipment, many farmers found the policy degrading.[77] The real bone of contention was limitations on road construction in the area, which hampered the farmers' access to their fields. The problem was never solved to the satisfaction of the local residents, and so it flared up time and again.

The events of the 1987 round were typical. In standard Israeli partisan fashion, Druze leaders pressured then Minister of Agriculture Arik Nehamkin to cancel all limitations on the lands inside the reserve and order the Nature Reserves Authority to desist in its inspection duties. With elections in the air and the Druze vote hanging in the balance, he acquiesced. The SPNI called the agreement illegal, and the Supreme Court concurred. But the Druze were not happy with the verdict, and the SPNI never succeeded in establishing a real dialogue with them about the full dimensions of their complaint.[78] The sides were back in court once again a decade later.

Beit Ja'an certainly had other problems, but it would be wrong to dismiss the Druze view about the Meron Reserve as simply cynical tactics. It did not expect special treatment, but a village that had contributed so significantly to Israel's security justifiably counted on sensitivity to the needs of its citizens. Beit Ja'an's residents felt that the Nature Reserve was suffocating them. The Nature Reserves Authority was tuned in to the complaints of the residents and tried to present the reserve as an economic resource with the potential for ecotourism. To help with the problem of unemployment and to give the village a stake in the reserve, the Authority hired a large number of Druze veterans to work as NRA rangers. Because of the sincerity of their efforts at accommodation, NRA leaders remain bewildered when their anti-Arab reputation is raised. Lands for the Meron nature reserve, after all, really were selected according to strict ecological criteria.[79] Yet although they understand that conflicting interests must learn to coexist, Arab environmentalists perceived the NRA's limitations as overzealous and paternalistic.

Things came to a head in the summer of 1997, when the Society for the Protection of Nature in Israel renewed a Supreme Court petition to enjoin road expansion by the residents. Beit Ja'an residents received the news of the Court's agreement with outrage, staging a violent demonstration in Jerusalem in which twenty-five people were hurt.[80] Beit Ja'an's mayor told the Court that he could not promise that his constituents would abide by the ruling. He explained to the press, "Only in the covenant of death do they remember us. What about the covenant of life?"[81] In the face of such a violent confrontation, the Nature Reserves Authority, under the pragmatic leadership of its new director, Aaron Vardi, forged a compromise.

Under the March 11, 1998, agreement, the SPNI agreed to cancel its Supreme Court action, and the Authority agreed to allow the route from Beit Ja'an to Chorpish to be redesignated as an agricultural road with the asphalting of certain steep sections. Landowners would be allowed to utilize heavy mechanical equipment in their agricultural work, and limitations on tree cutting and planting on private tracts would be removed.[82] Beit Ja'an agreed to forgo future claims and even agreed to launch a nature preservation education and public awareness campaign among its residents.

Local leaders in Beit Ja'an were satisfied with the agreement, asserting that hiking and love of nature are key elements in Druze heritage: "We are a law-abiding people as long as the law is fair and serves the public interest. But we couldn't accept a policy in which plants were deemed to be more important than people," explained Asad Dabur, the deputy head of the Local Council.[83] But scientists in the Authority did not share this romantic perception of traditional stewardship practices. Former chief scientist Aviva Rabinovich argued that aerial photographs of Beit Ja'an from the Mandate period are devoid of tree cover. All had been cut down except for a few groves that were deemed holy.[84] "Preservation was never an integral part of their culture," she claims.

This condescension is not lost on Israeli Arabs, who interpret preservation policies as driven by chauvinism rather than ecology. While Beit Ja'an residents may temporarily be satisfied with their agreement, decades of enmity are not easily erased.

THE GREEN PATROL

The bad feelings in Beit Ja'an pale alongside the legacy of hostility spawned by the NRA's Green Patrol in its dealings with Israel's Bedouin community. Here, the quest for nature preservation came into direct conflict with indigenous culture and claims to land ownership. A cursory review of Bedouin history in Israel suggests that, as in most of the serious dilemmas in conservation, there is no right or wrong party.

When the Roman Byzantine government in the Holy Land collapsed in the seventh century, Palestine's deserts—or *badia* in Arabic—became the domain of the Bedouin.[85] They were divided into seven tribes or confederations of tribes, spread across the southern half of what is now Israel, all of Jordan, and the Sinai. The Turks had a post office in Gaza until 1900 when it moved to Beer Sheva. For the most part, however, until 1948 the Bedouin of the Negev were on their own and wandered freely.

Each tribe was related to a common ancestor. Within a tribe, any member could use the pasture and water sources of his territory. But rival tribes were absolutely excluded, except in the case of agreement or alliance. Even without the trappings of modern maps and surveying, borders were clearly demarcated, and whoever wanted to pass through a Bedouin tribe's lands needed permission. Unlike some of the tribes east of the Arava, who only migrated, the Negev Bedouin would cultivate, especially during years of plentiful rain.[86] For extended periods, however, the land could be dormant.[87] To the Bedouin, this in no way meant that they had relinquished their claim. The calls by the Turks and later the British on the Bedouin to register these lands were largely ignored. Bedouin perceived it as merely a ruse for collecting taxes.[88]

All this changed when Israel assumed control of the Negev desert in 1949. The new nation had the jeeps and the national will to penetrate this inhospitable arid region. The Bedouin understood that the rules of the game had changed.[89] Of the fifty-three thousand Bedouin that the British registered as living in this area, all but 12,500 moved on.[90] They scattered to the West Bank, to Gaza, and mostly to Sinai and Jordan. Initially the Bedouin, like all Israeli Arabs, were considered to be a security threat. Israel set a policy of concentrating the Bedouin in the area east of Beer Sheva on some thousand square kilometers of land. After the 1948 War, the Bedouin district was declared a "closed area," and until the policy was canceled in 1966, Bedouin were not allowed to leave or take their flocks to other parts of the Negev without special permission. Ben-Gurion, of course, perceived the Negev as largely unsettled and as the ultimate land reserves for Jewish immigrants, who would continue to come.

After Israel conquered the West Bank and Sinai in 1967, connections between members of Bedouin tribes were renewed, and there were extensive migrations towards the Negev. In the aftermath of the cancellation of military restrictions, Israeli Bedouin began to wander throughout the country, reaching as far north as Rehovoth and as far south as Eilat. There were problems associated with this return to nomadic living. For instance, Bedouin would regularly break water pipelines to water their flocks. Yet the huge land reserves in Sinai overshadowed any concerns about control of Negev lands.

The Yom Kippur War and the subsequent peace negotiations with Egypt set in motion a process where the bulk of Israel's military training shifted from the Sinai desert to the Negev. There was increasing pressure to settle the Bedouin, whose numbers had grown dramatically.[91] This was hardly a new concept in the area. The Turks founded modern Beer Sheva in 1900 in an effort to settle the area's Bedouin (see Figure 31).

The Israeli government decided to establish seven cities in the northern Negev in which Bedouin tribes could live. The first of these, Tel Sheva, became residential in 1967. The conditions and the incentives offered to settle, by Israeli standards of the time, were attractive.[92] The Bedouin, of course, resisted.[93] Many were indignant at what they perceived as a violation of their rights and dignity. Others were less political but simply did not like the town or want to live in the proximity of rival tribes. Only about 40 percent of the Bedouin initially agreed to move into the new cities.[94] The homes provided for the Bedouin were by no means inferior by Israeli standards, and the towns were relatively spacious when compared to similar Bedouin villages created in Jordan.[95] Nonetheless, the cities' infrastructure and social services were perceived as inferior to that offered Jewish development towns.[96]

It was at this time, in 1977, that the Nature Reserves Authority established the "Green Patrol." The initiative brought together the four major territorial players in Israel: the Land Authority, the JNF, the Nature Reserves Authority, and the Ministry of Agriculture. Zvi Uzan, the Minister of Agriculture, was the most militant advocate for countering Bedouin proliferation, and he lobbied for a common initiative to address the issue. When a Bedouin herd passed through his Moshav farming settlement, defiling the local cemetery, he was sufficiently indignant to get a Green Patrol approved.[97] Technically the Patrol was established to protect the government-owned open spaces from a variety of illegal squatters. But clearly they were most alarmed by the Bedouin in the Negev, whose population had tripled during Israel's first thirty years.

NRA director Avram Yoffe "volunteered" the services of Alon Galili, a veteran ranger, and gave him three subordinate rangers and a jeep. Galili, who had recently become director of the southern region in the NRA, was unenthusiastic about leaving his new job. But General Yoffe had given an order, and Galili stayed at the helm for twelve years.[98] Although the Green Patrol is invariably associated with government efforts to expedite Bedouin settlement in the new urban centers, Galili claims that there were

Jewish farmers with whom they were also in constant conflict, as well as a sleazy underworld element, "who raised horses, grew drugs, and set up illegal dumps."[99]

The Green Patrol well understood that there were two sides to the issue. Many Bedouin had never bothered to register their lands when it had been possible during the Mandate and later were prevented from doing so by military orders. When the orders were lifted in 1966, they rushed out to stake claim on as much land as possible. Joint committees were created to deal with the problems of farmers (represented by Ministry of Agriculture personnel) and with the Bedouin (represented by a sheikh or a tribal leader). But it was not an effective format for dispute resolution. The Green Patrol was authorized to negotiate only with regard to the timing of the evacuation rather than the land rights themselves.

When discussion failed, the Green Patrol enforced the law—with an emphasis on force. Relying on the notion that "if you show them you are strong once, then the Bedouin will give you respect in the future," Galili nurtured the Patrol's tough image.[100] Indeed, Galili was frustrated with the police, who brought with them operating procedures he found too timid for the rules of the game required to confront the Bedouin in the field. When Arik Sharon became Minister of Agriculture in 1977, he identified with the Patrol's mission and expanded its ranks.

Systematically the Green Patrol began to remove Bedouin from State lands. Thirty armed inspectors would typically come to enforce an evacuation. It was not a pretty scene. The press reports were generally sympathetic toward the illegal squatters.[101]

Galili claims that no one was ever bodily removed without first attempting to resolve the matter amicably. The first three workers at the Patrol were all fluent in Arabic, and there was a disproportionately large percentage of Druze rangers. Non-Arabic speakers were sent to Arabic language courses. But frequently dialogue was fruitless, and there was a standoff. At this point the Patrol resorted to a variety of tactics to make its point. There was undoubtedly some exaggeration about the Green Patrol's brutality, and Galili is quick to point out that while there may have been a thousand complaints to the press, only twenty-three were filed with the police, and all of those were investigated and dismissed. Bedouin tell a completely different story.

Farkhan Shlebe might be the "model modern Bedouin" that Israel would like to put on display for the world. He runs a highly successful tourist operation and, like most energetic entrepreneurs, always seems to be on one of his three phone lines, the fax, or the computer. His Hebrew is almost accentless. He is a veteran of the Israeli army. He has a picture of the late Yitzhak Rabin on his wall. But he hates the Green Patrol, who he believes declared war on the Bedouin:

> Rumor had it that if you wanted to join the Green Patrol, Alon Galili would punch you in the chest. If you cried out, you weren't tough enough to make it. They would attach a jeep to a tent and just drive off. They would poke holes in our jerry cans so that we'd run out of water. Imagine how a man felt when he returned from the army to find his tent destroyed and his wife beaten. They shot our dogs even when they knew there weren't any rabies involved. They never bothered to ask if the dog might have been vaccinated against rabies, and they certainly could have. Maybe the collar had just fallen off? As a boy I remember I would see one of their jeeps in the distance and then would pull down the tent, hoping they wouldn't be able to see us. The very sight of their jeeps filled us with fear. I had a puppy and I would lie on it inside the tent just praying they wouldn't shoot my puppy.[102]

Green Patrol brutality and associated intimidation tactics have become almost a new chapter in local Bedouin folklore. The fact that many of the Bedouin were active in the military service did not sway the Green Patrol. Some Bedouin tried to use their connection with the military to attain permission to graze on certain closed military zones. But the Green Patrol was not inclined to allow such concessions. Beyond demoralization, for many families the policy was financially disastrous. They lost their most basic means of production. Any trespassing livestock was subject to immediate confiscation, and frequently Bedouin had to sell their animals at a fraction of the market price to opportunistic meat suppliers, mostly from the West Bank and Gaza.[103]

Although the conflict was over different values and cultures, much of the controversy centered around Alon Galili. Galili added fuel to the fire when he shot two thieves who were breaking into a local grocery store while he was on civil guard duty at his home at Sdeh Boqer. They turned out to be Bedouin. Galili warned them to stop, shot in self-defense, aimed for and hit them below the knees, and drove to the hospital the one who did not escape. But the incident was a cause célèbre for his many critics. The NRA management refused to abandon Galili as a scapegoat. Dan Perry, for example, believes that Galili's image is unfounded, caused by Bedouin exaggeration. The Green Patrol's strategy always emphasized respect (the Galili shooting incident he attributes to "bad luck").[104]

But it was lack of respect that seemed most to enrage the Bedouin, who place a premium on individual dignity. They point to Jordan as an example of a nation that knows how to treat Bedouin correctly. Jordan has its own way of inducing Bedouin to settle, and this was highlighted when Bedouin were uprooted from their traditional caves inside the historic site of Petra. Nonetheless, the Hashemite Kingdom does a better job of venerating Bedouin tradition, and although in decline, nomadic grazing continues.[105]

Ultimately, it is not the Green Patrol's tactics but its mission that remains contested. Many environmental leaders are unapologetic about framing the issue in nationalistic as well as ecological terms. "The Green Patrol performed an enormous mitzvah, a good deed," declares scientist Aviva Rabinovich. "The war over land is painful and difficult, and it always will be. We are fighting here. (There should be no illusions in this regard.) It's a battle for this land and our survival."[106] Rabinovich claims that the aerial photographs from 1917 to 1948 that she used as an expert witness in NRA court hearings clearly showed that the Bedouin never laid any real claim on most of the Negev.

Ecologically, Rabinovich sees the conflict as part of the general dynamics surrounding overpopulation in developing countries. "Before the creation of the State, Bedouin would have one or two children. Most of their children died. Now, they have ten or twenty children per family. There are those who cry about the loss of their culture. Well, that's the price of having twenty children instead of two. With antibiotics we changed their world. And I understand that a Bedouin may want to live like his grandfather did and enjoy the nomad's life. But the land simply can't sustain that population level of nomads. If you want to have twenty children, then live in a city."[107]

Israel's Bedouin do not deny the impact of overgrazing on the desert, but they believe that the division of land is grossly inequitable. They have accommodated the best that they can, switching from goats to sheep (which cause less erosion). They also have become better at utilizing the legal system to fight for their rights and at arguing ecologically. Dozens of cases were filed by Bedouin contesting expulsions; at the very least, the suits provided considerable delays. There is a ten-year average span from the time of filing until the final ruling.[108] One case dragged on for nineteen years.[109]

When removed from the charged nationalistic context, the debate over the actual impact of grazing herds on semiarid lands is a fascinating ecological issue. Goats, particularly aggressive in foraging, have traditionally been identified as agents of erosion and enemies of plant and tree rejuvenation. Indeed, no sooner had the British caught their breath and taken control of the Mandate after World War I than they issued an Ordinance prohibiting the grazing of goats on protected forestland in Palestine.[110]

Yet today many ecologists argue that the disturbance to the land caused by moderate doses of grazing may actually increase soil productivity. Because undisturbed soil is crusted by salinity and microphytic organisms (algae and lichens), seeds cannot take hold and germinate. In addition, the grazing animals leave behind droppings that provide an important source of nutrients. Grazing may also be beneficial for biodiversity, because it limits competitive exclusion. Dominant species that would otherwise overwhelm weaker ones are kept in check by the grazing animals.[111] The variety of plant types in the Mediterranean areas of Israel is four times higher than in regions with a similar climate in California. This can be explained by the regimen of human disturbance in the former region during the past thousands of years. Among these disturbances, the grazing of domestic animals may have an important place.[112]

With the major expulsions complete, an uneasy equilibrium in the political realm between the parties has also set in. NRA personnel try to coordinate grazing schedules for the forty thousand Bedouin who now live in Israel's deserts, primarily the northern Negev. The Bedouin tribes were allocated eight hundred thousand dunams for grazing. Although the quota allotted is one hundred twenty thousand goats, the actual number tolerated is closer to two hundred thousand. Between February and May the flocks are moved toward the western Negev. Then they drift eastward in summer.[113] The NRA monitors the movement and hurries the stragglers along. But Israeli Arabs hardly see it as a benevolent public servant.

The younger generation of Bedouin is increasingly comfortable with city life,[114] and more than three quarters of the population have settled.[115] Although scores of Bedouin still live in tents, very few truly nomadic Bedouin remain in Israel today. As Bedouin acclimated to their new lifestyle, attention gradually began to focus on social services and employment. There is much to improve.

For all the official lip service, as well as bona fide efforts, the Bedouin remain one of the poorest segments of Israeli society. Ironically, poverty and pervasive unemployment are reflected in sustainable environmental indicators. For example, Rahat, the largest Bedouin town in the Negev, has a car "motorization" rate only one-fifth of Tel Aviv's while producing thirty times less garbage per capita.[116] On the other hand, although the birthrate has begun to show signs of a slight decline, an average Bedouin family still has more than ten children.[117] As a result, the Israeli government recently agreed to expand the number of Bedouin cities from seven to fourteen, with a corresponding effort to upgrade conditions in them.

For a while it seemed as if the worst tensions were over. Then in August 1998, the newspapers reported that a NRA ranger shot and killed Suliman Abu-Jlidar.[118] Subsequent investigations revealed that the gunman was not an NRA ranger at all but a traveling companion of the victim, and criminal charges were dropped. Still, the incident reopened all the old scars and dashed any illusions of reconciliation. The Green Patrol may have won its war for desert lands, but in so doing, it lost the sympathies of Israel's Bedouin and Arab communities for the foreseeable future.

To offer a complete picture, it is important to mention that the Nature and National Parks Authority has begun to make a token effort to reach out to Israeli Arabs. In 1997, Giselle Hazzan, a vivacious and thoughtful Technion graduate, was given the job of organizing an educational initiative, based out of the Ein Afek reserve, that would target Israeli Arabs. Ein Afek is one of the last remaining pockets of wetlands in Israel, and it is surrounded to the east by Arab villages. Focusing on the younger generation, Hazzan's program brings roughly 25,000 school children from 50 Arab communities each year.[119] There they enjoy a stimulating dose of ecological wisdom, conservation biology, and the lumbering water buffalo that used to be so prevalent throughout the landscape. Any educational and managerial criteria would deem the Arab educational program a huge success. Yet it is still most conspicuously defined by its modest scope and the fact that Hazzan's remarkable achievements have been attained within the contraints of a half-time position.

AFFIRMATIVE ACTION AT THE MINISTRY OF THE ENVIRONMENT

There was perhaps no clearer case of institutional neglect and discrimination against Israel's Arab minority than the central government's investment in local environmental protection. While a few dozen Jewish municipalities and "Unions of Cities" enjoyed funding to support local environmental workers, before the 1990s, Arab communities, which needed help the most, were somehow overlooked.

When he took over as Minister of the Environment in 1992, Yossi Sarid brought his passion for social justice to the position, making the Arab sector a priority. Calling the present dynamic "a disgrace and a certificate of poverty for previous governments of Israel and for Israeli society in general," he prescribed a policy of affirmative action. Sarid called for establishing a disproportionately high percentage of local units among Israel's Arab sector to compensate for what he perceived to be systematic, historical discrimination.[120]

Sarid was as good as his word. He charged Dror Amir, a veteran Ministry administrator in the international realm, with the task, and nine environmental units began operating in Arab population centers.[121] Their staffs are small and frequently limited to a director, educational coordinator, and field technician. Yet the unit provides an important catalyst. For instance, in Sakhnin, Husain Tarabiah, a charismatic young engineer, took over as head of the unit. He began to pressure polluting tradesmen to clean up their act. He initiated construction of the first regional wastewater treatment and reuse facility for the city's wastes, converting it into an ecological education center for Arabs and Jews. Tarabiah even started a local nongovernmental organization—Naga Environmental Protection Society—to serve as a watchdog and bark at his own government unit—and perhaps expand environmental fund-raising opportunities.[122] Eventually senior Ministry of the Environment officials stopped referring to the Sakhnin program as the most effective local regulatory unit in Israel's Arab sector and began calling it the best in the entire country.

Arab environmentalists appreciated the new units that fostered education and generated more fastidious licensing procedures and new advocates for the issue within the system.[123] The success of each unit, however, was largely dependent on the degree of cooperation with the affiliated mayors, and here responsiveness is by no means uniform.[124] Much depended on the capabilities and resourcefulness of the unit director. Tarabiah, for example, succeeded in expanding his unit into a regional "union of cities." Yet, even though institutionally Arab municipalities have begun to catch up, it will take much longer for the environmental gap to close in the field. Solving pollution problems takes money, and Israeli Arabs have less to work with than their fellow Jewish countrymen. The units hover in perennial danger of closing due to the lack of matching municipal funding. In the meantime, however, Israel's Arab sector no longer lacks local agencies for addressing its many environmental woes.

OCCUPIED AND POLLUTED

Israeli Arabs maintained a distinct separation in their political and cultural identities from their cousins in the West Bank and especially in the Gaza Strip, but they were certainly familiar and empathetic with their ecological plight. If anything, environmental conditions there were so much worse than those inside of Israel that they offered a moderating sense of proportion. For the most part, Israel's small circle of Arab environmental activists did not become involved in ecological problems over the Green Line. There was too much work to do at home and, although Arab citizens of Israel were aware of Palestinian complaints, the problems were deemed insurmountable.

Regardless of one's view about the aftermath of the 1967 War, few would view Israel's occupation of the West Bank and Gaza as a glorious chapter in the country's environmental history. The adverse geopolitical circumstances tended to dominate the ecological reality on the ground. Many commentators are content to file the entire matter under the general heading of neocolonialism.[125] Yet, the unfolding of events had little to do with a premeditated exploitation of West Bank and Gazan resources. Rather it was an unintended consequence of the tragic diplomatic stalemate. The unnatural status of a military occupation, which was supposed to be a temporary arrangement, did not help.

Owing to Minister of Defense Moshe Dayan's powers of persuasion and the absence of feasible alternatives, Palestinian political and clan leaders in the West Bank and Gaza in 1967 were inclined to respond favorably to his appeal: "We don't ask that you love us; we ask only that you care for your own people and work with us in restoring normalcy." This was achieved by maintaining the status quo—legally, culturally, and economically—for the one million Arabs in the West Bank and Gaza. Israel chose not to annex the territories, whose residents formally kept their passports and their Jordanian and Egyptian national affiliations. For the first years of occupation, agriculture continued to be the central pillar of the local economy, and the West Bank farmers' traditional markets in Jordan were maintained (two days after the end of the Six-Day War, bridges over the Jordan River opened, and the flow of produce and people continued unencumbered).[126]

Initially the Palestinian standard of living improved dramatically under Israeli rule. Spurred by Israeli training and buoyed by Israeli markets, within five years, West Bank agricultural production increased by 100 percent, and per capita income was up 80 percent.[127] Even though they received the lowest wages in Israel, hundreds of thousands of Palestinian workers brought home salaries that were lucrative relative to pay in Jordan and Egypt. In the Gaza Strip, 98 percent employment was reported in 1973.[128] The Israeli administration also built piped water systems for hundreds of villages and initially allowed the drilling of some forty new deep wells. It even imported water from Israel's National Water Carrier to expand water supplies for Palestinian cities, leading to initial increases in per capita water usage in the West Bank.[129]

Yet as the occupation stretched on, the population swelled, and requests for new well and irrigation permits from the civil and military authorities were rarely granted. Water became a security issue. The amount of irrigated land in the West Bank remained static.[130] The Israeli civil administration did little to change previous Jordanian and Egyptian policies that had discouraged a serious Palestinian industrial base in the West Bank and Gaza. From an environmental perspective, the lack of industry may have been beneficial. Still, before long, water contamination involving sanitary rather than industrial wastes was a problem that could not be ignored.[131]

Most West Bank and Gaza homes were not connected to a sewage system in 1967. As the Palestinian economy stagnated, the sanitary infrastructure was the first thing to go, creating an exaggerated parallel to the situation in the Israeli Arab sector. The few sewage systems that were installed discharged the collected wastes without any treatment at all.[132] With water in short supply, many Palestinian farmers chose to use sewage as an irrigation source, health risks notwithstanding. Severe contamination became widespread,[133] and associated illnesses reached epidemic levels. In 1994 scientist Karen Assaf reported that waterborne diseases, in particular diarrheal illness, were second only to respiratory diseases in causing mortality and morbidity among children, who make up over 50 percent of the Palestinian population.[134]

As Jewish settlement expanded across the territories, the political as well as the environmental atmosphere soured. Although the settlers usually installed reasonable sanitary facilities, they brought with them the polluting habits of Israeli industry. The government supported an environmental protection unit to service the Jewish settlements in Samaria, run by a conscientious former NRA ranger, Yizhak Meir. Like most of the local environmental offices, however, its influence was limited. For instance, in the West Bank's largest industrial park, Barkan, fifteen of the ninety factories required sophisticated treatment of wastes. But a study in the 1990s showed that in practice, effluents streamed through a disabled treatment facility, polluting Wadi Rabba and the surrounding groundwater.[135] But it was not only the chemical contamination that bothered the Palestinians; they perceived the rectangular Jewish architecture as alien, another form of pollution on their landscape.[136]

International law prohibited the imposition of Israeli environmental legal norms on the local Arab population. This produced an extreme manifestation of the underlying paradox of environmental versus political rights. In contrast to political rights, which rely on a fundamentally passive government posture, the realization of environmental rights requires active government intervention. In the occupied territories, Israel's role was reversed. For the most part, Palestinians wanted to see the Israeli authorities as little as possible. Yet the environment, for some, was an exception, and Israel was resented for not imposing pollution standards.[137] Once the Intifada mushroomed into a full-fledged revolt in the 1980s, however, it was difficult to think about environmental oversight or progress. Still, in 1998, regulations were finally passed that allowed Israeli authorities to enforce environmental norms on the occupied West Bank.[138]

After the 1977 elections, the Likud government began expanding Jewish settlement exponentially, and the resource gap became an embarrassment. Twenty years later, Palestinian hydrologists claimed that 37 percent of the West Bank's Arab population still had no water piped to their homes. Ad hoc solutions such as rainwater collection, delivery by trucks, or rural wells presented their own health problems.[139] Even if their consumption was only half that of Californians (with a similar climate),[140] the 280 liters of water that Israelis used each day was four times higher than that allocated to Palestinians in the territories. During the summer months many West Bank cities experienced acute water shortages.

The image of "thirsty" Palestinian children, many of whom were without running water, staring resentfully across a valley at swimming pools in a neighboring Jewish town, was a powerful weapon in the ongoing propaganda war.[141] Meir Ben Meir, Israel's crusty two-time Water Commissioner, stressed Palestinian responsibility for such crises, citing the 40 percent loss of water to leaky pipes in cities such as Hebron. But many Israelis were embarrassed by such a callous technocratic response in the face of human privation.[142] Israel invariably responded to specific shortages on a humanitarian level by temporarily upping allocations.[143] The Israeli government, however, did little to assuage the Palestinian sense of injustice and resentment at the inequitable allocation of resources.

Comparisons to the quality of life inside Westernized Israel may have been inappropriate. But the state of the environment in the Gaza Strip was deplorable in absolute as well as relative terms. The wells that serviced the Strip in the southern tip of the Coastal Aquifer already suffered from alarmingly high salinity rates in 1967, because of chronic overpumping and saltwater intrusion during the years of Egyptian rule.[144] (Israeli studies suggested that the sustainable yield was 70 million cubic meters per year, but actual pumping exceeded 120![145]). But the squalid situation grew worse as the population doubled. With open sewage ponds festering in the center of refugee camps, dangerous nitrate levels rising in drinking water, withering aquifers, and ubiquitous litter, conditions became unbearable.[146]

Following the Madrid Peace Conference of 1991, the Middle East peace talks included special working groups for the environment and for water. The discussions produced a tentative optimism among observers that Green issues would offer a win-win basis for rapprochement. Yet progress with the Palestinians was slow. Figures collected by Meron Benvenisti documenting the enormous asymmetry between the parties explain why:

Two million Palestinians (one-third of Israel's population) controlled only 8 percent of its water resources and 13 percent of the land; they generated 5 percent of Israel's gross domestic product, with industrial production equivalent to that of a single medium-sized Israeli plant. Palestinian per capita income was one-tenth that of Israelis.[147] With little to lose, it is no wonder that the Palestinians were inflexible.

Much of the tension in negotiations centered around control of the Mountain Aquifer. Hydrologically, the aquifer includes three distinct units that offer the sole source of water for Palestinians on the West Bank and supply one-third of Israel's drinking water.[148] Together the subaquifers can produce a safe annual yield of roughly 630 million cubic meters of water. The problem was that as much as 90 percent of this recharge water originates as rainfall in the West Bank. The largest and highest-quality subaquifer (Yarkon-Taninim) then flows in a westerly direction, where it is most easily tapped in wells that spread throughout the central region of Israel.[149] By 1967, Israeli hydrologists knew that most of the aquifer's potential had already been tapped. Additional wells threatened to lead to massive salination, because of the aquifer's karstic limestone/dolomite geological composition. To the consternation of Palestinians, Israeli authorities drilled additional wells after 1967. These wells, however, added only 50 million to 65 million cubic meters per year, and most were located in the unexploited, saline eastern subaquifer.[150]

Israel claimed historical rights to wells in its territories under one theory of international law, while the Palestinians claimed riparian rights to the water under another.[151] Indeed, the Palestinian negotiators even demanded financial compensation for the years past when Israel utilized the groundwater. Yet most Israelis did not believe they should be expected to make any payment for resources they had won in a war started by past Arab aggression.[152]

The conflict provided material for countless academic gatherings and articles, as well as joint (and sometimes conflicting) reports.[153] The basic premise was easy to agree on: The hydrological interdependence between Israelis and Palestinians made "divorce" impossible.[154] A joint management strategy seemed inevitable. Yet academics had little influence on the zero-sum-game dynamic that characterized official rhetoric about water allocation.

Unwilling to embark on a drastic curtailment of Israeli agriculture,[155] Israelis argued that efforts had to focus on "expanding the pie," not just dividing it up. As things stood, water resources were already inadequate and would only become more scarce. Palestinians, whose baseline water quality and quantity were so much poorer than those of the Israelis, rolled their eyes as Israelis again brought up expensive high-tech options such as desalinization[156] and importation schemes.[157] It seemed that rather than make sacrifices for peace, Israel wanted to have its cake and eat it too.

At the same time, even those Israelis inclined to make concessions were worried that Palestinian control could lead to massive contamination. After Israel left the Gaza Strip to Palestinian control in 1994, there was a rash of unauthorized well drillings. These further exacerbated not only water contamination there but also Israeli concerns about Palestinian commitment to sustainable water management.[158] When Palestinians continued to blame deterioration of the Gaza aquifer—which is either a closed system[159] or only marginally connected to Israel's groundwater[160]—on external Israeli activities, it seemed especially disingenuous.

In the original 1994 Cairo accord that granted autonomy to Jericho and the Gaza Strip, negotiators had basically dodged the water issue.[161] So the 1995 "Oslo B" Interim Agreement surprised skeptics on both sides by taking meaningful steps toward compromise:[162]

- Israel recognized Palestinian water rights in the West Bank (although postponed defining them until negotiations over its permanent status). In the meantime Israel acknowledged future Palestinian needs to be 70 million to 80 million cubic meters annually and immediately made available an additional 28.6 million cubic meters each year for domestic use.
- Palestinians recognized the need to develop additional water resources for various uses as part of an ultimate solution.
- A Joint Water Committee, composed of Israeli and Palestinian representatives, was formed to coordinate management of water and sewage resources on the West Bank. In addition, joint supervision and enforcement teams were set up to monitor, supervise, enforce, and rectify problems arising from unauthorized drillings and connections as well as inappropriate water use.[163]

This spirit of optimism quickly diminished in 1996. As the peace process stalled, implementing the water agreement was not a priority for the new Netanyahu administration or Yasir Arafat's Palestinian National Authority. The joint committees ceased meeting, oversight was terminated, and the two sides began to exchange charges of intentional pollution and environmental mismanagement. Palestinian environmental negotiators became increasingly disenchanted when they faced one indignity after another as they made efforts to meet with Israeli counterparts or even cross from the West Bank into Gaza.[164] In December

1996 the Israel-Palestine Center for Research and Information held its third annual conference on bilateral environmental issues in Jerusalem. Practically no Palestinians were in attendance. Many had been denied permits to enter Israel.

Consequently, several Israeli proposals for joint projects were often ignored, deemed a marginal concern in the larger debate over territorial concession. For example, Israel's Ministry of the Environment was especially disappointed when an opportunity to receive funding from Germany for a West Bank sewage project fell through because the use-or-lose money went unutilized due to lack of Palestinian interest.[165] The only binational cooperation appeared to be between pirate smugglers of illegal toxic waste for dumping in the West Bank.[166]

Palestinian leaders were left with their daunting list of ecological problems and insufficient international aid to address them. Under the circumstances, it was still easiest to fall back on blaming Israel for ecological delinquency. For instance, in a 1998 summer symposium, Dr. Imad Sa'ad, a senior Palestinian official, blamed Israel for lack of infrastructure, the major obstacle to Palestinian environmental protection (of 648 Palestinian villages, 153 were still without drinking-water networks[167]).

Even when the broader political barriers are removed, a shared sense of priorities is likely to remain elusive. As the Israeli economy flourished, the Palestinian economic profile retreated.[168] For the foreseeable future, Palestine will be a developing country. With their children's health in real danger due to poor drinking-water quality, to Palestinians, many Israeli environmental concerns, such as radon concentrations in schools,[169] cellular phone antennas, or endangered-species repatriation, seem downright frivolous. When per capita GNP dropped 35 percent following the Oslo accords in 1993, Palestinian resources were more limited than ever.[170] If Western nations have a difficult time reaching a common environmental denominator with Third World countries, it is hard to imagine that Israelis and Palestinians, with all their additional political baggage, will find it easy to do so.

A DWINDLING INDIGENOUS ENVIRONMENTAL ETHOS

Because of the generally lower socioeconomic position of Arabs relative to Jews in Israel, it might seem objectively harder to enlist the public in environmental concerns. Yet the organic connection to the land, a most crucial part of Israeli Arab ideology, is a conviction to which environmentalists can appeal. There is certainly a strong religious foundation for environmentalism among the country's Muslims.

Islamic theology has been described by scholars as having almost an ecocentric perspective. The Koran contains a steady naturalist motif throughout, and the book is filled with ecologically relevant parables.[171] For instance, the prophet Solomon is moved by the warning of an ant to its fellows to get out of Solomon's way lest they be crushed.[172] It is interesting to note that, although *Sharia* is typically translated into English as "Islamic law," the word literally means "source of water." True to its hydrological etymology, the Sharia considers groundwater a public good that cannot be individually appropriated.[173]

It is not surprising, therefore, that a devout connection to local land and landscape is a highly romanticized part of traditional Palestinian identity and a common theme of exposition for poets and politicians.[174] Yet the connection has been intellectualized as a result of the ongoing trauma of exile and indignity. The environmental ethic of Arab communities in Palestine prior to its politicization was a nonintellectual intuitive impulse—or what has been called "premodern environmentalism."[175]

It is therefore ironic that Israeli Arabs have so quickly come to resemble the Jewish majority in the distinctly nonspiritual nature of their connection to nature and the natural world. The attachment to the land, instinctive in a society of fellaheen, is hard to find among the younger generation. Arab environmentalists are remarkably uniform in their assessment of the present situation: They believe that the problem can be traced to the appropriation of lands and the mass departure from agriculture.

The SPNI's Gazawi lays the blame for this alienation squarely at the feet of the Israeli government's preferential policies: "Arabs' movement to find work in the city was part of a policy. Water was withheld and given to the kibbutzim, and the rainfall was not enough. If we had gotten more water, we might have stayed more agrarian and connected to the land."[176] Whether or not his political diagnosis is correct, pointing the blame will not turn back the clock. Nor will it restore the sense of harmony born of a world that had fewer distractions and thus greater familiarity with nature and the land.

The changes go beyond the occupational profile or recreational preferences of Arab society, informing countless areas in its landscape. Arab architecture is still distinct from that found in Jewish settlements. Palestinian villages swelled after the initial displacement of 1948 was reinforced by an astonishing birthrate. As they turned into small cities, the villages lost much of their original, pastoral flavor. With memories of an agrarian lifestyle fading into the past, Western values and diet filled the cultural vacuum.[177]

This ideological drift was manifested in the ubiquitous trash found along the streets of many Arab settlements. "Until twenty years ago, you couldn't find much garbage," bemoans Ghattas. "There were hardly even any tractors to collect trash. Thirty years ago everything got used several times, even plastic bags. Whatever was left would be burned once a week. Today everything is thrown into the trash, and it is hard to get people to go back to the ways of grandmother and mother."[178] The same phenomenon is even more pronounced among Bedouin citizens, even though indigenous Bedouin culture offered a profound statement about simplicity, reuse, and material self-discipline.[179]

Some see historical continuity in the phenomenon. From the time of the draconian implementation of Ottoman land laws, "miri," or government-owned territories, were perceived by Palestinians as off-limits; the relationship was primarily one of alienation and disregard. The ubiquitous litter in public areas presumably constitutes a modern manifestation of this attitude.[180]

Even if the events of 1948 had gone differently, it is unlikely that the bucolic fellah existence would have lasted forever. Ultimately the nostalgic tune of Arab environmental leaders sounds very much like the sad refrain one hears old-time Israeli environmentalists repeat—mourning a lost society where communalism, a nonmaterialistic mission, and "the land" were the highest ideals.

The challenge ultimately is to salvage the most wholesome and nourishing aspects of traditional rural culture and transform them into a sustainable modern form. Integration of the olive harvest into the present calendars of Arab communities constitutes one such example. Even though most families no longer depend on the olive crop, schools are on vacation in November and October for the harvest. As families flock to relatives or small family plots to pick, an economic necessity was transformed into a celebration of the ancient covenant with the trees.[181]

The resonance of the SPNI message and the increasing number of Israeli Arabs who join the hordes of weekend hikers and picnickers (especially around springs and streams) suggest that if environmental passion is not always evident, it is certainly dormant there.[182] A 1996 review of environmental activity in Israel cited a variety of Arab community organizations—daycare centers, sports facilities, libraries, and after-school educational programs—that have begun to focus on environmental issues and projects.[183]

A JOINT ENVIRONMENTAL AGENDA

Conventional wisdom holds that environmental cooperation offers one of the best vehicles for building bridges between Arabs and Jews. Slogans put forward by both sides, such as "pollution knows no borders" and "our common environment," offer an optimistic prognosis for cooperation. The notion is that if environment truly drives the discussion, agreement can be reached. Even if they frequently make for better rhetoric than actual coalitions, such slogans offer an important first step for rapprochement. It also suggests that after all the historical settling of scores, among Israeli Arabs, self-reliance and pragmatism may be the more powerful impulse.

This has proven to be so in a few cases involving public-interest advocacy. Joint ventures between the Galilee Society and Adam Teva V'din—the Tel Aviv-based environmental advocacy group—were driven by the fact that both groups found themselves involved in the same case. One represented Arab communities and the other Jewish. So cooperation is a practical matter of expedience.

In one case that stretched throughout the 1990s, Adam Teva V'din argued on behalf of residents of the Galilee village of K'lil. It opposed approval of an industrial zone that threatened the integrity of the adjacent Kabri springs, one of the country's cleanest sources of drinking water. The development would also have impinged on the neighboring town of Sheikh Danun, which turned to the Galilee Society for help. A combination of lobbying, appeals to planning committees, and Supreme Court actions stymied the project.

In another case, consulting scientists at the Galilee Society considered the effect that a planned glass factory at the new Zipori industrial zone would have on the air quality in neighboring villages. Their analysis showed future violations of sulfur dioxide standards, but the insult was not only environmental. The industrial area had been given as a gift to the Jewish municipality of Natsaret Elit, located a few kilometers away, whose industrial-park space had become limited. Neighboring Arab (and Jewish) villages would be the first to feel the effect of the pollution but would not enjoy the associated property tax benefits. The organization filed suit. Lack of an appropriate impact statement and of "Best Available Technology" led to a parallel Supreme Court action by Adam Teva V'din.[184] The Court merged the two petitions, but the two organizations had already engaged the factory's Director General in talks about technical solutions. In the ensuing negotiations, the factory committed itself to installing emission control equipment far more stringent than that required by the Ministry of the Environment. The plant also agreed to bankroll independent air quality monitoring.[185]

Joining with an Arab organization to sue against Jewish development posed no dilemma for Adam Teva V'din. After all, the Galilee Society's environmental credentials and capabilities were above reproach. The Society had never hesitated to challenge Arab interests when they threatened the environment. For instance, the Society sued the town of Baka al-Biyah in the Supreme Court when the town left an entire neighborhood unconnected to the sewage grid. The Society later came to the brink of legal action against the city of Nazareth for opening a large slaughterhouse adjacent to a residential neighborhood.[186]

In the aftermath of the unrest and rioting that led to the shooting death of thirteen Arab citizens in October 2000, Life and Environment, the umbrella group for Israel's Green organizations, convened a meeting at the Arab town of Tira. The goal was to explore the role the environmental movement might have in bridging the social gaps and mutual alienation that had swelled between Jewish and Arab Israelis. Representatives from over twenty organizations launched a new initiative: "Room for Everyone." Its express purpose was to identify specific sites of shared environmental challenges that could bring together Arab and Jewish NGOs in a common campaign.[187] Its first major initiative was a joint week-long bike ride for Arab and Jewish cyclists along a route that highlighted common environmental hazards.

Yet it is simplistic to assume that this level of cooperation is always possible. Technical questions such as water quality criteria or the combination of particulate and flue gas controls are not divisive. Other ecological issues, on the other hand, are not free of politics. Environmental priorities reflect community values and by definition will diverge.

An example of an obstacle to forging a shared environmental agenda is the issue of open spaces. For Jewish environmentalists, preventing land-intensive development constitutes the highest ecological imperative of the day. All efforts are made to tighten zoning restrictions and contain suburban sprawl. But Israel's Arab community sees such prescriptions in the context of their own history of harassment and subjugation. They certainly have not forgotten to whom much of the land used to belong.[188] Indeed, this issue may be the single greatest catalyst for community solidarity. In 1975, the Rakah Arab-dominated Communist Party created a National Committee for the Defense of Arab Lands. The committee declared March 30 to be Land

Day, calling on Arabs to join them in a general strike to protest against proposed expropriations.[189] It soon became an unofficial day of solidarity for Israeli Arabs and a constant source of nervousness for Jewish Israelis. (For years Earth Day was not pushed as an eco-celebration by Israeli NGOs, because it was too close in name and time to this new Arab commemoration.)

While Israeli Arabs identify with the general aspiration to preserve the landscape for future generations, they feel that they have done more than their share in the area of forfeiting land resources. This is why the campaign against the Trans-Israel Highway resonated so strongly with Arab communities whose land holdings were lost to the asphalt, who forged their first real environmental alliance with Israel's Green organizations.[190]

Zoning, therefore, looms as a perennial source of tension. After the 1948 war, Arab villages swelled because of the influx of internal refugees. Today 20 percent of Israeli Arabs trace their lineage to those families that could not return to their homes after the 1948 war and flocked to other towns.[191] At the same time, peripheral lands around the villages were frequently nationalized. With the high birthrate and a rising standard of living, land shortages became chronic. Arab municipalities wishing to have their municipal boundaries expanded find themselves in a constant struggle with Jewish-dominated planning authorities. Attempts to enforce zoning regulations and destroy illegal construction by Arabs have become highly charged political propositions. New Jewish settlements dotting the Galilee hillsides literally and figuratively cast a dark shadow.

Israeli Arabs see their cities' congestion in terms of the ongoing battle for sovereignty and a pattern of discrimination. This is exacerbated by a cultural conflict. Planners argue that given Israel's territorial constraints, Arab construction needs to become more efficient and move to the more vertical high-rise structures found in Jewish settlements. Such a change runs counter to the architectural tradition and tastes of the Arab population. And, as it happens, most Arab villages are already densely populated.[192]

Here again, the key to protecting open spaces in the countryside is improving the quality of life in the city. Although residents of high-density Arab housing projects dislike the crowdedness in their apartments, one survey suggests that they have a higher satisfaction with health and community services than do residents in traditional neighborhoods. These and other environmental amenities were singled out as the principal advantage of their apartment lifestyles.[193]

Finally, no issue exacerbates mutual suspicions more than overpopulation. "Judaizing the Galilee" is not just a long-held Zionist axiom; it is the source of many ecologically ill-advised development projects. Not surprisingly, Israeli Arabs oppose a new regional master plan for the north of the country that is based on the settlement of an additional one million Jewish citizens. They have seen the impact of policies designed to keep them "in their place" for fifty years. Arab environmentalists have no trouble finding the ecological terms of reference in which to couch their case.[194] Many Israelis agree about the plan's effect on air and water quality, congestion, and open space decimation. Yet precisely because Israeli Arabs are honest about the full range of their motivations, mainstream Green organizations may be apprehensive about high-profile coalitions.

Repeated allusions by Palestinian political leaders to the high Arab birthrate as their secret weapon only make things worse. Indeed, it is not just rhetoric. In 1996, an average Jewish family had 2.6 children; Muslim families had 4.6.[195] The present dynamic constitutes a classic "prisoner's dilemma." Ideally, both Arab and Jewish environmentalists should join together in a call for policies to promote zero population growth in all sectors of Israeli society. But this would amount to heresy. For Jews, the ingathering of the Jewish exiles remains the raison d'être of the State. For Arabs, population growth holds the key to greater political influence and autonomy. Yet if demography continues to be a legitimate weapon in the battle for sovereignty between Arabs and Jews, both sectors, and the environment they profess to love, will be the losers.

11 Environmental Activism Hits Its Stride

A graduate student runs an air pollution monitoring station on her kibbutz and lobbies her Regional Council to fund an independent risk assessment of the effect of a toxic-waste incinerator at Ramat Hovav. Thousands attend a rally, sponsored by ultraorthodox communities in Jerusalem, hoping to stall expansion of high-tech factories in Jerusalem and reduce the amount of hazardous chemicals they store. A group of Eilat scientists publishes a full-page advertisement asking Israelis not to visit the town until sewage stops polluting the Red Sea. A suit filed by a neighborhood in the Arab city of Shfaram leads to the jailing of a polluting-factory owner for contempt of court when he flouts an injunction limiting work hours and noise. Ad hoc campaigns against solid-waste landfills that threaten quality of life are launched by various groups, from Jerusalem's Giloh neighborhood to Kibbutz Mishmar ha-Emeq to Beer Sheva. This is a small sampling of the environmental activism that rippled across Israel during the 1990s.[1] And it does not include the increasingly numerous—and often radical—animal rights groups that run a parallel network of activities.

Environmental history around the world has often been influenced by small groups of committed individuals who chose to make a difference.[2] Government obviously has a crucial role in eliminating the incentives to pollute and abating environmentally destructive activity, but it will never be able to solve all environmental problems. During the 1990s, more and more Israelis began taking matters into their own hands.

Mobilizing the public on any broad societal issue is not an easy task. Sociologist Stuart Schoenfeld explains that the dynamic constitutes a classic "social dilemma": an individual (or group) that invests resources in a collective problem finds itself relatively disadvantaged in competition with other similar individuals.[3] This dynamic is particularly salient in Israeli society, where leisure is so rare and the colloquial slur "freier" is so common. A "freier" in Hebrew slang refers to the sucker who always gets stuck pulling more than his share of the load. In more altruistic days this may have been an ideal, but today it is unmistakably pejorative.[4]

Yet Israel's society is sprinkled with individuals who chose involvement and decided to fight the environmental foe, be it in the form of pollution, corruption, apathy, or ignorance. They belong to a group of citizens that has been called Zionism's "new pioneers."[5] Many of their stories deserve mention—far more than one book, much less a chapter, can contain. Israel would be an uglier, dirtier, and unhealthier place had they not become involved. Their stories are also important for what they teach about the underlying motivation that impels ordinary people to translate concern for ecology or health into action. Presumably that impulse must be replicated if a broader societal commitment to environmental protection is to be nurtured and galvanized.

When the CRB Foundation commissioned a study about the state of environmental activism in Israel in 1996, eighty environmental organizations were identified throughout Israel.[6] In addition, many national organizations such as the Rotary Club and the Histadrut Labor Union, as well as several women's organizations—Na'amat, Hadassah, and WIZO—had started environmental initiatives.[7] Indeed, as early as 1975 there were already so many groups working nationally that Josef Tamir, the retiring Knesset eco-advocate, set up an umbrella organization, Life and Environment, to foster coordination and communication. By 2000, Life and Environment had almost eighty member organizations. In a comprehensive survey, Orr Karassin, its young attorney-director, estimated that there were probably twice as many identifiable organizational efforts that were environmental in character.[8]

In the CRB survey, forty-three of the environmental groups responded to researchers' questionnaires. The picture that emerged was encouraging: Over 80 percent of the organizations had been established during the previous decade.[9] The new Green players changed the face of the environmental movement. For instance, starting in the early 1990s, the radical "GreenAction" brought a refreshing abandon and penchant for civil disobedience to a somewhat staid environmental community. Greenpeace joined the local scene, plugging industrial effluent pipelines and pulling a variety of other high-profile stunts.[10] Green Course, the burgeoning student network, provided both the imagination and the shock troops on the ground to invigorate and inspire the entire environmental community.

Israel's environmental groups were never merely social clubs. About half claimed that enhancing awareness was their most important goal, but in fact they were predominantly activist in their orientation. Over 90 percent of organizations in the survey had met with government representatives to further their cause, and 41 percent did so on a regular basis. Most attempted to engage the media and write letters to decision makers. Over three-quarters reported participating in demonstrations. Almost as many claimed to have been involved in some sort of legal action. And although there are the usual turf battles and territorial tensions, these organizations are beginning to work together. Only a handful of groups had never been part of a broader alliance, and by the end of the century, groups banded into formal Green coalitions in all the major cities, led largely by revitalized SPNI branches.

As with nonprofit groups anywhere, money is perceived as the key obstacle to progress, but the financial resources of environmentalists as a movement have grown. In the 1996 CRB study, less than 20 percent of organizations surveyed reported budgets over 100,000 shekels ($30,000). Four years later, 28 percent had budgets over $100,000, and only 22 percent operated on less than five thosand dollars. Although only a third of the organizations impose dues on their members,[11] 70 percent of today's organizations operate with no assistance from the Diaspora or other international funding.[12] Still, the vast majority of Israeli environmental groups are "grassroots" in character, operating without offices, staff, or the typical trappings of nonprofit institutions.[13]

The chronic lack of organizational assets in Israeli environmental groups reflects a universal reality of activism. Generally a few key people stand behind an organizational initiative. The only real resource they can count on is their own passion and commitment. This is, however, exactly what makes their tales so inspiring. The following small sample reflects the sheer variety of the citizens who created the wave of environmentalism that swept through Israel during the 1990s. It also illustrates the range of obstacles they faced.

THE FIRST ENVIRONMENTAL ADVOCATES

For the record, it is important to note that the aggressive environmental advocacy model that is the prototype for many groups in Israel today was not invented in the 1990s. Indeed, as early as the 1960s, citizens' groups popped up in Tel Aviv and Haifa. The best known of these, Malraz, the Public Council against Noise and Air Pollution, was active for almost thirty years. Malraz was established soon after the Knesset passed the Kanovich antinuisance legislation in 1961. The group took upon itself the job of enforcing the new law, which forbade unreasonable noise, odors, and air pollution. Its early leaders included Kanovich's widow, Ganiah, and Antonio Feranio, the public-spirited American engineer from Haifa with a specialty in air pollution and Supreme Court litigation.

For its first few years, Malraz lacked a clear focus, and the organization was largely ineffective. All this changed in 1965 when the government's intention to expand the Tel Aviv Reading Power Station made air pollution an urgent political issue. With the help of a supportive Israeli press (in particular the ever-environmental *Ha-Aretz* daily), Malraz became the lightning rod that rallied the public while Josef Tamir led the parliamentary battle in the Knesset. Its petition against the project was signed by two hundred thousand people,[14] an Israeli environmental record to this day, even though the country's population has more than tripled.

Although technically he did not found the organization, Malraz is most closely associated with Yedidyah Be'eri, Israel's first public-interest environmental lawyer. Be'eri moved to Israel from Germany as a child just before the Third Reich closed its doors to Jewish emigration. Ten years later he lied about his age to enlist in the military. In 1949, as a soldier in the new Israeli Army, Be'eri was seriously wounded in combat and hospitalized for eight months. As part of his recuperation, he attended law school. After graduating, he opened his own private practice in 1958, dabbling on the side in the politics of Israel's (rightward) Liberal Party.

It was in 1965 that Be'eri became environmentally active, offering Malraz his legal services. He remembers his personal motivation for involvement as tied to a visceral disgust at the black smoke spewing from the Dan and Egged buses of the period. Yet he acknowledges that his "German" sense of civic order probably colored his worldview.

After the Government pushed its plan for the power station through the Knesset, a weary and discouraged Malraz decided to avail itself of Be'eri's pro bono offer as a last-ditch effort. In 1968 he filed suit against the Israel Electric Company on the grounds that the expanded plant would create a public nuisance and damage the health of residents living as far as six kilometers downwind. Preliminary motions by the Electric Company stalled a decision, but the magistrate judge, Yosef Charish, eventually ruled in the company's favor, holding that Be'eri had not shown the prima facie damage to an individual plaintiff that was needed to gain standing in public-nuisance cases. Behind the decision was the judge's preference for *private*-nuisance suits; he feared that recognizing vague *public*-nuisance actions would open the proverbial floodgates and allow citizens to harass Israeli industry at will.[15] Justice Charish reckoned that only the Attorney General should initiate public-nuisance actions.

The District and Supreme Courts felt differently, ruling in Malraz's favor with a liberal interpretation of standing in public-nuisance actions.[16] By the time of the final appeal, however, the station was up and running,[17] and the Supreme Court counseled Be'eri to take a wait-and-see approach regarding actual pollution damages.[18] The decision, like a few others that Be'eri would bring to the Supreme Court,[19] created an important precedent, even if it was a loser on the ground. In retrospect, though, Malraz's legal work and the case law it spawned opened the doors to public-interest suits and citizen enforcement in Israel.

By its own account, the lost battle exhausted Malraz's energies, and during 1968 and 1969 it lapsed into dormancy.[20] A small grant from the Ministry of Health, however, led to its subsequent revival. After its highly publicized battle against the government's energy policy, the organization's orientation changed. It took upon itself the more reparable micronuisances suffered by an increasingly beleaguered urban populace. Rapid development, crowded metropolitan conditions, and government apathy combined to produce considerable human discomfort. Malraz became the address for those seeking help. Within a few years its membership reached two thousand. Between 1974 and 1977 Be'eri served in Israel's Knesset and brought his environmental concerns to issues ranging from water quality in the Jordan River to the preservation of sand dunes in Rishon L'Tzion.

When Be'eri was not reelected to the Knesset, he returned to Malraz as a voluntary chairman, and for the next fifteen years he patiently assisted dozens of citizens with their problems.[21] The late 1970s were the golden years for the group. Malraz boasted branches in ten Israeli cities and responded to dozens of cries for help, where it championed the "little guy" against polluters or bureaucratic lethargy.[22] It also ran a number of campaigns, from antihonking to police awareness crusades.[23]

Although only a tiny minority of the public's complaints went to court, frequently Be'eri's florid, legalistic letters and the implicit threat of litigation were enough to expedite progress. A 1979 pamphlet describes a long list of nuisances where Malraz successfully intervened, from muffling a noisy refrigeration system in a Safed supermarket to shutting down a dusty concrete-block production plant in Hadera.[24]

Malraz laid the blame for Israel's urban environmental woes at the feet of municipal government. "The Ministry of the Environment has a tendency to shove all complaints from the public over to the municipality," opined Be'eri. "But I always ask: What happens when the municipality doesn't do its job? Who helps citizens then?"[25]

By the 1990s, the organization was in decline. Malraz had never generated its own funding, relying on support from modest government allocations. Be'eri, already close to seventy years old, acknowledged that fund raising had never been his strong suit. He had always relied on his personal connections in government to attain the monies needed for basic organizational operations. (Be'eri actually went so far as to cancel membership dues when he became chairman.) When the Ministry of the Environment began to phase out its annual stipend, Be'eri could not even afford to pay a part-time secretary to type out his letters or pay basic court costs. By then, other organizations had picked up the gauntlet, and it seemed that Malraz had faded away, only to be revived in 2001 by the creative energies of environmental regulator Hilik Rosenblum, and by a major grant to monitor mobile-source air pollution in Tel Aviv.

Malraz may not have changed long-term air quality trends or revolutionized the authorities' sensitivity to noise pollution, but the organization improved the quality of life for hundreds—if not thousands—of Israelis. Moreover, Malraz proved time and again that the Israeli public had the legal tools and moral edge to make a difference. What it lacked was confidence, patience, and will.

THE NUCLEAR NEMESIS

By the late 1980s, there were only two sizable nongovernmental organizations in Israel's environmental community. The Society for the Protection of Nature was still the predominant Green organization in the country, although there were many environmental issues that it did not address. The other group was the Council for a Beautiful Israel, established in 1968 by Josef Tamir, who chaired the Knesset's Interior and Environment Committee.[26] The Council benefited from the dedication and connections of its chairwoman, Aura Herzog, who later became Israel's First Lady. The Council's objectives emphasized aesthetics, with a strong focus on cleanliness.[27] Its orientation was primarily educational, and the organization worked intensively with teachers. Its approach to activism in its nine local chapters typically did not go far beyond community beautification projects.[28]

In this institutional landscape, it is not surprising that a number of small groups and countless individual initiatives sprang up to fill the vacant ecological niches. The sheer variety of characters who became environmental activists in Israel during the period made for a fascinating, if occasionally unruly, community. EcoNet was one of the first such successes. It chose to focus on the taboo subject of nuclear power and radiation. No Israeli environmental group of that period better reflected the spirit, aptitude, and limitations of its "founder" or set off as many unanticipated ripples.

Shirley Rose came to Israel from California during the 1970s. Living in Jerusalem, she found employment as an English teacher and enjoyed the large and friendly English-speaking Anglo-Saxon community there. She had once worked for the American Civil Liberties Union, and she soon became restless for something more provocative than teaching English. Coming from the United States in the post–Three Mile Island era, she was astonished at the degree of local indifference to nuclear issues.[29] Israel's Atomic Energy Commission was directly overseen by the Prime Minister's Office, and it was highly furtive about its activities. Israelis generally relegated questions involving nuclear power, exposures, and wastes to the broader—and inaccessible—category of "security." Indeed, Israel's fringe Communist Party was the only Knesset faction willing to make it a political issue.[30] Environmental groups had "more pressing" and less controversial topics on their agenda. But Shirley Rose did not. When she met and married Herschell Benyamin, she had a partner in her antinuclear crusade (see Figure 32).

Herschell Benyamin was among the many distinguished British Israelis who brought their World War II experience to Palestine to fight for the Zionist cause. Benyamin was quickly put to work as an artillery officer. But unlike many of the foreign volunteers, he was smitten by the young country and stayed on after the cease-fire. (Though he was by nature a man of pacific character, he ended up fighting as a reservist in all four of Israel's subsequent wars.) An avid gardener, Benyamin was a natural person to assume responsibility for the landscaping and grounds at Israel's only golf course, in Caesarea. The position gave him unusual access to Israel's small upper-class community. This elite circle of acquaintances widened after he trained as a tourist guide and became popular among an exclusive clientele for his understated wit and charm.[31] By the mid-1980s Herschell Benyamin was well into his seventies and would have been quite content to retire quietly to a small farm in the rural town of Karkur and tend to his avocado trees and sundry other horticulture ventures. But he had married the wrong woman. Shirley Benyamin had decided to do something to stop Israel from going down the nuclear power pathway, which she saw as a danger to the country's and the planet's future.

As a third coconspirator she called on her friend D'vora Ben Shaul. Dr. Ben Shaul had long since quit her post at the Nature Reserves Authority. Yet her work as Israel's first full-time environmental journalist at the *Jerusalem Post* still left her ample leisure for the new project, and she was happy to serve as editor of a quarterly communiqué, *Israel Nuclear News*. Working out of the Karkur farmhouse, they tried to establish the Israel Agency for Nuclear Information. They hoped to create a clearinghouse that could offer the public reliable information about the perils of radiation.[32] Although over 150 people sent donations after receiving the first newsletters,[33] the Ministry of the Interior refused to register the organization as a nonprofit "*ahmutah*." In the wake of the Vanunu affair, where a disgruntled former worker sold revealing photographs of Dimona's nuclear weapons facility to an English paper, it may have seen something perfidious in an antinuclear group. But 1986 was also the year of the Chernobyl disaster, and the group pressed on.

In 1989, Shirley Benyamin decided to modify her tactics, changing the organization's name to EcoNet.[34] Her rationale was that because a nuclear power plant was not imminent, it was prudent to expand the base of the organization's supporters. When the nuclear monster eventually reared its head, she would be ready to field a formidable opposition.[35] After stating much broader environmental goals, EcoNet's request to register as a nonprofit organization was finally approved. Herschell was formally the chairman, with his many personal connections and his driver's license. But Shirley was the driving force.

Although she was a woman past retirement age, Shirley Benyamin showed the energy of a college activist. The organization found funds to study pesticide exposures and sponsored over a dozen student research theses in environmental health.[36] As the Soviet immigrants began to pour in, the group recognized that it had a unique opportunity to assess the health impact of the Chernobyl accident. Benyamin supported research that monitored and advised Israeli Chernobyl survivors. In 1993 EcoNet helped sponsor a conference on the subject.[37] When concerns about the proposed Voice of America radio transmitter arose, much of the information about nonionizing radiation came from the burgeoning EcoNet library.[38] Ben Shaul's quarterly newsletter, now entitled *EcoNet News,* tackled a different issue each month. The bulletin ultimately reached 150,000 recipients, including all Knesset members and government decision makers.[39] And the group continued to hammer away at the nuclear-power option.[40]

Most of Shirley Benyamin's time, however, was taken up with an environmental hot line. Israel did not have a Freedom of Information Act until 1998, and obtaining data about the environment from Israel's government was not a trivial task. Sometimes the call was for expertise or interpretation of an environmental situation, rather than hard numbers. Relying on a growing pool of friends in academia, she became a Green matchmaker, linking people who had problems to the appropriate expert or journal article. The number of requests for free advice and information grew. Shirley Benyamin presided over her network of 130, mostly retired, volunteers, who helped translate and research, as well as lick envelopes.

Shirley Benyamin saw her group as a catalyst that worked behind the scenes. Besides pursuing the research initiatives, she would, like a fairy godmother, seek out innovative initiatives and share what little resources she had: a computer or a printer for a deserving activist who did not have one, or funding for a part-time staffer to get a promising group off the ground. In 1990, a twelve-thousand-dollar EcoNet seed grant to the author, along with much encouragement, led to the creation of Adam Teva V'din, a public-interest environmental law group that is now Israel's second-largest environmental organization.

Many government officials dismissed the pesky septuagenarian as "that crazy American" (even though her Hebrew was fine, Shirley preferred to speak in English, annoying government officials in a variety of forums with her frank and merciless attacks). Slowly but surely, though, Israelis began to talk about nuclear issues.[41] The process culminated with a televised documentary about health impacts on workers at the Dimona nuclear facility and the environmental Minister's high-profile fact-finding visit to the reactor area in 1994.[42] That year the Knesset found EcoNet's work impressive enough to grant the Benyamins the coveted Knesset Prize for their contribution to Israeli society.[43]

ASTHMA AND ACTIVISM

There is an exceptionally high percentage of immigrants in Israel, in particular from North America, who have heeded the call for societal involvement.[44] Their disproportionate participation on behalf of various causes has led to occasional stereotyping. Referring to the heavy American presence in West Bank settlements, one real estate developer sniped that Americans who come to Israel either move to Hebron or join environmental groups such as Adam Teva V'din.[45] It was true that for the first half of the 1990s, all of the environmental law courses at Israeli universities were taught by American-Israelis. The campaign for clean air by Haifa residents is an exceptionally successful case of community activism. Not surprisingly it was spearheaded by an American-Israeli.

When she moved to Israel in 1982, nothing would have led Lynn Golumbic to think she was about to become an environmental leader. She and her husband Marty brought their daughters to Haifa from New Jersey. They both focused on their jobs at IBM—he as a mathematician and she as a marketing specialist. Golumbic recalls that their small apartment at the Absorption Center north of the city faced Haifa. During the autumn season, when the ventilating winds dissipated, they could see a black cloud sitting on the Carmel hills. They could also feel the pollution. Marty Golumbic immediately developed respiratory problems and had difficulty sleeping; the doctors told him he had asthma. Lynn had always had allergies, but three years later these too blossomed into asthma. When their fourth daughter, Adina, joined her sisters in showing similar symptoms, they knew that the cause could be found in Haifa's air.[46]

Initially, Golumbic was too absorbed in the day-to-day exigencies of adapting to a new country, meeting professional challenges, and raising a family of six to explore the air pollution issue.[47] She did, however, become a member in the local branch of the Association of Americans and Canadians in Israel (AACI), an immigrant support group, and a 1987 maternity leave gave her the time to become more active. By chance she took a phone call in the AACI Haifa office from a newly formed grassroots group: ENZA—Citizens against Air Pollution.

ENZA was founded a year earlier by a group of disgruntled SPNI activists in Haifa, who felt that the nature organization did not devote sufficient attention to the city's air quality. Those were the days when Haifa's sulfur dioxide levels reached perilously high levels—two and three times above the legal ceiling.[48] Golumbic was sufficiently impressed with the presentation of the ENZA volunteers that she wrote an article about Haifa's air pollution for the AACI newsletter. Overnight it turned her into the group's resident environmental expert.[49]

Although there were two thousand factories in Haifa Bay, the lion's share of the sulfur dioxide came from the giant oil refineries and the power plant. The Israel Electric Company cooperated with the Environmental Protection Service, but the Haifa Oil Refineries took a more defiant line. Moshe Shachal, as Minister of Energy, was responsible for the refineries. He was also a Haifa resident. The ENZA activists organized a rally in front of his home and called on Golumbic to bring along her American troops. Upon their arrival, the environmentalists were stunned to discover that they were a minority. The oil refineries had bused in their workers to organize a counterdemonstration. The workers had come to take on the Green zealots who, they believed, had come to close down their company.[50] It was not clear whether the disinformation was intentional or not. Golumbic convinced the workers that she and her fifty environmental demonstrators had not come to demolish their factory but to improve air quality for all of their children. They were so impressed with her that they lent her their megaphone and allowed her to address the crowd.[51]

Exhaustive discussions with experts and a review of scientific literature confirmed empirically what Golumbic had sensed intuitively for some time. Haifa's air was making her family sick. A 1989 survey of Haifa's schoolchildren by the Ministry of the Environment statistically confirmed significantly impaired lung function among people residing in areas of the city with relatively high levels of exposure.[52] Eventually sulfur dioxide levels were printed in the daily press, and health warnings were issued for people with respiratory problems, advising them to stay indoors. Golumbic claims that she always knew when a pollution episode hit the city anyway, because the school would call to have her take one or another of her girls home. During the Gulf War, when Haifa became a target for Saddam Hussein's Scuds, Golumbic's asthmatic children gasped under their gas masks. For her, motivation was never a problem.

The salient issue on the local activists' agenda soon became the location of a new power plant. Israel needed to augment its energy production, and the powerful Haifa branch of the electric company's union wanted the facility expanded and the additional jobs. The environmentalists argued that Haifa already had the most polluted air in Israel, and that the power plant in Hadera, fifty kilometers south, offered a more logical location. Proponents, including Minister of Energy Moshe Shachal and Haifa Mayor Ariyeh Gurel, countered that a new Haifa station would somehow be different from the existing polluting plant.[53] This time they would demand scrubbers and other emissions control technologies.[54] The decision was formally in the hands of the National Planning Council, a conglomeration of diverse governmental and nongovernmental interests.

When ENZA activists began to squabble among themselves, Golumbic took the lead and forged a coalition with Shoshi Perry, who ran the SPNI's Haifa branch. They distributed bumper stickers whose slogan soon became ubiquitous throughout the city: "Air Pollution Is Destroying the City." The year 1988 was an election year. Golumbic made sure to plant environmental activists at every political meeting where candidates appeared to lobby against the Haifa option. Radio talk shows were also targeted. It seemed that every time the Minister of Energy came home to Haifa from Jerusalem he was greeted by a demonstration. Booklets about air quality were published and were distributed in the schools. Research by Haifa University economist Motti Shechter translated the costs of Haifa's air pollution into monetary terms of over twenty million dollars. His study provided another advocacy tool. And forty thousand people signed a petition promoted by the environmental coalition.[55]

The environmental effort was not without opposition. Haifa Mayor Ariyeh Gurel was openly prodevelopment but feared that the pollution issue would hurt him politically. When Gurel canceled a permit at the last minute, Golumbic had to postpone an air quality demonstration complete with pop singer Arik Sinai. The headlines about the mysterious cancellation were much more damaging to the mayor than the rally would have been. In a classic move of co-optation, Gurel did an about-face and offered a seat on his ruling Labor Party's City Council list to an environmental representative. Technion medical professor Noam Gavrielli was happy to answer the call. An expert in respiratory physiology, Gavrielli had been a particularly valuable member of ENZA. The mayor was disappointed when the environmental coalition still chose not to work actively on his behalf. Although ENZA claimed that it was politically neutral, Golumbic admits that its hesitancy to join the political campaign also was driven by a classic activist fear: It might reveal the thinness of their actual numbers. Once elected to the City Council, the energetic Gavrielli was appointed chairman of the Haifa-area Environmental Protection Union; he served in this position for nine years, a significant environmental achievement in its own right.

But the most significant victory came when Hadera was selected over Haifa as the site for the expanded power plant. This would have been inconceivable even a year earlier.[56] The Israel Electric Company did not seem to be overly upset by the turnaround. Along with the Haifa Oil Refineries, it protested the politicization of the environmental issue, claiming that this turned pollution from a technical question into a motherhood and apple pie issue. Acknowledging that their plants were not a "rose garden," they asked rhetorically, "How much is the State willing to invest in clean air?"[57] With the millions of shekels in past profits and a functional monopoly, the question was a little disingenuous. Yet after the Haifa campaign, the answer was clearly, "A lot more than it was in the past."

Golumbic continued her environmental activities for another three years. In this capacity she participated in a suit in local court against the oil refineries and in Supreme Court litigation that hastened tougher emissions standards. But with time she began to lose steam. The time demands became prohibitive. At the same time, Haifa's air was also significantly cleaner; instances of sulfur dioxide rates that were above the ambient standard became increasingly rare, and her family became healthier. It was time to move on. In retrospect she feels that the Haifa campaign made a national contribution. Certainly, air quality in Ashdod benefited as its oil refineries and electric plant ratcheted emissions down to the new standard set in Haifa.

The best indication of both Golumbic's and Benyamin's relative contribution is their organizational epilogues. No sooner did they bow out than both EcoNet and ENZA essentially closed up shop. The finite financial resources, attention spans, and energies of volunteer activists have always been an obstacle to environmental progress. Professionalization, salaries, and support staff can extend an activist's life cycle but often take the joy out of the cause. When a new environmental group sought advice from veteran environmental reporter Elli El Ad, he distilled his message down to one word: "Persist." Golumbic and Benyamin did so long enough for Israel to benefit from their efforts.

POLITICS AND CEMENT

Air quality provided the impetus for community activism in another city—and one that could not be further removed from Haifa in its sociological and ethnic makeup. Beit Shemesh was one of the many development towns that planners scattered around Israel during the 1950s to absorb new immigrants. A high percentage of immigrants from North Africa settled there, and it soon became associated with Sephardic socioeconomic disadvantage. When the Nesher Company built a cement factory in the Har Tuv industrial park north of Beit Shemesh, it rescued the local economy. Instantly it became the town's largest employer—and polluter. During the 1950s, cement factories were nasty operations. But in those days, the Har Tuv plant sat far away from the closest cottages, and people were thankful for the work.[58]

As the city expanded northward, filling the open spaces along the hillsides opposite the Nesher factory, the pollution problem worsened. The porches, cars, trees, and lawns of the northern neighborhoods were constantly covered with a thin white powder. When it became damp, the dust would harden. Such is the nature of cement. Clearing a car windshield in the morning literally became a grinding ritual. The particles also affected the lungs of the local residents. A 1994 government health survey confirmed two decades of respiratory complaints. Schoolchildren in Beit Shemesh were three times more likely to suffer from chronic colds and coughing episodes than a comparable control group in nearby Givat Sharet.[59]

Elli Vanunu had come to Beit Shemesh from Morocco with his parents in 1963, at age three. As an adult, he commuted to work in a print shop in Tel Aviv but was committed to the Beit Shemesh community. Vanunu represented his neighborhood in the town's urban renewal projects and did not hide his political ambitions as an activist in the Likud Party. When Vanunu bought a spacious new apartment for his family in 1988, it seemed as if his standard of living had taken a big step forward. But the apartment was deep in the town's northern dust belt. After speaking to his new neighbors and hearing repeated stories about sick, coughing children, his real estate investment did not seem quite as attractive.[60] He soon recognized that he had found an issue.

The Nesher factory had a long history of disputes with regulators. Many of the big factories in Israel underwent air quality regulation during the 1970s and 1980s, and yet the Beit Shemesh facility managed to elude a personal decree (under the Kanovich law) from the Ministers of Health and Interior.

Vanunu immediately demanded that the Ministry of the Environment call the factory to task for its massive releases of dust into the air. After long negotiations, Prime Minister (and environmental Minister stand-in) Yitzhak Shamir finally signed a personal decree on November 18, 1990.[61] But expectations for improvement were soon dashed. The Ministry of the Environment decided to show flexibility and agreed to accommodate the factory's request for "gradual controls." This gave the factory several years' grace before it would be required to install an electrostatic precipitator in the central smokestack of the kiln. The Ministry had also done shoddy technical work: Several of the smokestacks that were the main source of dust in the complex did not appear on the map that accompanied the permit.

The factory interpreted the omission as a loophole and refused to negotiate over further controls, even though the unmapped sources may have been the major source of emissions. Under the permit, the factory was required to measure particulate levels, but the management dragged its feet on continuous monitoring of the chimney and skipped several of the mandatory ambient measurements around the plant. When the incomplete data were submitted, they still showed a pattern of air quality violations.

Vanunu emerged as a formidable local political force. He had no trouble persuading the Beit Shemesh newspaper to adopt air quality as a lead issue. He sought and received legal assistance from Adam Teva V'din and, through public-interest attorney Tirtseh Keinan, he began to bombard the Ministry, the factory, local environmental officials, and the Mayor with demands to enforce the environmental permit, in particular its monitoring requirements. Ministry personnel who may have been dismissive of the brash young activist began to take him seriously. So did his neighbors.

Vanunu's efforts struck a chord with the citizens. He was a natural organizer. He set up a march of citizens to the factory gates, demanding better emission controls. He videotaped pollution episodes. Many of the factory workers were hesitant to join the ranks of activists, but most were highly supportive, and several quietly provided invaluable intelligence.[62] When he called an emergency meeting at the neighborhood community center to vent frustrations over air pollution in the city, several hundred people showed up.[63] A nervous Mayor Shalom Fadida (from the Labor Party) appeared, an hour late, and made promises to the angry crowd about his commitment to his city's health. Eventually Vanunu filed a private criminal suit against the plant and its manager. Although the resulting indictment did not produce results, because of an unsympathetic judge, it maintained the sense of momentum.

Publicly the factory was defiant, arguing that the dust that was choking Beit Shemesh residents actually came from nearby quarries. It tried to show its environmental commitment by planting trees inside the factory grounds. But the Nesher plant was feeling the pressure and made some adjustments: Ambient air measurements around the plant became much more regular, and when the new electrostatic precipitator was finally installed on the central chimney in 1993, the main source of emissions was virtually eliminated.[64] The factory still had "a long way to go" and periodically caused air quality violations that left Vanunu grumbling about the inadequate number of continuous monitors.

But at least he had a front-row seat from which to observe. The municipal elections of 1994 had catapulted him into the City Council. "I simply reached the conclusion that if we won't sit in the places of power, where the real decisions are made, then we won't make any meaningful progress," explained Vanunu. "The political campaign was purely environmental."[65] Beit Shemesh's new Likud mayor was only too happy to grant him oversight authority over the same local environmental protection unit that had previously defined him as persona non grata. And although there was no love lost between Vanunu and the Nesher management, it began to cooperate.

Vanunu's triumph disproved assertions that the environment was an elitist issue. Few middle-class Ashkenazi neighborhoods ever put together as fierce and effective a campaign as had the Beit Shemesh activists. Vanunu scoffs at attempts to infer ethnic or environmental-justice dimensions from his efforts. ("Ashkenazic children get just as sick as Sephardic ones," he joked.[66]) Rather, the Beit Shemesh experience is instructive regarding the power of uncompromising activism and the potency of health impacts for politicizing air pollution campaigns. Areas adjacent to cement plants are never truly clean, and Vanunu's neighborhood is no exception. But after a decade of pressure by a politically astute and ambitious local, Beit Shemesh's air quality had improved.

A MARSH AND MIGRATION

Some of the more dramatic Green episodes around the world involve accidental heroes, whose environmental cause runs up against powerful and sometimes violent interests. Chico Mendes, the defender (and in 1988 the martyr) of the Brazilian rain forest rubber-tapper movement, may be the most famous such case.[67] As land shortages in Israel lead to windfall profits in real estate ventures, greed seems to have become an increasing factor in turning environmental conflicts ugly. That such vicious intimidation tactics would surface in Eilat's seemingly remote bird sanctuary, of all places, suggests the pervasiveness of the phenomenon.

One of the most common misconceptions about nature is that birds migrate because they cannot endure the cold. In fact, their feathers are sufficiently warm to allow them to maintain their high body temperature and survive the winter. Rather, birds fly south to find food. When temperatures drop below ten degrees Celsius, insects are barely active, and the birds must seek them elsewhere. Hundreds of millions of insect-eating birds make their move from Europe to Africa, and predators such as hawks, eagles, and several dozen raptor species naturally follow their prey. Once they get to Africa though, the birds find it unwise to stay there beyond the winter. The competition for food is simply too keen to find the extra calories needed for breeding. Their instincts guide them back to northern latitudes. Once defrosted, the land is again crawling with insects, offering the birds sumptuous dining.[68]

The round-trip flight, however, is perilous. Only about 40 percent of the half billion birds that attempt it each year survive the thirty-five-hundred-kilometer trek through the Syro-African Rift.[69] The Rift is the long, narrow valley connecting Lake Victoria with Anatolia and is the only land bridge to fuse three continents. On the return trip in the spring, when the birds finally cross the Sahara Desert, they have only made it halfway to their summer home. Hovering on the verge of exhaustion, they are depleted of the energy they need to complete the trip to their northern breeding grounds. "Refueling" takes place in a few staging areas—isolated swamps in Morocco, Algeria, Tunisia, and the Nile Delta.

No pit stop is more popular, however, than the salt marshes north of the Red Sea.[70] Eons before a city named Eilat was a twinkling in Ben-Gurion's eye, twice a year, 283 species of migrating birds quite literally "dropped in."[71] When bird expert Hadurum Shirihai made a list of the species that are dependent on this specific salt marsh as a staging area, he counted eighty.[72] But he did not include raptors, water fowl, or pelagics (seabirds), so the list may be twice that long.

Aerial photographs taken in 1948 show that the entire twelve-square-kilometer complex of marshes on the eastern edges of the Eilat municipal border was intact. Salt marshes usually connote knee-deep swamps, but the arid Eilat marshes looked more like an overgrown prairie. A sink for most of the surrounding rivulets and ravines when they overflow in winter, the land quickly dries. Yet the saline soils supported copious quantities of sea blite and to a lesser extent Nile tamarisk and insects to snack on.[73] In the town's early years the ecosystem was largely unaffected, despite Eilat's rapid growth, from five hundred residents in 1955 to thirteen thousand in 1972. It was about this time, however, that Europeans discovered the winter sunshine along the Red Sea, the ease of direct Scandinavia-Eilat charter flights, and the surrounding red-black magmatic mountains. Even though Eilat sits in one of the world's most active geological faults, the Israeli government subsidized a third of tourism-related construction costs for foreign investors. High-rise hotels became the rage. The dry prairie leading north from the city's coastal beach was carved up into an artificial lagoon, agricultural fields, commercial algae ponds, high-rise five-star resorts, and access roads. By the 1990s virtually nothing was left of the natural marshlands.[74]

Eilat is renowned not only for its avian diversity but for its exotic human population as well. It is probably the only place in Israel where a man as unconventional as Shmulik Tagar could be a figure in local politics. An early settler who has lived in Eilat since the 1950s, Tagar has held a succession of jobs—from keeper of a now-defunct zoo (where he was once chased by a runaway lion) to deputy mayor. Lanky, with long, flowing white hair, a cigarette always lit, and dogmatic opinions about every imaginable subject, he fits right into the Eilat scene. And like many of the old-timers, he loves the birds that come to visit twice a year. Despite the pressure from developers, he was influential enough to get an old municipal garbage dump designated as a Bird Park.

From 1955 until 1977 Eilat's trash was unceremoniously heaped in an ugly wasteland a few kilometers north of the city's beaches. It was the last corner of the salt marsh. Tagar saw that no one had yet grasped the real-estate value of the dump; it was still perceived as a nuisance. His plan was to cover the rubbish with 1.5 to 2 meters of the original salty, dry soil that was being bulldozed to prepare the next generation of hotel foundations. Then indigenous plant species could be planted and the marsh restored. Local politicians thought Tagar's idea a bit crazy but saw no reason to stand in his way. Although he harbored grand visions, Tagar was pragmatic enough to find the only Israeli scientist who might be able to pull off the rehabilitation.

Reuven Yosef (Figure 33) does not fit any typical Israeli stereotypes either. Raised in India, he moved to Israel as an adolescent. Although he was a newcomer, his powerful intellect, field skills, and determination helped him rise through the ranks in one of Israel's crack military units. He was unable to reconcile himself to Israel's Lebanon War, and in 1982 he exchanged his officer's bars for a student card and returned to his studies. It did not take him long to complete degrees at Haifa and Ben-Gurion Universities and breeze through a doctorate at Ohio State and Cornell Universities in conservation biology. His research at the Archibald Biological Station in Florida was proceeding well when in 1993 he was called back to Israel to run the bird sanctuary project.[75]

Yosef brought considerable entrepreneurial skills to the initiative, and the garbage dump was completely reclaimed. It was only a six-hundred-dunam site, 5 percent of the original marsh, but it contained a huge seed bank in the ground that immediately started to attract ants. A few years after it opened, a Hungarian researcher estimated that its biomass was four times higher than that of the surrounding area. It made for a unique educational resource.[76]

Yosef raised research funds as well as monies for the restoration work itself.[77] College students came from abroad to study with him, and to ring, tag, and monitor the birds. The local International Birding Center became a focal point for bird-watching tourists. And Yosef, the ornithologist, generated a prodigious quantity of scholarly publications on subjects from flamingoes[78] and herons[79] to warblers[80] and hawks.[81] Most of the locals, however, were apathetic about the Bird Park and its efforts to restore an indigenous ecosystem. Despite his exciting presentations, Yosef never fully succeeded in conveying the importance of preserving the key link with the natural heritage of the region, or in saving one of the last urban open spaces for residents' enjoyment. The project did not fit into Eilat's commercialized concept of recreation.

These political problems were weathered with relative ease. The more violent enemies were more trying. As the last swatches of available land around the Red Sea shoreline were converted into hotels, developers set their eyes on the Bird Park, which still did not enjoy formal statutory protection. A campaign of intimidation against Yosef and his family began. It was clear that without his passion, the bird park initiative would fall apart. In 1996 Yosef began to receive threatening telephone calls. Soon thereafter, the Bird Center's jeep was vandalized. His wife's car was smeared with eggs. Then seedlings in the sanctuary were uprooted.[82] The Yosef family dog was hanged outside their home. Finally, in the spring of 1998, things got even nastier: A fire was started in the reserve, destroying much of the infrastructure and all the research equipment Yosef had accumulated.[83] The Eilat police tried to attribute the acts to juvenile delinquency and vandalism, but the persistence of the campaign cast a much more insidious shadow. Yosef was convinced that a few local developers, who had sent hooligans to do their bidding, were behind the acts.[84]

Many times during the ordeal he was tempted to take lucrative academic jobs abroad, but Yosef stayed on. It was not in Reuven Yosef's nature to retreat. He got an unlisted phone number, and raised a 3.5-meter barbed-wire fence around his center. But there was not much else he could do, except hold on. Yosef saw his own efforts in the context of the central ecological question confronting the country: "We're pouring millions of shekels into a captive-breeding program for birds like the lappet-faced vulture. But why waste money on the program if we can never release them in the wild because we don't protect their habitat?"[85]

At the start of the twenty-first century, Eilat was a bustling resort town of forty thousand people. It had ten thousand hotel rooms and at least as many more planned. The city is host to European guests throughout the winter and Israeli tourists in the spring and summer. Whether or not it can also find room to host the seasonal avian visitors remains unclear. But the birds would have no chance at all were it not for a small salt marsh that survives thanks to human inspiration and courage.

PUBLIC-INTEREST ENVIRONMENTAL LAW TAKES HOLD: ADAM TEVA V'DIN

Most of the above stories share a common element: With the exception of the International Bird Center, all of the initiatives mentioned in this chapter utilized the legal or scientific services of Adam Teva V'din—the Israel Union for Environmental Defense.[86] The Association for Civil Rights in Israel had remarkable success in expanding the protection of human rights in Israel through its masterful utilization of the courts. Adam Teva V'din was established by the author in 1991 as a public-interest law (and later science) organization to emulate this success in the environmental realm. Starting with the EcoNet seed money and donations from Toronto-based philanthropists Linda and David Bronfman, office space was found near the old port on Tel Aviv's Ha-Yarkon Street.

Although the fuse box at the organization's office smoldered too much to run air conditioners in summer and the roof leaked in winter, there were two phone lines. They were enough to allow the group to set about its business of hounding polluters. Adam Teva V'din began by initiating a lawsuit against the city of Eilat, whose illegal sewage discharges upset the delicate nutrient balance in the Red Sea.[87] Dozens of legal actions would soon follow. Ruth Yaffe, the first attorney hired by the organization, describes the initial challenges:

> When I joined the staff I immediately noticed that environmental laws in Israel were somewhat anachronistic and sparse. When you contrasted the environmental legislation to the volumes of environmental laws in Europe or the United States, it was clear that legal tools existed but not at all clear how to use them. We wanted to be pioneers and were ready to be creative, but ultimately decided to be pragmatic, going down a better-tread path of focusing on specific problems rather than policy questions.[88]

The organization benefited from several advantages that Malraz had not enjoyed thirty years earlier. There was now a Ministry of the Environment, and it knew how to funnel useful information discreetly to nongovernmental watchdogs, to help them bark up the right tree. The press had become interested in environmental cases, providing additional pressure points. Money was the ultimate cause of Malraz's demise. But during the 1990s, the international Jewish community was beginning to take an interest in Israel's environment, and generous funding became available. Foundations such as the Goldman Fund, the Moriah Fund, the Nathan Cummings Foundation, the CRB Foundation, the Conanima Foundation, the Bracha Fund, and the

Porter Fund provided critical and sustained support for Israeli NGOs. Still, the primary advantage of the 1990s was legal.

Israel's activist Supreme Court had lowered its barriers to public-interest petitioners, allowing for almost unfettered access to the Court's administrative powers of oversight. More important was a series of environmental laws and amendments that opened the lower courts' gates to environmental litigants. In 1991, Meretz party Knesset member Dedi Zucker sponsored a Citizens' Suit bill that had been drafted by Jerusalem lawyers Reuven Laster and Bruce Terris. To the astonishment and delight of environmentalists, a year later, just prior to the Knesset's adjournment, he slipped the law by the Knesset's Constitution and Law Committee. The statute was carved up in the process and fell far short of the original vision that would have rewarded lawyers handsomely for winning pollution cases, but still it created a broad range of civil actions against polluters for offenses that now included hazardous materials, solid wastes, and radiation.[89] Four years later, Zucker's colleague, Benny Temkin, amended the statute and the Water Law,[90] removing a number of annoying procedural barriers.[91] Finally, in 1997, sweeping legislation absolved selected public-interest groups of the vexatious requirement to prove actual damages to their members before filing most environmental suits.[92]

Adam Teva V'din also benefited from the proliferation of grassroots organizations that lacked technical expertise. It was a natural match: a national organization that needed local constituents and local environmental groups that needed professional assistance. Within two years of its inception, over twenty organizations signed power-of-attorney agreements with the Adam Teva V'din legal staff.[93] They often needed reassurance as much as they needed representation. For example, in the ultraorthodox neighborhood of B'nei Brak, a local group suffered from an array of dreadful hazards and eventually went door to door to raise funds to pay for the air quality monitoring required for the suit Yaffe eventually filed.[94]

Adam Teva V'din's independent status, combined with its tendency to open dialogues with polluters by making implicit or explicit legal threats, enabled small citizen groups to engage powerful interests. Sewage is just one example of a seemingly intractable ecological quagmire where public-interest advocacy has helped. The City of Ramla's sewage had long been a problem, flowing untreated into the Ayalon Stream. From there it continued to the Yarkon River. In a meeting with the public-interest attorneys, the mayor of Ramla actually invited legal action that might force the central government to provide some of the financial support for which he had long been asking.[95]

A Supreme Court petition by Adam Teva V'din against the city called for imposing a construction schedule for building the plant. The Ministry of the Interior found special funds for the project. Eight other cities were soon confronted, with most of the cases ending in legal actions that pushed cities such as Ra'ananah and Zefat toward a treatment solution.[96]

Once word got out about the availability of free legal assistance for environmental problems, the telephone rang continually. Irit Sappir, who originally was hired to be an office manager, spent most of her time giving environmental first aid for people's problems and eventually went back to school for a graduate degree in environmental studies. Tel Aviv University Law School offered a seminar in environmental legal aid, in which students worked as interns at Adam Teva V'din, helping citizens with their problems.

As was the case for any enforcement agency, Adam Teva V'din's first hurdle was obtaining evidence, in the form of quantitative measurements; such evidence was very expensive to collect. Even if government agencies had been willing to share information freely (the Ministry of the Environment generally was open, the Ministry of Health generally was not), they frequently lacked critical data because of their limited monitoring capacity. So money was raised to open a small "citizens' laboratory." Mouna Noufi had no sooner submitted her Ph.D. dissertation in analytical chemistry at the Technion than she was hired as chief scientist and began to acquire the equipment for measuring pollution levels. The U.S. EPA hosted Noufi on a summer training visit, after which she began to test the effluents released by factories, using a variety of methods. The results of her scientific reports provided the basis for press releases and, in several cases, evidence in litigation.

The most important of these was Adam Teva V'din's private criminal suit against a pair of Haifa petrochemical plants, Deshanim Ltd. and the megapolluting Haifa Chemicals, which were dumping highly acidic effluents into the Kishon River. Danny Fisch left a Haifa law firm to join Adam Teva V'din's legal team in 1994, bringing with him a pragmatic but unyielding legal style. The Kishon cases were among his first in a long string of successful suits. Under the Water Law, the evidence showing flagrant violations of the Water Commissioner's standards was not by itself enough to file suit. Fisch found a group of Haifa fishermen whose boats had been damaged by the acidic water; this met the standing requirements for filing suits that existed prior to the Water Law amendments.

The legal battle that ensued was stormy, as Fisch faced a team of Israel's most high-powered attorneys (including the legendary Amnon Goldenburg and his large Tel Aviv law firm, along with Haifa-based Solomon Lipshitz). For two years they launched a steady barrage of procedural missiles at the suit (as well as ongoing attempts to influence the Attorney General to cancel the case).[97] Time and again, Haifa Magistrate Judge Yitzhak Dar ruled in favor of Adam Teva V'din, and testimony finally began to unfold. The evidence was compelling, and Haifa Chemicals relented. The company agreed to sign a consent agreement that the judge formally approved. Its central component was a timetable to build a seventeen-million-dollar, state-of-the-art effluent treatment facility. In addition, for the first time in Israeli legal history, a polluter created a $250,000 environmental protection fund as part of a legal settlement, covered the full costs of the damage to fishermen's boats, and paid full legal expenses.[98]

As open-space preservation took center stage in the environmental agenda of the 1990s, the organization turned its attention to planning issues. The stakes were usually much higher in these cases than in previous ones, forcing the organization to test the boundaries of environmental precedent. This culminated in a petition to enjoin construction of the Trans-Israel Highway until a comprehensive environmental impact statement was prepared.

To support its case, the organization tapped a dozen international and Israeli experts for help. Their affidavits argued that the impact statements submitted for different segments of the highway did not capture the cumulative or indirect effect of an eight-lane freeway. The experts' opinions confirmed that the highway would gobble up open spaces and exacerbate air pollution and traffic fatalities.[99] The case dragged on for almost two years until Justice Misha Cheshin rejected the petition, offering a narrow interpretation of Israel's 1982 Environmental Impact Statement regulations.[100] Paradoxically, as a private attorney, just five years earlier Cheshin had represented local residents who demanded a better impact statement for the Voice of America transmitter. His decision to opt for judicial conservatism, rather than a creative interpretation of the regulation's language that supported environmental interests, was particularly disappointing. But there would be other cases where environmental interests would win.

THE BATTLE FOR ISRAEL'S COASTS

Adam Teva V'din fared far better when it challenged the rash of construction projects that began to threaten Israel's shoreline in the mid-1990s. The country's Mediterranean coast stretches for 190 kilometers, and the adjacent plains embrace more than 70 percent of the country's population and contain the bulk of the country's economic infrastructure. When one subtracts the military installations and the lands neutralized by industrial installations (four power stations, two oil refineries, two ports, and so forth), only about half of the beaches are left for the Israeli public. This comes to less than two centimeters of shoreline per person. In 1970, the National Planning Council decided to draft a special master plan to manage coastal-zone development. Eleven years later, National Master Plan Number 13 was finally proposed and was approved by the government in 1983.[101]

At first glance the Master Plan, with its highly protective tone, was a triumph for conservation. Development intensity levels were assigned to beaches on the basis of visitor capacity. The plan designated 98 kilometers for beaches; 47 kilometers for beach reserves; 30 kilometers for national parks, reserves, and public open spaces; and 15 kilometers for infrastructure.[102] It required that developers prepare an impact statement prior to building, and it banned any sort of residential housing along the coast. The plan was more lenient with hotel construction, presumably because tourism was an economic value of sufficient importance to override conservation interests. Significantly, it imposed a one-hundred-meter setback of structures from most of the Mediterranean waterline, preventing construction from encroaching on Israel's beaches. This level of protection set Israel apart as a regional leader, ahead of Cyprus's sixty-meter setback line or Turkey's ten-to-thirty-meter proscription.[103]

But the plan had several Achilles' heels. These included fourteen marinas that were to monopolize huge chunks of city shorelines. Not only did marinas obstruct public access to the seashore, but they also decimated the contiguous beaches to the north. The sand on Israel's shores comes from sediment that originates in the Nile Delta. Deposition had already been reduced 90 percent by construction of the Aswan High Dam.[104] The wave breakers that enclose marinas serve as a physical barrier, trapping the sand and preventing replenishment of the beach. The massive erosion that resulted was particularly unfortunate, because there was no real demand for anything approaching the yacht capacity that fourteen marinas provided. Yet marinas offered a boom for the local tax base, since developers offered to pay the cities' costs of construction on the water in return for permission to build on the land behind it. When the one-hundred-meter barrier was broken, it was open season on the most lucrative real estate in the country.[105] Once environmental Minister Yossi Sarid learned of the impact of the marinas, he tried to stem the tide, but mayors ignored his call to freeze marina development.[106]

Government impotence sparked a sequence of successful litigation by Adam Teva V'din attorneys. Some lawsuits were conducted in conjunction with local groups (such as the lively Residents for the Seashore—Nahariah) and some independently. Danny Fisch, a former naval officer, was particularly committed to coastal preservation. When an SPNI activist in Ashdod asked him in 1995 to help confront the marina there, he did not hesitate, even though it was already under construction. The Southern District Court in Beer Sheva quickly held hearings on the injunction request. Negotiations ended in compromise, with the city agreeing to sponsor an independent report on coastal erosion and to repair any coastal damage caused by the project.[107]

At the same time, the City of Tel Aviv proposed its second municipal marina, to be located at the outlet of the Yarkon River. As no plan had yet been formally submitted for public scrutiny, it provided an opportunity for setting a more meaningful precedent. Gila Stopler had been a lawyer for less than a year when she appeared on behalf of Adam Teva V'din, before Sara Serota, a seasoned judge in the Tel Aviv District Court. She asked Serota to enjoin the Regional Planning Committee from submitting the marina plan for public consideration until an environmental impact statement could be prepared.[108] Initially the Adam Teva V'din team was surprised by the judge's unfriendly reception: *"Bubeleh,"* Justice Serota condescended to the youthful-looking Stopler, "Why don't you save yourselves the court costs? It's a fairly unusual remedy you're asking for. Your case is premature; first let them submit the plan, and then file an objection if you like."

Stopler did not back down, insisting that the public did not have the necessary information to file objections and that it could find such material only in the contents of an impact statement. The environmental Ministry's Regional Director, Dalia Be'eri, happened to be in attendance. She asked to speak to the court (a request that was completely out of order) and proceeded to criticize the government attorneys' position. When Be'eri described the damage caused by Herzliyah's marina, Justice Serota changed her mind. In the decision against the city and the developers, she wrote: "It is incumbent on us to leave something for our children. The tendency to derive financial benefit from public lands ought to be secondary to the benefits public lands bring the public."[109]

The next Adam Teva V'din coastal achievement was a decision by District Court Judge Gabrielle Kling, canceling the development plan behind the Herzliyah marina.[110] Soon thereafter, legal action derailed a marina (again replete with luxury housing units) that was planned in Haifa.[111] The marina would have changed the face of the scenic coastal landscape there, with bombastic high-rise compounds on seaside landfills, breaking the gentle slope of the green Carmel hillsides into Haifa Bay.[112]

The enormous profits for coastal developers guaranteed that sooner or later the legal battles would turn ugly. So it was when Adam Teva V'din turned its attention to the Carmel Towers. A high-rise complex had been approved for Haifa's western coastline in 1978, prior to the passage of the Coastal Master Plan with its protective policies. The six towers were to hold 1200 luxury resort apartments and 800 hotel rooms as well as conference centers. It was not until 1993, however, that construction on the first tower began.[113] When Adam Teva V'din attorney Elli Ben-Ari went to check the plans, he was astonished to find enormous discrepancies between the blueprint and the city's building permit. The approved structure was supposed to be nine stories high and fifty-six meters wide. Instead, the developers were building eighteen stories that covered one hundred and thirty-four meters of beachfront.[114]

Ben-Ari, a Haifa native, was appalled when he saw that, if the construction continued, a virtual wall of high-rise towers would block off a full kilometer of beach—a seventh of his hometown's coastline. The only beneficiaries would be the luxury apartment owners and the developers. The Adam Teva V'din Board of Directors was convinced that the organization could not stand idly by, despite the low likelihood of success in court.

When the matter came before the Haifa District Court in April 1996, Judge Dan Bayne agreed to grant a temporary injunction, implying that the building approval was not as "kosher" as the developers insisted. But Adam Teva V'din could not afford the bond that he imposed as a condition for receiving the injunction. The organization was also wary of a precedent that would impose huge financial risks on future public-interest litigants. For most of the next year, the judge's decision alone proved enough to chill the construction on the second tower. In the meantime the developers counterattacked.

The hotel project was sponsored by a group of investors that included Itzhak Tshuva, one of Israel's most prosperous contractors. They claimed that the stalled construction was costing them hundreds of thousands of dollars per week. The Carmel Towers attorney (and partial owner), Itzhak Segev, dashed off letters to Adam Teva V'din's managers and board, informing them of their personal liability for losses caused by delays. The personal threats were elevated on March 3, 1997, by a more formal round of letters threatening that suits were imminent.[115]

The harassment did not stop there. Prior to the organization's general meeting in 1997, last-minute applications for membership were faxed from Tshuva's Tashluz construction headquarters. Several more arrived suspiciously at the Adam Teva V'din office via courier. Newspaper reports later confirmed that at least one of the applicants had been asked to join the organization by "senior Tashluz management."[116] Although the membership forms were rejected, it was impossible to seal a democratic organization hermetically. At the general meeting itself, Dan Marom, a hitherto anonymous member, created an uproar when he sent his attorney to represent him. Marom owned an apartment in the first Carmel Tower, and his lawyer warned the membership that they could each be financially imperiled if the organization continued with the suit.[117] Those in attendance unanimously voted to discontinue Marom's membership, but he promptly filed suit against the organization, calling his dismissal illegal.

Meanwhile, in the cramped Haifa courtroom, the developers were particularly bellicose. They drafted over fifty apartment owners to be co-defendants, imposing the considerable logistical headache associated with serving papers. (Environmentalists closed ranks, and the Society for the Protection of Nature offered its machines for the copious quantities of photocopying needed.) Although technically the suit was filed against the government (specifically, planning agencies and the city of Haifa), government attorneys were conspicuously uninterested throughout the trial and frequently did not bother to attend court sessions. This left the phalanx of lawyers for the apartment owners an open field to drive the defense strategy. They chose to go on the offensive, putting Adam Teva V'din's legitimacy and its management's good faith on trial.

In an attempt to establish the groundwork for the personal threats against the organization's leaders, their attorneys hammered away at the legality of the decision by the Adam Teva V'din board to file suit. Complaints were filed with the National Registrar of Nonprofit Organizations, calling for disqualification of past decisions by the general meeting on technical grounds. The organization was painted as corrupt and disorganized; the press was told that the suit was filed only to bolster the property value of the Adam Teva V'din Chairman's parents, whose view would be obstructed (although at the time they actually owned no real estate in Haifa at all).[118] Despite the constant objections of Adam Teva V'din attorney Elli Ben-Ari, the judge allowed cross-examination to continue for weeks.

Danny Fisch was appointed Adam Teva V'din Director in 1996, and founding director Alon Tal became the volunteer chairman. Fisch fielded each of the attacks with steely composure. Dan Marom, the disgruntled member, had the bad luck to find his case ruled on by the same Justice Serota who had blocked the Tel Aviv marina. She quickly sent him packing. The organization filed a complaint with the Israel Bar Association against Carmel Towers attorneys for harassing witnesses in the middle of their testimony with the threat of a lawsuit. The threatening letters suddenly stopped. The nonprofit registry was ultimately appeased by the group's explanations. The organization even passed a mysterious audit by tax authorities The harassment may have temporarily put Adam Teva V'din on the defensive, but it did not stop the organization from filing other legal actions against coastal projects.

When the Haifa court finally ruled on the Carmel Towers suit, both sides claimed victory.[119] Justice Bayne did not agree to tear down the two buildings already up, yet he ruled that contracts to sell them as apartments were illegal and that the one-hundred-meter line to the sea had ostensibly been breached. An appeal will decide who ultimately won, but the case had the immediate effect of changing the second building into a far less imposing, skinnier tower, and it also put the financial future of the project's four remaining towers in doubt. The cumulative impact of the marina litigation temporarily chilled some of the excitement for additional coastal development. But it was a piecemeal conservation strategy. Adam Teva V'din recognized that the country needed comprehensive legislation to tighten coastal management. The organization's attorneys wrote a strict law and began lobbying to pass it.

The hundred-odd court cases and countless more environmental matters that Adam Teva V'din lawyers advanced during the 1990s proved that Israel's legal system was a good equalizer. The unlevel playing field created by inadequate information, a weak Environmental Ministry, and prodevelopment policies could be tipped back in the environment's favor, especially if a case fell to a sympathetic judge. Developers and polluters began to pay more attention to legal requirements, and the public was on occasion roused from a fatalistic resignation that victory by economic interests was ineluctable. But not every environmental problem had a legal solution, and there were far more cases than there were public-interest lawyers. The judicial system could provide only part of the solution. Political empowerment would also be needed.

GREEN PARTIES

Throughout the Knesset's history there have been legislators who cared about environmental issues. The honor roll began with Shimon Kanovich, Yizhar Smilansky, and Josef Tamir and includes Yedidyah Be'eri, Yael

Dayan, Uzi Landau, Boaz Moav, Yehudith Naot, Mossi Raz, Benny Temkin, Rachel Zabari, Dedi Zucker, and most recently an environmental economist from the Russian Yisrael b'Aliyah party, Michael Nudleman. Well aware of the impact of the Green Party in Germany and other countries, environmentalists have for twenty years wondered whether the time was ripe in Israel for a comparable effort. The answer invariably was "No."

The reasons were more practical than ideological. A Green Party could not muster the 1.5 percent threshold of the popular vote required to capture a seat in the Knesset. Other issues, in particular the question of security and territorial concession (and, more recently, ethnic patronage), dominated the Israeli voter's consciousness. Yedidyah Be'eri toyed with the idea of trying to return to the Knesset as part of a Green Party but ultimately rejected it. Israeli voters cared about the environment but not enough to cast their vote for a political party that made it the only issue.

Furthermore, despite the conventional wisdom that ecology is apolitical, it turns out that not all Israeli voters feel the same about the environment. Political scientist Avner De-Shalit found a remarkable correlation between Israelis' political opinions regarding the peace process and their sensitivity to environmental issues. Voters for leftist "pro-peace" parties showed the highest interest in environmental issues; levels of commitment dropped with voters on the right side of the political spectrum. De-Shalit also found that 68 percent of all activists and workers at the Society for the Protection of Nature voted for the Meretz or Labor Parties, which are identified as the "pro-peace" camp. Some 90 percent of the survey respondents who expressed biocentric views were leftist Meretz voters.[120] This led him to the conclusion that Israel already had a Green party: Meretz.

Beyond electability, the idea of a Green party gets sticky when one considers the range of issues that divide Israeli society at present. Anybody who even considered establishing a Green party naturally had tried to enlist the help of SPNI founder and media personality Azariah Alon. But he always demured: "I always ask: 'What's a party?' Josef Tamir and I can agree about the environment, but what about civil marriages or the West Bank? How could we be on the same side? In Israel, one or two people can actually decide who is running the government, and you'd have no idea of whom you were working with."[121]

This pessimism changed overnight in the municipal elections of 1998. Shmuel Gilbert, a Haifa architect, headed an unaffiliated list, "Our Haifa," for the city government. The campaign focused on the development excesses of Mayor Amram Mitzna but was going nowhere until

Gilbert decided to add the subtitle "the Greens" to the party name. He was delighted when he was rewarded with five seats in the local city council.[122] Pe'er Weisner registered a national "Green Party," which then took two seats in the Tel Aviv city council elections. Other Green-affiliated parties won seats in Yehud, Mivaseret Zion, and Ashdod.[123] Once in the city councils, the new cadre of Green politicians became a lightning rod for environmental concerns, with varying degrees of radicalism. The Tel Aviv Greens offered a take-no-prisoners antiestablishment approach that attacked Tel Aviv's air pollution problem and the proliferation of cellular phone antennas. Haifa's thoughtful and charming Gilbert preferred a professional and uncompromising focus on sustainable urban planning.

The political success in the local elections did not escape the attention of Nehama Ronen, the ambitious Director General of the Ministry of the Environment. In January 1999 she resigned her post to form the Voice of the Environment Party and run for Knesset. The press conference announcing the move was broadcast nationally, and Ronen's candidacy enjoyed considerable media coverage.[124] Recognizing her electoral potential, Israel's Center Party offered her seventh place on their candidate slate (a "guaranteed" Knesset spot), and she brought her platform and ecological concerns to what was then Israel's third-largest political party.

The Tel Aviv Greens, who had attacked Ronen unceasingly during the campaign, continued with their own national efforts. The party ultimately attracted MK Dedi Zucker to sit at its helm, after he was shunned in the Meretz Party primaries, as well as Tel Aviv night club and trance music impresario Hagay Ayad.[125] Ultimately Israel's Green Party ran an unfocused and poorly funded campaign, receiving a pitiful thirteen thousand votes. When the Center Party chairman, Yitzhak Mordechai, resigned in disgrace after a conviction for sexual harassment, Ronen took her seat in Israel's parliament and soon formed a lobby of 22 Knesset members with a professed environmental commitment. These were only pilot ventures, but they left little doubt that Green politics in Israel had entered a new era.

NEW AGENDAS AND PARTNERS

After fifty years of independence, Israeli environmentalists needed to take stock. As is the case with any healthy ecosystem, the diversity in the shades of Green gives the environmental movement strength and stability, despite the headaches of coordinating overlapping efforts. The many serious and creative environmental institutions that have taken root and blossomed in this small country alongside the dominant SPNI suggest that the ground is fertile for ecological commitment. Yet after looking beyond their own immediate microsuccesses, honest environmentalists in Israel must admit that it has not been enough. Financial constraints and unfavorable geopolitical circumstances join a host of other excuses to explain why. But Israel's environment deserves results, not excuses.

Environmental history will not be changed by merely toughening emissions standards or defeating yet another destructive development plan. The symptoms of environmental degradation have their origins in broader societal phenomena. Not only governmental but also environmental nongovernmental organizations now recognize that environmental problems have been framed too narrowly. If clean air and water are to be attained and the Israeli landscape to be preserved, environmentalists today must make the difficult foray into the hitherto untraveled worlds of transportation, consumerism, architecture, and family planning. These issues have long been on the menu of environmental groups around the world. Altering strategies may in fact prove easier than adjusting tactics; indeed, changes in orientation have already begun.

Because of its direct link to air quality, transportation was the first new discipline that environmentalists began to study. Since systematic monitoring began, air pollution emissions in Israel have doubled every ten years. During the same period, the number of vehicles on the roads did too. Despite 100 percent import taxes on vehicles, growing societal affluence propelled car ownership rates from 6 cars per thousand people in 1950 to 198 in 1995.[126] The average daily distance traveled by the vehicles of an increasingly suburbanized population also grew, from 43.6 kilometers in 1980 to 47.1 kilometers in 1994.[127]

Despite the optimism at the Ministry of the Environment over catalytic converters, they have never been a panacea. It did not take long for many of the converters to stop functioning after they were tainted by the low-grade fuel sold in Israel, and, with import taxes of 84 percent of the catalog price, replacements were expensive.[128] By 1998, a third of Israeli cars reportedly failed to meet the new emissions standards, and noncompliance among older vehicles was especially rampant.[129] Although a tougher inspection and maintenance program could have led to their replacement, the Ministry of Transportation clung to relatively lenient European standards and was not inclined to beef up its oversight of emissions testing or roadside inspetions.[130]

Without the clear legal authority to do something about the problem, the Ministry of the Environment could only offer scientific explanations for the steady increase in the concentrations of oxides of nitrogen in the air, which kept breaking records and choking Israel's cities.[131] There was no reason to believe that a 1993 prediction made by Israel's leading air quality modeler, Menahem Luria, would not come true. Luria projected that by the year 2010, smog levels in Jerusalem would be comparable to those in Mexico City, a frightening place where children's health is endangered merely by playing outside in summer.[132]

As congestion became unbearable and rush hour turned into a prolonged ordeal, public transportation was more a reflection of government neglect than a government service. By the end of the 1990s, the country had over fourteen thousand kilometers of roads, but only 596 kilometers of railway tracks. The number of seats available on buses per inhabitant had not increased since 1970.[133] With only a minimal network of designated bus lanes, passengers were penalized twice. After walking to their bus stop and waiting, they faced the same delays as all other drivers. Consumer response was swift. Between 1989 and 1994, public transport ridership decreased 13 percent in Tel Aviv.[134]

Environmentalists also came to recognize that open-space preservation was linked to a sustainable transportation policy. By 1997, 5 percent of the country's land was covered in roads.[135] If the number of vehicles continued to rise, even more land would be sacrificed on the altar of mobility. New low-density bedroom communities were built around a two-car-family reality and the resignation that public transport would not be able to serve them. At the same, the inadequacy of the transportation infrastructure posed one of the primary barriers to stepping up density in cities like Tel Aviv. The underground parking garages proposed for the new generation of skyscrapers only exacerbated congestion. Without a subway or a monorail that enabled people to move quickly and easily across Tel Aviv, Jerusalem, and Haifa, it was unrealistic to think about doubling urban population. In the meantime, development continued to spill into the countryside.

The expansion of the environmental agenda enabled groups to form new coalitions with partners that in the past had been overlooked. There are many examples. Transportation, for example, opened the door to powerful allies, such as the accident prevention community, the elderly, the disabled community, and even youth movements. Connections with ultra-orthodox advocates for preserving the sanctity of ancient Jewish graves had a tentative beginning in the Trans-Israel Highway campaign. (The Israeli Arab community, whose lands were to be appropriated by the road, proved to be far better environmental allies.[136]) At the same time, in 1992 a group from farm communities, dedicated to preserving the lifestyle that the highway threatened to disrupt, registered as a nonprofit organization.[137] Indeed, by the end of the 1990s, the Forum for Public Transportation, a broad but highly effective coalition, brought together these disparate interests. A new organization, "Transportation—Today and Tomorrow," provided public-interest expertise.

These new constituencies brought with them a tactical component lacking in previous environmental endeavors. It fell under the broad heading of "outreach." Because of NGO reliance on volunteer experts, chronic insulation undermined effective advocacy. Ever since the SPNI rode the success of Azariah Alon's radio broadcasts in the 1950s and 1960s, its following became increasingly predictable, and it failed to embrace new communities. For instance, in a survey conducted during the 2000 holiday season, only 3 percent of visitors in JNF forests were immigrants from North Africa, and only 3 percent identified themselves as ultra-orthodox—rates far below the actual census levels.[138] The inability to field their own transportation experts willing and able to back up the environmental agenda in public forums left enviros looking like amateurs.

If fundamental aspects of Israeli lifestyles are to be influenced, traditional communications methods need to be reexamined. Television, the most powerful media tool, rarely gave the environment the sympathetic coverage offered so freely by newspapers and radio. Environmentalists needed to find out why no commercial films made the environment a theme, or why the evening news reported only the most outrageous eco-gimmick or disaster, neglecting the more arcane chronic problems. After dogged lobbying, in 2000, Life and Environment secured a place for an environmental representative on the Board of the Israel Broadcast Authority. Public-interest scientist Dr. Noam Gressel embraced the challenge and began to push for better environmental coverage on both radio and television. The growing cable network generated modest, local ecological programs on the popular National Geographic and other channels. The time had come to learn from other countries about making the broader environmental agenda a salient election topic and how to induce political parties to engage in a competitive, Green one-upmanship.

Other potential partners certainly existed who may have been willing to join hands in environmental matters. For example, Greens in Israel never connected with the powerful labor movement in the country—allowing themselves to become trapped by the false dilemma of "jobs versus ecology." With the exception of the work of attorney Richard Laster and physician Ellihu Richter against particularly egregious exposures to workers in a nickel-cadmium battery plant, an asbestos factory, and the Dimona nuclear facility, there were practically no crossovers into the occupational realm. Yet workers often serve as an early warning—the proverbial canary in the mine—that can uncover a polluting factory.

Once Israel began the reconciliation process with its neighbors, there was no shortage of environmental outreach efforts over the borders. It may have been the euphoria sparked by the glimpse of a peace that had been elusive for so long. It may also have been a shot of opportunism and the expectation of funding for joint Arab-Israeli cooperative ventures. In either case, after the peace process heated up, Israeli nongovernmental environmentalists were among the first to seek out their Arab colleagues. They began umbrella groups such as EcoPeace or the Palestine Israel Environmental Secretariat.[139] Joint educational programs were started, such as the Arava Institute for Environmental Studies, which brought Arab and Israeli university students together in an intensive interdisciplinary environmental leadership program.[140] Within four years, dozens of the institute's graduates had assumed key environmental positions or launched public-interest start-up initiatives in Israel, Jordan, and the Palestinian Authority.[141] And there was a handful of remarkable projects, including Yossi Leshem's bird migration center, whose satellite hookup allowed schoolchildren to track bird migrations through computer systems in their respective countries.[142]

Environmentalists in the Middle East's "peace region" quickly became acquainted with one other. In many areas, however, a healthy dose of realism soon tempered expectations. In some areas, such as water quality protection and nature preservation, cooperation could be vital. In others it seemed a waste of time. The Jordanian, Egyptian, and Palestinian economies were completely different from that of Israel, producing a very different set of environmental challenges. Transportation was not yet deemed an ecological priority for Arabs. Drinking-water quality did not top the list in Israel. The bottom-line natural-resource balance sheet was unclear: Israel's dwindling sand reserves stood to benefit from imports from its neighbors; water scarcity would grow worse as negotiations provided more equitable allocations; wildlife might benefit from binational reserves.

Contacts also suffered from a conspicuous lack of reciprocity. Environmental organizations, in particular in Jordan, were hesitant to embark on high-profile collaborative ventures. Palestinian enthusiasm initially was greater but vanished when the escalated violence of the Intifada came to dominate Palestinian-Israeli relations after the autumn of 2000. Here, Israeli environmentalists had little choice but to heed their new friends' suggestion for greater patience.

THE PUBLIC'S HEARTS AND MINDS

Eilon Schwartz, an environmental education expert from Hebrew University, served as the first Chairman at Adam Teva V'din. In addition to his views about the appropriate tactical approach for any given case, he often raised deeper "strategic" questions. Victories would not be sustainable if the general public that the organization purported to represent did not share the organization's underlying environmental values. During the debate over the Trans-Israel Highway he reminded advocates that even if public transportation or bike lanes were dramatically improved through litigation or lobbying, most Israelis would still prefer their cars.

This need to address the intellectual origins of Israel's environmental problems drove Schwartz to join Jewish studies scholar Jeremy Benstein to establish the Abraham Heschel Center for Environmental Learning and Leadership in 1994. The Center is named after the great Conservative rabbi who taught so profoundly about traditional Judaism's "radical amazement" toward the natural world. The overarching goal behind the Center's curricula, teacher seminars, and leadership training programs is to confront the environmental crisis as a crisis of values. The Center's mission statement read: "Only a renewed appreciation for the wonder of the world in which we live and a renewed commitment to basic human responsibility for that world can lead to long-lasting environmental protection." Schwartz taught a perspective that was both universal and consciously Jewish.[143] The message appealed to a variety of governmental and nongovernmental bodies, including the SPNI and the Ministry of the Environment, which frequently used the Heschel Center on a consulting basis. By the end of the 1990s the Heschel Center was Israel's fastest-growing environmental initiative.

Reaching out to Israeli hearts and minds and infusing them with a renewed commitment to environmental ethics is a daunting task. Simpler educational challenges have repeatedly failed. Despite years of campaigns, gimmicks, and slogans, 78 percent of Israelis polled in a government survey admitted to littering.[144] The environmental ideal runs counter to the relentless message blasted across sixty cable television stations, suburban shopping malls, and an increasingly hedonistic urban culture. Despite a multimillion-dollar water conservation media campaign after the third consecutive dry winter in 2001, water consumption among the Israeli public dropped by only 1.5 percent. It is not clear whether Israelis will be able to leave their cellular phones long enough to listen to the natural wonders that might inspire them to take an alternative route. (Israel is truly a world leader in cell-phone usage: Between 1996 and 1998, Cellcom, Israel's second cellular phone company, saw its customer base jump from five hundred thousand subscribers to one million.[145])

Lest the clash with the prevailing cultural bent appear too discouraging, it is worth noting that in some cases Israelis have listened to public-interest messages. For instance, Israelis used to be among the heaviest smokers in the world. During the country's first thirty-five years, theaters, buses, and offices reeked of tar and nicotine. Yet during the 1990s, smoking rates plummeted. As a result, today's 28 percent smoking rate is only 4 percent higher than that of the United States.[146] More surprising was the commitment to indoor air quality. Without any discernible enforcement efforts by the government, a tough 1983 ban on smoking in public areas and buses was widely honored.[147] Even El Al airlines prohibited smoking on its flights.

It is also well to remember that all Western societies are not the same. Values, and not simply transportation policies, inspire millions of Dutch citizens to get on their bicycles and pedal to work.[148] By the late 1990s, organizations such as Tel Aviv for Bicycles began to add an environmental message to their more traditional recreational appeal and were eventually recognized by *Time* magazine for doing so. But the demarcation of bicycle lanes has been delayed again and again, and the rare cyclist remains a brave pioneer amidst the tumult of Tel Aviv traffic. The virtues of civic duty, caring for the natural world, and concern for public health are hardly alien to Israeli society. But it was not clear how to tap into these amorphous commitments and harness them to change many popular but environmentally unfriendly aspects of modern living.

One source of encouragement was Israel's school system and universities, which expanded their environmental tracks during the 1990s. A 1997 survey reported that the subject of environmental studies was taught in half of the country's elementary school classes. At the secondary level, 150 out of Israel's 450 high schools offer a final *bagrut,* or matriculation exam, about the environment—up from only three programs in 1991.[149] Furthermore, a 1996 study commissioned by the Ministries of Education and Environment showed a relatively high environmental literacy rate among students who participated in formal studies programs. It also confirmed a correlation between the level of knowledge and commitment to environmental values.

Over the long run, formal education can offer crucial *content* to the Israeli public. Yet it remains an unlikely forum for changing *consciousness*: convincing Israelis to forgo a large family, a suburban development, or a second car. Private actions like these are at the heart of Israel's most troubling ecological issues. Getting the public to reconsider them is a challenge that must be addressed by informal educational outreach as well as government policy.

Perhaps the ultimate value of grassroots expressions of environmentalism, therefore, is that they offer a compelling experience that flies in the face of the pervasive consumer culture. People really do learn best by doing. Hiking through a pristine valley may no longer awaken the same commitments for many as it did when Israel was less metropolitan. Sadly, many of the old natural treasures are tarnished. In 1996 journalist Meir Shalev wrote a column about his regret at having participated in trail-marking campaigns during his youth. Returning to the sites, he found the treks covered in plastic wrappers and beverage cans. Shalev wondered whether only an elite group of bushwhackers, who had to work to find the scenic routes, was worthy of hiking them.

Israel may be more urban than ever, but that does not mean that people have changed. Through activism (as shown in Figure 34), scores of citizens have come to rediscover their local communities and to connect to the basic human need to belong somewhere. Precisely because Israel is such a young country, with half its residents still foreign-born, many citizens suffer from shallow roots. Enlisting the public in efforts to enhance the physical fabric of their common urban experience offers an important tool for deepening these ties.

And aesthetics and beautification have taken their rightful place on the country's environmental agenda, as environmentalists began to think about ethics. Life expectancy for Israeli males, now standing at seventy-seven years, has consistently been among the highest in the world, and Israeli women live two and a half years longer on average![150] Statistically Israelis' "quantity of life" is doing fine. It is quality of life that needs attention.

The ancients had a keen sense of the ability of greenways to transform and edify drab urban life. This awareness can be found in pastures mandated for Levite towns in the books of Numbers and Leviticus.[151] The medieval rabbi and physician, Moses Maimonides, prescribed open spaces as a tonic for city congestion. These considerations were secondary in the strategic planning of Israel's cities.[152] But there is a growing pride in local parks and beaches, or just clean, attractive neighborhoods, that energized many of Israel's ad hoc groups during the 1990s. The exceptional coalition that rose up to fight to preserve the last three hundred acres of the Jerusalem Forest became a model.[153]

It has been argued that this sense of place may be the most powerful, identifiable force behind environmental commitment.[154] Growing numbers of Israelis were seeking to safeguard those places they called home. The vast majority of grassroots environmental campaigns attempted to preserve a specific cherished resource. Often a group's only purpose was to stop an undesirable landfill, factory, or road. Even in Hebrew, this phenomenon has been dubbed by centralist environmental regulators as "NIMBY"—the pejorative English acronym for "not in my back yard." It is disparaged as a disingenuous, selfish desire to deflect the necessary environmental price of progress onto some other, weaker segment of society. This egoism presumably leads to less than optimal environmental results.

But from the vantage point of a sustainable society, NIMBY actually represents a positive phenomenon. The same citizens—who refuse to breathe toxic air, to endanger drinking-water sources, or to concede the open spaces that offer them a bridge to their natural world—may be the harbingers of a mass environmental movement in Israel. This movement understands intuitively that only a sense of place, community, continuity, and responsibility can return the harmony between humans and their environment.[155] The educational challenge is to expand the public's perception of its "backyard." In a country as intimate as Israel, the micro for many is hard to distinguish from the macro, and this educational mission may prove easier to overcome than in larger nations.

Environmentalists are often discouraged. Even when the ecological situation is good, it is their job to worry about how easily things could turn bad. But the record of local and national activism in Israel during the 1990s offers some basis for optimism. More and more Israelis were waking up to their ecological problems and to their own power to improve them. Beyond the many resources that were saved and health hazards that were mitigated, activists may have infused their ranks and communities with a higher sense of purpose: that the healthier environments they protected not only bestowed benefits to their bodies but to their spirits.

12 Toward a Sustainable Future?

It was the eve of Rosh ha-Shanah (the Jewish New Year) 1995 when Yisrael Peleg got a phone call from Dr. Dureid Mohasneh in Jordan. Peleg, then Director General of the Ministry of Environment, had become friendly with Mohasneh when they met at the negotiating table during the Middle East multilateral environmental peace talks.[1] Mohasneh, director of the Port of Aqaba, was calling from his home with some bad news: "There's been a major oil spill here outside our port," he reported.[2]

Water usually flows in a counterclockwise direction in the Gulf of Aqaba. This means that Jordanian pollution directly affects Israel, whose pollution in turn directly affects Egypt. The proximity and ecological interdependence were precisely the reason that the first important environmental agreement signed as part of the Middle East peace process was a pact for cooperation in the Gulf. Despite the concern about the ecological damage, the best tiding of the New Year was that the new system seemed to be working.

Peleg already knew about the problem. Elli Varberg, a burly former Navy officer and head of the Ministry's marine protection unit in Eilat (Figure 35), had already alerted him to the situation. Within two hours the Israeli Emergency Response squad was on its way. The squad's five inspectors were indeed a seasoned team. In addition to their considerable experience containing spills in Eilat,[3] they had already been through exercises with their Jordanian counterparts in a joint European workshop. The cooperative procedures were to be tested for the first time.

A six-million-dollar Japanese-funded oil spill station was not yet running in the Gulf of Aqaba, so initially the Israelis had to do most of the work. They brought with them their snakelike booms for containment and quickly sealed off the area, confining the spill. Then, using skimmers and pumps, they began the tedious process of soaking up the oil out of the black water. The team worked well into the night. It soon became clear that the original estimate by the Jordanians of an inconsequential two-ton spill had been optimistic. It was closer to a hundred tons, and much of it was hiding under the piers of the Aqaba port.[4]

Varberg had to sail back to Eilat to pick up heavier equipment. It took three days and the help of hundreds of Egyptian day laborers to bring the situation under control. But the oil spill did not spread, and a catastrophe for the Jordanian coral reef was averted. Removing the thick, gooey mess from the equipment in Eilat took several weeks more. But what Varberg remembered most from the incident was the warm welcome offered the Israeli pollution prevention team by the Jordanian hosts. Not only did they feed them lavish catered meals, but they also made sure that the offending ship's insurance company paid the Israeli unit in full for its work.[5]

Because it happened during the New Year's holiday, the Israeli press did not pick up the story, and the Jordanian press, never particularly enthusiastic about highlighting the peace process, did not report Jordanian reliance on Israeli assistance. Nonetheless, it was a landmark event—one about which peace negotiators had every right to be proud. The joint effort stood in contrast to the somewhat fatuous environmental euphoria associated with many of the amorphous and ultimately empty proclamations and regional codes of conduct that had been signed.[6] The environment was finally benefiting from the Gulf of Aqaba treaty.

Even so, bilateral emergency response operations are not a panacea for the Gulf's problems. Rather, they serve only to underscore the inadequacy of present strategies. The diplomatic significance of the cooperation was at least as great as any immediate ecological gain. By definition, oil spill emergency responses are not proactive. Little energy went into eliminating the cause of the spill or developing long-term plans to ensure that the unique coral reefs were not devastated by a full-scale, future oil tanker disaster. This time the habitats had been lucky. The three marine protection stations established in Aqaba, Eilat, and Egyptian Nueiba were equipped to deal with only small (two-hundred-ton) events.[7] They could not begin to counter the damage from a spill caused by the Egyptian ship that regularly brought 160,000 tons to Eilat from the Abu Rodeis oil fields.[8]

In March 1995 EcoPeace, a nascent regional environmental umbrella group, convened Egyptian, Israeli, Jordanian, and Palestinian environmental organizations in Cairo to talk about the Gulf. Participants unanimously called for making the Gulf a tanker-free zone.[9] Most of the exported oil was bound for Europe anyway, so it was much safer to transfer it by pipeline to the Mediterranean and cheaper in the long run.[10] The proposal was a shout in the desert. This sort of shared commitment to environmental protection might impose real expenses on the parties. As such, it was not a realistic subject for negotiation between Middle Eastern governments.

At any rate, such a bold step might ultimately contribute little to the preservation of the remarkable coral ecosystem in the Gulf of Aqaba. In 2001, the Israeli government commissioned a report from an international team of scientific experts to address the growing controversy surrounding the contribution of mariculture and fish cages on the northern edge of the Gulf to the sudden rise in coral loss in Eilat. They cited a discouraging list of pollution sources to explain the massive 70 percent death rate: phosphate dust from the port, sewage discharges, siltation, the municipal marina, and port ballast waters, as well as the fish cages. But the most insidious problem was the hardest to regulate. People, not tankers, have brought large areas of the reef to the brink of extinction. The cumulative impact of well-meaning swimmers and divers probably constitutes the single greatest cause of coral destruction. Divers' flippers raise the sediments, which then cover and suffocate the delicate reef. Direct contact between coral and hands, diving fins, hoses, or scuba tanks adds injury to insult.

At least on a theoretical level, government negotiators truly cared about their shared aquatic resource. The deep, tropical, salty blue waters of the Gulf support 230 species of coral and nearly a thousand species of fish, thrilling laymen and marine biologists around the world. To preserve this wealth of diversity requires an immediate reduction in the population pressures along the Gulf shore. The subject of carrying capacity, however, was not on the official agenda; in fact, it ran counter to it. The one thing that the governments in the region could truly agree on was that they could make a lot more money and create many more jobs by expanding the respective tourist industries in the Gulf of Aqaba. Israel announced plans for twenty thousand Eilat hotel rooms, Jordan spoke of four thousand, and Egypt envisions a whopping one hundred six thousand—albeit dispersed along its much longer Sinai coastline.[12]

Marine biologist Nanette Chadwick-Furman studied the impact of diving on coral reefs in the U.S. Virgin Islands before bringing her expertise to the Inter-University Institute in Eilat. In collaboration with David Zakai, a biologist at the Nature Reserve Authority, she analyzed empirical observations for the first time, confirming what has long been known by seasoned divers in Eilat. The research showed a perfect correlation between the number of dives and the extent of the damage to coral. The Caves Reef, an enchanting labyrinth of marine mystery, is one of the most popular diving sites along Israel's Red Sea coast. It is also the most devastated. More than 60 percent of its total coral has been damaged. In contrast, the Japanese Gardens, with only a quarter of the Cave's divers, is in much better shape, showing only 10 percent losses.[13] Chadwick-Furman then began working with the Nature Reserves Authority to limit the number of visitors who can utilize the one-kilometer-long Coral Beach reserve, even though most of them come to sunbathe and swim, rather than snorkel and dive.

Divers and dive clubs bristle at the notion of reduced access to the Red Sea. Yet the question facing decision makers in the Gulf of Aqaba is simple: Will strict limitations be placed on access to the coral now—while there is still some coral left—or will the public continue to enjoy unrestricted diving until it wipes out all that remains? Sadly, despite growing concern about the predicament, no city council, legislature, or Ministry has ever seriously deliberated the issue. Preliminary steps, such as bans on diving classes in the reefs until novices develop buoyancy control, have only now reached the proposal stage.[14]

This question of sustainability, or carrying capacity, does not seem to interest government decision makers in any of the Gulf countries. It took almost fifty years for Eilat to build its first eight thousand hotel rooms. Its own plans call for a doubling of rooms within four years' time.[15] This aggressive economic strategy may ultimately backfire, as the infrastructure certainly has not kept pace with the growth.

Visitors to Eilat will certainly not forget the summer of 1998, the most scorching in Israel's history. As much as the heat, they may recall the collapse of the city's main sewage line, when the northern and central sections of the beach were engulfed in sewage. With bacteria levels fifty times above permissible levels, the Ministry of Health ordered everyone out of the water and relegated tourists to hotel swimming pools.[16] Presumably they will find a more pleasant environment for their next vacation. Even if the city finds the emergency funds necessary to handle the sewage glut, and even if clever promotion can gloss over the fiasco, glib explanations will not correct the ecological reality.

The case of Eilat serves as a microcosm for the environmental dilemmas facing Israel.[17] Eilat is a remarkable success by most of the criteria that guided the nation's founders. Even when geopolitical turmoil has decimated Israel's tourist industry, the city has provided nearly full employment for its forty thousand residents.[18] It attracts billions of dollars in foreign currency from its steady stream of international visitors. The city has a music conservatory, a small college, a tennis center, one of Israel's best professional basketball teams, and some of the finest restaurants in the country. But unless there is a radical change, Eilat will not have any coral to speak of. Little will remain of the diverse, luminescent fish populations that its reef once sustained. If one wishes to see exotic marine creatures, one will have to visit the town's impressive but pricey aquarium and observatory. Or one can always go see what remains of the magnificent marine fortunes of Egypt or Jordan.

Do the Eilat coral reef and the Yarkon River offer the environmental bottom line for the Zionist dream? In a reversal of the Biblical admonition, have the inhabitants unwittingly devoured the land they came to redeem? After a century of Jewish settlement and over fifty years of Israeli sovereignty, present environmental trends give little basis for encouragement. As recently as the 1950s, Israel's air, water, and land were generally clean. Today they are not. The earnest efforts and frequent successes forged by Israeli institutions and individuals described thus far have not been able to stem the larger deterioration, reflected in a variety of bleak physical parameters.

In a critical retrospective of Zionism's first hundred years, Ari Shavit, a leading Israeli journalist, wrote:

> Zionism is destined to continue to cause an ecological catastrophe in a substantial part of the small and sensitive land that many of the planet's inhabitants still consider *Terra Sancta*. . . . In a certain sense Zionism was a grandiose engineering scheme drafted in the final years of the nineteenth century and implemented during the twentieth. A somewhat arrogant scheme whose implementation hinged on a wide-reaching act of development. An act that could not have been nonviolent. An act that would finally, after one hundred years, succeed in submerging most of the valleys and plains and hills of the biblical land under the asphalt and cement of a megalopolis state—thriving, commercialized, and very loud. . . . The task of settling five or six million immigrants on this narrow piece of land, which measures only twenty thousand square kilometers, required the rape of the land. Ironically it was the success of Zionism that wreaked irreparable environmental damage on Zion.[19]

Shavit's pessimistic analysis fails to mention that Israel's environmental movement has also evolved and come of age. It can yet galvanize Israeli society to confront its environmental responsibilities at last—before it is too late. What must be done then to set the country and Zionism on a sustainable track? What new issues, approaches, and vision are required, and is it a realistic mission?

LIFE IN THE THIRD JEWISH COMMONWEALTH

The land of Israel witnessed astonishing changes during the turbulent twentieth century, when Zionism forged the third Jewish State. Battles were fought; empires twice unseated; millions of refugees found shelter and others fled; a democratic state was born. The land and its people were transformed as never before. Once sparsely populated and undeveloped, Israel now teemed with humans, who sought to fill every available space. One-fifth of the entire countryside (not counting the Negev desert) was covered by construction, while only a fifth of the land around cities remained as unbuilt open spaces. Trees made a comeback and covered a tenth of the hills and valleys. Although it has retreated as of late from its most prosperous periods, farming still dominated the scenery as never before. At the start of the century, regardless of their origins, most people were indigent, living at subsistence levels. Now, by global standards, Israelis are wealthy. An agrarian economy has been supplanted by industry—at first heavy and more recently high-tech. But an environmental tragedy has accompanied the inspiring drama of modern Zionism.

Culturally Israel perceives itself as a Western nation; its best basketball team successfully competes in Europe, and its entertainers occasionally win pop-music contests there. Israeli orchestras, fast food, theaters, websites, arcades, hospitals, cable television stations, pubs, universities, magazines, and tattoo and sushi bars may have a Hebrew flair but unquestionably evoke the West. English-language fluency may be higher than it was during the Mandate. Yet the dynamics of environmental protection in Israel are fundamentally different from those of other First World countries.

In his best-selling and controversial 1995 tome *A Moment on the Earth: The Coming Age of Environmental Optimism,* Gregg Easterbrook posited that most Western nations go through an environmental cycle. Relying on trends rather than snapshots, Easterbrook argued that the majority of Western nations have already passed through their most polluted periods and are beginning to exhibit ecological improvement.[20]Empirically the numbers in Israel tell a different story.

Unfortunately pollution has not yet peaked in Israel. The air gets dirtier every day, with lead and sulfur dioxide the exceptions that prove the rule. Since 1980, for example, oxides of nitrogen (NO_x) emissions doubled twice.[21] Estimates of annual air-pollution-related premature deaths in Israel consistently range between three and four digits. Quibbling over whether the actual mortality caused by air pollution is in the hundreds or in the thousands, however, misses the broader public-health issue.[22] Asthma, once relatively rare among Israeli children, is an increasingly common trauma for Israeli families. (Between 1980 and 1989 incidence among children increased from 5.6 to 11.2 percent[23] and today reaches 17 percent,[24] bringing respiratory illness to epidemic levels.[25]) Breast cancer incidence among Israeli women ranks among the highest in the world. One in eight women ultimately develops breast cancer, with four thousand cases and eight hundred deaths per year.[26] The 32 percent increase in incidence during the 1990s may partly be linked to early diagnosis, but the especially high breast cancer rates in Haifa (32 percent above the national average) are certainly a scary reflection of the many years of environmental neglect.[27] Apparently the environment is getting worse.

The largely irreversible deterioration of groundwater continues, although contamination rates may have begun to stabilize.[28] There is much talk of river restoration,[29] but the Australian experience in the Yarkon's toxic waters speaks for itself. Even the JNF, after a high profile embracing of river restoration, has made hasty retreat, apparently finding its restoration projects on the banks futile given the quality of the water. The number of fatal accidents involving hazardous materials remains alarming.[30] As natural habitats dwindle, dozens of mammal and reptile species are now listed as endangered.[31] Of the 185 bird species that bred regularly at the turn of the century, fourteen have become extinct and fifty-eight are threatened.[32] Israel's urban cat and crow population, however, may be the densest in the world.[33]

The annals of Israel in the twentieth century offer some obvious explanation for the legacy of continued ecological deterioration. A population that grew sixfold in fifty years, and with a ravenous ambition for economic development, generated residuals. At the same time, because it was so small, the country did not enjoy the margin of error that allows larger nations to make mistakes with relative ecological impunity. Israel never had any spare aquifers to tap. Neighborhoods quickly reached the borders of garbage dumps and dirty industrial centers. For many of the animals, there was literally no place to hide.

During the second half of the century, the environment may have been abused, but it was not forgotten. Even when faced with unparalleled military threats and relentless waves of immigration, the country still found the resources to make environmental history. Its afforestation efforts began early, picked up momentum, and, within a largely semiarid context, remain unprecedented. Its nature reserves managed to stem the tide of most extinctions and even return a few almost-lost species to nature. The Society for the Protection of Nature emerged as an activist conservation group, attracting a large public membership a full decade before the modern environmental movement emerged in the West. And almost everyone showered with water heated by the sun.

The environmental successes and failures are in fact opposite sides of the same coin. Both were symptoms of the powerful patriotism that characterized the young State. Israelis gave up a national pastime of picking wildflowers; they invested in water infrastructure, ceded considerable public lands to preserve nature, and planted trees, because they were perceived as national objectives. Like the levying of outrageous taxes or the rationing of food supplies, it was part of the price for restoring the ancient homeland, and most citizens willingly paid. But building the State also meant sparing no expense to create jobs, mining the Dead Sea, granting factories carte blanche to maximize profits, and meeting agriculture's unquenchable thirst for water. The attendant environmental price was perceived as part of the same package.

Conservation and environmental groups conducted countless battles along with their government colleagues, who eventually, languidly, took the lead. Initially these efforts hardly had any effect at all on the larger war over what should constitute Israel's true national interests. Yet the efforts had a slow but cumulative effect that was continuously strengthened by the international environmental awakening. The shift in this ideological clash probably began in 1961, when Dr. Kanovich's Abatement of Nuisances Law specifically branded unreasonable noise and air pollution as crimes. By the start of the twenty-first century Israelis were aware that their environmental problems would wait no longer. Even industry recognized that no national interest was served by production that made air unfit for children to breathe or by budgets that ignored the contamination of a finite supply of water.[34] Perpetrators of such acts pursued individual or parochial interests rather than legitimate societal objectives. This transition has set the stage for a possible reversal in the way Israel treats its environmental resources.

Israel's environmental history certainly teaches that economic progress has a price and that environmental progress is not automatic. Most areas marked by improvement were the results of clear government policies, supported by an unshakable societal commitment. Polluting facilities such as Israel's electric power stations,[35] oil refineries,[36] and cement plants are cleaner because they were forced to clean up. Looking ahead, there is no reason that a thoughtful but insistent government leadership with a strong supporting role by citizens' groups cannot bring about similar breakthroughs in other troublesome areas. Incremental pollution mitigation, however, may not be enough for many threatened resources and species. There is such a thing as being too late. Large sections of Israel's aquifers testify to the finality of irreversibility.

GETTING TOUGH AND SMART WITH POLLUTERS

Conventional pollution problems are still acute, causing severe health and ecological damage. They cannot be abandoned as environmentalists broaden their menu of concerns. The task is especially formidable because of the change in the country's pollution profile. Reaching the multitudes of non-point-source polluters in their cars, buses, or dry cleaners is quite a regulatory challenge. Yet if government law enforcers finally decide to get tough on polluters, there can be many reasons for cautious optimism.

For most of Israel's history, the "polluter pays" principle was an empty slogan. Almost without exception, pollution paid very handsomely for the scores of industries that knowingly shirked environmental responsibility while continuing to enjoy an economic advantage.[37] Israel's police force, attorney general, district attorney, and judges quite simply had other priorities. After forty years, a Ministry of the Environment began to make inroads. By the second half of the 1990s, the Ministry filed close to two hundred criminal prosecutions per year. Yet the majority were directed at small polluters and litterbugs, whereas larger factories managed to keep regulators at bay.[38] There were many excuses for the gap between the Ministry's stated priorities and its enforcement program—most conspicuously, the professional backgrounds of enforcement personnel. The very progress, however, of which it could justifiably be proud when it did enforce the law (such as the closure of dozens of illegal garbage dumps) suggested that the Ministry of the Environment needed to focus its enforcement energies more effectively. At the same time Israel's Water Commission needed to make "enforcement" part of its vocabulary.

It is not just toughness that is required. Israel's enforcement policy must become more sophisticated in seeking compliance. Enormous technical advances in pollution monitoring—from analytical chemistry field kits to remote and infrared sensing—have yet to become tools used by most enforcement agents. Technology is in fact an ally, not an enemy, in the effort to improve air quality. Cars all over the planet are cleaner than ever and will continue to be so.[39] A recent analysis suggests that when Tel Aviv switches to electric buses, it will save the lives of sixteen people a year.[40] Some even argue that only a few years remain for environmental advocates to use air pollution as a compelling argument for improvements in rapid transport. Although they are not always cheap, technological fixes exist for most of

Israel's air pollution problems. During the same ten years (1987–1996) that ambient nitrogen oxide levels doubled, reaching crisis proportions in Israeli cities, they dropped by 10 percent in the United States.[41] Many local patriots bemoan the "Americanization" of Israeli society, but this is surely one area worth emulating.

It is now possible to imagine marshaling the societal resources to make this kind of transition. Money is no longer the problem it once was. Israel was a poor country when it was established. Since then, the standard of living has increased tenfold.[42] From 1989 to 1997, for example, gross domestic product rose by 50 percent, whereas population growth was only 29 percent. In 1995 Israel's per capita GDP stood at $17,200.[43] This compared reasonably with Sweden ($18,400), France ($20,000), or Canada ($21,000) and favorably with Spain, which had a per capita GDP of only $14,300.[44] Having for so long gotten the short end of the Israeli economic miracle, environmental protection should begin to enjoy the benefits of prosperity as well. Even though the tax burden that funds high-tech military capabilities is still considerable, there is ample wealth to sport a First World environmental portfolio. Tourism and high-tech industries, the leading engines of economic growth, offer their own ecological challenges, but each can be leveraged in the service of the environment.

The crisis has too often been framed as a clash between profits for polluters and the health of the polluted. Yet, many industrialists have begun to recognize the simple fact that they and their children also breathe the air and drink the water. Moreover, powerful signals have been broadcast by the global economy, which expects environmental performance from industry. International standards such as the ISO 14000 production codes[45] are motivating factories to surpass Israeli government regulatory controls and voluntarily seek improved environmental performance. Some fifty major Israeli factories have joined the voluntary ISO management initiative.[46] German packaging laws, for example, changed the way Israeli exporters wrap things. European rather than Israeli pesticide residue standards and monitoring drove farmer awareness of this issue.[47] Technology is improving, and Israeli factories will become cleaner, because they need to keep up.

As more factories comply with domestic and international limits, it will be easier to identify and punish the polluting operations that ignore them. The fact that environmental violations are prosecuted as criminal infractions may be the single most important difference between Israeli pollution control law and that of other Western countries.[48] Given pollution's impact on health and resources, such a proceeding is not without justification. Yet there have increasingly been calls, on pragmatic grounds, to downgrade environmental infractions to civil or administrative offenses. Based on the success of American litigators in securing substantial penalties for environmental violations, civil penalties might finally produce the badly needed deterrence against polluting activities.[49]

Israel came into being only because for the first half of the century Zionist leaders systematically broke the law. They bequeathed to the present generation a fast-and-loose attitude toward normative constraints; a self-reliant, self-righteous anarchism remains a characteristic part of Israeli chutzpah. Some nations, such as Japan and Switzerland, may be able to rely on a more submissive civic culture for environmental successes (for years Japan did not even have a recycling law, yet it buried only 20 percent of its trash). Israelis expect their government to set limits and are keenly aware of follow-up. When enforcement is feeble, it sends an insidious message, rewarding lawlessness and punishing conscientious industries who do the right thing.

In 1997 the Knesset passed a law that jacked up the fines for environmental infractions of five laws by two orders of magnitude.[50] The press made a point of telling the public that they could now be fined thousands of shekels for throwing a cigarette butt on the ground. Negligence with hazardous chemicals could theoretically cost an offending industry over a million shekels. Yet average fines for successful Ministry prosecutions were typically around ten thousand shekels at the end of the century. Despite a growing criminal-prosecution program, the cumulative amount of penalties collected has never exceeded three million shekels in any given year.[51] If Israeli judges, police, and prosecutors do not share a commitment to going the distance with environmental offenders, having such laws on the books will not help.[52] In addition, enforcement agents in the field must have better biological and chemical backgrounds if they are to enforce standards that are based on biological and chemical parameters.

Over two hundred years ago, Italian philosopher Cesare Beccaria first pointed out that it is not the severity of the penalty but the likelihood of getting caught that drives compliance levels.[53] The hundreds of industrial plant operators in Israel who do not bother to transport their hazardous wastes to the Ramat Hovav disposal site know this intuitively. It is astonishing that, even at the end of 2001, the Ramat Hovav management still reports that only 18 to 19 percent of the toxic waste that should be delivered to the country's only hazardous-waste facility actually arrives there for treatment or incineration.[54]

Environmental policy-makers and regulators must never lose sight of the fact that their goal should be not to catch violators but to create deterrence that will prevent violations in the first place. An efficacious compliance program should go beyond enforcement and "self-reporting" techniques to include implementing new standards, launching innovative economic schemes, and trying experimental technology. Treating pollution sources as "clients" at first, providing them with technical assistance in meeting environmental standards, should be the proverbial carrot that precedes the stick.

Finally, a stronger commitment must be made to restoration and repair. Not all past damage can be fixed, but much can be. In 1978 Israel created a fund for restoring land damaged by quarrying activities.[55] Promulgated under the old Mining Ordinance, the fund is fed by a surcharge paid by all quarry owners as a proportion of the materials they extract. The money is used to reimburse the expenses incurred in restoring the quarry lands after they are no longer productive.[56] Although the fund has not grown beyond 150 million shekels, and has rehabilitated only a third of the four hundred sites that require attention,[57] nonetheless some old scars on the land are already starting to heal. For example, in 1993 a quarry that dated from the Second Temple period (operating two thousand years ago) was restored.[58]

Presumably it will not take that long to repair other sites. Writing timetables into permits, and backing them up with surcharges and designated funds, can ensure that garbage dumps are converted into parks after they close. International experience teaches that although it may take time, given real powers and budgets, river basin authorities can bring Israel's weary streams back to life. Animals whose absence was felt for many years have begun to reclaim their natural habitat when given a little respite in reasonably sized reserves. It is well within Israel's capabilities to lighten the heavy imprint of human progress.

DISAPPEARING SPACE

The majority of Israelis have always lived in cities and towns. Between 1936 and 1948, 70 percent of the Jewish population lived in either Tel Aviv, Haifa, or Jerusalem.[59] In the most ambitious agricultural visions set forward in the Sharon plan of 1951, the farming population never exceeded 22 percent, and in reality most people remained concentrated around the big cities. Except for the scenic stony vistas of Jerusalem, however, cities were never idealized in Zionist culture. The dense, high-rise metropolis was almost an anathema. Planners in the 1950s clung to the inexpensive and flexible model of cottages in garden village settings.[60] As these neighborhoods became more polluted, crowded, and noisy, more and more Israelis looked for a way out.

Environmentalists were quick to point out the disastrous consequences for open spaces. At the same time, they avoided the more difficult task of addressing the underlying motivation behind the changing geographic inclinations. In fact, it should not have been hard to analyze. Environmental leaders' kibbutz affiliations were by now a thing of the past for the most part, yet most Green leaders continue to opt out of the predominant urban lifestyle. Some Israelis were moved by a residual "ruralist" impulse inculcated by youth movements or school trips and nourished during outdoor military training, and which later came to dominate their leisure activities. Others sought the suburbs because they just wanted a little more privacy, peace, and quiet. Sprawl, after all, is an international phenomenon.

Although tougher zoning and more efficient cities can be called an ecological imperative, many Israelis see them as ecological tyranny. Without an attractive urban alternative, limits on development will not last for long in a democratic society. So quality of life in the cities has to be part of an open-space strategy. Immediate measures are also necessary.

People are typically unable to detect gradual environmental harm. The sudden disappearance of vast swatches of land during the 1990s, however, was dramatic enough to jolt the general public. Entire areas, such as the fertile Sharon region in central Israel or the scenic Judean hills, seemed to be swallowed by the deluge of concrete, asphalt, and sprawl.[61] A recent inventory suggests that available open space in the greater Tel Aviv region is down to a parsimonious 0.05 dunam per person, and that 23 percent of the land in the center of the country is already fully built.[62] It was not just the swiftness of the phenomenon but its irreversibility that shocked. Entire natural vistas, which had once moved ancient prophets and psalmists as well as more modern painters and poets, simply ceased to exist. In a country dotted with archaeological sites, from prehistoric dwellings to Roman roads or Nabataean cities, Israelis know that construction does not easily disappear even when it is shoddy.

Some experts believe that more-detailed and better-quality statutory plans constitute the most effective strategy for preserving Israel's open spaces.[63] A key first step involves ranking the aesthetically and biologically unique terrain and demarcating large areas that are to be spared the onslaught of development. National Master Plan Number 35, which is being finalized at the time of this writing, presumably will do a better job of drawing such a line in the sand. The plan mandates the continued physical separation of different urban enclaves, as well as a "green corridor" that would extend from the Golan to the Negev. Yet a central lesson of the 1990s is that statutory protection is easily circumvented when there is sufficient panic about housing shortages, windfalls for developers, or new tax bases for cities. Enforcement of building codes and zoning is often more farcical than enforcement of emissions standards. It is not just high-rise hotels that make their own rules. Meron Chomesh, from the Israel Lands Administration, acknowledges that he does not know of a single quarry that has not pushed out beyond the limits of its licensed zone.[64] The enormous profits to be made from breaking the rules, when coupled with the complicity of prodevelopment government entities, offer a temptation that is difficult to overcome. Here again, the rare fine inevitably costs a small fraction of the profit.

Saving Israel's open spaces requires a change of perspective. Since the Israeli government administers 92 percent of the nation's lands, it must decide what sort of landlord it wishes to be. Does it want to maximize short-term profits, or to protect its assets for the future?[65] It is extremely unfortunate that during the modern age of privatization, the Israel Lands Authority is overseen by a Housing Ministry committed to development. Israel's government needs to reconfirm its commitment to preserving land reserves, categorically reject the idea of selling to the highest bidder. It needs to change the institutional bias, which reflects the orientation of politicians who seem to be automatically swayed by even poor-quality economic analysis.

The signs are so severe that the crisis is even starting to register with senior members of Israel's government. Thus, Shimon Peres, a Ben-Gurion protégé who may be Israel's most durable politician, has come to embrace the importance of land preservation. Peres has gone as far as calling the new city of Modi'in, located at the rocky ascent to the Judean hills, "a mistake" and eventually came to oppose massive roadway expansion, such as the Trans-Israel Highway.[66] The passage of the Forestry Master Plan No. 22, or Master Plan No. 31 for Immigrant Absorption, suggests that the professional planning community is committed. Getting maps approved, however, may be the easy part. Catching and punishing violators is harder. Keeping the constraints in place will probably be the toughest challenge of all.

It is important to remember that top-heavy policies, in a realm that affects such a key aspect of ordinary life, have limitations. Government can move only slightly faster than its own citizens. This dynamic was evident in one failed effort to bolster plans for the greater Beer Sheva metropolitan area. Because of the northern Negev's unexploited carrying capacity and land reserves, the city is envisioned as supporting one to two million more people, relieving the pressure in Israel's crowded northern half. To expand employment opportunities, the Israeli Cabinet tried to force a government corporation, Israel Chemicals, to move its headquarters south to Beer Sheva. Faced with an open rebellion by its Director General Victor Medina and the management, who insisted on staying in Tel Aviv, the government backed down.[67]

Had a high-speed, half-hour train ride to the center of the country been available, the move might have seemed more appealing. At the same time, it may not be an altogether negative change that citizens today cannot be moved about by government fiat as they were during the 1950s and 1960s. If people choose their domiciles, presumably they will stay there and work to make their homes a better place to live. And if the same people choose to make recreation in open spaces part of their lifestyle, presumably they will sacrifice to save these lands.

THE LAND IS FULL

Almost by definition, Israel's environmental community lives with paradoxes. As environmentalists, they know that nuclear arms production is a nasty business, with waste disposal a conundrum that has yet to be solved. But as Israelis, most understand that without a nuclear deterrent their country may not survive. As Israelis, they long for peace. But as environmentalists, they know that development in a tranquil Middle East can bring ecological ruin to resources that would otherwise be safe in times of war.[68] As environmentalists, they believe that if sustainable development is a questionable concept,[69] sustainable growth is an oxymoron.[70] But as Israelis, they are ideologically committed to embracing any of their tribe's exiles who are willing and able to join their ranks. Contradictions are inevitable and often help define the personalities of individuals as well as societies. Some can be brushed aside as the price of living in a complex world. Others can be ignored only at a nation's peril. The population paradox is a case in point.

The most critical single factor in understanding the downside of Israel's environmental history is population pressure. With an average increase of one million people a decade, the land soon became very crowded. Israel's population density is now roughly 270 people per square kilometer. This is greater than that of India and is fast approaching the density of Japan, which has 327 people per square kilometer. If one subtracts the generally uninhabited Negev desert, the population density per square kilometer rises to over five hundred, making Israel the most densely populated Western country in the world. It will not be long, though, before today's levels of suffocation will be remembered as the good old days. At present rates of growth, conventional estimates foresee Israel's population reaching 8.7 million people by the year 2020.[71] Other estimates suggest that when the Palestinian population is included, the number could be as high as fourteen million between the Jordan and the Mediterranean Sea.[72] By that time, northern Israel will reach the hair-raising density of eight hundred people per square kilometer—twice that of Holland, the most populated country in the West.[73]

Almost all the environmental clashes described in this book share a common characteristic. A growing population needed more electricity, more highway capacity, more neighborhoods, more water, and more jobs. Indeed, one longitudinal study analyzing environmental conflicts in Israel showed that they were often the outcome of the residential invasion of once-outlying facilities such as airports, landfills, or treatment plants.[74] A country with abundant land resources might have been able to find a reasonable balance and accommodate competing needs. In a tiny nation like Israel, the result was environmental crisis.

It took a while for environmentalists to recognize that they had a population problem.[75] But now, at the turn of the century, among themselves, there is growing agreement that if something does not slow demographic growth soon, Green efforts will have been for naught. Population pressure promises to undermine even the most optimistic scenarios. Yet publicly, even nongovernmental organizations remain uncharacteristically timid in expressing their concerns. At most they call for debate about the issue.[76] A few brave souls, however, have gone public with the message.[77]

Israel's population growth has two engines: immigration and an unusually high birth rate. Ever since David Ben-Gurion called for "internal immigration" and for subsidized fertility treatment through a special fund in the Prime Minister's office, Israelis have been encouraged to have children.[78] The typical Israeli family has 2.9 children, but this does not reflect the geometric growth in certain sectors such as the Muslim community, where the rate is 4.6.[79] Average family size among ultraorthodox Israelis is 6.9 people and growing; in some sectors ten has become the norm. Religious couples in their eighties often boast a progeny of over a hundred people.[80]

Both immigration and birthrates are linked to broader cultural and philosophical issues. Indeed, Jewish immigration touches the very heart of modern Israel's raison d'être. Similarly, calls for zero population growth clash with Muslim, Catholic, and Orthodox Jewish traditions. The modern Hebrew expression for a large family is "a family that is blessed with children." When Minister of Justice Yossi Beilin questioned the validity of the term, an angry public debate ensued about just how great the blessing was.

Public policies certainly contribute to the problem. Israel subsidizes large families with tax breaks and grants the fathers relief from military reserve duty. Part of the historic rationale for the policy was to provide enough able bodies to defend the country against the overwhelming numerical superiority of the Arab countries. (A desire to replace the six million Jews slaughtered in Europe also remains, at least subconsciously, an important factor.) Yet even if the peace process sputters, high-tech weaponry has revolutionized the military dynamic completely. Moreover, the Arab and ultraorthodox sectors with the highest birthrates do not contribute appreciably to local defense capabilities. Other policy areas also affect the demographic equation. For instance, a tightening of criteria led to a massive drop in the number of requests to abort unwanted pregnancies—from 44,000 in 1980 to 19,500 in 1996. (Presumably, some abortions took place outside the official system and went uncounted.)

Some politicians began to seize on the issue and play on their public's resentment of perceived ultraorthodox parasitism. Avraham Poraz, a Knesset Member from the radically secular Shinui Party, wrote voters a direct mail appeal, blasting the bias of Israel's National Insurance stipends, which reward large families, most of whom do not serve in the military. A family with ten children, regardless of income, was automatically subsidized by a government stipend of over $1200, as opposed to a family of two, which received around $80.[81] Such attacks may not be the most constructive contribution to the population debate; indeed, they may be counterproductive. The human misery and hunger caused by a cancellation of support for large families—many of whom are completely indigent—would certainly be politically unacceptable in a Jewish state that retains its fundamental decency and sensitive collective social conscience. Yet grandfathering present benefits in, while in future eliminating the bonuses granted for the fourth child onward, may prove to be an acceptable alternative. Unfortunately, it is not at all clear that the cancellation of child support would alter the commitment of religious families to produce a big brood. The decision is ideologically driven and reinforced by powerful sociological pressures and community services.

A frank presentation of the sociological implications may offer a more compelling case than the narrow environmental arguments. There are documented impacts on equality of opportunity and quality of life. A recent study by sociologists Ilana Broch and Yohanan Peres concludes that offspring from large Israeli families suffer more than economic disadvantage. The data of Broch and Peres indicate that children from Jewish families with more than six children have an average IQ of 90, as opposed to average levels of 99 among children from smaller nuclear families. Furthermore, levels of professional aspiration, as measured in sociometric tests, proved far lower among children from large families.[82] There is probably some basis for finding a cultural bias in such test results. Yet the very fact that these studies made the front page of *Ha-Aretz* suggests that the secular public is at least ready for a discussion about the full implications of the present levels of procreation.

Eventually the issue will have to be breached theologically. The first commandment in Genesis certainly calls on humans to "be fruitful and multiply and fill the land." But the rabbis who interpreted this dictate in the Talmud stated that two children were enough (if one was a son) to meet the religious duty.[83] Moreover, in times of famine, Jews are expected to show self-restraint. Environmentalists have been largely unwilling to make the scarce-resources analogy, proclaim that the land is already full, and suggest moving on to the other 612 biblical commandments that are less enthusiastically fulfilled.[84] Future environmental historians may well look back and identify the present deadlock as one of Israel's greatest tragedies: Attempts to raise the population issue were doomed to be received cynically, as a transparent diversionary tactic in the openly acknowledged cultural war between secular and religious Israelis.

Yet ecology will not wait for a religious modus vivendi. With each passing year, the likelihood of reversing the population policy diminishes. Charles Galton Darwin (the grandson of the eminent geneticist) explained that purely voluntary population controls were bound to fail because they rewarded individuals and cultures with a proclivity for large families.[85] If it is serious about solving problems, Israel's environmental movement must embrace the population issue without hesitation while politically there is a chance to change public policy, and let the chips fall where they may.

Immigration is an equally thorny issue that cannot be ignored if Zionism is to forge a new environmental ethic. The negation of the

Diaspora was an integral part of the *Weltanschauung* for Israel's founding political elite. The Zionist solution for the Jewish problem was in direct competition with alternative ideologies, from Communism and Bundism to the American Dream. This pushed the pioneers and their leaders to pose Palestine as the superior choice, if not the exclusive alternative. But as the year 2000 has now come and gone, a reality check is desperately needed. With a solemn, post-Holocaust resolve, Israel for fifty years sought to liberate Jewish communities suffering from persecution around the globe. Today, however, the last Jews who wanted to have already left Arab countries such as Yemen, Iraq, and Morocco. There is an open-door policy among former Eastern-bloc Communist nations. Zionists should be delighted that anti-Semitism is no longer a compelling reason to move to Israel.

Demographic trends show that because of assimilation and a low birthrate among North American and other Western Jewish communities, very soon Israel will boast the largest Jewish population on the planet. It is time to bury the old insecurities about Israel's standing in the Jewish world and the obsessive need to validate the Israeli option versus the American, Argentinean, or English alternative. Most mainstream Jewish organizations around the world recognize the centrality of Israel in Jewish life and the importance of the "homeland" for strengthening identity, especially among Jewish youth. Indeed, almost all forums and institutions in which Israeli and Diaspora Jews work together speak of a new, more "mature" relationship, based on mutual respect and equality. This is reflected in initiatives with names like "Partnership 2000."

If Israel is serious about accepting the legitimacy of Judaism in the Diaspora, it is not clear why it needs to send emissaries to the ends of the earth with a mission of convincing people to immigrate. Sending Jewish educators to support the burgeoning religious school systems would be a far more constructive expression of Jewish solidarity. If Jews feel that moving to Israel is spiritually or professionally edifying, they should be welcomed. Talented individuals can certainly make important contributions to Israeli society. But the land of Israel no longer needs more people.

Eventually, economics, water resources, noise, and the general dysfunction caused by unbearable density will push Israel into a confrontation with advocates of large families and mass immigration. While Dan Perry was Director of the Nature Reserves Authority, his call for assessing options to modify immigration and family size policies was met with outrage from Israel's religious community. Yitzhak Levy, then a member of the Knesset and later Minister of Education, called for Perry's dismissal. Perry was upset that the press had misinterpreted his comments. It was not to preserve nature that he called for population policies but to protect humans and save the quality of Israeli life.[86] Environmentalists, however, should not be afraid to speak on behalf of the many natural treasures that will otherwise be decimated by the crowds; flora and fauna are the first to pay for human encroachment on shrinking habitat. Future generations will certainly have reason to resent today's deafening silence about Israel's excessive population growth.

There are any number of explanations for the present ostrichlike myopia among decision makers regarding Israel's ultimate carrying capacity. Religious Jews can fall back on the Talmud, which explains that unlike settlement in other countries, population growth in Israel is comparable to the skin of a gazelle. As a young doe, the animal's skin is stretched so tightly that it seems as if it would simply burst if the body underneath expanded at all. But sure enough, as the gazelle grows, its skin grows with it. And so it is that the Almighty provides for the land of Israel, guaranteeing space and support for all comers.[87] On the secular side, there may have been residual distrust among Zionist leaders who felt that the British Mandate authorities disingenuously used carrying capacity as a tool for stifling Zionist aspirations. Today's Zionists ultimately are fervent technological optimists. It is no wonder that carrying capacity remains a nonissue.

The Ministry of the Environment, for the foreseeable future, remains unwilling to define overpopulation as an environmental priority, maintaining that Israel's population limits have yet to be determined. The official view remains that there is probably room for many millions more in the Negev. Turning Beer Sheva into a Phoenix is an easier operational goal than defining the long-term impacts of unencumbered geometric demographic growth and adopting policies to stay it.[88]

Some environmentalists argue that it is "consumption" rather than overpopulation that constitutes the preonderant cause of Israel's environmental deterioration. A recent analysis, based on car ownership, water, and garbage production, concluded that consumption is growing faster than Israel's population is, and that poor "large" families may consume less than rich "small" families do.[89] Clearly, if the analysis had chosen different indicators (such as disposable diapers, BOD loadings into sewage systems, or even heating oil), the results would have been different. Nor does the "snapshot" analysis attempt to evaluate what the long-term implications of a high birth rate might be. Nonetheless, the choice of indicator is gratuitous, because clearly, both overpopulation and wasteful consumption patterns are critical environmental issues that deserve much greater attention.

TECHNOLOGICAL OPTIMISM

The great American environmentalist David Brower liked to explain that his adversaries were wrong when they claimed he was opposed to everything. Rather he was in favor of clean air, clean water, and nature. That makes for good rhetoric, but the critique underscores the need for environmentalists to articulate a positive vision that can capture the public's imagination. Rather than always saying "No," Greens need to initiate ideas to which they can unhesitatingly say "Yes." Given recent experience with products from DDT to aerosol sprays, a healthy suspicion of technology is well deserved. But this does not mean that technology cannot be part of an Israeli environmental vision.

In Israel at least, it is too late for Luddites, who would quickly find themselves in a hungry, dirty, and desolate country. Rather than stand aside, skeptically seeking out the holes in new scientific innovations, it would be well for environmentalists to seize the initiative and encourage projects that offer promising ecological outcomes. People often forget that after documenting the hazards of pesticides, Rachel Carson went on in the second half of *Silent Spring* to describe the alternative biological controls that she so strongly advocated. Although conservation and pollution prevention must be the cornerstones of any environmental policy, it is also true that bold and unorthodox technological solutions will be needed.

Energy towers are one such possible innovation. After leaving the Water Commission, Dan Zaslavsky led a sizable Technion engineering team that designed a potentially revolutionary new source of electricity. Their energy tower is a machine that takes advantage of dry climates to produce huge amounts of power carried by wind. Water is sprayed at the top of a 1.2-kilometer-high and four-hundred-meter-wide tower. The water then evaporates, cooling the air. Because the cool air is denser, it descends down the tower, the opposite of the usual air flow (as in a chimney). This inversion can create wind speeds of eighty kilometers per hour. When run through an extensive network of turbines, such winds could run generators powerful enough to produce 17 percent of Israel's electricity needs—without any emissions.[90] The tower might be the answer to prayers for those concerned about global warming. The facility is also being designed to desalinate several hundred million cubic meters of water a year at less than half of present costs.[91]

There are two serious drawbacks to the project: One is its 1.3-billion-dollar price tag; the other is the way it looks.[92] Even proponents agree that although it may be clean, it would be a monstrosity. The tower would be visible from distances of twenty kilometers and would break up the grandeur of the Arava desert plains. The knee-jerk environmentalist response in the past would typically be opposition in a case such as this, where unproven technology is weighed against certain damage to the landscape. Yet, Zaslavsky's Green credentials and exhaustive attention to ecological details stood him in good stead. Local environmentalists were convinced that the benefits of copious quantities of emission-free electricity and fresh water trumped the troubling aesthetics. Whether or not funding for energy towers ever materializes, planners have penciled them into regional development plans in the south.[93]

Large desalination installations will undoubtedly become part of the local water strategy. When he was Foreign Minister, Arik Sharon made a regional desalination vision the sole subject of his presentation when speaking to his fellow Ministers at the United Nations in 1999. Already, 60 percent of the world's desalination capacity comes from the Middle East, primarily from huge water factories in the Persian Gulf.[94] In Israel's Arava desert, for almost two decades, houses have had three faucets: hot, cold, and desalinized drinking water. This offers a vision of the region's future. Indeed, an important opportunity was missed during Israel's housing boom of the 1990s, when seperate drinking-water systems were not installed in new neighborhoods.

Water Commissioner Meir Ben-Meir, a relatively recent convert to the desalination camp, ultimately resigned his post because of the government's foot-dragging in the area. Invoking the "precautionary principle," he called the consequences of delaying "catastrophic" to both the urban and the agricultural sectors.[95] At last, after decades of marshaling cost-benefit evidence against investment in desalination infrastructure, Israel's Ministry of Finance conceded the urgency of the situation in 2000 and supported a government decision calling for facilities that were to produce 50 million cubic meters by 2004. Much more will be needed.

Environmentalists need to think ahead and ensure that ecological concerns are duly noted. Reverse osmosis, the most common desalination process, involving diffusion of salty water through membranes at a high pressure, is not without environmental impacts. Beyond the copious amounts of electricity (and, presumably, greenhouse gases) required to squeeze out the salts, the residual brine is typically discharged back into the sea around the plant. The increase in salinity can be extremely detrimental to adjacent ecosystems, as was shown to be the case after five years of desalination plant operation in Qatar Bay.[96] Security interests will certainly ensure that the dangers of military sabotage to desalination facilities are minimized, but will sabotage to the environment receive the same degree of attention?

Environmentally friendly technologies, especially those that are not fully cost effective, can help only as much as public policy allows them to do so. This is one of the key lessons of Israel's disappointing experience with solar power—a lesson that has not really been learned. For many years Israeli solar activities were honored around the world. Israel led the world in per capita hot-water heating from the sun. To this day solar water heaters save 620 kilowatt hours nationally a year, approximately 3.2 percent of the country's total electricity usage. But public policies, largely responsive to oil price drops, did little to bolster expansion of the solar industry.[97] It is one of Israel's great ecological ironies that the largest solar-power projects ever assembled were built in California during the 1980s by Lutz, an Israeli company. Although the technology came from Jerusalem, Lutz never produced a single watt of electricity in Israel and eventually went bankrupt.[98]

The peace process spawned a series of large regional economic gatherings. All countries in the peace region came with large projects to sell. Jordan's energy projects included photovoltaic solar facilities; Israel's did not.[99] Here again, even as the world rediscovered wind turbines and biomass, the Israeli environmental movement's agenda did not seriously push clean, renewable electricity sources. Even if there is no perceived need to preempt the nuclear option, energy remains a crucial factor for both long- and short-term air quality.

As he barnstormed the country, promoting his Master Plan for the year 2020, Technion planning professor Adam Mazor shaped the thinking of governmental and nongovernmental environmentalists alike.[100] One easy adjustment he recommended was multiple-level zoning, where land could find three different designations: underground, on the surface, and in the air. New marginal lands would need to be developed, and the contaminated "brownfields" of the inner city must be restored.[101] Environmentalists already agree that the desert, in particular the northern semiarid regions, needed to be transformed to accommodate millions more people. There is some comfort in the reassurance of ecological experts who insist that words such as "pristine" and "wilderness" are misnomers in the context of a land that for millennia has been continuously inhabited. Still, when developers began to propose artificial islands within Israeli territorial waters, even the broadest environmental visions were challenged.

Ever since Mark Twain quipped that "land is the one thing they don't make more of," the finite nature of this commodity is an operational assumption in most real estate markets. Yet, even in Twain's day, reclaiming land from the sea had already begun. Less famous than Holland's dikes and drainage were its artificial islands. Dejima was built off the Nagasaki coast in 1636 and for three centuries served as a trading post and home for Dutch citizens.[102] Artificial islands are now home to the airports of Osaka and Hong Kong; their areas are in excess of twenty square kilometers apiece—half the size of Tel Aviv's municipal boundaries.[103] In western Australia, small artificial islands have been utilized for purely ecological ends—facilitating the return of fish species, such as grouper and queen snapper, that had disappeared.[104] Japan alone has eighty artificial islands, proving that land prices in coastal cities can reach levels high enough to make the creation of urban enclaves cost effective.

The idea is proposed with sufficient frequency in Israel to suggest that it may well come to be. Architecture students' assignments include design of islands off the Tel Aviv coast. The more ambitious designs visualize islands that would be located two kilometers off the sea and offer housing for tens of thousands of people as well as a great deal of retail and office space. Other island proposals are built around a Venice-like matrix of canals.[105] As early as 1997, Ariel Sharon, Israel's Minister of Infrastructure, proclaimed the feasibility of the venture, and Dutch and Israeli companies began preparing specific plans.[106] A typical island is projected to cost a billion dollars. It would provide housing for twenty thousand people, employ ten thousand, and attract twenty thousand more for tourism and business.[107]

The very idea of moving Israel's international airport to a yet-to-be-built artificial Mediterranean island fills the beleaguered residents of Tel Aviv suburbs Holon and Bat Yam with euphoria. For years they and the airport's rural neighbors have suffered from Ben-Gurion Airport's noisy air traffic. Moving the airport to an island would also liberate thousands of dunams in the heart of Israel from their present fate as an asphalt runway.

The Ministry of the Environment has taken a cautious approach to the proposals, making its support for a limited number of artificial islands contingent on the blessings of attendant environmental impact statements.[108] Considerable questions remain about a range of technical problems. The two most significant issues are the effect of mining the prodigious quantities of materials required for the islands' underwater bulk and the flow of sediments from the Nile basin to Israel's Mediterranean beaches. Some environmentalists have assailed the aesthetics of an obstructed horizon.[109]

Israel need only look south to the unintended ecological consequences of Egypt's Aswan High Dam to justify mixing in a teaspoon of humility when conjuring up grandiose projects. Yet the world is full of examples of land reclamation projects that today are an accepted, and even cherished, part of the landscape. Years after the pioneer swamp-draining fervor, large sections of Israel's own heartland testify to the possibility of a salubrious transformation, and the potential for humans to be positive actors in the process.

HEALING THE PROMISED LAND

Half of the people who call themselves Israelis are not native, having decided at some point in their lives to move to an alternative homeland. With emigration a very real option, the majority of native-born citizens also consciously choose to make their home in Israel. During the 1990s, the number of Israelis living abroad who returned to Israel began to surpass the number leaving. Each year, newspaper surveys report that 91 percent of Israelis are quite satisfied with their country.[110]

The satisfaction is linked to a general sense of quality of life. There are many places where it is easier to become rich, but the ancient land still casts a spell over newcomer and veteran alike. The stimulating lifestyle offers advantages that trump military insecurities, overzealous tax collection, and a host of other social and economic maladies. Besides leading the world in registered synagogues and institutions of Jewish religious study, Israel also ranks first in areas as diverse as orchestras per capita, theater subscription rates, in vitro fertilization rates, and dairy cattle productivity. And although the society is not perfect, for most Jews it is still family.

It is important, therefore, to temper any ire toward Israel's ecological negligence with an honest recognition of the material and basic comforts that the Zionist revolution produced. There is a tendency to glorify the first half of the century in a romantic vision of milk and honey—when crystal rivers babbled through ancient wadis to a sewage-free seashore, and fresh air ventilated the soul, unencumbered by lead and nitrogen dioxide emissions. But Palestine was also a land where scorching summers without air conditioners (or even electric fans) were almost unbearable, where insects and malaria prowled mercilessly, where even a spartan diet was often unattainable, and where plumbing (including indoor toilets) was a rarity. Indeed, the pristine rivers of the land dried up during the summer and became perennial only when fed by a steady source of sewage. Even those environmentalists fully aware of the ecological consequences of the progress made would be disinclined to turn the clock back and forgo the astounding advancements of the twentieth century.

It is precisely because quality of life holds such an essential place in most citizens' values, however, that Israelis are now able to turn their attention to their relationship to their environment. Zionism has completed a crucial stage. There are many good reasons for believing that it can evolve to address the new ecological challenges spawned by its own success. The first is environmental consciousness.

Israelis are beginning to look at the entire planet with new, greener eyes. During summer heat waves, newspapers are suddenly writing editorials about the greenhouse effect, acknowledging that Green calls for concern, once dismissed as alarmism, are compelling. The Kyoto climate convention does not require "developing countries" such as Israel to reduce their greenhouse gas emissions. Nonetheless, by the century's end, the Ministry of the Environment had lobbied other Ministries to support a ceiling on carbon dioxide emissions, both because it was good business and "the moral thing to do."[111]

Yossi Sarid is the politician who best represented this new level of environmental awareness. He believes that information holds the key to reconciling the Zionist dream and the Green dream. "In the early years of the State there were mistakes made in all areas, not just the environment. (In those days they thought that wars would solve problems. That's why we had five wars.) Even the serious mistakes were understandable during the first stages of the country. People just didn't know. Do you think that Ben-Gurion knew anything about ecology? I can tell you that Levi Eshkol didn't know a thing about it. They didn't know because there was no one to explain it to them. The opposite was the case. They told them, you have to build more and more." [112]

Sarid is therefore willing to absolve Zionism for many of the ecological follies that were perpetrated in its name during the 1950s and 1960s. He is, however, unforgiving toward political leaders of the past twenty-five years who knowingly did the wrong things in the name of the public interest. This public interest has clearly changed: "During my last year as Minister of the Environment, I stopped all development around Lake Kinneret," Sarid explains. "I saw it as one of my most important achievements. Once, they didn't understand the problems caused by building around the Kinneret. Today building there is actually anti-Zionist. It's not just that the country needs to be a home; it needs to be a home that you can live in."[113]

Not all Israelis have made this leap in understanding. But a growing number have, and they have the power to help the rest along. When a law banning smoking in public places was enacted in 1983, many Israelis laughed at it, given the pervasiveness of smoking on buses and in restaurants and public halls.[114] Today, public areas are smoke-free. It was a silent majority that cared quite deeply about the air it breathed, not the government, that enforced the new norms. In addition, the law pushed ahead a consciousness that was already evolving. Now other areas of Israel's environment are ripe for such a push. Israelis are looking beyond their physical survival for a better way to live.

There has always been a spiritual dimension to Israeli patriotism. It has taken a new form in the growing number of secular Israelis who seek a renewed connection with Judaism, in a variety of nontraditional forms. The best place to find spiritual renewal will probably always be in the natural world, just outside the window.

Israel in the next millennium will still be home to rocky hills and verdant valleys, sandy coasts and a desert that is both empty and teeming with life. The Israeli instinctively returns to this natural world, finding rest from the cacophony of city life, much as the prophets did when seeking inspiration in days of old. There is no better place than in the natural world to purge defeatist impulses from the human heart—and indeed, the earth is full of examples of successful cures for each and every Israeli ecological ailment.

Adopting a new environmental ethos can open the public's minds to available solutions, but it cannot minimize the sacrifice required for the common good and for the good of future generations. Clean-ups cost money. The requisite self-discipline in transportation, consumption, and housing is not always convenient. Israel's ultimate national challenge may therefore be to regain the higher sense of purpose that comes from giving to a society that aspires to be a "light unto the nations." Ecology needs to be a central component of this renewed Zionist dream.

As the twentieth century closed, Israeli society had become more individualistic and divided over core questions about its collective identity.[115] A growing number of citizens embraced self-indulgence and material acquisition over ideological satisfaction, shirked military reserve duty, and voted for narrow sectarian concerns. Yet relative to other countries, Israel still rests solidly on a deep reservoir of common purpose and willingness to sacrifice that has shown little signs of depletion. It is there in the deference to the elderly on the buses or to the wildflowers in the fields. It is there as yet another round of terrorism tests a nation's resolve. Once people are convinced of the severity of the environmental situation, this spirit will provide the requisite commitment to preserve the water, earth, and air. Regardless of when they actually arrived, most Israelis' ties to the land are deep.

Before he went to his lonely death, Moses wrote a farewell letter to the people of Israel that is otherwise known as the book of Deuteronomy. He offered them the options of a blessing or a curse. The choice has been woven into Judaism's central prayer, the Shema. According to the Scriptures: "The land that you go to possess is not like the land of Egypt that you just left. There you had to sow your seed and water with your foot as in a vegetable garden. No, the land you will pass through and inherit is filled with hills and valleys, and it drinks the water of the rains of heaven. It is a land which the Lord your God cares for. Indeed, the eyes of God are always on this land, from the start until the end of the year. And I will give it rains in their due season, the first rain and the last rain. And you shall gather your grain and wine and oil. And I will send grass."[116]

If one allows oneself to lapse into a puerile conception of divinity, it is instructive to imagine what the Creator might see, peering down from on high upon this promised land four thousand years later. In the most recent cycle of dispersion and redemption, Jews returned en masse to their ancestral homeland, making much of it a greener place. The renewed forests, the abundant agriculture, and the reserves set aside for the other creatures would surely be a source of happiness. Yet other omens would be troubling. Watering so many fields has sapped and sullied many of the reservoirs. The toll that the farmers' and the factories' chemicals took would also not be overlooked. Asphalt strips crisscrossing the hills and plains added little to the landscape, except perhaps the acrid exhaust of cars. And all those people—indeed as numerous as the stars in the sky and the sands along the seashore.

Even an omniscient Creator might not have anticipated the mischief that civilization was capable of once it forgot that with dominion came responsibility. Who would expect the people to turn the good rivers of their land into streams of squalor? Who could imagine producing such copious quantities of garbage and toxins with no serious plan about how to dispose of them? Or allowing a land so venerated to be paved and built into submission? If one took the biblical admonitions seriously, it would almost seem that the conditional curse offered a few verses later was coming to pass: "And the stranger that comes to your land from afar and sees the plagues upon the land and the sickness that the Lord has laid on it—sulfur and salt and burning, where the land will not be sown and will not bear any grass. . ."[117]

Rationalizations would be duly noted. There had been poverty and refugees and wars to overcome; a modern state was created. But these cannot temper a deep sense of sorrow at what has been lost. The trying circumstances serve only to amplify the merits of those who successfully worked to preserve and heal the land. Still, if any dispensation was due to the Holy Land's ecological saints (and sinners), it would have to wait for a world to come. While on this earth, much more was expected.

And so it is important to remember that it is not divine decree, but human ambition, myopia, negligence, and sometimes greed that brought these curses to the land. Precisely because the people of Israel created their many environmental problems, they were also blessed with the collective wisdom, wealth, laws, technologies, and passion to solve them. The same Zionist zeal that allowed an ancient nation to defy all odds for an entire century can be harnessed to confront the newest national challenge. More than any of their ancestors, the present generation stands at an ecological crossroads—offered the choice of life and good, or death and evil. This "last chance" to preserve a healthy Promised Land for posterity is a weighty privilege indeed. Surely as it writes the next chapters in its environmental history, Israel will once again choose life.

Notes

CHAPTER 1: THE PATHOLOGY OF A POLLUTED RIVER

1. "Two Dead and Sixty-five Injured in Maccabiah Opening Ceremonies," *Yediot Ahronot,* July 15, 1997, pp. 1–9.

2. Yehudit Yehezkel, Moshe Sheinman, and Hilah Elroy, "The Bridge Is Falling, the Bridge Is Falling, Help! Help!" *Yediot Ahronot,* July 15, 1997, p. 2.

3. Yigael Sarena and Guy Leshem, "Something in the Water Is Killing People," *Yediot Ahronot,* August 22, 1997, p. 17.

4. Dafna Linzer, "Two Die in Bridge Collapse: 64 People Injured When Temporary Structure Falls During Opening Ceremonies at Maccabiah Games in Israel," *Associated Press Report,* July 15, 1997.

5. Sarena and Leshem, *op. cit.,* p. 18.

6. Tova Tzimuki and Yoram Yarkoni, "Five Indictments Will Be Filed against Suspects in the Bridge Collapse," *Yediot Ahronot,* January 12, 1998, p. 5.

7. "A Sense of Shame," *Jerusalem Post* editorial, at www.jpost.com, July 16, 1997.

8. Brian Schiff, "Tragedy and Triumph," *Jewish News of Greater Phoenix,* August 8, 1997, http://www.jewishaz.com/jewishnews/970808/athletes.shtml.

9. Tzimuki and Yarkoni, *op. cit.,* p. 5.

10. Anat Midan, "This Isn't How I Imagined It," *Yediot Ahronot—7 Yamim,* January 16, 1998, p. 18.

11. Noami Segal, "Fourth Australian Dies Due to Bridge Collapse; Polluted Water Cited," *Jewish Telegraphic Agency,* August 1997.

12. Lina Deodhar, Priya Miskeen, Al Kirpalani, Jinni Jagose, and Shubhu Sarkar, "*Pseudallescheria boydii* Brain Abscess in an Immunocompromised Host," *Bombay Hospital Journal* 42, no. 1 (January 2000),http://www.bhj.org/journal/2000_4201_jan00/case_180.htm.

13. *Infections in the Lung Transplant Recipient,* http://path.upmc.edu/divisions/pulmpath/infe03.html, University of Pittsburgh, 1997 at f.

14. Midan, *op. cit.*

15. Sasha Elterman's medical condition can be monitored on a special website dedicated to the Maccabiah victims: http://www.tfta.com.au/maccabi_bridge_victims. She is available to receive email messages through the site.

16. Midan, *op. cit.*, p. 8.

17. Sarena and Leshem, *op. cit*, p. 18.

18. Interview with Nehama Ronen, Director General, Israel Ministry of the Environment, Tel Aviv, December 18, 1997.

19. David Pargament, Director, Yarkon Streams Authority, oral presentation, Kibbutz Ketura, October 26, 1997.

20. Yeshayahu Bar Or, Director of the Water and Streams Division, *Summary of the Laboratory Findings of Water and Sludge Samples Taken from the Yarkon River, Following the Maccabiah Bridge Collapse Accident*, (Jerusalem: Ministry of the Environment, August 18, 1997), copy on file with author.

21. Haim Watzman, "Rivers of Death," *New Scientist*, August 16, 1997.

22. Interview with Dror Avisar, hydrologist, Adam Teva V'din (Israel Union for Environmental Defense), Tel Aviv, January 10, 1998.

23. Daliah Mazori, "It's Final: Industrial Pollution and Machine Oils Caused the Death of the Athletes at the Maccabia," *Maariv*, August 27, 1997, p. 2.

24. Interview with Sol Benheim, Kibbutz Ketura, October 26, 1997.

25. Abraham Moshe Lunts, *A Guide to the Land of Israel* (1891), as quoted in Ofer Regev, *Forty Years of Blossoming* (Tel Aviv: Society for Protection of Nature, 1993), p. 96.

26. Arie Rahamimoff and Amos Brandeiss, "The Yarkon as a National and Regional Resource," *Preservation of Our World in the Wake of Change* (Jerusalem: Israel Society for Ecology and Environmental Quality Services, 1996), pp. 240–244.

27. Shoshana Gabbay, *The Environment in Israel* (Jerusalem: State of Israel, Ministry of the Environment, 1994), p. 29.

28. Rahamimoff and Brandeiss, "The Yarkon as a National and Regional Resource," pp. 240–244.

29. *Hashachar Shiron* (New York: Young Judaea, 1985).

30. The Public Health Regulations, Determination of Standards for Effluent Waters, 1992. K.T. 9440.

31. The Yarkon River Authority, *The Yarkon River Master Plan*, 1996, p. 13.

32. David Pargament, oral presentation, Kibbutz Ketura, October 26, 1997.

33. Tsvi Sonovsky, "The Survival Limitations in Difficult Conditions for Fish Used in Mosquito Extermination," *Biosphera*, May/June 1994, pp. 25–31. Also, Gila Shneider and Adam Kanark, "Transplanted Fish in the Sands, Mosquito Extermination with the Help of *Gambusia* Fish," *Teva V'Aretz*, 263 (1994): 20.

34. The oil increases the water's surface tension, causing a mosquito larva to slip underwater, extend its breathing "pipe," and ingest the oil. This physically clogs its air supply, and the mosquito suffocates.

35. Revital Bracha, "Thousands of Fish Discovered Dead Again in the Yarkon: Fear of Additional Kills," *Ha-Aretz*, July 28, 1993. Also, Liat Collins, "Probe into Death of Yarkon Fish," *Jerusalem Post*, July 29, 1993.

36. Nir Kipnis, "The Yarkon Isn't Restored At All," *Zman Tel Aviv*, August 6, 1993.

37. The Yarkon River Authority, *The Yarkon River Master Plan*, 1996, pp. 6–9.

38. Gabbay, *op. cit.*, p. 29.

39. Ford Pearce, "Back to the Days of the Deadly Smogs," *New Scientist*, December 5, 1992, pp. 25–28.

40. Andrew Hoffman, "An Uneasy Rebirth at Love Canal," *Environment*, 1995, pp. 5–31.

41. Rachel Adam, personal communication, May 23, 1998.

42. Lisa Beyer, "The Filthy Holy Land," *Time*, September 1, 1997, p. 27. Also, Lisa Beyer, "Trashing the Holy Land," *Time*, September 7, 1998, p. 54.

43. Mazori, *op. cit.*, p. 2.

44. Rachel Adam, personal communication, May 23, 1998.

45. Ruth Rotenberg, personal communication, May 1, 2000.

46. Interview with Nehama Ronen, Director General, Israel Ministry of the Environment, Tel Aviv, December 18, 1997.

47. David Pargament, personal communication, October 17, 2001.

48. Andra Jackson, "Thirteen Million Dollars for Maccabiah Games Victims, " *The Age News*, May 10, 2001.

49. Israel requires that sewage effluent meet a "20-30 standard" (20 mg/L BOD, 30 mg/L suspended solids). Public Health Regulations (Establishing Standards for Effluent Waters), 1992.

50. Baruch Weber, Deputy Director of Water and Streams Division, Ministry of the Environment, personal communication, Tel Aviv, December 18, 1997.

51. Rachel Adam, personal communication, May 23, 1998.

52. Interview with Dror Avisar, Tel Aviv, January 10, 1998.

53. Rachel Adam, personal communication, May 23, 1998.

54. Interview with Menahem Kantor, Kibbutz Ma'agan Michael, November 20, 1997.

55. Letter from Shalom Goldberger to Noah Rabinovich, Ministry of Education, September 22, 1997. Copy on file with author.

56. Sarena and Leshem, *op. cit.*, p. 18.

57. David Pargament, personal communication, December 24, 1998.

58. Iris Cohen and Dror Avisar, "One Year After the Maccabiah Disaster—and the Yarkon Is Still Polluted," pamphlet published by Adam Teva V'din and the Council for a Beautiful Israel, Summer 1998, p. 1, and "The Yarkon Is Again a Disappointing River," *Green, Blue and White* (August-September 1998): 8–9.

59. Noam Gavrielli, "New Findings on Additional Mortality as a Result of Particulate Air Pollutants," *Particulate Air Pollution* (Proceedings) (Haifa: Technion, 1995).

60. "Report of the Committee to Assess the Water Pollution in the Krayot," *Biosphera* (July 1985): 5–16.

61. Michael Graber, "Emissions and Concentrations of Air Pollutants in Israel," *Biosphera* (January 1995): 13–14.

62. Joyce Whitman, *The Environment in Israel* (Jerusalem: Ministry of the Interior, 1988), p. 162.

63. Yossi Sarid, personal interview, December 30, 1997.

64. Al Gore, *Earth in the Balance, Ecology and the Human Spirit* (New York: Plume, 1992).

65. "Musaf L'Chag," *Yediot Ahronot,* October 1, 1997, p. 1.

66. "Public Opinion Survey on the Subject of Environmental Quality," *Biosphera* (October-November 1994): p. 2.

67. Interview with Yizhar Smilansky, January 13, 1998.

68. Eran Hadas, "Listening Survey, An Impressive Increase to 'Reshet Gimmel'—The Move to Broadcasting Israeli Music Took a Bite in Listeners from the Military and Regional Stations," *Yediot Ahronot—24 Hours,* November 11, 1997.

69. Michael Cohen, personal communication, March 10, 1998.

CHAPTER 2: RECLAIMING A HOMELAND

1. Translated in Howard Sachar, *A History of Israel, From the Rise of Zionism to Our Time* (New York: Alfred A. Knopf, 1976), p. 8.

2. Walter Laqueur, *A History of Zionism* (New York: Holt, 1972).

3. David Ben-Gurion, *Maarechet Sinai* (Tel Aviv: Am Oved, 1964), p. 130.

4. Stuart Schoenfeld, personal communication, June 3, 1998.

5. Oz Almog, *The Sabra—A Profile* (Tel Aviv: Am Oved, 1997), p. 255.

6. See: L. White, "The Historical Roots of Our Ecological Crises," *Science* 155 (1967): 1203–1207.

7. Manfred Gerstenfeld, "Jewish Attitudes to the Environment in the Bible's Narrative," in *Judaism, Environmentalism and the Environment* (Jerusalem: Institute for Israel Studies, 1999), 163–207.

8. Yoav Sagi in *The Zionist Dimension of Nature Preservation* (Jerusalem: Ministry of Education, 1981), p. 4.

9. Aaron David Gordon, *The Human and Nature* (Jerusalem: Zionist Press, 1951), p. 44.

10. "Biographical Sketches," *Eretz Israel* (New York: Jewish National Fund, 1932), pp. 111–112.

11. Rachel Blubstein, "Sham Harei Golan," *Rachel's Poems* (Tel Aviv: Davar, 1978), translation by William Slott, 1998.

12. R. Y. Ben Zvi, *We Immigrate* (Jerusalem: Am Oved, 1975), p. 22, as cited in Izhak Schnell, "Nature and Environment in the Socialist-Zionist Pioneers' Perceptions: A Sense of Desolation," *Ecumene* 4, no. 1 (1995): 69.

13. R. Avraham Yitzhak Hakohen Kook, *A Vision of Vegetarianism and Peace* (Jerusalem: Lahai Ro'i, 1961), p. 207.

14. Josef Tamir, "A Chosen Land or a Lousy Land," *Green, Blue and White* (April 1998), p. 36.

15. R. Aryeh Levine, in Simcha Raz, *A Tzadik in Our Times* (Jerusalem, 1976), pp. 108–109.

16. Avner De-Shalit, "From the Political to the Objective: The Dialectics of Zionism and the Environment," *Environmental Politics* 4 (1995): 70–87.

17. Interview with Eilon Schwartz, Tel Aviv, January 12, 1998.

18. Michael Bar-Zohar, *Ben-Gurion, A Biography* (New York: Delacorte Press, 1977), pp. 6–12.

19. Ibid., p. 21.

20. David Ben-Gurion, *Divrei ha-Knesset*, 1963, p. 331.

21. David Ben-Gurion (1962) as quoted in *Science and Technology* (Jerusalem: Israel Information Center, 1997), 1.

22. Tom Segev, *1949, The First Israelis* (New York: Free Press, 1986), p. 150.

23. Interview with Eilon Schwartz, Tel Aviv, January 12, 1998.

24. Stuart Schoenfeld, personal communication, June 3, 1998.

25. Avner De-Shalit, "From the Political to the Objective: The Dialectics of Zionism and the Environment," *Environmental Politics* 4 (1995): 75–76.

26. Interview with Yizhak Shamir, Tel Aviv, November 24, 1997.

27. Avram Burg, lecture at the Annual Meeting, Society for Protection of Nature, Tel Aviv, January 13, 1998.

28. Meron Benvenisti, "An Image of a Homeland," *Conflicts and Contradictions* (New York: Eshel, 1989), 19.

29. Einat Ramon, "The Zionist Myth of the Mother: The Land of Israel in the Thought of A. D. Gordon," presented at the Annual Conference, Association of Jewish Studies, Boston, December 14, 1993, p. 14 (copy with author).

30. Sachar, *op. cit.*, pp. 154–155.

31. Amos Oz, "On Loving the Land," in *The Zionist Dimension of Nature Preservation* (Jerusalem: Ministry of Education, 1981), p. 14.

32. H. P. Simonsen, ed. *The Significance of the Frontier in American History* (New York: Ungar, 1963).

33. Adam Werbach, keynote address, Eco Zionism Conference, Marin County, California, March 1997.

34. Bar-Zohar, *op. cit.*, p. 14. Ben-Gurion's numbers have no clear empirical basis, and many experts believe them to be far smaller. Gideon Biger, personal communication, March 5, 1999.

35. Ibid., p. 16.

36. Meir Dizengoff, *On Tel Aviv and Its Life Styles* (Tel Aviv: Yediot Iriyat, 1934), p. 2.

37. Almog, *op. cit.*, p. 266.

38. Interview with Ra'anan Weitz, Jerusalem, January 12, 1998.

39. Amos Keinan, "To Understand the Land," in *The Zionist Dimension of Nature Preservation* (Jerusalem: Ministry of Education, 1981), p. 20.

40. Iris Milner, "The Rabbit's Foot," *Ha-Aretz*, August 22, 1998.

41. Heinrich Mendelssohn, personal interview, Tel Aviv University, August 27, 1997.

42. See generally Benvenisti, *op. cit.*, pp. 19–23.

43. Eliezer Livneh, Yosef Nedbeh, and Yoram Efrati, *NILI* (Jerusalem: Shoken, 1980), pp. 61–62. Three years later, twenty-eight-year-old Feinberg was killed on a mission for the pro-British NILI spy ring.

44. Almog, *op. cit.*, p. 255.

45. Azariah Alon, personal interview, Kibbutz Beit ha-Shita, September 15, 1997.

46. Azariah Alon, "By the Side of the Abandoned Train Bank," *Teva V' Aretz*, vol. 254, p. 46.

47. Almog, *op. cit.*, p. 259.

48. Ibid., p. 268.

49. Ibid., p. 260.

50. Uri Marinov, presentation, September 19, 1997, Anglo-Israel Colloquium, Suffolk, England.

51. Noah Efron, personal communication, April 5, 1998.

52. *The Blue Mountain* is the title of the English translation of *Roman Russi* (Russian Novel). Meir Shalev, personal communication, November 11, 1998.

53. Meir Shalev, oral presentation, Kibbutz Ketura, Purim, 1997.

54. Meron Benvenisti, "Part II," *Moment*, January-February 1986, p. 26.

55. Benvenisti, "An Image of a Homeland," pp. 17–47.

56. Benvenisti, "Part II," p. 24.

57. Avner De-Shalit, "Where Do Environmentalists Hide?" in *Our Shared Environment—The Conference 1994*, ed. Robin Twite and Robin Menczek (Jerusalem: Israel-Palestine Center for Research and Information, 1994), pp. 274–276.

58. Azariah Alon, "Nature Preservation versus Zionism," in *The Zionist Dimension of Nature Preservation* (Jerusalem: Ministry of Education, 1981), p. 27.

59. Oz, *op. cit.*, pp. 16–17.

CHAPTER 3: PALESTINE'S ENVIRONMENT, 1900–1949

1. Efraim Orni and Elisha Efrat, *Geography of Israel* (Jerusalem: Israel Universities Press, 1971), p. 96.

2. Gideon Biger and Amnon Kartin, "The Emeq Yizrael Swamps—Legend or Reality," *Cathadra on the History of Eretz Israel and Its Settlement*, vol. 30, 1984, pp. 179–182; also see recollections of Akiba Ettinger, "The Development of the National Fund in Ten Year Periods," *Eretz Israel* (New York: Jewish National Fund, 1932), pp. 46–47.

3. Zeev Carmi, "Description of a Journey, 1910," as quoted in Adam Ackerman, *The Deeds of The Pioneers in Eretz Israel*, 1840–1940 (Jerusalem: Good Times, 1982), p. 25.

4. Gertrude Bell, letters of 1905; see Ackerman, *op. cit.*, p. 25.

5. Nahum Benari, diary as quoted in Ackerman, *op. cit.*, p. 35.

6. In this chapter, the "land of Israel" refers to the geographic boundaries of the British Mandate, specifically from the Mediterranean to the Jordan River and from the Galilee to the Sinai desert's edge.

7. *Statistical Abstract of Palestine*, 1944–1945, p. 17.

8. Paul Ehrlich and John Holdren, "The Impact of Population Growth," *Science* 171 (1971): 1212–1217; see also Paul Ehrlich and Anne Ehrlich, *Healing the Planet* (Reading, Mass: Addison-Wesley, 1991), p. 7.

9. Shoshana Gabbay, *The Environment in Israel* (Jerusalem: Ministry of the Environment, 1994), p. 3.

10. Yosef Weitz, *Ha-Ya'ar v'ha-Yi'ur b'Yisrael* (Ramat Gan: Masada, 1970), particularly Chapter 1, "Forests in Biblical Times."

11. H. Weiss, M. A. Courty, W. Wetterstrom, F. Guichard, L. Senior, R. Meadow, and A. Curnow, "The Genesis and Collapse of Third Millennium North Mesopotamian Civilization," *Science* 20 (August 1993): 995–1004.

12. Jared Diamond, *Guns, Germs, and Steel: The Fates of Human Societies* (New York: Norton, 1997), p. 312.

13. Arnold Blumberg, *Zion before Zionism* (Syracuse: Syracuse University Press, 1985), p. 158.

14. Yehoshua Ben-Ariyeh, "Population of the Land of Israel and Its Settlements on the Eve of the Zionist Settlement," *Research in the Geography and Settlement History of the Land of Israel*, ed. Y. Katz, Y. Ben-Ariyeh, and Y. Kaniel (Jerusalem: Ben Tzvi Institute, 1991), pp. 1–14.

15. Hazem Zaki Nuseibeh, *Palestine and the United Nations* (New York: Quartet Books, 1981), p. 15.

16. Edward Said, *The Question of Palestine* (London: Routledge and Kegan Paul, 1980), pp. 79–81.

17. For example, Laurence Oliphant, *Life in Modern Palestine* (Edinburgh: Blackwood and Sons, 1887); Gertrude Bell, "The Wilderness of Judea" (1903) in *The Sierra Club Desert Reader*, ed. Greg McNamee (San Francisco: Sierra Club Books, 1995).

18. Yoram Yom-Tov and Heinrich Mendelssohn, "Movement and Distribution of Vertebrates in Israel in the 20th Century," in *The Plants and Animals of the Land of Israel*, ed. Azariah Alon (Tel Aviv: Ministry of Defense Press, 1990), p. 67.

19. Interview with Gideon Biger, Tel Aviv, April 29, 1998.

20. Uzi Paz, *Eretz ha-Tsvi v'ha-Yael* (Land of the Gazelle and the Ibex), *I* (Givataim: Masada, 1981), p. 21.

21. Interview with Gideon Biger, Tel Aviv, April 29, 1998.

22. Yom-Tov and Mendelssohn, *op. cit.*, p. 67.

23. Mark Twain, *The Innocents Abroad* (New York: Oxford Press, 1996), p. 495.

24. "Woods and Forest Ordinance," *Laws of Palestine*, October 1920, pp. 92–102; see also Richard Laster, "Israel," *Environmental Law* (Deventer, The Netherlands: Kluwer Publishers, 1993), p. 84.

25. Interview with Gideon Biger, Tel Aviv, April 29, 1998.

26. Shaul Ephraim Cohen, *The Politics of Planting* (Chicago: University of Chicago Press, 1995), pp. 44–47.

27. Paz, *op. cit.*, p. 85.

28. J. V. Thirgood, *Man and the Mediterranean Forest* (London: Academic Press, 1981), p. 114, as quoted by Cohen, *op. cit.*

29. Yom-Tov and Mendelssohn, *op. cit.*, p. 64.

30. Paz, *op. cit.*, p. 22.

31. Blumberg, *op. cit.*, p. 3.

32. Ackerman, *op. cit.*, p. 50.

33. M. Lilian and H. Shuval, *Ten Years of Sanitation in Israel* (Jerusalem: Ministry of Health, 1959), p. 15.

34. Theodor Herzl, *Diary*, Jerusalem, October 31, 1898, http://www.cet.ac.il/~history/herzl/yoman17.htm.

35. Sneier Levenberg, "History," in *Israel*, ed. Muriel Manuel (London: St. James Press, 1971), p. 15.

36. Ronald Storrs, *Orientations* (London: Nicholson and Watson, 1945), pp. 286–310.

37. Ackerman, *op. cit.*, p. 46.

38. Ibid, p. 46.

39. A. J. Sherman, *Mandate Days* (New York: Thames and Hudson, 1997), p. 74.

40. Ibid, pp. 42–43.

41. Nachum T. Gross, *The Economic Policy of the Mandatory Government in Palestine* (Jerusalem: Falk Institute, 1982), p. 4.

42. John Marlowe, *The Seat of Pilate: An Account of the Palestine Mandate* (London: Cresset Press, 1959), p. 108.

43. Storrs, *op. cit.*, p. 417.

44. See Eli Shaltiel, *Pinchas Rutenberg, 1879–1942: Life and Times* (Tel Aviv: Am Oved, 1990).

45. Tony Travis, personal communication, March 3, 1998.

46. Alon Tal, "Citizen Suits to Improve Sewage Treatment in the Poleg River: The Limitations and Potential of Legal Actions," *Ecology and Environment* 2 (1995): 151–158; see also Alon Tal, "Law of the Environment," *Israel Law and Business Guide*, ed. Alon Kaplan (Boston: Kluwer Publishers, 1994), p. 341.

47. Laster, *op. cit.*, pp. 23–37.

48. Hillel Shuval, *The Inspection of Environmental Sanitation in Israel* (Jerusalem: Ministry of Health, 1960), pp. 3–4.

49. *Palestine Gazette 1065*, December 20, 1940, p. 191.

50. *Palestine Gazette*, vol. I, 1936, p. 74.

51. Dr. A. Y. Levi, *Sanitation: A Guide for Food and Beverage and for Installing Comfortable and Healthy Conditions at Home and Outside* (Tel Aviv: Ahiabar, 1936).

52. Lilian and Shuval, *op. cit.*

53. *Palestine Gazette*, 1937, p. 667.

54. *Laws of Palestine*, vol. III, Rotenburg, Tel Aviv, 1933, pp. 852–860.

55. *Laws of Palestine*, vol. 1, p. 710, amended in the *Palestine Gazette*, no. 600, January 22, 1937, p. 1.

56. *Palestine Gazette*, 1940 II, p. 66.

57. Simcha Blass, *Water in Strife and Action* (Givataim: Masada, 1973), p. 136.

58. Laster, *op. cit.*, p. 6.

59. A primitive series of industrial effluent standards were promulgated as part of the licensing of business system but never really enforced. Indeed, sewage treatment itself was hardly regulated. The Municipal Corporations (Sewerage, Drainage and Water) Ordinance of 1936 (*Palestine Gazette*, 1936, p. 560) essentially threw the problem at local governments.

60. Laster, *op. cit.*, pp. 7–8.

61. Howard Sachar, *A History of Israel from the Rise of Zionism to Our Time* (New York: Alfred A. Knopf, 1975), p. 230.

62. Walter Clay Lowdermilk, *Palestine, Land of Promise* (New York: Harper and Brothers, 1944), p. 90.

63. Sachar, *op. cit.*, p. 230.

64. "Woods and Forest Ordinance," *Laws of Palestine*, October 1920 (sec. 14), p. 96. The ordinance was amended in 1926, but the above activities still were allowed only after receiving a permit.

65. *Laws of Palestine*, vol. 1, p. 710, amended in the *Palestine Gazette*, no. 600, January 22, 1937, p. 1.

66. "Review of the Agricultural Situation in Palestine," p. 54, as quoted in Weitz, *op. cit.*, p. 44.

67. Weitz, *op. cit.*, pp. 44–45.

68. Nili Lipshitz and Gideon Biger, "Afforestation Policy of the British Government in Land of Israel," *Ofakim B'Giographiah* 40–41 (1994): 5–16.

69. Nili Lipshitz and Gideon Biger, *The Rise and Fall of the Jerusalem Pine as the Primary Tree in the Land of Israel* (Jerusalem: Jewish National Fund, 1994), pp. 13–14.

70. For instance, in March 2000, local activists prevented the conversion of the Shaked Forest, an old British forest reserve, into a military base that was to be moved as part of the peace agreement.

71. A. Y. Goor, *Forest Reservations in Palestine December* 31, 1946, ISA A403/F/26/12/4, sheet 1, as quoted in Cohen, *op. cit.*, p. 53.

72. Government of Palestine, *Annual Report for the Year* 1947, *Department of Forestry*, as cited in Lipshitz and Biger, *The Rise and Fall of the Jerusalem Pine*, pp. 13–14.

73. Department of Forests, *Report of the Period* 1936–39, ISA AF/41/39, p. 11, as quoted in Cohen, *op. cit.*, p. 57.

74. Gideon Biger and Nili Lipshitz, "Protected Trees and Flowers in the Mandate's Land of Israel," *Ariel, A Journal of Israeli Geography* (April 1994): 242–246.

75. Heinrich Mendelssohn, personal interview, Tel Aviv University, August 27, 1997.

76. Paz, *op. cit.*, p. 22.

77. Iris Milner, "The Rabbit's Foot," *Ha-Aretz*, August 22, 1998.

78. Heinrich Mendelssohn, personal interview, Tel Aviv University, August 27, 1997.

79. Ibid.

80. Laurence Oliphant, *Life in Modern Palestine* (Edinburgh: Blackwood and Sons, 1887).

81. David Gilmour, *Dispossessed: The Ordeal of the Palestinians, 1917–1980* (London: Sidgwick and Jackson, 1980), p. 33.

82. Evelyn Abel, "Righting the Record," *JNF Illustrated*, Spring 1994, pp. 30–33.

83. Baruch Kimmerling and Joel Migdal, *Palestinians: The Making of a People* (New York: Free Press, 1993), p. 23.

84. Ibid., p. 31.

85. "Although much slower than the tractor, the horse-pulled plow does not cause the stirring up and erosion of the shallow hilly soil in the rocky and sloped land of today's West Bank." Said Assaf, "Overview of Some Traditional Agricultural Practices Used by Palestinians in the Protection of the Environment," *Our Shared Environment—The Conference 1994*, ed. Robin Twite and Robin Menczel (Jerusalem: Israel-Palestine Center for Research and Information, 1995), pp. 12–13.

86. *Palestine: Report on Immigration, Land Settlement and Development* (London: HMSO, 1930).

87. Adolf Reitenberg, translated by Cyril Leonard Whittles, *The Soils of Palestine* (London: T. Murby, 1947), p. 161.

88. Kimmerling and Migdal, *op. cit.*, p. 33.

89. Reitenberg, *op. cit.*, p. 158.

90. ISA 672/AG/38/1/2, November 10, 1941, as quoted in Kimmerling and Migdal, *op. cit.*, p. 52.

91. Ibid.

92. Ibid., p. 52.

93. Kimmerling and Migdal, *op. cit.*, p. 28.

94. Lowdermilk, *op. cit.*, p. 152.

95. Gilmour, *op. cit.*, p. 45.

96. Ibid.

97. Tom Segev, *1949, The First Israelis* (New York: Free Press, 1986), p. 78.

98. Reitenberg, *op. cit.*, p. 157.

99. Kimmerling and Migdal, *op. cit.*, p. 117.

100. *Report of the Royal Commission* (London: HMSO, 1937), p. 127.

101. Yoram Yom-Tov and Heinrich Mendelssohn, "Altered Landscapes," *Eretz Magazine*, November 1985, p. 55.

102. Josef Tamir, personal communication, June 3, 1998.

103. Kimmerling and Migdal, *op. cit.*, p. 329.

104. Reitenberg, *op. cit.*, p. 166.
105. Kimmerling and Migdal, *op. cit.*, p. 329.
106. Reitenberg, *op. cit.*, pp. 162–163.
107. Kimmerling and Migdal, *op. cit.*, p. 329.
108. Reitenberg, *op. cit.*, p. 164, Table 87.
109. Abraham Mercado, "The Coastal Aquifer in Israel: Some Quality Aspects of Groundwater Management," in *Water Quality Management under Conditions of Scarcity: Israel as a Case Study*, ed. Hillel Shuval (New York: Academic Press, 1980), p. 99.
110. Interview with Ra'anan Weitz, Jerusalem, January 12, 1998.
111. Ibid.
112. Efraim Talmi and Menahem Talmi, "Yizhak Elazari-Volcani," *Lexicon of Zionism* (Tel Aviv: Maariv Library, 1982), p. 21.
113. Interview with Ra'anan Weitz, Jerusalem, January 12, 1998.
114. Lowdermilk, *op. cit.*, p. 90.
115. Simcha Blass, *op. cit.*, p. 136.
116. Richard Laster, *The Legal Framework for the Prevention and Control of Water Pollution in Israel* (Jerusalem: Ministry of the Interior, 1976), p. 26.
117. Paz, *op. cit.*, pp. 21–22.
118. Lowdermilk, *op. cit.*, p. 219.
119. Ibid., p. 6.
120. "A Man Who Cared for the Earth," *The Jerusalem Post*, July 5, 1976.
121. Assaf, *op. cit.*
122. Interview with Gideon Biger, Tel Aviv, April 29, 1998.
123. Paz, *op. cit.*, pp. 94–95.
124. Tamar Keinan, Director of Information Systems, Israel Water Commission, personal communication, March 5, 1998.
125. Ackerman, *op. cit.*, p. 49.
126. "The Year in Palestine," *Eretz Israel* (New York: Jewish National Fund, 1932), p. 93.
127. Sachar, *op. cit.*, p. 190.
128. Josef Tamir, personal communication, June 3, 1998.
129. Sachar, *op. cit.*, p. 190.
130. "The Year in Palestine," p. 101.
131. Lowdermilk, *op. cit.*, p. 3.
132. Ackerman, *op. cit.*, p. 58.
133. Marlowe, *op. cit.*, p. 109.
134. Aaron Baroway, "Palestine's Mineral Wealth," in *Eretz Israel* (New York: Jewish National Fund,1932), p. 93.
135. Sachar, *op. cit.*, p. 190.
136. EcoPeace, *Dead Sea Challenges: Final Report* (Jerusalem: EcoPeace, 1996), p. 7.
137. Alon Tal, "Methyl Bromide and the Ozone Hole," *Eichut ha-Sviva*, 1994.
138. Meir Shalev, *Primarily about Love* (Tel Aviv: Am Oved, 1995), p. 31.

139. Numbers 13:32.

140. World Health Organization, web site, www.who.ch/programmes/ctd/diseases/mala/maladis.htm, 1997.

141. Ibid.

142. T. Talitranick, "Malaria and Stages of Its Eradication in Our Land," *Public Health* (1963): p. 356.

143. H. Joffe, "Campagne antipaludéenne en Galilee, 1914," p. 2, as quoted in T. Talitranick, *op. cit.*

144. Talitranick, *op. cit.*

145. I. Klinger, *Epidemiology and Control of Malaria in Palestine* (Chicago: University of Chicago Press, 1930).

146. Talitranick, *op. cit.*, p. 362.

147. Meir Dizengoff, *On Tel Aviv and Its Life Styles* (Tel Aviv: Yediot Iriyat, 1934), pp. 3–4.

148. Ibid. pp. 3–6.

149. Dror Avisar, personal communication, March 5, 1998.

150. Tom Segev, *The Seventh Million: The Israelis and the Holocaust* (New York: Hill and Wang, 1993).

151. Josef Tamir, *Haver Knesset* (Jerusalem: Ahiabar, 1987).

152. David Heller and Azariah Alon, "Nature Researchers of the Land of Israel," in *The Plants and Animals of the Land of Israel*, ed. Azariah Alon (Tel Aviv: Ministry of Defense Press, 1990), pp. 55–63.

153. Ofer Regev, *Forty Years of Blossoming* (Tel Aviv: Society for Protection of Nature in Israel, 1993), pp. 8–9.

154. As quoted in Paz, *op. cit.*, p. 41.

155. Heller and Alon, *op. cit.*, pp. 60–61.

156. A. Brotzkos, "The Possibilities and Functions of National Planning," *Ha-Binyan* (1938).

157. Regev, *op. cit.*, pp. 9–10.

158. Heinrich Mendelssohn, personal interview, Tel Aviv University, August 27, 1997.

159. Hillel Shuval, "Israel's Impending Water Crisis," *Selected Papers on the Environment in Israel*, no. 9 (Jerusalem: Israel Ministry of the Interior, 1982), p. 47.

160. Interview with Gideon Biger, Tel Aviv, April 29, 1998.

161. Benny Morris, *The Birth of the Palestinian Refugee Problem* (Cambridge: Cambridge University Press, 1987).

162. The United Nations estimated the number of refugees in 1949 to be between 716,000 and 726,000, but Palestinians claim the number was higher. See George Kossaifi, "Demographic Characteristics of the Arab Palestinian People," *The Sociology of the Palestinians*, ed. Khalil Nakhleh and Elia Zureik, (London: Croom Helm, 1980), pp. 18–26.

163. Storrs, *op. cit.*, chapter 15.

164. Lowdermilk, *op. cit.*, p. 174.

165. Lipshitz and Biger, *Rise and Fall*, p. 15.

166. "Israel 2020, A National Master Plan for the Year Two Thousand," *Biosphera* 24 (1994): 6.

167. Avner De-Shalit and Moti Talias, "Green or Blue and White? Environmental Controversies in Israel," *Environmental Politics* 3 (1994): 278–280.

168. EcoPeace, *op. cit.*, p. 4.

169. Interview with Gideon Biger, Tel Aviv, April 29, 1998.

170. For example, Yizhak Bar-Yosef, "The Military Industries Are Polluting the Water in Ramat ha-Sharon," *Yediot Ahronot*, March 3, 1998.

171. The National Parks, Nature Reserves, Memorial Sites, and National Sites Law, 1992, *Sefer ha-Hokim*, 1397 (sec. 23), p. 230.

CHAPTER 4: THE FOREST'S MANY SHADES OF GREEN

1. Uri Marinov, presentation, September 19, 1997, Anglo-Israel Colloquium, Suffolk, England.

2. Interview with Aviva Rabinovich, Kibbutz Kabri, January 11, 1998.

3. Interview with Henrich Mendelssohn, Tel Aviv University, August 27, 1997.

4. Walter Lehn, *The Jewish National Fund* (London: Kegan Paul International, 1988), pp. 16–18.

5. Yisrael Cloyzner, *Land and Spirit: The Life and Activities of Professor Tsvi Herman Schapira* (Jerusalem: Hebrew University, 1966), p. 12.

6. Maximilian Hurwitz, "The Father of the National Fund," in *Eretz Israel* (New York: Jewish National Fund for America, 1932), pp. 24–30.

7. Leib Yafeh, "Zvi Schapira," in *Megilat ha-Adamah* (Jerusalem: Ha-Gesher, 1951), pp. 15–20.

8. Cloyzner, *op. cit.*, p. 61.

9. Lehn, *op cit.*, p. 18.

10. Shlomo Shva, *One Day and Ninety Years: The Story of the Jewish National Fund* (Jerusalem: JNF, 1991), p. 26.

11. Ibid., pp. 17–18.

12. Ibid.

13. Menahem Ussishkin, Memorandum to National Committees of the Keren Kayemet L'Yisrael, December 24, 1922, Zionist Archive Number, KKL 10/5111-514.

14. Keren Kayemet L'Yisrael, Budget for the Year 1997, Jerusalem, December 1996.

15. Shva, *op. cit.*, p. 26.

16. Simon Schama, *The Two Rothschilds and the Land of Israel* (London: Collins, 1978), pp. 21–22.

17. Yaakov Tahon, "Yehoshua Chankin," *Megilat ha-Adamah* (Jerusalem: Ha-Gesher, 1951), p. 150.

18. Ibid., pp. 147–149.

19. "Infertile Women Seek Redemption at the Grave of Chankin, The Land Redeemer," *Yediot Ahronot*, December 12, 1997.

20. Berl Katznelson, "Menahem Ussishkin," in *Megilat ha-Adamah* (Jerusalem: Ha-Gesher, 1951), p. 116. This is a eulogy written by the Labor Party Leader.

21. Shva, *op. cit.*, pp. 35–36.

22. *Jewish Advocate*, May 10, 1941.

23. Katznelson, *op. cit.*, p. 115.

24. Ronald Storrs, *Orientations* (London: Nicholson and Watson, 1945), p. 417.

25. Avraham Gronovsky, "Nehemiah Di-Lieme," in *Megilat ha-Adamah* (Jerusalem: Ha-Gesher, 1951), pp. 122–130.

26. Shva, *op. cit.*, p. 42.

27. Avshalom Rokach and Haim Zaban, "Forty Years of Land Development and Afforestation in Israel," JNF reprint from *Ariel, A Review of Arts and Letters in Israel,* no. 75 (1989): 4.

28. Efraim Orni, *Land in Israel* (Jerusalem: Ha-Makor, 1981), p. 27.

29. Katznelson, *op. cit.*, p. 109.

30. Interview with Yerahmiel Kaplan, Rehovoth, May 5, 1999.

31. Interview with Chaim Blass, Herzliyah Pituach, September 3, 1997.

32. Interview with Shimon Ben Shemesh, Jerusalem, January 12, 1998.

33. Shva, *op. cit.*, p. 52.

34. For example, *The Australian Jewish News,* October 17, 1941, p. 1.

35. Interview with Yerahmiel Kaplan, Achovoth, Israel, May 5, 1999.

36. Shva, *op. cit.*, p. 26.

37. Shaul Ephraim Cohen, *The Politics of Planting* (Chicago: University of Chicago Press, 1995), p. 47.

38. Shva, *op. cit.*, p. 26.

39. "The primary claim of the JNF is in expanding the national lands. . . and foresting them only has value insofar as it provides the requisite occupation for establishing ownership of the land." Ussishkin, *op. cit.*

40. Some JNF publications suggest that a full half of the 45,000 planted trees in 1948 were JNF in origin (Rokach and Zaban, *op. cit.*). Most other estimates are set between 10,000 and 12,000. See Nili Lipshitz and Gideon Biger, *The Rise and Fall of the Jerusalem Pine as the Main Tree in the Land of Israel* (Jerusalem: JNF, 1994). Also see Azariah Alon, "This Is Not the Forest We Sought," *Green, Blue and White* 8 (May-June 1996): 36.

41. Yosef Weitz, *Ha-Ya'ar V'ha-Yi'ur B'Yisrael* (Ramat Gan: Masada, 1970), pp. 82–84.

42. Ibid, pp. 85–86. See also Efraim Orni, *Afforestation in Israel* (Jerusalem: Sivan Press, 1969), p. 24.

43."Rishon L'Zion Is Seventy," pamphlet, pub. N. Tiberski (1952), pp. 97–98, as quoted in Weitz, *op.cit*, p. 94.

44. Lipshitz and Biger, *op. cit.*, p. 3.

45. Weitz, *op. cit.*, p. 69.

46. "Naked and barefoot, you spent all day in the mud, amidst a thicket of bushes. The sun heated the swamp and created bubbles until you became dizzy

and fainted. Then a committee of mosquitoes of every type and family, malaria-bearing mosquitoes, conduct a group escort, buzzing after you step after step and there was no escape." Shimon Gorden, one of the first foresters from Hadera, as quoted in Weitz, *op. cit.*, pp. 70–71.

47. Ibid., p. 71.

48. Ibid.

49. Lipshitz and Biger, *op. cit.*, p. 3.

50. David Brower, personal communication, Eugene, Oregon, March 15, 1993.

51. Interview with Chaim Blass, Herzliyah Pituach, September 3, 1997.

52. Ibid.

53. Golda Meir, *My Life* (London: Weidenfeld and Nicolson, 1975), p. 67.

54. Weitz, *op. cit.*, p. 99.

55. Orni, *Afforestation in Israel*, p. 25.

56. Interview with Chaim Blass, Herzliyah Pituach, September 3, 1997.

57. Interview with Heinrich Mendelssohn, Tel Aviv University, August 27, 1997.

58. Interview with Azariah Alon, Kibbutz Beit ha-Shita, September 15, 1997.

59. Interview with Mordechai Ruach, Beit Zayit, September 9, 1997.

60. Interview with Shimon Ben Shemesh, Jerusalem, January 12, 1998.

61. Tom Segev, *1949, The First Israelis* (New York: Free Press, 1986), pp. 29–30.

62. Benny Morris, "Yosef Weitz and the Transfer Committees, 1948–1949," 1948 *and After: Israel and the Palestinians* (Oxford: Clarendon, 1990), pp. 89–149.

63. Interview with Ra'anan Weitz, Jerusalem, January 12, 1998.

64. Weitz, *op. cit.*, pp. 140–141.

65. Yosef Weitz, journal entry from June 26, 1946, published in *Tlamim Ahronim* (Jerusalem: Keren Kayemet, 1974), pp. 24–25.

66. Nili Lipshitz and Gideon Biger, "Forestry Policy of the British Government in the Land of Israel," *Geographic Horizons* 40 (1994): 5–16.

67. Orni, *Afforestation in Israel*, pp. 27–28.

68. The indigenousness of the brutia pine tree is the subject of some debate. Senior foresters of the JNF insist that it is an exotic tree, first appearing in Israel in 1924. Yerahmiel Kaplan and Rene Karshon, personal communication, February 8, 1999.

69. Interviews with Yerahmiel Kaplan and Rene Karshon, Rehovoth, May 5, 1999.

70. Michael Zahari, *Geobotanica* (Tel Aviv: 1959), pp. 342–345, as quoted in Lipshitz and Biger, *Rise and Fall*, footnote 2.

71. Lipshitz and Biger, *Rise and Fall*, p. 3.

72. Ibid., p. 3.

73. Ibid., pp. 4–5.

74. Ibid., p. 7.

75. Ibid.

76. Interview with Mordechai Ruach, Beit Zayit, September 9, 1997.

77. Imanuel Noy-Meir, "An Ecological Viewpoint on Afforestation in Israel: Past and Future," *Allgemeine Forst Zeitschrift* 24–26 (1989): 615.

78. Interview with Mordechai Ruach, Beit Zayit, September 9, 1997.

79. Orni, *Afforestation in Israel*, p. 64.

80. Interview with Azariah Alon, Kibbutz Beit ha-Shita, September 15, 1997.

81. Lipshitz and Biger, *Rise and Fall*, p. 15.

82. Orni, *op. cit.*, p. 27.

83. Cohen, *op. cit.*, p. 58.

84. Interview with Gideon Biger, Tel Aviv, March 5, 1999.

85. Zwi Mendel, "Major Pests of Man-Made Forests in Israel: Origin, Biology, Damage, and Control, *Phytoparasiticia* 15, no. 2 (1987): 683.

86. Lipshitz and Biger, *Rise and Fall*, p. 17.

87. Oppenheimer report, October 30, 1939, as quoted in Lipshitz and Biger, *Rise and Fall*, p. 19.

88. Orni, *Afforestation in Israel*, pp. 38–40.

89. Rokach and Zaban, *op. cit.*, p. 6.

90. JNF report to the Twenty-Third Zionist Congress as cited in Lehn, *op. cit.*, p. 132.

91. Letter from Dorothy de Rothschild to David Ben-Gurion, July 15, 1957, reprinted in Schama, *op. cit*, pp. 327–329.

92. Lehn, *op. cit.*, pp. 131–133.

93. Keren Kayemet L'Yisrael Budget for the Year 1997, Jerusalem, December 1996, p. 6.

94. Eric Greenberg, "JNF Probe Expands," *The Jewish Week*, October 11, 1996; Seth Gitell, "Lone Rangers Try to Lasso Herzl's Fund," *Forward*, September 20, 1996.

95. "JNF Moves to Slash Operations, Tighten Budget," *St. Louis Jewish Light*, September 24, 1997, p. 33.

96. Ibid., p. 8.

97. Lehn, *op. cit.*, p. 122.

98. "The Jewish National Fund: Its Origin, Object, History, and Achievements," *Eretz Israel* (New York: Jewish National Fund for America, 1932), p. 31.

99. Weitz, *Ha-Ya'ar v'ha-Yi'ur b'Yisrael*, p. 300.

100. In fact Goor's promotion to head of a Mandate department was remarkable for a non-British Jew. It was made possible when the previous forest "conservator," G. N. Sale, became embroiled in a broadly publicized sex scandal. Interview with Yerahmiel Kaplan, May 5, 1999.

101. Interview with Chaim Blass, Herzliyah Pituach, September 3, 1997.

102. Weitz, *op. cit.*, p. 302.

103. *Adam Teva V'din and Others v. the Ministry of Interior and Others*, Bagatz 288/00, August 29, 2001 (text on: www.jnfreform.org).

104. Abraham Granot, a former aide to Nehemiah Di Lieme and Menachem Ussishkin and himself the JNF chairman from 1944 to 1961, is credited with getting the government to adopt this JNF axiom. Efraim Orni, *Land in Israel*, p. 30.

105. Yosef Weitz, *The Struggle for the Land* (Tel Aviv: Tubersky, 1950), p. 336.

106. Segev, *op. cit.*, pp. 117–129.

107. Orni, *Afforestation in Israel*, p. 33.

108. Shva, *op. cit.*, pp. 93, 98.

109. Interview with Chaim Blass, Herzliyah Pituach, September 3, 1997.

110. Journal entry from August 18, 1949, Yosef Weitz, *My Diary and Letters to My Sons*, vol. IV (Ramat Gan: Masada, 1965), p. 48.

111. Interview with Mordechai Ruach, Beit Zayit, September 9, 1997; Weitz, *Ha-Ya'ar v'ha-Yi'ur b'Yisrael*, p. 295.

112. Moshe Kolar, "The Forest Department of the JNF," *Allgemeine Forst Zeitschrift* (June 1989): 600.

113. Interview with Shimon Ben Shemesh, Jerusalem, January 12, 1998.

114. Weitz, *Ha-Ya'ar v'ha-Yi'ur b'Yisrael*, p. 300.

115. Ibid., p. 302.

116. Rokach and Zaban, *op. cit.*, p. 22.

117. Ibid., pp. 9–10.

118. Ibid., p. 11.

119. Interview with Chaim Blass, Herzliyah Pituach, September 3, 1997.

120. Interview with Mordechai Ruach, Beit Zayit, September 9, 1997.

121. Interview with Chaim Blass, Herzliyah Pituach, September 3, 1997.

122. Orni, *Afforestation in Israel*, p. 10

123. Interview with Mordechai Ruach, Beit Zayit, September 9, 1997.

124. Interview with Menahem Sachs, Director of JNF Forestry Department, October 28, 1997.

125. Interview with Iris Bernstein, Planner for JNF Central Region, Eshtaol, October 28, 1997.

126. Land Development Authority, *Annual Report 1996* (Jerusalem: Keren Kayemet L'Yisrael, July 1997), p. 18.

127. Ofer Regev, *Arbaim Shnot Pricha* (Tel Aviv: Society for the Protection of Nature, 1993), p. 46.

128. Ibid., p. 16.

129. Alon, *op. cit.*, 338–340.

130. Heinrich Mendelssohn, personal communication, November 19, 1998.

131. *Divrei ha-Knesset*, December 11, 1962, p. 413.

132. Uzi Paz, *Eretz ha-Tsvi v'ha-Yael* (Land of the Gazelle and the Ibex),*–I* (Givataim: Masada, 1981), p. 92.

133. Orni, *Afforestation in Israel*, p. 57.

134. Omri Boneh and John Woodcock, "Cherishing Precious Resources," *JNF Illustrated* (Spring 1994): 3.

135. Extension Toxicology Network, Oregon State University, "Simazine," http://ace.orst.edu/info/extoxnet/pips/simazine.html, June 1996.

136. Aviva Rabinovich, "Evaluation of the Damages Caused to the Soil and Groundwater as a Result of Spraying with Simazine in Open Spaces," as quoted in *Adam Teva V'din and Others v. the Ministry of Interior and Others,* Bagatz 288/00, *op. cit.*

137. Mendel, *op. cit.*, p. 2.

138. Extension Toxicology Network, Oregon State University, "Endosulfan," http://ace.orst.edu/info/extoxnet/pips/endosulf.htm, June 1996.

139. Boneh and Woodcock, *op. cit.*, p. 2.

140. Mendel, *op. cit.*

141. Mendel, "Integration of Management of Forests and Preservation of Forest Habitats as a Way to Address Damaging Insects," lecture at JNF Research Symposium, Hebrew University, Agriculture Faculty, Rehovoth, December, 19, 2001.

142. Orni, *Afforestation in Israel*, p. 61.

143. Land Development Authority, *op. cit.*

144. Interview with Azariah Alon, Kibbutz Beit ha-Shita, September 15, 1997.

145. Meir Shalev, *Primarily about Love* (Tel Aviv: Am Oved, 1991), p. 34.

146. Interview with Chaim Blass, Herzliyah Pituach, September 3, 1997.

147. Mendel, "Major Pests."

148. Land Development Authority, *op. cit.*

149. Avraham Weinstein, "The Species of *Pinus* and *Eucalyptus* Used for Afforestation in Israel," *Allgemeine Forst Zeitschrift*, June 1989, p. 627.

150. *JNF Mini-Manual* (Jerusalem: JNF, 1973), p. 34.

151. The percentage of conifer trees planted is actually higher, owing to their relatively high density, reaching as high as 65 percent. Zvika Avni, head of JNF Forestry Department, personal communication, March 12, 2000.

152. Mordechai Ruach, "Organization and Activities of the Forest Department," *Allgemeine Forst Zeitschrift*, June 1989, p. 604.

153. Interview with Yerahmiel Kaplan, Rehovoth, May 5, 1999.

154. Interview with Iris Bernstein, Eshtaol, October 28, 1997.

155. Danny Orenstein, oral presentation, JNF Educational Department, December 8, 1997.

156. Interview with Y'nass Maalachi, Yatir Forest, December 8, 1997.

157. Noam Gressel, personal communication, May 2000.

158. Noy-Meir, *op. cit.*

159. Interview with Menahem Sachs, Director of JNF Afforestation Department, October 28, 1997.

160. Noy-Meir, *op. cit.*

161. Ibid.

162. Land Development Authority, *op. cit.*

163. Ibid., p. 33.

164. *Planning and Building Law, 1965—National Master Plan for Forests and Forestry*, submitted to the Government, Ministry of the Interior, Jerusalem, August 1995. Published in *Yalkut Pirsumim*, no. 4363, December 19, 1995.

165. "Keren Kayemet: Twelve Million People Can't Be Wrong," *Green, Blue and White* (May 1996): 39.

166. Avi Goren, "Hands Off the Forest," *Green, Blue and White* 3 (March 1995): 34–35.

167. Interview with Iris Bernstein, Eshtaol, October 28, 1997.

168. *Adam Teva V'din and Others v. the Ministry of Interior and Others*, Bagatz 288/00, August 29, 2001 (text on www.jnfreform.org).

169. Ibid.

170. Zafrir Rinat, "Grounds for Suspicion," *Ha-Aretz*, August 14, 1998.

171. Land Development Authority, *op. cit.*, p. 6.

172. David Blougrund, *The Jewish National Fund Policy Study No.* 49, The Institute for Advanced Strategic and Political Studies, September 2001. Also, Arnold Regular, "'Destructive and Dangerous' Management: The Report of the Comptroller of the Jewish Agency Regarding the JNF," *Kol ha-Ir*, February 9, 2001.

173. Lehn, *op. cit.*, p. 147.

174. Shva, *op. cit.*, p. 93.

175. Keren Kayemet L'Yisrael Budget for the Year 1997 (Jerusalem, December 1996), p. Zayin-2.

176. Land Development Authority, *op. cit.*, p. 45.

177. Interview with Dr. Benny Shalmon, Kibbutz Ketura, January 14, 1998.

178. J. Kaplan, R. Karschon, and M. Kolar, "Israel," in *Afforestation in Arid Zones*, ed. R. N. Kaul (The Hague: Junk, N.V. 1970), pp. 151–152.

179. Jewish National Fund, *Savannization—An Ecological Answer to Desertification* (Jerusalem: JNF, 1994).

180. Danny Orenstein, oral presentation, JNF Educational Department, December 8, 1997.

181. Patricia Golan, "Redeeming the Desert: Successful Experiment in Reversing the Process of Desertification in Israel's Negev," *Israel Environmental Bulletin* 13, no. 3 (1990): 21–22.

182. Dr. Burt Kotler, Ben-Gurion University of the Negev, personal communication, May 5, 1998.

183. Dr. Uriel Safriel as quoted in Janine Zacharia, "The Final Frontier," *The Jerusalem Report*, May 1, 1997, p. 14.

184. Elchanan Josephy, "The Natural Forest of Israel," *Allgemeine Forst Zeitschrift*, June 1989, p. 663.

185. Ibid.

186. Amit Shapira, Director of the Environment and Nature Preservation Department, Society for Protection of Nature in Israel, personal communication, April 27, 1998.

187. Irit Sappir-Gildor, "Environmental Attitudes among Visitors to JNF Forests," master's thesis, Tel Aviv University, Department of Geography 2001.

188. Ibid.

189. Interview with Menahem Sachs, October 28, 1997.

190. World Commission on Environment and Development (Brundtland Commission), *Our Common Future* (Oxford: Oxford University Press, 1987).

191. Mike McCloskey, "The Emperor Has No Clothes: The Conundrum of Sustainable Development," *Duke Environmental Law and Policy Forum* 9 (1999): p. 153.

192. Noy-Meir, *op. cit.*, p. 618.

193. Sappir Gildor, *op. cit.*

194. M. Schechter, N. Zaitsev, B. Reiser, and S. Nevo, "On Valuing Natural Resources Damages," *Preservation of the World in the Wake of Change*, ed. Y. Steinberger (Jerusalem: Israel Society for Ecology and Environmental Quality Sciences, Pub., 1996), pp. 343–349.

195. Zafrir Rinat, "There Once Was a Forest," *Ha-Aretz* (English Internet edition), December 2, 1997.

196. Meir Barzilai, personal communication, Jerusalem, May 1, 2000.

197. Keren Kayemet L'Yisrael Budget (Jerusalem: December 1996), p. "F."

198. "Keren Kayemet: Twelve Million People Can't Be Wrong," p. 39.

199. Rokach and Zaban, *op. cit.*, p. 29.

200. *Adam Teva V'din and Others v. the Ministry of Interior and Others*, Bagatz 288/00, August 29, 2001 (text on: www.jnfreform.org).

201. Blougrund, *op. cit.*, pp. 6, 11.

CHAPTER 5: THE EMERGENCE OF AN ISRAELI ENVIRONMENTAL MOVEMENT

1. Interview with Danny Rabinowitz, Maoz Aviv, September 30, 1997.

2. Azariah Alon, "Position Paper of the Society for the Protection of Nature in Israel Regarding 'Nesher's' Demand to Establish a Quarry In the Heart of Carmel National Park," *Biosphera* 73, no. 2 (1972): 1.

3. Azariah Alon, "Nature Protection in Israel: Activities and Campaigns of the Society for the Protection of Nature in Israel," unpublished manuscript, on file with author, 1997.

4. Environmental Protection Service, "Conditions for a Quarrying Permit in Tamra-Kabul," *Environmental Quality* (Jerusalem: Ministry of the Interior, 1977), p. 192.

5. Yosef Lapid, "The Pelican Survey," *Maariv*, August 20, 1992, Yoman Ma'ariv section.

6. Eitan Gidalizon, personal communication, April 27, 2000; also Israel State Comptroller, "The Society for the Protection of Nature: Field Schools," in "Ministry of Education, Culture and Sport," *The 47th Report of Israel's State Comptroller*, 1997, p. 415.

7. Shoshana Gabbay, *The Environment in Israel* (Jerusalem: Ministry of the Environment, 1995), p. 240.

8. Brock Evans, personal communication, May 23, 1998.

9. Ofer Regev, *Forty Years of Blossoming* (Tel Aviv: Society for the Protection of Nature in Israel, 1993), p. 23.
10. Interview with Uzi Paz, Ramat Efal, September 14, 1997.
11. Interview with Amotz Zahavi, Tel Aviv University, December 18, 1997.
12. Yosef Weitz, "The Redemption and Settlement of the Huleh Valley," in *The Huleh: An Anthology* (Jerusalem: World Zionist Organization, 1954), pp. 114, 119.
13. A. Zigelman and M. Gershuni, "The Flora and Fauna of the Huleh Valley," in *The Huleh: An Anthology* (Jerusalem: World Zionist Organization, 1954), pp. 62–63.
14. Ibid., pp. 62–99.
15. Interview with Amotz Zahavi, Tel Aviv University, December 18, 1997.
16. Regev, *op. cit.*, p. 18.
17. Interview with Uzi Paz, Ramat Efal, September 14, 1997.
18. Interview with Heinrich Mendelssohn, Tel Aviv University, August 27, 1997.
19. Haim Finkel, "The Hula Lake Project," *Jerusalem Post,* June 1993.
20. Regev, *op. cit.*, p. 18.
21. Alon, *op. cit.*
22. In his memoirs, Simcha Blass claims that it was Pinhas Lavon, who served briefly as Minister of Agriculture, who made the fateful decision to go ahead with the project, explaining, "There's no way to leave the swamp when it's in our hands." Simcha Blass, *Water in Strife and Action* (Givataim: Masada, 1973), p. 161.
23. Interview with Heinrich Mendelssohn, Tel Aviv University, August 27, 1997.
24. Ibid.
25. See, for example, Boomi Toran's account of the February 25, 1953, meeting in Linda Zach, *Boomi: To Collect Golden Moments,* Publication of Kibbutz Mabarot, 1997, pp. 173–174.
26. Alon, *op. cit.*
27. Ibid.
28. Interview with Amotz Zahavi, Tel Aviv University, December 18, 1997.
29. Ibid.
30. Regev, *op. cit.*, p. 24.
31. Ibid.; interview with Amotz Zahavi, Tel Aviv University, December 18, 1997. A decade later, after it joined forces with the natural history journal *Teva V'Aretz,* the number reached fifteen thousand.
32. Zach, *op. cit.*, pp. 173–174.
33. Interview with Yoav Sagi, Tel Aviv, November 19, 1998.
34. Regev, *op. cit.*, p. 27.
35. Uzi Paz, *Eretz ha-Tsvi v'ha-Yael* (Land of the Gazelle and the Ibex), *I* (Givataim: Masada, 1981), p. 42.
36. Regev, *op. cit.*, p. 27.
37. Paz, *op. cit.*, p. 42.

38. Interview with Uriel Safriel, Sdeh Boqer, January 6, 1998.

39. Ibid.

40. Ibid.

41. Interview with Azariah Alon, Kibbutz Beit ha-Shita, September 15, 1997.

42. Interview with Uzi Paz, Ramat Efal, September 14, 1997.

43. Interview with Amotz Zahavi, Tel Aviv University, December 18, 1997.

44. Regev, *op. cit.*, pp. 29–30.

45. Shmuel Amir and Reuven Stern, "Involvement of Citizen Groups in Environmental Quality Protection," *Biosphera* (1975): 8–12.

46. Regev, *op. cit.*, p. 31.

47. Interview with Adir Shapira, Ramat Gan, December 18, 1997.

48. Yizhar Smilansky, *Divrei ha-Knesset*, December 11, 1962, p. 419.

49. Support is delivered mainly in the form of direct Ministry of Education payments to field school staff for educational trips. The figure is actually an understatement and does not include indirect subsidies, such as those arising from the use of the schools themselves by the SPNI. Israel State Comptroller, *op. cit.*, p. 413.

50. Interview with Benny Shalmon, Kibbutz Ketura, January 14, 1997.

51. Ibid.

52. Ibid.

53. Interview with Uzi Paz, Ramat Efal, September 14, 1997.

54. Interview with Azariah Alon, Kibbutz Beit ha-Shita, September 15, 1997.

55. Alon, *op. cit.*

56. Zach, *op. cit.*, pp. 173–174.

57. In a 1962 Knesset debate Ben-Gurion proclaimed: "We only have one Carmel in this land. It is necessary, in may opinion, to use the Knesset's authority to do for the Carmel what we must to safeguard the character of the land." *Divrei ha-Knesset*, December 11, 1962, p. 474.

58. Areas traditionally owned by neighboring Druze villages of Daliat al-Karmel and Usifiyah lay outside of the Park's boundaries, as were the lands of Kibbutz Beit Oren, a small sanitarium, military bases, and even a quarry run by the Nesher Cement plant. But the remaining 75 percent of the area in question was protected.

59. Yoav Sagi, personal communication, January 7, 1999.

60. Interview with Yoav Sagi, Tel Aviv, November 19, 1998.

61. "Environmental Prize to Yoav Sagi," *Israel Environmental Bulletin* 13, no. 3 (1990): 2–3.

62. Interview with Danny Rabinowitz, Maoz Aviv, September 30, 1997.

63. "Activities of the Society for the Protection of Nature Regarding the Power Station on the Sharon Coast," *Biosphera* 72, no. 8 (1972): 15–17.

64. Interview with Danny Morganstern, Tel Aviv, March 31, 1996.

65. "Power Station, Nahal Taninim or Hadera," *Biosphera* 72, no. 7 (1972): 5.

66. Interview with Yoav Sagi, Tel Aviv, November 19, 1998.

67. Ibid.

68. Ibid.
69. Regev, *op. cit.*, p. 69.
70. Interview with Adir Shapira, Ramat Gan, December 18, 1997.
71. Danny Rabinowitz, "The Bird Man Flies Again," *Ha-Aretz*, October 11, 1991, p. B-4.
72. Orit Nevo, "Nature Knows No Borders," *Eretz v'Teva* (May 1994): 8.
73. Yossi Leshem, "Birds without Borders," *Teva ha-Dvarim* 24 (October 1997): 40–68.
74. Paz, *op. cit., II* p. 282.
75. Regev, *op. cit.*, p. 89.
76. Interview with Yossi Leshem, London, September 21, 1997.
77. Regev, *op. cit.*, p. 90.
78. Interview with Yossi Leshem, London, September 21, 1997.
79. Regev, *op. cit.*, p. 90.
80. Interview with Yoav Sagi, Tel Aviv, November 19, 1998.
81. Meron Benvenisti, *Moment Magazine*, February 1986, p. 24.
82. Nahum Barnea, "The Fall of the Giants," *Yediot Ahronot*, special section, "A New Land," Independence Day edition, 1998, p. 10. It was only because of his personal distaste for many of the Greater Israel movement's leadership (born of historical political rivalries) that Alon did not become more deeply involved.
83. Interview with Yossi Leshem, London, September 21, 1997.
84. Avner De-Shalit, "Where Environmentalists Hide," in *Our Shared Environment—The Conference 1994*, ed. Robin Twite and Robin Menczel (Jerusalem: Israel-Palestine Center for Research and Information, 1995), p. 273.
85. Yoav Sagi, "Militevah," *Teva v'Aretz* (May 1992): p. 30.
86. Interview with Danny Rabinowitz, Maoz Aviv, September 30, 1997.
87. Interview with Amotz Zahavi, Tel Aviv University, December 18, 1997.
88. SPNI Director General's Office, *Summary of Activities in the Society for the Protection of Nature, 1997* (Tel Aviv: SPNI, 1998).
89. Amit Shapira, personal communication, May 28, 1998.
90. Yoav Sagi, personal communication, January 13, 1999.
91. Gabbay, *op. cit.*, p. 240.
92. Regev, *op. cit.*, p. 108.
93. Interview with Lynn Golumbic, Haifa, November 24, 1997.
94. Interview with Azariah Alon, Kibbutz Beit ha-Shita, September 15, 1997.
95. Yoav Sagi, keynote address at the International Conference on the Role of Nongovernmental Organizations in Protecting the Environment, Eilat, Israel, March 21, 1994.
96. SPNI Director General's Office, *op. cit.*, p. 13.
97. Interview with Lev Fishelson, Tel Aviv, September 28, 1997.
98. Interview with Uriel Safriel, Sdeh Boqer, January 6, 1998.
99. Yisrael Caspi, personal communication, December 13, 1997.
100. Mickey Lipshitz, oral presentation, Law School Class, Tel Aviv University, December 11,1997.

101. Ariella Ringle-Hoffman, Mordechai Gilat, and Meli Kritz, "They Don't Stop on Green," *Yediot Ahronot*, December 12, 1997, Musaf Shabbat, pp. 12–15; Ariella Ringle-Hoffman, "Recorded Evidence: This Is How They Will Take Over the Society for the Protection of Nature," *Yediot Ahronot*, December 26, 1997, pp. 1, 6; "Hostile Takeover Threat," *Ha-Aretz*, December 15, 1997.

102. Yisrael Caspi, personal communication, December 13, 1997.

103. Emily Silverman, personal communication, May 28, 1998.

104. Interview with Emily Silverman, Tel Aviv, September 14, 1997.

105. Ibid.

106. Ibid.

107. CRB Foundation, *Catalogue of Environmental Organizations in Israel* (Tel Aviv: Adam Teva V'din, 1997), pp. 19–20.

108. Eitan Gidalizon, personal communication, April 27, 2000.

109. Eitan Gidalizon, personal communication, December 10, 1998.

110. Orit Nevo, "170 Men and Women Guides in the Society for the Protection of Nature," *Teva v'Aretz* 262 (1993): 12.

111. Israel State Comptroller, *op. cit.*, p. 423.

112. Ibid. On average, SPNI guides were underemployed, idle 25 percent of the time, even though they were never expected to work more than 160 days a year. Field schools were often deserted on weekends.

113. Part of the financial problem stems from the SPNI's substantive integrity. Schools prefer a lighter, more attraction-packed experience. The SPNI policy of two guides per bus rather than one is pedagogically sound, but twice as expensive.

114. Israel State Comptroller, *op. cit.*, p. 414.

115. Eitan Gidalizon, personal communication, April 27, 2000; see also Israel State Comptroller, *op. cit.*, p. 415.

116. Ibid., p. 415.

117. Interview with Lev Fishelson, Tel Aviv, September 28, 1997.

118. Israel State Comptroller, *op. cit.*, p. 413.

119. Eitan Gidalizon, personal communication, April 1997.

120. Corri Gottesman, *Case Study of the Carmiel-Tefen Road: Management of an Environmental Conflict* (Jerusalem: Environmental Protection Service, 1988).

121. Interview with Valerie Brachya, Jerusalem, February 17, 1999.

122. Tal Setzmeski, "The Voice of America Station in the Arava: Summary of Events," unpublished paper, for the Arava Institute for Environmental Studies, January 1997.

123. Avner De-Shalit and Moti Talias, "Green or Blue and White? Environmental Controversies in Israel," *Environmental Politics* 3 (Summer 1994): 274.

124. Reuven Yosef, "Clues to the Migratory Routes of the Eastern Flyway of the Western Palearctics: Ringing Recoveries at Eilat, Israel," *Die Vogelwarte* 39 (1997): 131–140.

125. Letter from Yoav Sagi to Shimon Peres, May 13, 1985, on file with Yoav Sagi, Tel Aviv.

126. Interview with Uriel Safriel, Sdeh Boqer, January 6, 1998.

127. Yaakov Skolnik, "Save the Ashdod Sands," *Eretz Magazine,* Winter 1991, pp. 10–12.

128. Interview with Bilha Givon, Tel Aviv, November 25, 1997.

129. *The Central Arava Regional Council et al. v. the National Planning Board and the Tomer Project Council, Piskei Din,* vol. 47, section 1 (1990), pp. 610–618.

130. Interview with Bilha Givon, Tel Aviv, November 25, 1997.

131. Azariah Alon and Yoav Sagi, "Cancel the Voice of America in the Arava," SPNI flyer, January 1990.

132. Yosef Lapid, "The Pelican Survey," *Maariv,* August 2, 1992.

133. *The Central Arava Regional Council et al. v. the National Planning Board and the Tomer Project Council, Piskei Din,* vol. 47, section 1 (1990), pp. 610–618.

134. Interview with Valerie Brachya, Jerusalem, February 17, 1999.

135. Interview with Yoav Sagi, Tel Aviv, November 19, 1998.

136. "After reviewing the written materials that were submitted to us and the arguments of the attorneys of the [two] sides, we are convinced that the following subjects were not checked in a sufficiently serious fashion by the National Planning Board: A) The dangers facing the birds from the nets of the Voice of America Transmission Station of the Arava. . . B) The change in the location of the IDF's firing zones. . . C) The possible noise and explosions from the location of the IDF firing zone." Bagatz 3476/90, May 20, 1991, pp. 1–2.

137. Berry Pinshow, personal communication, June 16, 1998.

138. Idit Vitman, "Green Birds in the World," *Globes,* September 4, 1992.

139. Berry Pinshow and Ron Frumkin, *Bird Migration Survey, Spring 1992, Ground Survey Report,* Mitrani Center for Desert Ecology, Ben-Gurion University, Sdeh Boqer Campus, 1992.

140. Elli Elad, "Migrating Bird Survey in the Negev Determines That Only a Few Individuals Will Be Injured by the Voice of America," *Ha-Aretz,* August 16, 1992.

141. Ben Caspit, "Forbes Is Coming to the Arava," *Maariv,* October 27, 1992.

142. "They Demonstrated in Jerusalem Against the Voice of America: The Land of Israel is Blossoming, We Won't Let It Go Bald, " *Davar,* August 25, 1992; "The Air Force Is Liable to Accidentally Bomb Arava Elementary Schools," *El ha-Mishmar,* November 11, 1992.

143. Elli Elad, "Protection of Nature: The Tomer Council Forged a Report About the Voice of America," *Ha-Aretz,* October 16, 1992.

144. "Namir: In the Question of the Voice of America, the Matter of Employment Can't Be Ignored," *Yediot Ahronot,* August 19, 1992.

145. Amir Rosenblit, "The Society for the Protection of Nature to Rabin: Don't Ignore the Supreme Court's Decisions," *Davar,* December 29, 1992.

146. "The Senate and Congress: To Freeze the Funds for the Voice of America," *Maariv*, October 1, 1992.

147. Interview with Bilha Givon, Tel Aviv, November 25, 1997.

148. SPNI press release, May 3, 1993.

149. Interview with Danny Morganstern, Tel Aviv, March 31, 1996.

150. "Last Hope for the Jordan," *World Rivers* (September/October 1991): pp. 13–14.

151. Eran Solamon and Yoram Gabizon, "A Money Station at K'far ha-Nasi," *Ha-Aretz*, June 3, 1992, p. C-1.

152. De-Shalit and Talias, *op. cit.*, p. 274.

153. Yoav Sagi, personal communication, January 12, 1999.

154. *The Movement for Quality Government in Israel and Others v. the National Planning and Building Council and Others, Piskei Din*, vol. 45 III, 1991, pp. 680–689.

155. Ibid., pp. 680–689.

156. De-Shalit and Talias, *op. cit.*, p. 274.

157. Interview with Danny Rabinowitz, Maoz Aviv, September 30, 1997.

158. Interview with Mickey Lipshitz, Tel Aviv, January 8, 1998.

159. Interview with Yoav Sagi, Tel Aviv, November 19, 1998.

160. *The Movement for Quality Government in Israel and Others v. the National Planning and Building Council and Others, Piskei Din*, vol. 45 III, 1991, p. 688.

161. David Rudge, "Electric Plant Inauguration Sparks Protest," *The Jerusalem Post*, May 21, 1992.

162. De-Shalit and Talias, *op. cit.*, pp. 281–282.

163. Zur Sheyzaf, "The River against the World," *Ha-Aretz*, May 31, 1991.

164. Elisha Efrat, personal communication, March 1994.

165. Trans-Israel Highway Company, "1993 Was an Important Year for Drivers," promotional material, 1999.

166. Daniel Morganstern, "Don't Want Trans-Israel Highway," *Green, Blue and White* (August 1999): p. 11.

167. Adam Teva V'din, "Cross-Israel Highway or Cross-Israel Railway," 1995.

168. Yossi Brachman, Ilan Salomon, Eran Feitelson, and Danny Shefer, *The Place of the Train in the Transportation System in Israel* (Jerusalem: Jerusalem Institute for Israel Studies, 1993); see also Ilan Salomon and Eran Feitelson, "Israel: Transport on a Small Turbulent 'Island State,'" *A Billion Trips a Day*, ed. I. Salomon, I. P. Bovy, and J. Orfeuil (Kluwer: Academic Publishers, The Hague, 1993).

169. "Do We Need the Trans-Israel Highway?" debate between Alon Tal and Ilan Salomon, *The Jerusalem Report*, January 3, 2000, p. 56.

170. Letter from Yoav Sagi to Alon Tal, July 1, 1993.

171. DESHE Transportation Team, *Summary of the Staff Position Regarding the Trans-Israel Highway (Highway 6)*, Revised and shortened version approved by DESHE Transportation Committee, June 1993.

172. Initially, *Adam Teva V'din and Others v. The National Planning Commission and Others*, Bagatz 2920/94 *Piskei Din* 50(III)446; in 1999,

Adam Teva V'din v. Minister of Treasury and Others, Bagatz, 4119/99 99(3) p. 793.

173. Amit Ashkenazi, "How to Direct Traffic to a Highway and Still Get Stuck," *Telegraph,* October, 7, 1994, Dividend Supplement, p. 1.

174. Interview with Yossi Sarid, Jerusalem, December 30, 1997.

175. Orit Nevo, "More against Trans-Israel Highway," *Eretz v'Teva* 47 (1997): 17.

176. SPNI Director General's Office, *op. cit.*

177. Ruth Sinai, "With a Tie and a Hat Against the Bulldozers," *Ha-Aretz,* December 5, 1999, p. B-3.

178. Bilha Givon, "A Law unto Itself," *Eretz Magazine* (Winter 1993): p. 70; Seffi Ben Yosef, "Improvisers Out: The Dead Sea Yet Lives," *Teva v'Aretz* 255 (1993), pp. 6–9.

179. Yehudit Goren and Rinat Klein, "Severe Pollution in the Golan Streams," *Maariv,* August 10, 1993; see also Alon Tal, "Pollution Prevention in the Kinneret Watershed through the Regulation of Animal Feedlots," Proceedings from Pollution Prevention NGO Conference in Sfax, Tunisia, September 1994.

180. Orit Nevo, "Down by the Seaside, Sifting Sand," *Eretz Magazine* (Autumn 1963): p. 70; see also "Beach Clean Up Campaign," *Biosphera* 22 no. 8–9 (1993): 31.

181. Orit Nevo, "The Forum to Promote Trash Recycling," *Teva v'Aretz* 256 (1993): 9.

182. Avi Goren, "Hands off the Forest," *Green, Blue and White* 3 (March 1995): 34–36.

183. SPNI Director General's Office, *op. cit.*, p. 7.

184. "The Society for Preservation of Nature to the Supreme Court: The Water Commissioner Is Endangering the Kinneret," *Shomrei ha-Sviva* 27 (October 2001): 4.

185. Interview with Emily Silverman, Tel Aviv, September 14, 1997.

186. Yoav Sagi offers a somewhat different perspective on the strategic approach of SPNI during the 1980s and 1990s in his article "Militevah," *Teva v'Aretz* (May 1992): 28–32.

187. Interview with Amotz Zahavi, Tel Aviv University, December 18, 1997.

CHAPTER 6: A GENERAL LAUNCHES A WAR FOR WILDLIFE

1. Interview with Avinoam Finkleman, October 28, 2001.

2. Giora Ilani, *Zoogeographical, Ecological Survey of Predators, in Israel, the Golan, Judea and Samaria, and Sinai* (Jerusalem: Nature Reserve Authority, November 1979), pp. 125–128.

3. H. B. Tristam, "The Land of Israel," in *A Journal of Travels in Palestine* (London: Society for Promoting Christian Knowledge, 1866), as quoted in Giora Ilani, *op. cit.*

4. Yoram Yom-Tov and Heinrich Mendelssohn, "Altered Landscapes," *Eretz Magazine* (November 1995): 52.

5. Interview with Azariah Alon, Beit ha-Shita, September 15, 1997.

6. Shoshana Gabbay, "Human Activity and Wildlife Protection: Conflicts and Challenges," *Israel Environmental Bulletin* 20, no. 1 (1997): 15–16.

7. R. Nathan, U. Safriel, and H. Shirihai, "Extinction and Vulnerability to Extinction at Distribution Peripheries: An Analysis of the Israeli Breeding Fauna, *Israel Journal of Zoology* 42 (1996): pp. 361–383.

8. Uzi Paz, *Eretz ha-Tsvi v'ha-Yael* (Land of the Gazelle and the Ibex), *I* (Givataim: Masada, 1981).

9. Moshe Sneh, in *Divrei ha-Knesset*, December 19, 1962, p. 469.

10. Theodor Herzl, "The Jewish State," in *The Zionist Idea*, ed. Arthur Hertzberg (Athenum, New York: Temple, 1981), p. 221.

11. Reuven Yosef and Rony Malka, "Avian Conservation in Israel," unpublished manuscript, 1998.

12. Tal Shavit, "All-Terrain Vehicles on You Israel," *Maariv*, June 18, 1995.

13. Yom-Tov and Mendelssohn, *op. cit.*, p. 57.

14. Uzi Paz, personal communication, November 11, 1998.

15. Heinrich Mendelssohn, "Nature Protection in Israel," *Biosphera* 72, no. 3 (1972): 3.

16. Interview with Uzi Paz, Ramat Efal, September 14, 1997.

17. Paz, *op. cit.*, p. 42.

18. Interview with Alon Galili, Sdeh Boqer, January 3, 1998.

19. Ibid.

20. Ibid.

21. Paz, *op. cit.*, p. 70.

22. Mendelssohn, *op. cit.*, p. 4.

23. Benny Shalmon, personal communication, July 13, 1998.

24. Uzi Paz, "Nature Preservation in Israel," in *The Plants and Animals of the Land of Israel*, ed. Azariah Alon (Tel Aviv: Ministry of Defense Press, 1990), pp. 71–72.

25. Interview with Uzi Paz, Ramat Efal, September 14, 1997.

26. Mendelssohn, *op cit.*, p. 3.

27. *Sefer Chokim*, 1956, p. 10.

28. Iris Millner, "The Rabbit's Foot," *Ha-Aretz*, August 22, 1998.

29. Interview with Uzi Paz, Ramat Efal, September 14, 1997.

30. Interview with Reuven Ortal, January 13, 1998.

31. Uzi Paz, "Thus It Began," *Teva v'Aretz* 264 (1994): 62.

32. Ibid.

33. Paz, *Eretz ha-Tsvi, I*, p. 43.

34. Paz, "Nature Preservation in Israel," p. 70.

35. Paz, *Eretz ha-Tsvi, I*, p. 41.

36. Ofer Regev, *Forty Years of Blossoming* (Tel Aviv: Society for the Protection of Nature in Israel, 1993).

37. Ibid.

38. Interview with Uzi Paz, Ramat Efal, September 14, 1997.

39. David Ben-Gurion, Knesset debate, *Divrei ha-Knesset,* December 3, 1962, p. 331.

40. Interview with Uzi Paz, Ramat Efal, September 14, 1997.

41. Paz, "Thus It Began," p. 62.

42. Proposed law, National Parks and Nature Reserves Laws, 1962.

43. Yosef Weitz, "To Sit Narrowly—On Landscapes and Nature Reserves," speech from Weitz's JNF archives, dated second day of Hanukkah, 1963.

44. Interview with Uzi Paz, Ramat Efal, September 14, 1997.

45. *Divrei ha-Knesset,* December 3, 1962, p. 331.

46. Ibid., p. 413.

47. Ibid., p. 419.

48. Ibid., p. 473.

49. Ibid., p. 471.

50. In the Knesset, a bill must be read and voted on at least three times before it becomes law. The bill goes back to committee for revisions between the first and second readings.

51. Paz, *Eretz ha-Tsvi, I,* p. 44.

52. Paz, "Thus It Began," p. 63.

53. *Divrei ha-Knesset,* December 11, 1962, p. 413.

54. Uzi Paz, personal communication, November 27, 1998.

55. Interview with Uzi Paz, Ramat Efal, September 14, 1997.

56. Paz, "Thus It Began," p. 63.

57. Ben-Gurion was suspicious of the rival Palmach's "barefoot" image as well as their leftist Mapam political ideology.

58. "The Zimmerman Prize for the Environment—Judges' Explanation," *Biosphera* G, no. 8 (1978): 2.

59. Regev, *op. cit,* p. 38.

60. An English officer who established a reconnaissance unit of Jews from the Yishuv prior to and during World War II.

61. Interview with Danny Yoffe, Tel Aviv, November 15, 1997.

62. Danny Yoffe, personal communication, November 17, 1998.

63. Interview with Danny Yoffe, Tel Aviv, November 15, 1997.

64. "In 1957 I was out observing gazelles for the annual count, and Yoffe showed up hunting. He argued that he loved animals, and he underwent a metamorphosis when he became head of the NRA. But before that, there were no limits for him." Interview with Giora Ilani, Beer Sheva, September 9, 1997.

65. Interview with Batyah Paz, Ramat Efal, September 14, 1997.

66. Interview with Uzi Paz, Ramat Efal, September 14, 1997.

67. Paz, "Nature Preservation in Israel," p. 73.

68. Interview with Danny Yoffe, Tel Aviv, November 15, 1997.

69. Iris Millner, "The Rabbit's Foot," *Ha-Aretz,* August 22, 1998.

70. Interview with Leah Rabin, Tel Aviv, January 7, 1998.

71. Ibid.

72. Interview with Dan Perry, Tel Aviv, September 4, 1997.

73. Ibid.

74. National Parks, Nature Reserves, Memorial Sites and National Sites Law, 1992, sec. 22 and 34.

75. Paz, *Eretz ha-Tsvi, I,* p. 48.

76. Paz, "Nature Preservation in Israel," p. 70.

77. Paz, *Eretz ha-Tsvi, II,* p. 101.

78. Ibid., p. 48.

79. Azariah Alon, "The Nature Reserve Authority," in *The Plants and Animals of the Land of Israel,* ed. Azariah Alon (Tel Aviv: Ministry of Defense Press, 1990), p. 77.

80. Yom-Tov and Mendelssohn, *op. cit.,* p. 58.

81. Interview with Dan Perry, Tel Aviv, September 4, 1997.

82. Ibid.

83. Interview with D'vora Ben Shaul, Rosh Pina, January 11, 1997.

84. Ibid.

85. Azariah Alon, "After the Establishment of the Nature Reserve Authority," in *The Plants and Animals of the Land of Israel,* ed. Azariah Alon (Tel Aviv: Ministry of Defense Press, 1990), p. 74.

86. Interview with Dan Perry, Tel Aviv, September 4, 1997.

87. Ibid.

88. *The Legislation of the State of Israel* (Jerusalem, updated 1991), pp. 3045–3047.

89. SPNI, "Educating for Israel's Environment," promotional pamphlet, 1994; see also Uzi Paz, "A Success Story," in *Eretz ha-Tsvi, I,* pp. 76–80.

90. National Parks, Nature Reserves, Memorial Sites and National Sites Law, 1992, sec. 41(d).

91. Avraham Yoffe, "The Nature Reserve Authority," *Biosphera* 72, no. 3 (1972): 6.

92. Paz, *Eretz ha-Tsvi, I,* pp. 78–80.

93. Interview with Reuven Ortal, January 13, 1998.

94. Uzi Paz, letter to Eitan Gidalizon, March 19, 1998.

95. Paz, *Eretz ha-Tsvi, I,* pp. 76–80.

96. Mendelssohn, *op. cit.,* p. 4.

97. Paz, *Eretz ha-Tsvi, I,* p. 48.

98. Interview with Reuven Ortal, January 13, 1998.

99. Ibid.

100. Israel State Comptroller, *Twenty-First Annual Report,* Jerusalem, 1971, p. 236.

101. Paz, *Eretz ha-Tsvi, I,* p. 70.

102. Asaf Gotfeld, "The Judea and Samaria Team," *Yedion,* no. 45 (1993): p. 6.

103. Dina Weinstein, "Danger: Rabies," *Eretz v' Teva* 47 (1997): 79.

104. Oded Shani, personal communication, May 1997.

105. Interview with D'vora Ben Shaul, Rosh Pina, January 11, 1997.

106. Ibid.

107. "From the Activities of the Nature Reserve Authority," *Biosphera* D (1974): 7.

108. Heinrich Mendelssohn, "The Impact of Pesticides on Bird Life in Israel," *ICBP Bulletin* 11 (1972), pp. 75–104.

109. Interview with Aviva Rabinovich, Kibbutz Kabri, January 11, 1998.

110. Ibid.

111. Israel State Comptroller, *op. cit.*, p. 237.

112. Interview with Danny Yoffe, Tel Aviv, November 15, 1997.

113. Interview with D'vora Ben Shaul, Rosh Pina, January 11, 1997.

114. Interview with Mordechai Ruach, Beit Zayit, September 9, 1997.

115. Interview with Reuven Ortal, January 13, 1998.

116. Israel State Comptroller, *op. cit.*, pp. 234–242.

117. "Sea and Coasts," in *Environmental Quality in Israel, 1979–80* (Jerusalem: Ministry of the Interior, 1981), p. 102; see also Y. Shlezenger, "Monitoring for Oil and Phosphates in the Eilat Coast," *Biosphera* E, no. 10 (1979): 8–9.

118. Interview with Adir Shapira, Ramat Gan, December 18, 1997.

119. Interview with Ariyeh Cohen, Kibbutz Ketura, September 11, 1997.

120. Paz, *Eretz ha-Tsvi, I*, p. 341.

121. Interview with Leah Rabin, Tel Aviv, January 7, 1998; interview with Ariyeh Cohen, Kibbutz Ketura, September 11, 1997.

122. Interview with Dan Perry, Tel Aviv, September 4, 1997.

123. Interview with Alon Galili, Sdeh Boqer, January 3, 1998.

124. Interview with Dan Perry, Tel Aviv, September 4, 1997.

125. Interview with Alon Galili, Sdeh Boqer, January 3, 1998.

126. Interview with Reuven Ortal, January 13, 1998.

127. Interview with Lev Fishelson, Tel Aviv, September 28, 1997.

128. Aviva Rabinovich, personal communication, November 11, 1998.

129. Interview with Uri Safriel, Sdeh Boqer, January 6, 1998.

130. Gabbay, *op. cit.*, pp. 6–7.

131. Daphna Levi, "Gazelle Count in the Arava, October 1991," *Yedion*, (NRA, September 1992), pp. 39–44.

132. Interview with Uzi Paz, Ramat Efal, September 14, 1997.

133. Interview with Lev Fishelson, Tel Aviv, September 28, 1997.

134. Israel State Comptroller, *op. cit.*, p. 238.

135. Hagai Agmon-Snir, "Nature Preservation as Nostalgic Greed," *Teva v'Aretz* 260 (1993): 45.

136. Miriam Vamosh and Mike Livneh, "Effervescent Eden," *Eretz Magazine* (October 1995): 52.

137. Edward Abbey, the conservation novelist and philosopher, felt otherwise, as passionately expressed in his collection of essays, *Desert Solitaire: A Season in the Wilderness* (New York: Ballantine, 1968).

138. National Parks, Nature Reserves, Memorial Sites and National Sites Law, 1992, sec. 37.

139. Interview with Dan Perry, Tel Aviv, September 4, 1997.

140. Shoshana Ashkenazi, "Biosphere Reserves—Their Meaning, Value in Nature Preservation, and Function," *Ecology and Environment* 4, no. 3 (1996):

207–218. The Carmel and Meron areas have also been declared international biospheric reserves, although it is not yet clear what the implications are.

141. Interview with Dan Perry, Tel Aviv, September 4, 1997.

142. "Seven years after I had left the Authority, people who had criticized us approached me and said, 'If it were not for you, there would not be a reserve left today.'" Interview with Adir Shapira, Ramat Gan, December 18, 1997.

143. Noam Meshi, personal communication, May 1997.

144. There is a disagreement among zoologists regarding the taxonomical breakdown of Middle Eastern leopards. Today the predominant opinion among experts is that all subspecies in the Middle East, from Arabia through Sinai-Israel to Turkey and Iran, are considered to belong to a single subspecies: *Panthera pardus saxicolor.* Benny Shalmon, personal communication, July 13, 1998.

145. Ilani, *op. cit.,* pp. 125–128.

146. David Heller and Azariah Alon, "Nature Researchers of the Land of Israel," in *The Plants and Animals of the Land of Israel,* ed. Azariah Alon (Tel Aviv: Ministry of Defense Press, 1990), p. 57.

147. Tristam, *op. cit.,* p. 128.

148. Ilani, *op. cit.*

149. Interview with Giori Ilani, Beer Sheva, September 9, 1997.

150. Interview with Danny Yoffe, Tel Aviv, November 15, 1997.

151. Interview with Giori Ilani, Beer Sheva, September 9, 1997.

152. Ibid.

153. Ibid.

154. Ilani, *op. cit.,* p. 9.

155. Interview with Giori Ilani, Beer Sheva, September 9, 1997.

156. Interview with Danny Yoffe, Tel Aviv, November 15, 1997.

157. Amatzia Shochat, as quoted in Yehoshua Eliash, "Man and Leopard: Can the Two Live Together?", unpublished honors paper, 2001, p. 72.

158. Interview with Giori Ilani, Beer Sheva, September 9, 1997.

159. Shmulik Shapira, "Notices from the Southern Region," *Yedion,* (NRA) no. 28 (1986), p. 5.

160. For instance, in 1987 Ilani reported "Herod" mounting his mother, "Shlomzion." Giora Ilani, "Fauna News," *Teva v'Aretz* 31, no. 3 (December 1998): pp. 28–29.

161. Ariel Ben Avraham, "Desert of Leopards," *Teva v'Aretz* 254 (March 1993): pp. 31–34.

162. Interview with Uzi Paz, Ramat Efal, September 14, 1997.

163. Interview with Lev Fishelson, Tel Aviv, September 28, 1997.

164. Benny Shalmon, personal communication, July 13, 1998.

165. Interview with Dan Perry, Tel Aviv, September 4, 1997.

166. David Quammen, *The Song of the Dodo, Island Biogeography in an Age of Extinction* (Touchstone: New York, 1997), pp. 512–519.

167. Gabbay, *op. cit.,* p. 19.

168. Eitan Glickman, "Vulture Campaign," *Yediot Ahronot—24 Hours,* July 7, 1998, p. 10.

169. Ronen Tal, "A Leopard Who Came to the Movies," *Yediot Ahronot*, December 23, 2001, p. 23.

170. Israel State Comptroller, *op. cit.*, p. 236.

171. Interview with Alon Galili, Sdeh Boqer, January 3, 1998.

172. David Saltz, personal communication, June 14, 1998.

173. Ibid.

174. Yaakov Skolnik, "Back to Nature," *Eretz Magazine*, December 1997, p. 41.

175. Bill Clark, "Desert Compatible," *Eretz Magazine*, January 1995, p. 27.

176. Uzi Paz, "The Hai Bar Reserve," in *Eretz ha-Tsvi, II*, pp. 337–339.

177. Interview with Ariyeh Cohen, Kibbutz Ketura, September 11, 1997.

178. Benny Shalmon, personal communication, July 13, 1998.

179. Interview with D'vora Ben Shaul, Rosh Pina, January 11, 1997.

180. Adnan Budieri, personal communication, June 1996.

181. Interview with Uri Safriel, Sdeh Boqer, January 6, 1998.

182. "If today I was asked to start setting up a Hai Bar, I would be against it. However, the Hai Bars already exist, and unless the reintroductions are carried out, the chances of them ever closing down would be very slim. There is now a long-range program accepted by the Authority that the Hai Bars will shut down in eight years. Another positive point is that these projects raise quite a bit of money and are flagship projects." David Saltz, personal communication, June 14, 1998.

183. Interview with Ariyeh Cohen, Kibbutz Ketura, September 9, 1997.

184. Ibid.

185. Interview with Giora Ilani, Beer Sheva, September 9, 1997.

186. Arje Cohen, *The Pere Back in Nature* (The Netherlands, Valenzuela, 1994).

187. Skolnik, *op. cit.*, p. 45.

188. Shirli Bar David, personal communication, October 29, 1998.

189. Roni King, personal communication, January 10, 1997.

190. Barbara Sofer, "Israel's Wild Side," *Hadassah Magazine*, 1990.

191. Interview with Yossi Sarid, Jerusalem, December 30, 1997.

192. Salah Tarif, who was the Chairman of the Interior and Environment Committee, is Druze. With his long history of displeasure with the NRA over the Mount Meron situation and no love lost for the new Minister of the Environment, Raful Eitan, he grandstanded but eventually gave in.

193. A. Sofer and R. Finkel, *The Mitzpim in the Galilee: Goals, Achievements, Lessons* (Rehovoth: Center for Rural and Urban Settlement, 1986).

194. Interview with Dan Perry, Tel Aviv, September 4, 1997.

195. Zafrir Rinat, "The Minister of Environment Delays Allocation of the Budget for the Nature and Parks Authority," *Ha-Aretz*, October 26, 2001.

196. For instance, the Zipori National Park had three hundred dunams of breathtaking archaeological sites and sixteen thousand dunams of open spaces and indigenous forests that it did little to manage and monitor.

197. Interview with Aaron Vardi, Tel Aviv, December 3, 1998.

198. Levi, *op. cit.*, pp. 39–44.

199. Orit Nevo, "Counting Bats," *Eretz v' Teva* 47 (1997): 17.

200. Elli Elad, "The Departure of the Arava Gazelle," *Ha-Aretz*, October 15, 1993.

201. Interview with Giora Ilani, Beer Sheva, September 9, 1997.

202. Yom-Tov and Mendelssohn, *op. cit.*, p. 58.

203. Orit Nevo, "They Blossom and Are Picked," *Teva v' Aretz* 256 (1993): p. 9.

204. Yom-Tov and Mendelssohn, *op. cit.*, p. 57.

205. Interview with Alon Galili, Sdeh Boqer, January 3, 1998.

206. Ibid.

207. Zafrir Rinat, "Tire Tracks Scar Israel's Dunes," *Ha-Aretz*, April 25, 2000, p. A5.

208. Gabi Baron, "Thais Have Destroyed 90% of the Gazelles in the Golan Heights," *Yediot Ahronot*, April 6, 2000, p. 22.

209. Menahem Abadi, "Deterioration in the Southern Nature Reserves," *Yedion*, no. 44 (1992): pp. 28–34.

210. Ibid.

211. Shmuel Yaakov, "As a Result of the Article," *Yedion*, no. 44 (1992): p. 34.

212. Interview with Reuven Ortal, January 13, 1998.

213. Seffi Ben Yosef, "The Three Percent Problem, The Case of the Off-the-Record-Reserve," *Eretz Magazine* (March 1999): pp. 21–27.

214. Zafrir Rinat, "Ibex without Borders," *Ha-Aretz*, November 23, 1994.

215. Munir Adgham, "Non-Governmental Environmental Organizations in the Gulf of Aqaba-Bordering States: A Current Appraisal," *Protecting the Gulf of Aqaba: A Regional Environmental Challenge*, ed. P. Warburg (Washington: Environmental Law Institute, 1993), pp. 479–489.

216. Yaakov Skolnik, "Running with the Pack," *Eretz Magazine*, September 1996, pp. 28–30.

CHAPTER 7: THE QUANTITY AND QUALITY OF ISRAEL'S WATER RESOURCES

1. Daniel Hillel, *Rivers of Eden: The Struggle for Water and the Quest for Peace in the Middle East* (New York: Oxford , 1994), p. 26.

2. Itzhak Galnoor, "Water Policy Making in Israel," *Water Quality Management Under Conditions of Scarcity: Israel as a Case Study*, ed. Hillel Shuval, (New York: Academic Press, 1980), p. 293.

3. David Ben-Gurion, *Southbound* (Tel Aviv: Ahdut Press, 1956), p. 305.

4. Galnoor, *op. cit.*, p. 295.

5. Ibid., pp. 290–298.

6. Simcha Blass, *Water in Strife and Action* (Givataim: Masada, 1973), pp. 113–115.

7. Ibid., p. 157.

8. Ibid., p. 158.

9. Ibid., p. 159.

10. Joshua Schwarz, "Management of Israel's Water Resources," *Water and Peace in the Middle East,* ed. J. Issac and H. Shuval (Amsterdam: Elsevier, 1994), p. 69.

11. Interview with Menahem Kantor, Kibbutz Ma'agan Michael, November 20, 1997. A cubic meter is equal to 1000 liters, or roughly 250 gallons.

12. Schwarz, *op. cit.,* p. 69.

13. Tom Segev, *Nineteen Forty Nine: The First Israelis* (New York: Free Press, 1986) p. 95.

14. Schwarz, *op. cit.,* p. 70.

15. Yohanan Boneh, "The Historical Development of Underground Water Supply," *Water in Israel: Selection of Articles,* Part I (Tel Aviv: Water Commissioner, 1973), p. 44.

16. Blass, *op. cit.,* pp. 170–171.

17. "Tel Aviv and the surrounding cities were notable for their overpumping. From its first days it never installed water meters, and every citizen paid what he paid, without any connection to how much he consumed." Blass, *op. cit.,* p. 171.

18. Richard Laster, "Legal Aspects of Water Quality," in *Water Quality Management, op. cit.,* p. 279.

19. Water Law (Notice of Agreement), *Yalkut Pirsumim* 842, p. 1206, as reported in Laster, *op. cit.,* pp. 26, 34.

20. Interview with Hillel Shuval, Jerusalem, December 30, 1997.

21. Ibid.

22. M. Lilian and H. Shuval, *Ten Years of Sanitation in Israel* (Jerusalem: Ministry of Health, 1959), pp. 5–6.

23. Ibid., p. 10; see also "Environmental Sanitation," *Health Services* (Jerusalem: Ministry of Health, 1957), p. 12.

24. Blass, *op. cit.,* p. 166.

25. Blass, *op. cit.,* pp. 160–164.

26. Richard Laster, *The Legal Framework for the Prevention and Control of Water Pollution in Israel* (Jerusalem: Ministry of the Interior, 1976), p. 27.

27. Blass, *op. cit.,* p. 165.

28. Blass, *op. cit.,* pp. 165–166.

29. Laster, "Legal Aspects of Water Quality," p. 289.

30. Yehudah Karmon, *The Land of Israel: Geography of the Land and the Region,* 3rd ed. (Tel Aviv: Yavneh, 1976), p. 131.

31. Blass, *op. cit.,* pp. 141–143.

32. Efraim Orni and Elisha Efrat, *Geography of Israel* (Jerusalem: Israel Universities Press, 1973), p. 444.

33. Interview with Menahem Kantor, Kibbutz Ma'agan Michael, November 20, 1997.

34. Orni and Efrat, *op. cit.,* p. 446.

35. Blass, *op. cit.,* p. 168.

36. Lilian and Shuval, *op. cit.,* p. 17.

37. Interview with Hillel Shuval, Jerusalem, December 30, 1997.

38. Based on the geographical origins of the tribe of Dan, the "Dan block" is generally used to denote the Greater Tel Aviv metropolitan area.

39. Lilian and Shuval, *op. cit.*, p. 15.

40. Laster, "Legal Aspects of Water Quality," p. 277.

41. Hillel Shuval, "Public Health Aspects of Waste Water Utilization in Israel," *Proceedings of the 17th Industrial Wastes Conference*, Purdue University, 1962, pp. 650–664.

42. Yael Shoham and Ofra Sarig, *The National Water Carrier* (Sapir: Mekorot, 1995).

43. Walter Clay Lowdermilk, "The Jordan Valley Authority—A Counterpart of TVA in Palestine," in *Palestine, Land of Promise* (New York: Harper and Brothers 1944), pp. 168–179.

44. Blass, *op. cit.*, p. 135.

45. J. B. Hayes, *TVA on the Jordan, Proposals for Irrigation and Hydro-Electric Development in Palestine* (Washington, D.C: Public Affairs Press, 1949).

46. Shoham and Sarig, *op. cit.*, p. 7.

47. Blass, *op. cit.*, p. 161.

48. Interview with Hillel Shuval, Jerusalem, December 30, 1997.

49. Blass, *op. cit.*, p. 167.

50. Howard M. Sachar, *A History of Israel* (New York: Knopf, 1976), p. 519.

51. Shoham and Sarig, *op. cit.*, p. 6.

52. Blass, *op. cit.*, pp. 184–185.

53. Ibid., pp. 188–190.

54. Ibid., p. 190.

55. Ibid., pp. 195–207.

56. Arnon Sofer, "The Relevance of the Johnston Plan to the Reality of 1993 and Beyond," in *Water and Peace in the Middle East*, pp. 110–112.

57. Sachar, *op. cit.*, p. 458.

58. Yehudah Goldsmid, "Water Quality Management of the Israel National Water System," in *Developments in Water Quality Research* (Ann Arbor, Michigan: Ann Arbor Science Publishers, 1971), p. 9.

59. Ibid.

60. Orni and Efrat, *op. cit.*, p. 156.

61. Shoham and Sarig, *op. cit.*, p. 9.

62. Galnoor, *op. cit.*, p. 296.

63. Shoham and Sarig, *op. cit.*, pp. 1, 16–24.

64. Sachar, *op. cit.*, pp. 618–619.

65. Meir Ben Meir, personal interview, Tel Aviv, November 19, 1998.

66. Joyce Whitman, *The Environment in Israel* (Jerusalem: Ministry of the Interior, 1988), p. 136.

67. Shoham and Sarig, *op. cit.*, pp. 8, 11.

68. Hillel Shuval, "Drinking Water and Sewage Disposal," *Public Health* 8, no. 2 (1965): p. 475.

69. Blass, *op. cit.*, pp. 237–248.

70. Goldsmid, *op. cit.*, p. 8.

71. Richard E. Laster, "Lake Kinneret and the Law," *Israel Law Review* 12, no. 3 (1977): p. 296.

72. Blass, *op. cit.*, p. 23.

73. Goldsmid, *op. cit.*, p. 9.

74. B. E. Butterworth, et al. "The Role of Regenerative Cell Proliferation in Chloroform Induced Cancer," *Toxicological Letters* (1995), pp. 23–26.

75. Hillel Shuval, personal communication, November 20, 1998.

76. Yoram Avnimelech, "Irrigation with Sewage Effluents: The Israeli Experience," *Environmental Science and Technology* 27, no. 7 (1993), p. 1280.

77. Interview with Hillel Shuval, Jerusalem, December 30, 1997.

78. Goldsmid, *op. cit.*, p. 8.

79. Nessen's contribution to New York City's water system was substantial enough that the laboratory at one of the city's eighteen reservoirs was named in his honor. Cristina Manos, City of New York Department of Environmental Protection, personal communication, October 9, 1998.

80. Today aluminum sulfate is added at the Eshkol reservoir; it adsorbs to the sediments and turns them into larger flocs that settle in the settlement basin. Chlorine dioxide and chloramine are added to the water prior to its introduction into the municipal water systems. Interview with Hillel Shuval, Jerusalem, December 30, 1997.

81. Goldsmid, *op. cit.*, p. 9.

82. Shoham and Sarig, *op. cit.*, pp. 26–27.

83. EcoPeace, *Dead Sea Challenges: Final Report*, (Jerusalem: EcoPeace, 1996).

84. Whitman, *op. cit.*, p. 136.

85. C. Serruya and T. Berman, "The Evolution of Nitrogen Compounds in Lake Kinneret," in *Developments in Water Quality Research* (Ann Arbor, Mich.: Ann Arbor Science Publishers, 1971), pp. 73–78.

86. Goldsmid, *op. cit.*, p. 9.

87. Blass, *op. cit.*, p. 185.

88. "For years I had stood in opposition to the water legislation that the Mandate was about to legislate. It took time until I could get over my own principled hostility to a Water Law. " Ibid.

89. *Sefer ha-Hokim*, 1959, p. 169.

90. Water Law, Notice of Agreement, *Yalkut Pirsumim*, 842, p. 1206.

91. See generally: Laster, *The Legal Framework*.

92. Interview with Menahem Kantor, Kibbutz Ma'agan Michael, November 20, 1997.

93. Ibid.

94. Laster, "Legal Aspects of Water Quality," p. 280.

95. Dan ha-Levy, "Mekorot versus the Water Commissioner," *Davar*, November 22, 1994.

96. Daniel Hillel, "Water," in *The Negev: Land, Water and Life in a Desert Environment* (New York: Praeger, 1982), p. 32; see also Rachel Carson, *The Sea around Us* (New York: Oxford University Press, 1950).

97. Ben-Gurion, *op. cit.*, pp. 305–306.

98. Oded Lipshitz, "Agriculture, Water, and Much Adrenaline," *El ha-Mishmar, Musaf,* August 25, 1992.

99. Galnoor, *op. cit.*, p. 299.

100. Shoshana Gabbay, *The Environment in Israel* (Jerusalem: Ministry of the Environment, 1994), p. 21.

101. Zafrir Rinat, "Altering Mother Nature's Recipe: To Get Water, Just Add Salt," *ha-Aretz,* March 23, 1999.

102. Hillel Shuval, "Quality Management Aspects of Wastewater Reuse in Israel," *Water Quality Management,* p. 213.

103. Uri Marinov, "How Israel Handles the Environment and Development," *Environmental Science and Technology* 27, no. 7 (1993), p. 1253.

104. Interview with Hillel Shuval, Jerusalem, December 30, 1997.

105. Hillel Shuval, "The Problem of Sewage and Waste Discharges in Israel," lecture to sanitary engineers, February 2, 1967, Tzrifin, Israel (manuscript available with author).

106. Ibid.

107. Interview with Hillel Shuval, Jerusalem, December 30, 1997.

108. Yaakov Zak, "Water Quality in Israel's Streams," *Biosphera* F, no. 9 (June 1977): 4; see also Yaakov Raveh, "Sources of Pollution in the Streams of Israel," *Biosphera* 5 (1971): 9–12.

109. S. Alfi, "The Alexander Stream: Conclusions of a Field Study for the Ministry of Health," *Biosphera* 72, no. 7 (1972): 3.

110. Avnimelech, *op. cit.*, p. 1279.

111. Shuval, "Public Health Aspects of Waste Water Utilization in Israel," pp. 651–652.

112. Alberto M. Wachs, "The Outlook for Wastewater Utilization in Israel," *Developments in Water Quality Research,* ed. Hillel Shuval (Ann Arbor, Mich.: Ann Arbor Science Publishers, 1971), pp. 109–111.

113. Uri Marinov and Eitan Harel, *The Environment in Israel* (Jerusalem: National Council for Research and Development, 1972), p. 31.

114. Avnimelech, *op. cit.*

115. Laster, "Legal Aspects of Water Quality," p. 277.

116. Whitman, *op. cit.*, pp. 140–141.

117. Emanuel Idelovich, "Can We Get Water at Drinking Water Quality from the Dan Plant?" *Biosphera* E, no. 1 (November 1975): 14–15, and response by Hillel Shuval, p. 15.

118. Batiah Yadin, *Wastewater Reuse* (Tel Aviv: Mekorot, 1993).

119. Whitman, *op. cit.*, p. 142.

120. Hillel Shuval, "Waste Water Utilization in Agriculture and Hygenic Problems," *Journal of the Engineers and Architects Association of Israel* 9 (August 1951), pp. 38–40.

121. Hillel Shuval, "Waste Water Utilization in Israel," in *Proceedings of an International Seminar on Soil and Water Utilization* (South Dakota State College, Brookings, 1962), p. 40.

122. Badri Fattal and Hillel Shuval, "Historical Prospective Epidemiological Study of Wastewater Utilization in Kibbutzim in Israel, 1974–77," in *Developments in Arid Zone Ecology and Environmental Quality,* Ed. Hillel Shuval (Philadelphia, Pa.: Balaban ISS, 1981), pp. 333–343.

123. J. Cohen et al., "The Endemicity of Gastrointestinal Infections," in *Public Health* (Jerusalem: Ministry of Health, 1971), p. 3.

124. B. Gerichter and D. Cahan, "Laboratory Investigations During the Cholera Outbreak in Jerusalem and Gaza, 1970," *Public Health* (Jerusalem: Ministry of Health), pp. 26–35. T. A. Schwartz, "The Jerusalem Cholera Outbreak: The Course of the Epidemiological Investigation," ibid., p. 13.

125. Reuben Klasmer, "Primary Investigation and Chemoprophylaxis of Cholera," *Public Health* (Jerusalem: Ministry of Health, 1971), p. 19.

126. Z. Imre et al., "The Cholera Outbreak with *Vibrio cholerae* in the Gaza Area in 1970," *Public Health* (Jerusalem: Ministry of Health 1971), pp. 39–40.

127. Laster, "Legal Aspects of Water Quality," p. 274.

128. Ministry of Health Public Health Principles, 1981, *Kovetz Takanot,* no. 1357, p. 718.

129. T. Naff and R. C. Matson, *Water in the Middle East: Conflict or Cooperation* (Boulder, Colo.: Westview Press, 1984).

130. David Salick, "A Lebanese Time Bomb," *Green, Blue and White* (August–September 1998): 14–15.

131. Arie Issar, "Fossil Water under the Sinai-Negev Peninsula," *Scientific American* 253 (1985, 107–108): see also E. M. Adar et al., "Quantitative Assessment of the Flow Pattern in the Southern Arava Valley (Israel) by Environmental Tracers and a Mixing Cell Model," *Journal of Hydrology* 136 (1992): 333–352.

132. Cardiological problems are primarily associated with the sodium element in salts. See *Drinking Water and Health,* vol. 3 (Washington, D.C.: National Academy of Science Press, 1980), pp. 243–247.

133. "We pour about two million tons of salt into our water each year. There's a billion cubic meters of water with a hundred milligrams per liter of chlorine. And everyone adds 100 grams of sewage a day. Getting the salt out. That is our real environmental problem." Interview with Menahem Kantor, Kibbutz Ma'agan Michael, November 20, 1997. Other experts dispute this level of salinity infiltration, arguing that the amount is closer to one hundred thousand tons. Dan Zaslavsky, personal communication, September 22, 1998.

134. Baruch Weber, *Reduction of Salinity in Urban Effluents in Israel* (Jerusalem: Ministry of the Environment, 1996), p. 32.

135. Avnimelech, *op. cit.,* p. 1279.

136. Interview with Dan Zaslavsky, Haifa, September 29, 1997.

137. Ibid.

138. Ibid.

139. Israel State Comptroller, *Report on Water Management in Israel* (Jerusalem: State Comptroller's Office, 1990), pp. 22–23.

140. Ministry of the Environment, "Water Quality," in *Environmental Quality in Israel,* nos. 17–18, *1990–91* (Jerusalem: Ministry of the Environment, 1991), pp. 269–270.

141. Alon Rosenthal (Tal), "Nitrates in Drinking Water," *Harnessing Science for Environmental Regulation,* ed. John Graham (New York: Praeger, 1988), pp. 159–179.

142. S. Wago, "Two Cases of Methemoglobinemia in Infants," *Ha-Refuah* 5, no. 2 (1955): 35–36.

143. M. Morales-Suarez-Varela et al., "Nitrates and Stomach Cancer," *Cancer Detection and Prevention* 20 (1996): 5.

144. Shuval, "The Problem of Sewage and Waste Discharges in Israel."

145. Interview with Hillel Shuval, Jerusalem, December 30, 1997.

146. Abraham Mercado, "The Coastal Aquifer in Israel: Some Quality Aspects of Groundwater Management," in *Water Quality Management under Conditions of Scarcity: Israel as a Case Study,* ed. Hillel Shuval (New York: Academic Press, 1980), p. 99.

147. Hillel Shuval, "The Problem of Nitrates in Drinking Water in Israel," VIBAS conference proceedings, Tel Aviv, 1972.

148. C. Soliternick, Y. Kanovich, and Y. Shevach, "Sources of Ground Water Pollution by Nitrogen Compounds," reprinted in *Biosphera 72,* no. 9 (1972): 1–6.

149. Shuval, "Waste Water Utilization in Israel."

150. Y. Kanovich and D. Blank, "Groundwater Pollution by Nitrates in Israel," *Biosphera* I (1973): 7–8.

151. Hillel Shuval, "Utilization of Sewage Water in Agriculture and Hygiene Problems," manuscript available with author.

152. Meir Ben Meir, personal interview, Tel Aviv, November 19, 1998.

153. Lazi Shelef, personal communication, November 26, 1998. Nitron Purification Systems, which built Israel's first major facility of this type in Rishon L'Tzion, reports nitrate reductions from 90 to 40 milligrams per liter, with an additional benefit of cutting the chloride levels in half—from 120 to 60 milligrams per liter,"Technological Advantages of Ultra Filtration," Nitron Purification Systems promotional material, 1999.

154. Interview with Menahem Kantor, Kibbutz Ma'agan Michael, November 20, 1997.

155. Alon Tal, "Enforceable Standards to Abate Agricultural Pollution: The Potential of Regulatory Policies in the Israeli Context," *Tel Aviv Studies in Law* 14 (1998), pp. 223–286.

156. Joyce Starr, "The Quest for Water, from Biblical Times to the Present," *Environmental Science and Technology* 27, no. 7 (1993): p. 1265.

157. Blass, *op. cit.,* p. 330. A talmudic story (Mesechta Gittin 56b) relates that after the Roman Emperor Titus destroyed the Temple in Jerusalem, a mosquito flew into his nose and entered his brain, destroying it from within over seven years.

158. Shaul Arlosoroff, "Managing Scarce Water: Recent Israeli Experience," *Between War and Peace: Dilemmas of Israeli Security* (London: Frank Çass, 1996), p. 242.

159. Hillel, *op. cit.*, p. 221.

160. Amnon Greenberg, Southern Arava Agricultural Research Center, personal communication, 1998.

161. Blass, *op. cit.*, p. 351.

162. Netafim web site, http://www.netafim.co.il.

163. *Netafim*, promotional film (undated).

164. Ori Nir, "In California the Grapes—and the Grass—Are Greener," *Jerusalem Report*, July 24, 1997, pp. 30–31.

165. Hillel, *op. cit.*, p. 221.

166. Avnimelech, *op. cit.*, p. 1280.

167. Arlosoroff, *op. cit.*, p. 246.

168. Sandra Postel, *Last Oasis: Facing Water Scarcity* (Washington D.C.: World Watch, 1992).

169. *Sefer ha-Hokim*, 1972, p. 8.

170. Rachel Adam and Uri Marinov, "Problems in Enforcing the Water Law, 1959," *Water and Irrigation* 305 (1992): p. 235.

171. Laster, "Legal Aspects of Water Quality," pp. 272–273.

172. Adam Teva V'din (Israel Union for Environmental Defense), *Annual Report, 1995* (Tel Aviv: Adam Teva V'din, 1996), p. 5.

173. Zadok Yehezkel and Anat Tal-Shir, "The Navy Leaves the Water," *Yediot Ahronot*, July 27, 1001, pp. 1,3.

174. Water Regulations (Prohibition of Hard Detergents), 1974, *Kovetz Takanot*, no. 3208, p. 1621.

175. M. Ravid, "Hard Detergents Are Marketed in Israel in Contravention to the Law," *Biosphera* 6, no. 12, (1977): 1–3.

176. Avnimelech, *op. cit.*, p. 1279.

177. Mercado, *op. cit.*, p. 134.

178. Leah Muszkot, "First Results of Research That Examined Groundwater Point to Pollution by Hazardous Organic Materials," *Biosphera* (October 1988): 15; see also Leah Muszkot et al., "Large Scale Contamination of Deep Groundwaters by Organic Pollutants," *Advances in Mass Spectrometry* 11B (1990): 1628.

179. Leah Muszkot, "Groundwater Quality, Problems, and Solutions," *Our Shared Environment—The Conference 1994*, ed. R. Twite and R. Menczel (Jerusalem: Israel-Palestine Conference for Research and Information, 1995), pp. 70–86.

180. Muszkot, "First Results," p. 15.

181. Israel Standards Institute, *Gilayaon Hadracha, No. 193*, as reported in "The Ministry of Health, 1959–1964" *Public Health* 8, no. 2, (1965): 474–475.

182. Lilian and Shuval, *op. cit.*, p. 11.

183. Shuval, "Drinking Water and Sewage Disposal," p. 475.

184. Lilian and Shuval, *op. cit.*, p. 12.

185. The Water Regulations (The Sanitary Quality of Drinking Water), 1974, Regulation 7(e), *Kovetz Tekanot*, no. 3117, p. 556.

186. *Sefer ha-Hokim*, 1962, p. 96.

187. Uri Marinov and Deborah Sandler, "The Status of Environmental Management in Israel," *Environmental Science and Technology* 27, no. 7 (1993): 1258.

188. Alon Tal, "Six Reasons Behind Israel's Environmental Crisis," *Politica* 47 (1993): 48.

189. Laster, "Lake Kinneret and the Law," p. 303.

190. C. Serruya, "The Nutrient Load of Lake Kinneret," in Merinov and Harel, *op. cit.*, p. 41.

191. Laster, "Lake Kinneret and the Law," pp. 307–308.

192. Ibid., p. 303.

193. R. J. Davis, "Investigation of the Pollution Problems of Lake Kinneret," 1971, as cited in Laster, "Lake Kinneret and the Law," note 17.

194. Richard Laster, personal communication, January 30, 2002.

195. Laster, "Lake Kinneret and the Law," p. 311.

196. Whitman, *op. cit.*, p. 135.

197. Gabbay, *op. cit.*, pp. 27–28.

198. Rachel Adam, personal communication, July 13, 1998.

199. Israel State Comptroller, *op. cit.*

200. Israel Water Commission, *The Chemical Quality of Ground Water in the Coastal Aquifer of Israel, Report No. 76/1* (Tel Aviv: Ministry of Agriculture, 1976) and accompanying letter to Shaul Alazaroff, Deputy Water Commissioner, from authors Yaakov Kanfi and Daniel Ronen.

201. Hillel Shuval, "The Impending Water Crisis in Israel," *Developments in Arid Zone Ecology and Environmental Quality*, ed. Hillel Shuval (Philadelphia: Balaban ISS, 1981), pp. 101–114.

202. Israel State Comptroller, *op. cit.*, p. 9.

203. Oded Lipshitz, "Agriculture, Water and Much Adrenaline," *El ha-Mishmar, Musaf*, August 25, 1992.

204. Based on data from the 1980s, two Canadian experts wrote, "The State of Israel is commonly regarded as presenting a model of sound water management. The reality is, however, different from the image." Stephen C. Lonergan and David B. Brooks, *The Economic, Ecological and Geopolitical Dimensions of Water in Israel* (Victoria, British Columbia: Center for Sustainable Regional Development, 1992).

205. Hillel, *op. cit.*, p. 39.

206. *Pocket World in Figures: 2000 Edition* (London: Economist, 1999), p. 1.

207. Interview with Dan Zaslavsky, Haifa, September 29, 1997.

208. Ibid.

209. Ibid.

210. Dan Zaslavsky, personal communication, September 22, 1998.

211. Interview with Dan Zaslavsky, Haifa, September 29, 1997.

212. Interview with Yitzhak Shamir, Tel Aviv, November 24, 1997.

213. David Rudge, "Sewage Flows into Kinneret," *Jerusalem Post*, May 3, 1992.

214. H. Nahtomi, "Improvement in the Water Levels in the Aquifer and Southern Region," *Mabat l'Kalkalah v'l'Chevrah*, November 11, 1994.

215. Interview with Dan Zaslavsky, Haifa, September 29, 1997.

216. David Amiran, *Rainfall and Water Policy in Israel: Series of Dry and Rainy Years and Their Implications for Policy* (Jerusalem: Jerusalem Institute for Israel Studies, 1994).

217. See *Ha-Aretz* newspaper special edition on the water crisis, July 15, 2001, available electronically at: http://www2.haaretz.co.il/special/water-e/.

218. Zafrir Rinat, "Ex-Water Chiefs Report: Treasury Was Opposed to Desalination Plans," *Ha-Aretz*, July 10, 2001.

219. Shimon Tal, personal communication, Tel Aviv, January 1, 2002.

CHAPTER 8: ISRAEL'S URBAN ENVIRONMENT, 1948–1988

1. Golda Meir, *My Life* (London: Weidenfeld and Nicolson, 1975), pp. 315–316.

2. "The Jewish stamp on the face of the abandoned Galilee will ultimately be placed through Jewish rural settlements, rather than a large population in a single city." Yigael Alon, *A Nation and Its Land* (Tel Aviv: Am Oved, 1963), p. 79.

3. Yigael Alon, "International Environment Day," *Biosphera* 73, no. 7 (1973).

4. A. Donagi, "Sources of Air Pollution in Israel," *Humans in a Hostile Environment*, Conference Proceedings (Jerusalem: VIBAS, 1971), p. 18, Table 3.

5. Ibid., p. 13.

6. Tirza Yuval, "2020 Vision," *Eretz Magazine*, January–February 1995, p. 43.

7. Elisha Efrat, "Fathers and Sons in the Physical Planning of Israel," *Karka* (1995).

8. Ariyeh Sharon, *Physical Planning for Israel* (Jerusalem: Israel Government Printer, 1951).

9. "Immigrant and transit camps, housing projects and settlements, all planned and built in haste, will remain as social and economic blots on the landscape and may be succeeded by even worse blemishes later on." Ariyeh Sharon, *op. cit.*, reprinted in "Planning for the New State," *Eretz Magazine*, January 1995, p. 51.

10. Meron Benvenisti, *Conflicts and Contradictions* (New York: Eshel, 1986), pp. 59–60.

11. Adam Mazor, *Israel in the Years of* 2000 (Tel Aviv: Haim U'Sviva, 1993), p. 4.

12. Efrat, *op. cit.*

13. Ibid.

14. Ephraim Shlain and Eran Feitelson, *The Formation, Institutionalization and Decline of Farmland Protection Policies in Israel* (Jerusalem: Floersheimer Institute, 1996), pp. 10–11.

15. P. Rosen, "The Breeding of Flies in Tel Aviv Municipal Garbage," *Tavruah*, August 1957.

16. Uri Marinov and Eitan Harel, *The Environment in Israel* (Jerusalem: National Council for Research and Development, 1972), p. 54.

17. Hillel Shuval, "Composting Municipal Garbage in Israel," unpublished manuscript.

18. M. Lilian and H. Shuval, "Ten Years of Sanitation in Israel" (Jerusalem: Ministry of Health, 1959), p. 20.

19. Shuval, *op. cit.*

20. Yossi Leshem and Nehama Ronen, "Removing Hiriya Garbage Dump, Israel–A Test Case," presented to the International Bird Strike Committee Conference, Slovakia, September 14, 1998.

21. Hillel Shuval, "Israel Is Tackling Her Composting with a Will," *Municipal Engineering* (March 6, 1959): 241.

22. The Hiriyah plant used a windrow process. Shuval, "Israel Is Tackling Her Composting," p. 241.

23. Shuval, "Composting Municipal Garbage in Israel."

24. Shuval, "Israel Is Tackling Her Composting," 241.

25. Marinov and Harel, *op. cit.*, p. 54.

26. Lilian and Shuval, *op. cit.*, p. 19.

27. Richard E. Laster, "Lake Kinneret and the Law," *Israel Law Review* 12, no. 3 (1977): 303.

28. Mishna Baba Batra 2:9.

29. Moses Maimonides, "The Preservation of Youth," in Daniel Fink, "The Environment in Halacha," in *Judaism and Ecology* (New York: Hadassah, 1993), p. 42.

30. Lilian and Shuval, *op. cit.*, p. 22.

31. "Dr. Shimon (Zigfried) Kanovich," *Encyclopedia of Founders of the State* (Tel Aviv: Tedhar, 1975), p. 4068.

32. Ernest Katin and Mordechai Virshubski, "Environmental Law and Administration in Israel," *Tel Aviv University Studies in Law* 1 (1975): 210.

33. *Divrei ha-Knesset*, 1960, p. 580.

34. Proposed Prevention of Nuisances Law, 1961, sec. 6(2), *Proposed Laws, 1961*, p. 67.

35. Ibid., sec. 10.

36. Prevention of Nuisances Law, sec. 11, 1961, *Sefer ha-Hokim* 32, p. 58.

37. Prevention of Nuisances Regulations (Vehicular Air Pollution) (Hartridge Test Standard), 1963, *Kovetz Takanot*, no. 1506, p. 92. A Hartridge test measures the opacity of smoke emitted from a tailpipe, with a relatively high degree of precision.

38. Alex Donagi, "Air Pollution Prevention in Israel," *Public Health* 8, no.2 (1965), p. 506.

39. Ibid.

40. Lilian and Shuval, *op. cit.*, p. 18.

41. Yizhak Zamir, personal communication, January 13, 2002.

42. Bagatz 295/65, *Hillel Oppenheimer and Others v. Ministers of Interior and Health*, 1966, PADI 20 I, 309–338.

43. Prevention of Nuisances Regulations (Air Quality), 1971, *Kovetz Takanot*, no. 1633, p. 380.

44. "Hope for Recycling: Supreme Court Orders Ministry of Environment to Prepare Recycling Regulations within Three Months," *Law and Environment* (Fall, 1997), p. 3.

45. Josef Tamir, *Haver Knesset* (Jerusalem: Ahiabar, 1987).

46. Victor Shem-Tov, *Divrei ha-Knesset*, May 31, 1971, p. 2559.

47. Y. Yaron, "Laws Disregarded," *Israel Law Review* 6 (1971): 188.

48. *Sefer ha-Hokim, 1965*, p. 307.

49. For a description of Israel's planning system and the environment, see Valerie Brachya, "Environmental Management through Land Use Planning," *Our Shared Environment*, ed. Robin Twite and Jad Isaac (Jerusalem: Israel-Palestine Center for Research and Information, 1994), pp. 339–354.

50. Israel Electric Company study cited in speech of Yehudah Sha'ari, *Divrei ha-Knesset*, August 8, 1967, p. 2905.

51. Yehudah Sha'ari, *Divrei ha-Knesset*, August 8, 1967, p. 2905.

52. Josef Tamir, *Divrei ha-Knesset*, August 8, 1967, p. 2892.

53. Richard Laster, "Reading D: Planning and Building or Building and Then Planning," *Israel Law Review* 8 (1973): p. 482.

54. Interview with Yedidyah Be'eri, Tel Aviv, September 5, 1997.

55. Tamir, *Haver Knesset*, p. 360.

56. Ibid., pp. 361–362.

57. Tel Aviv Power Station Law, 1967, *Sefer ha-Hokim*, 1967, p. 141, sec. 1.

58. Ibid.

59. Laster, "Reading D," p. 492.

60. Shlomo Lorenz, *Divrei ha-Knesset*, August 8, 1967, p. 2893.

61. Tamir, *Divrei ha-Knesset*, August 8, 1967, p. 2891.

62. *Divrei ha-Knesset*, August 8, 1967, p. 2907.

63. Tamir, *Haver Knesset*, p. 358

64. Laster, "Reading D," p. 488.

65. Ibid., p. 489.

66. Ibid.

67. Lisa Perlman, "Threat from the Air," *Jerusalem Post Magazine*, February 24, 1989, p. 4.

68. Tamir, *Haver Knesset*, p. 358.

69. Laster, "Reading D," p. 490.

70. Lev Fishelson, personal communication, 1994.

71. Yosef Waksman, "Frutarom Factory Expansion Rejected," *Maariv*, February 14, 1973.

72. G. Kelner, "The Epidemiology of Cancer in Israel," *Public Health* (in Hebrew) 8, 1965, pp. 115–129.

73. Meir, *op. cit.*

74. "The U.N. Convention on the Human Environment, Stockholm," *Biosphera* 72, no. 7 (1972): 12.

75. Uri Marinov, personal communication, December 27, 1998.

76. "Eban in Stockholm: Air Pollution Endangers Humanity," *Yediot Ahronot*, June 7, 1972.

77. Interview with Uri Marinov, Winston House, Suffolk, England, September 19, 1997.

78. Ibid.

79. "The Life Science Branch of the National Council for Research and Development," *Biosphera* 70, no. 1 (May, 1970): 3.

80. "Editor's Word," *Biosphera* 71, no. 2 (June 1971): 2.

81. Ibid.

82. "I represented the Knesset at VIBAS. Except for academic arguments, the institution didn't contribute a thing. From its start it was a dud." Tamir, *Haver Knesset*, p. 169.

83. Protocol No. 12, the Kinneret Committee, p. 1, September 17, 1971, as quoted in Laster, "Lake Kinneret and the Law," p. 311.

84. "Haim Kubersky, Director General of the Ministry of the Interior, argued that "it is impossible in the foreseeable future to consider any proposal for a government ministry on the subject." "The Debate about Government Organization for Addressing Environmental Quality Subjects That Took Place in VIBAS," *Biosphera* 72, no. 11 (1972): 8.

85. Ibid.

86. "The Environmental Protection Authority," *Biosphera* 72, no. 11 (1972): 1–5.

87. Interview with Uri Marinov, Winston House, Suffolk, England, September 19, 1997.

88. Interview with Josef Tamir, Tel Aviv, July 1, 1997.

89. Government Decision No. 563, March 20, 1973.

90. Interview with Josef Tamir, Tel Aviv, July 1, 1997.

91. Interview with Uri Marinov, Winston House, Suffolk, England, September 19, 1997.

92. Uri Marinov, "The Service to the Ministry of Interior," *Biosphera E* (1 November 1975): 1.

93. Interview with Uri Marinov, Winston House, Suffolk, England, September 19, 1997.

94. Interview with Valerie Brachya, Jerusalem, February 17, 1999.

95. Ibid.

96. Interview with Richard Laster, Jerusalem, September 9, 1997.

97. Interview with Uri Marinov, Winston House, Suffolk, England, September 19, 1997.

98. Ibid.

99. Marinov, *op. cit.*, p. 2.

100. Ibid., p. 1.

101. Richard Laster, personal communication, November 18, 1998.

102. Interview with Uri Marinov, Winston House, Suffolk, England, September 19, 1997.

103. Ibid.

104. Interview with Richard Laster, Jerusalem, September 9, 1997.

105. Uri Marinov and Deborah Sandler, "The Status of Environmental Management in Israel," *Environmental Science and Technology* 27, no. 7 (1993): 1257.

106. Interview with Uri Marinov, Winston House, Suffolk, England, September 19, 1997.

107. Environmental Protection Service, *Environmental Quality in Israel, 1976* (Jerusalem: Ministry of the Interior, 1977), p. 69.

108. Planning and Building Regulations (Environmental Impact Statements), 1982, *Kovetz Takanot,* no. 4307, p. 502.

109. "Environmental Impact Statements," *Environmental Quality in Israel,* no. 11, *1983–84,* (Jerusalem: Ministry of the Interior, 1985): 259–261.

110. "Environmental Impact Assessment in Israel," *Israel Enviromental Bulletin* (Winter 1998): 23.

111. Interview with Valerie Brachya, Jerusalem, February 17, 1999.

112. Ibid.

113. Explanatory Notes, *National Master Plan for Garbage Disposal: TAMA 16* (Jerusalem Ministry of the Interior, 1980), p. ii.

114. Ibid.

115. The Master Plan explains, "In garbage burial sites, it will be possible to establish facilities for recycling of raw materials and facilities for collecting methane gas. Garbage disposal sites of a different type will allow the burning of trash in incinerators. . . . In certain cases, the sites for treating garbage will be attached to the transfer stations, as their primary activity is concentrated around the subject of sorting." *National Master Plan for Garbage Disposal: TAMA 16* (Jerusalem: Ministry of the Interior, 1989), p. 6.

116. The Israel State Comptroller took the government to task for this inordinate delay. See Israel State Comptroller's Report.

117. Danny Morganstern, "A Waste, Stinking to the Sky," *Ksafim,* March 15, 1989, pp. 25–29.

118. Ami Etinger, "Ben-Gurion Airport Will Close on Tuesday for Two Hours Because of Hiriyah," *Maariv,* January 16, 1998, p. 9.

119. D'vora Ben-Shaul, "Solutions That Go to Waste," *Jerusalem Post,* January 2, 1998, p. 15.

120. *National Master Plan for Garbage Disposal: TAMA 16* (Jerusalem: Ministry of the Interior, 1989), pp. 36–37.

121. Yossi Inbar, Deputy Director, Ministry of the Environment, lecture at the Third Conference of the Professional Forum for Solid Waste Management, Maaleh-ha Chamishah, January 9, 2002.

122. Micki Haran, lecture, Conference on Brownfields in Israel, Technion, December 1999.

123. Elli Liederman, Beer Sheva Environmental Unit, personal communication, January 4, 1998.

124. Elli Liederman, personal communication, January 13, 2002.

125. Y. N. Narkis and Y. Kornberg, "The Problem of Hazardous Waste in Israel," *Ecosystem Stability, Proceedings of the Fourth International Conference of the Israel Society for Ecology and Environmental Quality Sciences, Jerusalem, June* 4–8, 1989. Reprinted in *Israel Environmental Bulletin*, 1980, p. 17.

126. Joyce Whitman, *The Environment in Israel* (Jerusalem: Ministry of the Interior, 1988), pp. 208–209.

127. Ibid.

128. Avi Taub, presentation, Ramat Hovav, December 7, 1997.

129. Narkis and Kornberg, *op. cit.*

130. Licensing of Businesses Regulations (Disposal of Wastes from Hazardous Materials), 1990, *Kovetz Takanot*, no. 5298, p. 22.

131. "An inspection of the site revealed that that it runs in contravention to all the procedures.... These inappropriate activities create environmental risks that arise from leaks of wastes requiring immediate, high priced restoration activities." Quoted in Elli Elad, "Disposal of Wastes at Ramat Hovav Is Managed Negligently, Endangering the Environment," *Ha-Aretz*, June 4, 1991.

132. Avi Taub, presentation, Ramat Hovav, December 7, 1997.

133. Environmental Protection Service, *Environmental Quality in Israel, 1976* (Jerusalem: Ministry of the Interior, 1977), Introduction.

134. "Proposal to Create Environmental Quality Unit in the Local Authorities," *Biosphera* 6, no. 5 (1977): 6.

135. Uri Marinov, personal communication, December 27, 1998.

136. Interview with Uri Marinov, Winston House, Suffolk, England, September 19, 1997.

137. Uri Marinov, "Introduction," in *Environmental Quality in Israel, 1977* (Jerusalem: Ministry of Interior, 1978), p. d.

138. Fred Pearce, "Dead in the Water," *New Scientist*, February 4, 1995, pp. 26–31.

139. David Greenberg and Richard Helman, *Israel in the Mediterranean Ecosystem* (Jerusalem: Ha-Makor, 1978), p. 15.

140. Evangelos Raftopoulos, "The Mediterranean Action Plan: Appraisal of a Model for Regional Cooperation," in *Protecting the Gulf of Aqaba: A Regional Environmental Challenge*, ed. Philip Warburg (Washington: Environmental Law Institute, 1993), pp. 315–347; see generally: www.unepmap.org.

141. Convention for the Protection of the Mediterranean Sea against Pollution, February 16, 1976, 15 *ILM*, p. 290.

142. Marinov and Sandler, *op. cit.*, p. 1259.

143. See Alon Tal, "Preventing Pollution from Ports and Maritime Activity in the Gulf of Aqaba: Current Practices in Israel," in *Protecting the Gulf of Aqaba: A Regional Environmental Challenge*, ed. Philip Warburg (Washington: Environmental Law Institute, 1993), pp. 264–266.

144. The Abu Rodeis fields also supply the SOPED Suez-Mediterranean pipeline in Egypt. Danny Ben-Tal, "Peace and Pollution," *Green Pages at Ariga,* 1992.

145. Protocol Concerning Cooperation in Combating Pollution for the Mediterranean Sea by Oil and Other Harmful Substances in Cases of Emergency, February 16, 1976, 15 *ILM,* p. 306.

146. *Palestine Gazette,* no. 612, July 16, 1936, sec. 216–234.

147. Prevention of Seawater Pollution by Oil Ordinance (New Version), 1980, sec. 13–17, *Sefer ha-Hokim,* p. 630.

148. Interview with Uri Marinov, Winston House, Suffolk, England, September 19, 1997.

149. "Establishing the Marine Pollution Prevention Department in the Environmental Protection Service," in *Environmental Quality in Israel,* no. 11 (Jerusalem: Ministry of the Interior, 1985), p. 147.

150. Regulations for Prevention of Seawater Pollution by Oil, 1983, *Kovetz Takanot,* p. 1973.

151. "Establishing the Marine Pollution Prevention Department in the Environmental Protection Service," p. 150.

152. Marinov and Sandler, *op. cit.,* p. 1258.

153. Elik Adler, "The Marine Pollution Prevention Department: Review of the Past Two Years of Activities," *Biosphera* 18, no. 9 (1989): 16.

154. Whitman, *op. cit.,* p. 159.

155. Tal, "Preventing Pollution," pp. 260–261.

156. Elik Adler, personal communication, 1995.

157. Whitman, *op. cit.,* p. 162.

158. Tamar Ben-Yeshayahu, *Environmental Quality in Israel,* no. 16 (Jerusalem: Ministry of the Environment, 1990), p. 338.

159. Interview with Uri Marinov, Winston House, Suffolk, England, September 19, 1997.

160. Prevention of Nuisances Law, sec. 8, 1961, *Sefer ha-Hokim* 32, p. 58.

161. Whitman, *op. cit.,* p. 185.

162. Katin and Virshubski, *op. cit.,* p. 212.

163. *Engineer A. Feranio and the Haifa Public Council for Protecting Environmental Quality versus the Minister of Health and the Minister of the Interior,* Bagatz 372/71, *Piskei Din* 26 I 1972, pp. 809–811.

164. Public Health Regulations (Pollution Emissions from Vehicles), 1980, *Kovetz Takanot,* p. 1244.

165. The Ringleman standard defines air violations according to a scale of "opaqueness" or "blackness." Inspectors visually contrast the darkness of emissions to the levels of gray on their chart. Although it is criticized for being a very blunt instrument, it is also easy to implement. Public Health Regulations (Pollution Emissions from Vehicles), 1980, *Kovetz Takanot,* p. 1244. Reg. 2.

166. Announcement on Approval of a Decision under the Basic Law, the Government, May 4, 1982, reprinted in *Environmental Quality in Israel,* no. 11 (Jerusalem: Ministry of the Interior, 1985), p. 286.

167. Prevention of Nuisances Regulations (Air Quality) 1971, *Kovetz Takanot*, no. 2783, p. 380.

168. Prevention of Nuisances Regulations (Air Quality) 1992, *Kovetz Takanot*, no. 5435, p. 972.

169. "Air Quality," in *Environmental Quality in Israel*, no. 10 (Jerusalem: Ministry of the Interior, 1985), p. 14.

170. Interview with Ruth Rotenberg, Tel Aviv, December 11, 1997.

171. Ibid.

172. Ibid.

173. Interview with Bernanda Flicstein, Haifa, December 27, 1998.

174. Ibid.

175. Interview with Ruth Rotenberg, Tel Aviv, December 11, 1997.

176. Moshe Shachal, "The Ministry of Energy and Air Quality in the Haifa Region," speech by the Minister of Energy, reprinted in *Biosphera* 1 (October 1987).

177. "Amendment to the Personal Decree to the Haifa Oil Refineries," *Biosphera* 14, no. 11 (1985): 2–8.

178. Interview with Bernanda Flicstein, Haifa, December 27, 1998.

179. Haifa Criminal File 2004/87, 1987, on file at the Ministry of the Environment.

180. Interview with Bernanda Flicstein, Haifa, December 27, 1998.

181. Ibid.

182. Alon Tal, "The Economic Benefits of Noncompliance with Environmental Laws: The Role of Economic Analysis in Assessing Penalties for Polluters in Israel," *Ecology and Environment* (January–February 2000), p. 3.

183. Prevention of Nuisances Regulations (Unreasonable Noise) 1977, *Kovetz Takanot*, no. 365, p. 716; see also Alon Rosenthal (Tal), "Measuring Noise: Towards an Optimal Judicial Policy," *Israel Law Review* 20 (1985).

184. Meir Gazit, "A Law to Remove Advertising Signs on the Sides of Inter-City Roads Is Passed," *Biosphera* 9, no. 2 (1979): 5.

185. Yarkon River Authority Order, 1988, *Kovetz Takanot*, no. 5109, p. 855.

186. Streams and Springs Authorities Law, 1965, *Sefer ha-Hokim* 457, p. 150.

187. Uri Marinov, "The Service Doesn't Rest on Its Laurels,"*Biosphera* (December 1977): 2–4.

188. Plant Protection Law, 1956, *Sefer ha-Hokim* 206, p. 79.

189. Rachel Carson, *Ha-Aviv Ha-Domem* (Petah Tikva: Teva u'Briyut, 1966).

190. "The Use of Pesticides in Israel and Its Implications," *Biosphera* (November 1977).

191. Whitman, *op. cit.*, p. 212.

192. Israel State Comptroller, "The Use of Pesticides and the Control for Prevention of Toxic Impacts," *Annual Report* 37, Part A (Jerusalem: Government of Israel, 1986), p. 528.

193. Heinrich Mendelssohn and Yossi Leshem, "The Status and Conservation of Vultures in Israel," *Vulture Biology and Management*, ed. S. R. Wilbur and J. A. Jackson (Berkeley: University of California Press, 1983), pp. 86–98.

194. "The Impact of Pesticides on Fauna in Israel: The Situation at the Start of 1976," *Biosphera* F, no. 2 (1976).

195. "The Use of Pesticides in Israel and Its Implications."

196. Elihu Richter, "Sustainable Agriculture and Pesticide Programs: Perspectives and Programs," *Our Shared Environment*, ed. Robin Twite and Jad Isaac (Jerusalem: Israel-Palestine Center for Research and Information, 1994), p. 186.

197. "The Use of Pesticides in Israel and Its Implications."

198. Poison Control Center in Haifa, as cited in D'vora Ben Shaul, "A Pharmacy of Poisons," *Econet News* 7, no. 2 (1992) 1.

199. Y. Levy, F. Grauer, S. Levy, P. Chuwers, N. Gruener, J. Marzuk, and E. Richter, "Organophosphate Exposure and Symptoms in Farm Workers and Residents with Normal Cholinesterase," *in Environmental Quality* 4A (Jerusalem Israel Society for Ecology and Environmental Quality Sciences, 1991).

200. Aviva Kamar, personal communication, June 15, 1995.

201. The Pharmacist Regulations (Limitation of Pesticide and Chemical Spraying from Airplanes), 1979, *Kovetz Takanot*, no. 1114, p. 1003.

202. The Water Regulations (Prevention of Water Pollution) (Spraying near Water Sources), 1991, *Kovetz Takanot*, no. 5344, p. 776.

203. Shlomo Bravender, "The Polluter Pays: The Principle and Its Application," *Biosphera* 4 (1974): 2–3.

204. Tal, "The Economic Benefits of Noncompliance with Environmental Laws."

205. Tamir, *Haver Knesset*, p. 360.

206. Shmuel Amir, "Public Expenditures for Protecting Environmental Quality," *Environmental Quality in Israel*, no. 9, *1981* (Jerusalem: Ministry of the Interior, 1982), p. 281.

207. Interview with Uri Marinov, Winston House, Suffolk, England, September 19, 1997.

208. Protection of Cleanliness Law, 1984, *Sefer ha-Hokim*, p. 442.

209. It imposed legal presumptions of liability to expedite enforcement. For instance, a car owner was assumed to be driving his vehicle when trash was thrown from it, unless he could prove otherwise. The law contained a gimmick that allowed for citizens to be appointed "Cleanliness Trustees" and report anyone they caught littering in public places.

210. Interview with Ruth Rotenberg. Tel Aviv, December 11, 1997. Rotenberg emphasizes that subsequently other such sessions were well attended but that the EPS staff never forgot its first experience.

211. Alona Frankel, *Sir ha-Sirim* (Givataim: Masada, 1975).

212. Uri Marinov, personal communication, December 27, 1998.

CHAPTER 9: A MINISTRY OF THE ENVIRONMENT COMES OF AGE

1. Mark Landy, Marc Roberts, and Steve Thomas, *The Environmental Protection Agency: Asking the Wrong Questions*, 2d ed. (New York: Oxford, 1994).

2. P. Ali Memon, *Keeping New Zealand Green: Recent Environmental Reforms* (Dunedin: University of Otago Press, 1993), p. 48.

3. Interview with Uri Marinov, Winston House, Suffolk, England, September 19, 1997.

4. Ibid.

5. Interview with Ronni Miloh, Tel Aviv, January 19, 1999.

6. Shamir recalls the process: "There was never really any doubt that we had to create a Ministry of Environment. Except perhaps for the Third World, you can find one in every country today. I didn't think that sports needed a Minister." Yitzhak Shamir, personal interview, Tel Aviv, November 24, 1997.

7. David Vogel, "Israel Environmental Policy in Comparative Perspective," *Israel Affairs,* as quoted in Noga Morag-Levine, "The Politics of Imported Rights," in *Cause Lawyering and the State in a Global Era,* ed. Austin Sarat and Stuart Scheingold (London: Oxford University Press, 2000), p. 353.

8. Interview with Ronni Miloh, Tel Aviv, January 19, 1999.

9. Yosef Harish, personal communication, December 1989.

10. Abraham Atzmon and Yehezkel Dror, "Professional Opinion: Establishing the Ministry of Environment," in *Environmental Quality in Israel, 1990* (Jerusalem: Ministry of the Environment, 1991), p. 736.

11. Ibid.

12. Government Decision, no. 349, regarding the Authorities of the Ministry of the Environment, April 2, 1989, reprinted in *Environmental Quality in Israel, 1990* (Jerusalem: Ministry of the Environment, 1991), pp. 719–721.

13. Dror Amir, personal communication, December 22, 1998.

14. Government Decision, no. 525, regarding the Authorities of the Ministry of the Environment, May 28, 1989, reprinted in *Environmental Quality in Israel, 1990* (Jerusalem: Ministry of the Environment, 1991), pp. 721–722.

15. Shmuel Brenner, personal communication, Ra'ananah, October 25, 2001.

16. Continued Discussion, Government Decision, no. 1262, of January 21, 1990, reprinted in *Environmental Quality in Israel, 1990* (Jerusalem: Ministry of the Environment, 1991), pp. 725–727.

17. Josef Tamir, "The Embarrassing Budget," *Biosphera* 19, no. 5 (1990): 9.

18. Interview with Ronni Miloh, Tel Aviv, January 19, 1999.

19. Instructions for Abatement of Air Pollution Nuisances from the Castel Quarry according to the Abatement of Nuisances Law, 1961, May 2, 1989.

20. Prevention of Sea Pollution from Land-Based Sources Regulations, 1990, *Kovetz Takanot,* no. 5240, p. 250.

21. Abatement of Nuisances Regulations (Unreasonable Air Pollution and Odors from Solid-Waste Disposal Sites), 1990, *Kovetz Takanot,* no. 5250, p. 386.

22. Dror Amir, personal communication, December 22, 1998.

23. In practice, the planning advisors who worked for the EPS in the regions became the Regional Directors. Interview with Valerie Brachya, Jerusalem, February 17, 1999.

24. Interview with Uri Marinov, Winston House, Suffolk, England, September 19, 1997.

25. Amir Levine, "Inspection, Practice in the Field: The Ministry of the Environment Patrol," in *Environmental Quality in Israel, 1992* (Jerusalem: Ministry of the Environment, 1993), pp. 20–21.

26. Report of the Committee for Investigating Air Pollution Episodes in Haifa Bay on April 29 and 30, 1989, in *Environmental Quality in Israel, 1990* (Jerusalem: Ministry of the Environment, 1991), p. 53.

27. Ibid., pp. 54–56.

28. Eilan, *Note*, p. 10.

29. Prevention of Nuisances Regulations (Air Quality), 1971, *Kovetz Takanot*, p. 380.

30. Position paper summaries for ozone, particulates, and sulfur dioxide in "Environmental Air Quality Standards," *Environmental Quality in Israel, no. 11, 1985* (Jerusalem: Ministry of the Interior, 1985), pp. 44–54.

31. Yizhak Zamir, "Secondary Legislation: Procedure and Directives," *The Directives of the Attorney General* (Jerusalem: Ministry of Justice, November 1985).

32. "Instructions for Abatement of Air Pollution Nuisances from the Electric Company Station in Haifa (Amendment) according to the Abatement of Nuisances Law, 1961," and "Instructions for Abatement of Air Pollution Nuisances from the Petroleum Refineries in Haifa (Amendment) according to the Abatement of Nuisances Law, 1961," reprinted in *Environmental Quality in Israel, 1990* (Jerusalem: Ministry of the Environment, 1991), pp. 761–764.

33. Protocol of the Ministerial Committee regarding Abatement of Nuisances Regulations (Air Quality) November 14, 1989, Israeli Government.

34. Haim Harari, "Report Investigating the Subjects and Aspects Associated with the Controversy between the Minister of the Environment and the Minister of Energy Regarding Air Quality Standards and Associated Problems of the Fuel Industry," February 1990.

35. *Oil Refineries v. the Minister of the Environment and Others,* Bagatz 1201/90.

36. Interview with Ronni Miloh, Tel Aviv, January 19, 1999.

37. *Adam Teva V'din and Tsvi Levanson v. the Minister of the Environment (and Others),* Bagatz 2112/91, June 19, 1991 (unpublished).

38. Arik Meirovsky, "The Government Adopts the Recommendations of the Harari Committee for Air Quality Standards," *Kol Haifa,* July 5, 1991, p. 1.

39. *Adam Teva V'din and Others v. the Deputy Minister of the Environment and Others,* Bagatz 1183/92 (unpublished) on file with author.

40. Elli Elad, "New Orders Will Force the Oil Refineries and the Electric Company to Reduce Pollution," *Ha-Aretz,* April 15, 1992.

41. Alon Tal, "The Air Is Free," *Eretz v'Teva* (September-October 1999).

42. David Rudge, "Sewage Flows into Kinneret," *Jerusalem Post*, May 3, 1992.

43. Letter from Uri Marinov to Yosef Peretz, December 2, 1991.

44. Letter from Rachel Adam to Ruth Yaffe, April 28, 1992.

45. D'vora Ben Shaul, "No Excuse for Kinneret Sewage," *Jerusalem Post*, April 27, 1992; Rudge, *op. cit.*

46. Elli El Ad, "The State Attorney Has Filed an Indictment against the Sewage from the City of Tiberias and the City of Eilat," *Ha-Aretz*.

47. Interview with Ruth Rotenberg, Tel Aviv, December 11, 1997.

48. Criminal File 538/92, filed on May 26, 1992; see Deborah Sandler, "Environmental Law and Policy for the Gulf of Aqaba: An Israeli Perspective," *Protecting the Gulf of Aqaba: A Regional Environmental Challenge*, ed. Philip Warburg et al. (Washington, D.C.: Environmental Law Institute, 1993), p. 87.

49. Almost two hundred thousand new immigrants arrived in Israel in 1990 as opposed to ten thousand in 1988. Eran Feitelson, "Muddling toward Sustainability: The Transformation of Environmental Planning in Israel," in *Progress in Planning* 49, p. 1, ed. D. Diamond and B. H. Massam (London: Pergamon Press, 1998): 19.

50. *Sefer ha-Hokim*, 1990, p. 166, amended 1991, p. 8.

51. "Committees for Residential Construction."

52. Their express, but somewhat convoluted, mandate was "to approve building plans that offered urgent readiness for solutions to housing and employment needs in the country for absorbing immigration, young couples, homeless people and employment." The Planning and Building Procedures Law (Temporary Measures), 1990, sec. 1.

53. Ministry of the Environment, "Review of the Work of the Residential Building Committees from the Environmental Viewpoint," in *Environmental Quality in Israel*, nos. 17–18 (Jerusalem: Ministry of the Environment, 1992), pp. 89–90.

54. Ibid., pp. 91–92.

55. In 1988, twenty-two thousand residential units were under construction. The number rose to forty-three thousand by 1994. Feitelson, *op. cit.*, p. 23.

56. Itai Peleg, "The Knesset Approved the Valalim Law in the Second and Third Readings," *Telegraph*, November 23, 1994.

57. "Israel Is Elected to the Directorate of the Mediterranean Organization for the Environment," *Biosphera* 21, no. 1–2 (1991): 2–3.

58. "Ora Namir, The Minister of the Environment," *Biosphera* 21, no. 10 (1992): 2.

59. Yosef Hirshberg, "Environmental Quality Protection: From Service to Government Ministry," *Biosphera* 21, no. 10 (1992): 7.

60. "Dr. Yisrael Peleg, Director General of the Ministry of the Environment," *Biosphera* 21, no. 10 (1992): 4. Peleg, who calls himself "one of Peres's boys," was rewarded for his loyalty with a consular appointment to Philadelphia from 1988 to 1992.

61. Interview with Yisrael Peleg, Tel Aviv, September 30, 1997.

62. Uri Marinov, personal communication, September 18, 1997.

63. Shaikeh Ben-Porat, *Discussions with Yossi Sarid* (Tel Aviv: Sifriat Poalim, 1997), p. 28.

64. Ibid., p. 64.

65. Interview with Valerie Brachya, Jerusalem, February 17, 1999.

66. Uri Marinov, personal communication, December 27, 1998.

67. Ibid.

68. Interview with Mickey Lipshitz, Tel Aviv, January 8, 1998.

69. Rinat Klein, "Sarid in the Important Place," *Maariv,* August 10, 1993. Seven years later, Minister Daliah Itzik would cash in on the same photo opportunity. Liat Collins, "Itzik, Shahak Get a Whiff of Substandard Public Bathrooms," *Jerusalem Post,* April 18, 2000, p. 4.

70. A nature reserve that is home to one of three craters, or "machteshim," in the Negev desert.

71. Liat Collins, "Sarid: Country's Nuclear Reactors Pose No Threat," *Jerusalem Post,* March 31, 1993; Elli Elad, "Water, Air, Soil and Flora Samples from the Dimona Reactor to Be Passed Regularly to Minister Sarid," *Ha-Aretz,* March 31, 1993.

72. Liat Collins, "Sarid: Authorities Hushed Up Negev Nuclear Waste Leak," *Jerusalem Post,* April 15, 1993.

73. Amira Lem, "Yossileh: How Did It Happen?," *Kol ha-Ir,* July 23, 1993.

74. Liat Collins, "The Year of Public Awareness," *Jerusalem Post,* September 3, 1993.

75. Interview with Yossi Sarid, Jerusalem, December 30, 1997.

76. Dror Amir, personal communication, December 21, 1998.

77. Shoshana Gabbay, "We've Moved," *Israel Environmental Bulletin* 18, no. 2 (1995): 2.

78. Ben-Porat, *op. cit.,* pp. 94–95.

79. Uri Marinov and Deborah Sandler, "The Status of Environmental Management in Israel," *Environmental Science and Technology* 27, no. 7 (1993): 1258.

80. Interview with Yisrael Peleg, Tel Aviv, September 30, 1997.

81. Liat Collins, "Environmental Year Off to a Smelly Start," *Jerusalem Post,* September 7, 1993.

82. "The Opening of the 'Year of the Environment' at the President's Home," *Yom l'Yom,* September 8, 1993.

83. Yossi Sarid, "With the Opening of the Year of the Environment," *Biosphera* 22, no. 12 (1993): 6.

84. Liat Collins, "The Year of Public Awareness."

85. Gil Doron, "'94—The Year of the Environment: The Programs, the Events and the Campaigns," *Ha-Ir,* July 23, 1993.

86. Azariah Alon, "Environmental Quality—Not a Matter for Only a Year," *Teva v'Aretz,* June 1993, p. 7.

87. Rachel Adam, personal communication, July 2, 1998.

88. Lem, *op. cit.*

89. Orna Yosef, "Personal Decrees Issued to Directors of Quarries in the North That Require Presentation of Program to Correct Flaws That Constitute an Environmental Hazard," *Kol ha-Emeq v'ha-Galil,* October 7, 1994.

90. Uri Sharon, *Davar Rishon,* "The Oil Refineries in Haifa Will Change to Using Low Sulfur Fuel," *Davar Rishon,* December 8, 1995.

91. Amnon Greenberg, personal communication, March 1995.

92. United Nations Environment Programme, *Methyl Bromide: Its Atmospheric Science, Technology and Economics,* Synthesis Report of the Methyl Bromide Interim Scientific Assessment, June 1992, p. 1.

93. Rachel Carson, *The Sea around Us* (New York: Oxford, 1950).

94. Rinat Klein, "Israeli Bromide Endangers the World," *Maariv,* July 21, 1993, p. 1.

95. Rinat Klein, "Thus We Became Enemies of the World," *Maariv (Sof Shavua),* July 23, 1993, p. 1.

96. Michael Graber, personal communication, April 11, 1999.

97. Michael Graber, "Israel and the Montreal Protocol," *Eichut ha-Sviva,* 1993.

98. Michael Graber, personal communication, April 4, 1999.

99. Michael Shpiegelstein, "Methyl Bromide and the Ozone Layer," *Eichut ha-Sviva,* March 1994, pp. 15–18.

100. Bill Thomas, personal communication, August 4, 1998.

101. U.S. EPA, "The Montreal Protocol: Decisions at Copenhagen, 1992," fact sheet, 1993.

102. Alon Tal, "Damage to the Ozone Layer from Methyl Bromide," *Eichut ha-Sviva,* March 1994, pp. 20–22.

103. Liat Collins, "Sarid: We're Committed to Phase Out Methyl Bromide Use," *Jerusalem Post,* December 8, 1995.

104. Liat Collins, "Don't Stop Producing Methyl Bromide," *Jerusalem Post,* July 23, 1993.

105. Collins, "Sarid: We're Committed to Phase Out Methyl Bromide Use."

106. "Stopping Use of Methyl Bromide Will Destroy the Melon and Watermelon Branches," *Ha-Zofeh,* December 8, 1995.

107. Paul Horowitz, U.S. EPA, personal communication, August 6, 1998.

108. Ben-Porat, *op. cit.,* p. 137.

109. Eran Feitelson, "Protection of Open Spaces in Israel at a Turning Point," *Horizons in Geography* 42–43 (1995): 8.

110. Zafrir Rinat, "Once, Here Was a Tree," *Ha-Aretz, Weekend Supplement,* October, 27, 2000, p. B-12.

111. Shoshana Gabbay, "Israel 2020: A New Vision," *Israel Environmental Bulletin* 18, no. 2 (1995): 5.

112. During the 1980s, Israeli farmers faced a series of financing crises, largely associated with loans linked to triple-digit inflation that soon spiraled out of hand. With government assistance, most farms weathered the storm, but this did not alter the overriding economic reality: It was increasingly difficult to compete with the low-priced produce coming from Arab and other

Mediterranean countries. Ed Hofland, personal communication, April 22, 1999; see also Eran Feitelson, "Protection of Open Spaces in Israel at a Turning Point," *Horizons in Geography* 42–43 (1995): 9–13.

113. Feitelson, *op. cit.*

114. Ministry of the Interior, Planning Administration, "Report of the Valal reports," as printed in Feitelson, *op.cit.*, p. 14.

115. A 1968 amendment to Israel's planning law empowered the Committee to designate land for agricultural purposes. Its "agricultural default" claimed any soil that did not already have buildings on it. The Committee was, of course, stacked with agricultural and rural representatives. Feitelson, *op. cit.*, pp. 9–13.

116. Decision 533 (1990) of the Land Administration Council; later, Decision 611 (1993) allowed farmers to include water rights in the overall debt-restructuring and compensation package.

117. Interview with Valerie Brachya, Jerusalem, February 17, 1999.

118. Motti Kaplan, *Population Dispersion, Development, Construction and Preservation of Open Spaces* (Jerusalem: Ministry of the Environment, 1995).

119. Ibid.

120. EcoPeace, *Dead Sea Challenges* (Jerusalem: EcoPeace, 1996), p. 7.

121. Amir Rosenblit, "A Little Bit Salty Concession," *Davar,* August 20, 1993.

122. Society for the Protection of Nature in Israel, "The Law against the Law," information brochure, 1993.

123. Rosenblit, *op. cit.*

124. Greer Fay Cashman, "MK Proposes Amendment to Dead Sea Franchise Law," *Jerusalem Post,* March 5, 1994.

125. Rosenblit, *op. cit.*

126. Lem, *op. cit.*

127. Bagatz 2920/94, *Adam Teva V'din (Israel Union for Environmental Defense) and Others v. the National Planning Board and Others, Piskei Din,* 50 III pp. 443–470, 1996.

128. Ben-Porat, *op. cit.*, pp. 140–141.

129. Ruth Sinai, "With a Tie and a Hat against the Bulldozers," *Ha-Aretz,* December 5, 1999, p. B-3.

130. Interview with Dr. Basel Ghattas, Shfaram, December 9, 1997.

131. Terry Davies, *Comparing Environmental Risks: Tools for Setting Government Priorities* (Washington: Resources for the Future, 1996).

132. Cruelty to Animals Law (Protection of Animals), 1994, *Sefer ha-Hokim,* no. 56, p.304.

133. Zafrir Rinat, "Sarid Orders the Nature Reserve Authority to Prohibit Circus Performances in Israel That Involve Wild Animals," *Ha-Aretz,* December 28, 1995.

134. Gabi Baron and Yitzhak Bar Yosef, "Hundreds Will Be Injured in a Hazardous Materials Incident," *Yediot Ahronot,* November 15, 1994.

135. Hazardous Materials Law, 1993, *Sefer ha-Hokim,* no. 1408, p. 28.

136. Adam Teva V'din, "The Garbage Crisis in Israel," fact sheet, Tel Aviv, 1998.

137. Collection and Removal of Garbage for Recycling Law, 1993, *Sefer ha-Hokim*, p. 116.

138. Elli Elad, "Sarid Requests Cessation of Activities at Hundreds of Garbage Dumps across the Country," *Ha-Aretz*, March 23, 1993.

139. Ilan Nissim, personal communication, May 1996.

140. Alon Tal, "An Imperiled Promised Land: The Antecedents of Israel's Environmental Crises and Prospects for Progress," *Journal of Developing Societies* 13, ed. J. Jabra (1997): 116.

141. Kenneth E. Boulding, "The Economics of the Coming Spaceship Earth," *Environmental Quality in a Growing Economy*, ed. Henry Jarrett (Baltimore: Johns Hopkins Press, 1966).

142. "Symposium on the Recycling Law," *Environmental Quality*, November 1993, p. 6.

143. Yitzhak Bar Yosef, "Yossi Sarid: We Stopped the Recycling Momentum," *Yediot Ahronot*, September 26, 1994.

144. Yigael Avidan and Shai Shalev, "960 Tons of Plastic Garbage from Germany Will Arrive in Israel," *Hadashot*, March 31, 1993.

145. Yoram Avnimelech, personal communication, January 10, 1999.

146. Dov Rubin, "The Garbage Can Asks, the Ministry of the Environment Answers," *Sheva*, December 22, 1994.

147. Yossi Leshem and Nehama Ronen, "Removing Hiriya Garbage Dump: Israel—A Test Case," paper presented to *International Bird Strike Committee Conference*, Slovakia, September 14–18, 1998.

148. Larry Derfner, "Buried in Garbage," *Jerusalem Post*, January 2, 1998, p. 15.

149. Elli Khalifa, "The City of Tel Aviv Refuses to Transfer Its Trash to the Dudaim Site," *Sheva*, October 12, 1995.

150. "Petition Filed to the High Court against the Company That Won the Dudaim Tender," *City Fax*, Tel Aviv, October 6, 1994.

151. U.S. EPA, "Solid Waste and Emergency Response," January 1998, EPA530-K-98-008 (http://www.epa.gov/osw).

152. Orli Gil and Aliza Konter, "Caniel Produces Environmentally Friendly Cans," *Zomet ha-Sharon*, December 31, 1993.

153. Danny Morganstern, personal communication, January 9, 2002.

154. Derfner, *op. cit.*

155. Elli Elad, "The Route of Hospital Wastes," *Ha-Aretz*, April 20, 1993.

156. *Adam Teva V'din v. the Ministry of the Environment and Others*, Bagatz, 1997.

157. Interview with Yossi Sarid, Jerusalem, December 30, 1997.

158. Ultimately it was the residents who had to bring the unsuccessful suit to stop the buses. They did not succeed, and the "largest central bus station in the world" eventually opened, with the noise on neighboring porches exceeding an ear-splitting eighty-decibel level.

159. Interview with Yossi Sarid, Jerusalem, December 30, 1997.

160. Nehama Ronen, lecture at Hebrew University, May 17, 2000.

161. Liat Collins, "The Shadow Ministry," *Jerusalem Post Magazine*, December 4, 1998, pp. 15–17.

162. Sima Kadmon, "Raful Intends to Bequeath the Party to Nehama," *Yediot Ahronot, Sabbath edition*, November 27, 1998, p. 4.

163. Interview with Nehama Ronen, Tel Aviv, December 18, 1997.

164. Interview with Uri Marinov, Winston House, Suffolk, England, September 19, 1997; interview with Yossi Sarid, Jerusalem, December 30, 1997.

165. Interview with Nehama Ronen, Tel Aviv, December 18, 1997.

166. Ibid.

167. Elihu Richter, "Sustainable Agriculture and Pesticides: Problems, Perspectives and Programs," *Our Shared Environment*, eds. R. Twite and J. Isaac (Jerusalem: Israel-Palestine Center for Research and Information, 1994).

168. Alon Rosenthal (Tal), "State Agricultural Pollution Regulation: A Quantitative Assessment," *Water, Environment and Technology* 2, no. 8 (1990): 50.

169. Alon Tal, "Enforceable Standards to Abate Agricultural Pollution: The Potential of Regulatory Policies in the Israeli Context," *Tel Aviv University Studies in Law* 14 (1998).

170. Haim Watzman, "Israel Ministry Kicks Out the Greens," *New Scientist*, September 14, 1996, p. 6.

171. Binat Schwarz Millner, "Hula Plans," *Eretz Magazine*, March-April 1996, pp. 64–65.

172. Dan Alon, "Crane Reign," *Eretz Magazine*, January-February 1997, pp. 47–48.

173. Millner, *op. cit.*, p. 65.

174. Iris Millner, "The Rabbit's Foot," *Ha-Aretz*, August 22, 1998.

175. "In most countries, the Greens are in constant conflict with their ministries of environment. Environmental organizations should represent environmental interests and nothing else." Ronen in Watzman, *op. cit.*; also published in *Journal of Developing Societies* 13 (1997): 116.

176. Interview with Yossi Sarid, Jerusalem, December 30, 1997.

177. Yael Golan, personal communication, December 1997.

178. Nehama Ronen, lecture at Hebrew University, May 17, 2000.

179. Ibid.

180. Janine Zacharia, "The Wasteland," *Jerusalem Report*, August 21, 1997.

181. "I was wrong about this position. It's important and fascinating. It might not be considered important by most of the country, but I think it is prestigious. After all, it touches every aspect of our lives." Liat Collins, "Green and Fighting Fit," *Jerusalem Post Magazine*, April 7, 2000, p. 13.

182. Tamar Nahari and Liron Maroz, "The Minister of the Environment Votes against a Green Law," *Walla News*, November 7, 2001.

183. Gedaliah Shelef, "The Problem of Sewage, Treatment and Reuse as a Source of Water," *National Environmental Priorities in the Area of Israel's Environmental Quality* (Tel Aviv: Israel Economic Forum, 1999), chap. 6, p. 1.

184. Yehudah Bar, "Sewage Treatment as a Condition for River Reclamation," *Israel Rivers Conference*, 1995 (proceedings), ed. Rachel Einav and Avital Gazit (Tel Aviv: Tel Aviv University, 1995), pp. 13–14.

185. Alon Tal, "Reforms in the Prevention of Air Pollution in Vehicles: Towards the Era of Catalytic Converters," *Biosphera* 22, no. 7 (1992): 4–10.

186. Robin Dawes, *Social Dilemma* (1980), as quoted by Stuart Schoenfeld, "Negotiating the Environmental Social Dilemma: The Case of Greenpeace Canada," paper presented at the Jerusalem Conference on Canadian Studies, Hebrew University, July 1998.

187. Zafrir Rinat, "After the Fire, Yet Another Committee for Ramat Hovav," *Ha-Aretz*, August 4, 1998.

188. Amiram Cohen and Zafrir Rinat, "Israel—One of the Main Polluters of the Mediterranean," *Ha-Aretz*, November 24, 1997.

189. Gad Lior, "The 2002 Budget: 254.8 Billion Shekels," *Yediot Ahronot, Economics Supplement*, October 30, 2001, pp. 2–3.

CHAPTER 10: ISRAEL, ARABS, AND THE ENVIRONMENT

1. Interview with Dr. Basel Ghattas, Shfaram, December 9, 1997.

2. Basel Ghattas, "Planning Problems and Implementation of Solutions for Effluent Treatment in the Rural Sector from the Local Point of View," *Solutions for Disposal, Treatment and Reuse of Wastes in Rural Areas of Israel: Symposium Proceedings*, ed. Khatam K'naneh (Rama: Galilee Society, 1988), p. 46.

3. Ghattas, *op. cit.*

4. Galilee Society, *Health for All?* (Shefa Amr: Galilee Society, 1993), educational video.

5. Ghattas, *op. cit.*

6. Interview with Dr. Basel Ghattas, Shfaram, December 9, 1997.

7. Ibid.

8. Figures from *The Sociology of the Palestinians*, eds. Khalil Nakhleh and Elia Zureik (London: Croom Helm, 1980), p. 66.

9. Baruch Kimmerling and Joel S. Migdal, *Palestinians: The Making of a People* (New York: Free Press, 1993), p. 127.

10. The Israeli Arab population is located primarily in the Galilee and in the so-called triangle near the center of the country. Israel's Bedouin, who technically are Arab but who remain a distinct ethnic group with a particular identity (for instance, generally serving in Israel's army), are primarily concentrated in the south of Israel.

11. Howard Sachar, "The Seeds of Arab-Jewish Confrontation," in *A History of Israel* (New York: Knopf, 1976), pp. 163–167.

12. For example, Exodus 12:49, 20:10, 22:20, 23:9; Leviticus 19:10, 19:33–19:34, 23:22, 24:22; Deuteronomy 10:18–19, 27:19.

13. Kimmerling and Migdal, *op. cit.*, p. 161.

14. Peter Hirschberg, "Road Rage," *Jerusalem Report*, May 8, 2000, p. 24.

15. Magad el-Haj, "The Arab Village in Israel: General Lines," *Solutions for Disposal, Treatment and Reuse of Wastes in Rural Areas of Israel*, p. 8.

16. Kimmerling and Migdal, *op. cit.*, p. 174.

17. Ian Lustick, *Arabs in the Jewish State: Israel's Control of a National Minority* (Austin: University of Texas Press, 1980).

18. Meron Benvenisti, *Intimate Enemies: Jews and Arabs in a Shared Land* (Berkeley: University of California Press, 1995), p. 106.

19. Ibid, p. 165.

20. Elia T. Zureik, *The Palestinians in Israel: A Study in Internal Colonialism* (Boston: Rutledge and Kegan Paul, 1979), p. 178, as cited in Kimmerling and Migdal, *op. cit.*, p. 352.

21. Dan Rabinowitz, *Overlooking Nazareth: The Ethnology of Exclusion in Galilee* (Cambridge: Cambridge University Press, 1997), pp. 13–14.

22. David Kretzmer, *The Legal Status of the Arabs in Israel* (Boulder, Colorado: Westview, 1990).

23. *Environmental Justice: Issues, Policies, Solutions*, ed. Bunyant Briant (Washington, D.C.: Island Press, 1995); see also *Unequal Protection: Environmental Justice and Communities of Color*, ed. Robert D. Bullard (San Francisco: Sierra Club Books, 1994).

24. "The Alazama Tribe Members Petitioned the High Court of Justice against the Intent to Remove Them from Ramat Hovav," *Davar*, December 21, 1994.

25. Golda Meir, *My Life* (London: Weidenfeld and Nicolson, 1975), pp. 231–232.

26. Fuad Farah, "Problems of Sewage Reuse in the Arab Sector in Israel," *Solutions for Disposal, Treatment and Reuse of Wastes in Rural Areas of Israel*, p. 33.

27. Interview with Dr. Diab Shahir, Tamara, March 9, 1999. The odors and water contamination associated with the pig operations in the Ibalin region, for example, have been the focus of litigation for over a decade. Richard Laster, Hebrew University Law School, "Treatment of Piped Water from a Legal Perspective," 1988, unpublished manuscript.

28. el-Haj, *op. cit.*, p. 8.

29. Interview with Dror Amir, Tel Aviv, January 7, 1999.

30. Mahmud Ka'abiah, personal communication, July 12, 1993.

31. Yaakov Garb, "The Challenge of Sustainable Transport for Israel and Palestine," *Palestine-Israel Journal of Politics, Economics and Culture* 5, no. 1 (1998): p. 60.

32. Ministry of Health data provided in Meron Epstein, "Rise in Cancer Morbidity and Mortality," in *Israel: Vital Signs 2000* (Tel Aviv: World Watch/Heschel Center for Environmental Learning and Leadership, 2000), p. 159.

33. Gary Ginsburg, personal communication, August 28, 1998.

34. Lustick, *op. cit.*, p. 167.

35. Dalia Schwartz and Sharon Nussbaum, *Water in Israel: Consumption and Production*, 1996 (Tel Aviv: Israel Water Commission, 1998).

36. Interview with Dr. Fayad Sheabar, Kibbutz Ketura, December 1, 1997.

37. Adam Teva V'din, *Annual Report*, 1992 (Tel Aviv: Adam Teva V'din, 1993), p. 8.

38. Michael Sofer and Ram Gal, "Environmental Implications of Penetration of Non-agricultural Businesses to Moshavim," *Horizons in Geography* 42–43 (1995): 39–50.

39. "The Kadmon Report," officially known as "Additional Production Activities in the Family Farm, Planning, Physical and Organizational Aspects," submitted to the Minister of Agriculture, Tel Aviv, 1994.

40. Interview with Dr. Fayad Sheabar, Kibbutz Ketura, December 1, 1997.

41. State Comptroller's Office, *Report Number 46* (Jerusalem: State Comptroller's Office, 1996).

42. Yaakov Eran, "Health Aspects of Waste Treatment and Reuse in Village Regions," *Solutions for Disposal, Treatment and Reuse of Wastes in Rural Areas of Israel*, p. 40.

43. Interview with Dr. Basel Ghattas, Shfaram, December 9, 1997.

44. Farah, *op. cit.*, p. 33.

45. Interview with Mohamad Rabah, Director of Um el-Fahm Municipal Environment Unit, January 22, 2002.

46. Ibid.

47. Farah, *op. cit.*, p. 34.

48. Naim Daoud, *Survey Assessing the State of Arab Sewage in the Triangle* (Shefa-Amr: Galilee Society, 1997).

49. Farah, *op. cit.*, p. 34

50. David Gabrielli, "Engineering, Economic and Organizational Aspects of Planning Problems and Implementation of Solutions in Treatment of Wastes in Village Regions," lecture synopsis in *Solutions for Disposal, Treatment and Reuse of Wastes in Rural Areas of Israel*, p. 41.

51. Ghattas, *op. cit.*, p. 45.

52. Interview with Dr. Basel Ghattas, Shfaram, December 9, 1997.

53. Lustick, *op. cit.*, p. 105.

54. Galilee Society, *Health for All?* (Shefa Amr: Galilee Society, educational video, 1993).

55. Khatam K'naneh, "Human Health Savings Resulting from Central Sewage Systems in the Arab Village," *Solutions for Disposal, Treatment and Reuse of Wastes in Rural Areas of Israel*, pp. 15–16.

56. Report of the Committee to Assess the Water Pollution in the Krayot, *Biosphera* (July 1985): 5–16.

57. Shaul Ephraim Cohen, *The Politics of Planting* (Chicago: University of Chicago Press, 1995).

58. David Harris, "JNF Fears Arab Purchases under Proposal," *Jerusalem Post*, May 22, 1997, at www.jpost.com.

59. Interview with Dr. Basel Ghattas, Shfaram, December 9, 1997; see also Ziv Maor, "Israel Lands Authority and Jewish National Fund Are Finally to Separate," *Ha-Aretz,* November 4, 1998.

60. Cohen, *op. cit.;* Walter Lehn, *The Jewish National Fund* (London: Kegan Paul International, 1988), pp. 16–18.

61. Shirli Golan Meiri, "A Stroke of Fire," *Yediot Ahronot,* October 14, 1998, pp. 1–7.

62. Yael Danieli and Felix Frish, "The Foresters Chased the Arsonists and Discovered a Plan for Burning Forests," *Maariv,* October 29, 1998, p. 20.

63. Efraim Orni, *Afforestation in Israel* (Jerusalem: Sivan Press, 1969), p. 58.

64. Land Development Authority, *Annual Report,* 1996 (Jerusalem: Keren Kayemet L'Yisrael, July 1997), p. 25.

65. Avshalom Rokach and Haim Zaban, "Forty Years of Land Development and Afforestation in Israel," JNF reprint from *Ariel, a Review of Arts and Letters in Israel,* no. 75 (1989): 23.

66. Land Development Authority, *op. cit.,* p. 24.

67. Yael Danielli, "Record Response to the JNF and Ma'ariv Campaign to 'Return the Green to the Carmel,'" *Maariv,* October 20, 1998.

68. Interview with Mahmoud Gazawi, Tel Aviv, January 13, 1998.

69. Meir Barzilai, personal communication, December 3, 2001.

70. Mahmoud Gazawi, *SPNI News* (Spring 1996), p. 1.

71. Orit Nevo, "Conference of the Arab Sector," *Teva v'Aretz,* June 1993, p. 42.

72. Interview with Yossi Leshem, London, September 21, 1997.

73. Interview with Mahmoud Gazawi, Tel Aviv, January 13, 1998. Sunfrost is a leading Israeli frozen food maker.

74. Ibid.

75. Uzi Paz, *Nature Reserves in Israel* 2 (Givataim: Masada, 1981), p. 101.

76. Beit Ja'an Local Council, *Beit Ja'an,* promotional material, 1998.

77. Interview with Salah Raid, Beit Ja'an, October 21, 1998.

78. Yoav Sagi, "Enough of the Sad War," *Teva v'Aretz* 30 (August 1988): 11.

79. Interview with Dan Perry, Tel Aviv, September 4, 1997.

80. Batsheva Tsur, "Twenty-five Hurt in Supreme Court Protest," *Jerusalem Post,* July 21, 1997.

81. "Druze Land Day," *Maariv,* July 21, 1997.

82. Beit Ja'an Local Council and the Society for the Protection of Nature, Compromise Agreement between the Nature Reserve and Parks Authority, March 11, 1998.

83. Interview with Asad Dabur, Beit Ja'an, October 21, 1998.

84. Interview with Aviva Rabinovich, Kibbutz Kabri, January 11, 1998.

85. Daniel Hillel, "The Bedouin," in *The Negev: Land, Water, and Life in a Desert Environment* (New York: Praeger, 1982), p. 202.

86. "Because rainfall is sketchy, there was never enough pasture for the flocks in a single place. Bedouins had to seek pastures, so they migrated, which meant they had to travel light. The less they had, the better." Interview with Clinton Bailey, Kibbutz Ketura, December 1, 1997.

87. David Grossman, "The Felah and the Bedouin at the Edge of the Desert: Relations and Strategies for Survival," in *The Bedouin Settlement in Israel,* ed. David Grossman (Ramat Gan: Bar Ilan University Press, 1995), p. 31.

88. Interview with Clinton Bailey, Kibbutz Ketura, December 1, 1997.

89. T. Ashkenazi, *The Bedouin: Their Origins, Lives and Practices* (Jerusalem: Reuven Mas, 1974).

90. Eliyahu Babai, "The Situation of the Bedouins in Israel," *Karka* 42 (October 1996): 79.

91. Interview with Clinton Bailey, Kibbutz Ketura, December 1, 1997.

92. Babai, *op. cit,* pp. 75–80.

93. Yosef Ben-David, "Adaptation through Crisis: Social Aspects of Bedouin Urbanization in the Negev," in *The Bedouin Settlement in Israel,* pp. 48–49.

94. Interview with Clinton Bailey, Kibbutz Ketura, December 1, 1997.

95. A. Shmueli, "Bedouin Rural Settlement in Eretz-Israel," in *Geography in Israel,* ed. D. H. K. Amiran and Y. Ben-Arieh (Jerusalem, 1976), pp. 308–326.

96. Zvi Alush, "Kisifiyeah: 'We're Second Class Citizens,'" *Yediot Ahronot—24 Hour Supplement,* January 20, 1998, p. 6.

97. Interview with Alon Galili, Sdeh Boqer, January 3, 1998.

98. Ibid.

99. Ibid.

100. Ibid.

101. For example, one 1979 Jerusalem reporter spoke of ". . .bullying and intimidation by patrolmen carrying court evacuation edicts recurring with alarming frequency. Units have been known to swoop down on encampments, confiscating or scattering herds, destroying property, and threatening women and children with loaded pistols." Harry Wall, *Jerusalem Post,* February 23, 1979.

102. Interview with Farkhan Shlebe, Sfinat Midbar, January 6, 1998.

103. Interview with Clinton Bailey, Kibbutz Ketura, December 1, 1997.

104. Interview with Dan Perry, Tel Aviv, September 4, 1997.

105. "Shepherds and Wise Men," *New Scientist,* December 23/30, 1995, pp. 25–27.

106. Interview with Aviva Rabinovich, Kibbutz Kabri, January 11, 1998.

107. Ibid.

108. Babai, *op. cit.,* p. 80.

109. Interview with Alon Galili, Sdeh Boqer, January 3, 1998.

110. Woods and Forest Ordinance, 1920, sec. 14 (h), *Laws of Palestine,* 1925. The Ordinance, as well as its 1926 amendment, prohibited "pasturing or allowing cattle (e.g., sheep and goats) to trespass." The provision remains in force today.

111. For years the *sira kotzanit,* a thick and thorny plant, was controlled in the Galilee by grazing goats. Once the animals were removed, the plant overwhelmed weaker species, and wildlife had a hard time penetrating the thickets. Donny Orenstein, personal communication, December 8, 1997.

112. Avi Pervolotsky, "The Connection between Grazing and Nature Preservation in Israel," *Ecology and Environment* 3 (1997): 247.

113. Interview with Clinton Bailey, Kibbutz Ketura, December 1, 1997.

114. Yosef Ben-David, "Adaptation through Crisis, Social Aspects of Bedouin Urbanization in the Negev," in *The Bedouin Settlement in Israel,* p. 55.

115. Babai, *op. cit.,* pp. 76–80.

116. Dov Khenin, et al., in *Israel: Vital Signs 2000,* pp. 157, 171.

117. Avinoam Meir and Yosef Ben-David, "Demographic Trends among the Urbanizing Bedouins of the Negev," in *The Bedouin Settlement in Israel,* pp. 83, 86.

118. Aliza Arbeli, "Why Shouldn't They Vote for Shas?" *Ha-Aretz,* Friday, September 11, 1998.

119. Giselle Hazzan, presentation, Ein Afek, December 5, 2001.

120. Elli Elad, "The First Environmental Unit Is Dedicated in Sachnin," *Ha-Aretz,* April 20, 1993.

121. Boaz Kabel, personal communication, June 11, 2000.

122. Interview with Dror Amir, Tel Aviv, January 7, 1999.

123. Interview with Dr. Fayad Sheabar, Kibbutz Ketura, December 1, 1997.

124. Interview with Dr. Diab Shahir, Tamara, March 9, 1999.

125. Benvenisti, *op. cit.,* p. 85; see also Sarah Graham-Brown, "The Economic Consequences of the Occupation," in *Occupation: Israel over Palestine,* ed. Naseer Aruri (London: Zed Books, 1984), pp. 167–222; and Henry Cattan, *The Palestine Question* (London: Croom Helm, 1988).

126. Howard M. Sachar, *A History of Israel* (New York: Alfred Knopf, 1976), p. 672.

127. Vivian A. Bull, "The Agricultural Sector" and "General Economic Development of the West Bank Territory," in *The West Bank: Is It Viable?* (Lexington, Mass.: Lexington Books, 1975), pp. 37–89.

128. Sachar, *op. cit.,* pp. 687–688.

129. Hillel Shuval, "Towards Resolving Conflicts over Water between Israel and Its Neighbors: The Israeli-Palestinian Shared Use of the Mountain Aquifer as a Case Study," in *Between War and Peace: Dilemmas of Israeli Security* (London: Frank Cass, 1996), p. 226.

130. Yisrael Tomer, "Tumultuous Waters at the Economics Committee," *Yediot Ahronot,* October 20, 1993.

131. Alfred Abed-Rabbo and David Scarpa, "Assessing Pollution in the Mountain Aquifer and the Drinking Water It Supplies" (Bethlehem: Bethlehem University, 1995).

132. Jad Isaac, "Environmental Protection and Sustainable Development in Palestine," in *Our Shared Environment,* eds. R. Twite and J. Isaac (Jerusalem: Israel-Palestine Center for Research and Information, 1994), p. 17.

133. Alfred Abed-Rabbo, personal communication, December 1997.

134. Karen Assaf, "Palestinian Water Resources: Water Quality," in *Our Shared Environment,* eds. R. Twite and J. Isaac (Jerusalem: Israel-Palestine Center for Research and Information, 1994), p. 296.

135. Dror Avisar, *The Impact of Pollutants from Anthropogenic Sources within a Hydrologically Sensitive Area—Wadi Rabba Watershed—Upon Ground Water Quality* (Tel Aviv: Adam Teva V'din, 1996), p. 18.

136. Tariq Talahmeh, personal communication, May 14, 2000.

137. Karen Assaf, "Environmental Law and Its Enforcement in Israel, the West Bank, and Gaza," in *Our Shared Environment,* eds. R. Twite and J. Isaac (Jerusalem: Israel-Palestine Center for Research and Information, 1994), p. 115.

138. Liat Collins, "Environmental Law to Be Enforced in Judea and Samaria," *Jerusalem Post,* October 18, 1998.

139. Abed el-Rahman Tamimi, "Waste Water Options and Solutions," *Campaigning for Our Environment,* Eilat conference proceedings (Jerusalem: EcoPeace, 1997), p. 74.

140. Tomer, *op. cit.*

141. Amira Hess, "Water Pressure in the West Bank Rises, Sharon Says Palestinian Excuses Are All Wet," *Ha-Aretz,* August 19, 1998.

142. A newspaper editorial opined, "As far as the water wastage and leakage argument is concerned, Israel was responsible, over a period of 26 years, for the water systems in the territories, including the development budgets that were supposed to upgrade the water networks there." "Water Shortage in the Territories," editorial, *Ha-Aretz,* August 20, 1998.

143. Amira Hess, "Israel to Ease West Bank Water Shortage, Water Tanks to Be Placed in Driest Areas," *Ha-Aretz,* August 24, 1998.

144. Shuval, *op. cit.*

145. Oded Lipshitz, "Agriculture, Water and Much Adrenaline," *El ha-Mishmar, Musaf,* August 25, 1992.

146. Mohammed Ajjour, "Introduction to the Gaza Environmental Profile," in *Our Shared Environment—The Conference 1994,* ed. R. Twite and R. Menczel (Jerusalem: Israel-Palestine Center for Research and Information, 1995), pp. 121–126; see also Reitse Koopmans, "Environmental Problems in the Gaza Strip," ibid., pp. 126–137.

147. Benvenisti, *op. cit.*, p. 218.

148. These consist of the following: the Yarkon-Taninim Western Basin, the largest subaquifer, which spans the West Bank and the center of Israel; the Nablus-Gilboa Northeastern Basin, a subaquifer north of the Samaria mountains; and the Eastern Subaquifer, located almost entirely in the West Bank. See Shuval, *op. cit.*, pp. 216–219.

149. See Eyal Benvenisti and Haim Gvirtzman, "Harnessing International Law to Determine Israeli-Palestinian Water Rights: The Mountain Aquifer," *Natural Resources Journal* 33 (1993): 552–556.

150. Shuval, *op. cit.*, p. 220.

151. Deborah Housen-Couriel and Moshe Hirsch, "Aspects of the Law of International Water Resources," in *Environmental Quality and Ecosystem Stability,* ed. A. Adin, A. Gasith, and B. Fattal (Jerusalem: Israel Society for Ecology and Environmental Quality Sciences, 1992).

152. Dalia Mazori, "Unholy Waters," *Maariv,* October 22, 1993.

153. To name a few: Tony Allan, *The Middle East Water Question: Hydropolitics and the Global Economy* (London: I. B. Tauris, 2002); Sharif S. Elmusa, *Water Conflict: Economics, Politics, Law and the Palestinian-Israeli Water Resources* (Washington, D.C.: Institute for Palestine Studies, 1997); *Water and Peace in the Middle East,* ed. J. Isaac and H. Shuval (Amsterdam: Elsevier, 1994); Karen Assaf, Nader al-Khatib, Elisha Kallay, and Hillel Shuval, *A Proposal for the Development of a Regional Water Master Plan* (Jerusalem: Israel-Palestine Center for Research and Information, 1993); *Management of Shared Groundwater Resources: The Israeli-Palestinian Case with International Perspective,* ed. Eran Feitelson and Marwan Haddad (Ottawa: International Development Research Center, 2001); Arnon Soffer, "The Israeli-Palestinian Conflict over Water Resources," *Palestine-Israel Journal of Politics, Economics, and Culture* 5, no. 1 (1998): 43–54; and Deborah Housen-Couriel, *Legal and Administrative Aspects of the Establishment of a Regional Water Authority for the Jordan River Basin* (Tel Aviv: Armand Hammer Fund for Economic Cooperation in the Middle East, 1995).

154. Feitelson and Haddad, *op. cit.*

155. Soffer, *op. cit.*, p. 47.

156. Amiram Cohen, "Peres Will Present Projects in America for Expanding Water Sources for Israel and Jordan," *Ha-Aretz,* October 4, 1994.

157. David Chayon and Yaakov Ben Amir, "Shimon Peres: The Turkish Proposal for Supplying Water Is Being Checked Seriously; Yaakov Zur: It's Not Practical," *Globes,* October 20, 1993.

158. Jitzchak P. Alster, "Water in the Peace Process: Israel-Syria-Palestinians," *Justice,* no. 10 (September 1996): 6–8.

159. Haim Gvirtzman, "The Jewish-Arab Water Conflict: Hydrological and Jurisdictional Perspective," *Water* 10 (1995): 32–40.

160. Aharon Zohar and Yehoshua Schwartz, *The Water Problem in the Framework of the Arrangements between Israel and the Arab Countries* (Tel Aviv: Tahal, 1993).

161. The Gaza Strip and Jericho Agreement, May 4, 1994 (article 31 of Annex II); see also Moshe Hirsch, "Environmental Aspects of the Cairo Agreement on the Gaza Strip and the Jericho Area," *Israel Law Review* 28, no. 2–3 (1994): 374–401.

162. Alster, *op. cit.*, pp. 6–8.

163. Ibid.

164. Mohammed Said al-Hmaidi, personal communication, February 1997.

165. Aaron Lerner, "Dr. Mohammed Ajjour, Palestinian Authority Environmental Planning and Israeli Response," *Independent Media Review and Analysis,* December 1, 1996.

166. Zafrir Rinat, "Israelis Prefer Dumping Their Trash over the Green Line," *Ha-Aretz,* June 6, 1998.

167. Imad Sa'ad, presentation at Symposium on the Environment as a Casualty of the Israeli-Palestinian Conflict, Tel Aviv, EcoPeace—Friends of the Earth Middle East, August 10, 1998.

168. Benvenisti, *op. cit.*, p. 218.

169. Zafrir Rinat and Or Kashti, "Jerusalem Schools with High Radon Concentrations Will Remain Closed until the Classroom Measurements Results Are In," *Ha-Aretz*, December 28, 1995.

170. "Keeping People in Their Place," *Economist*, September 1998, p. 48.

171. See generally Mawil Izzi Dien, *The Environmental Dimension of Islam* (Cambridge: Lutterworth, 2000).

172. Sura XXVII: 18–19.

173. Joyce Starr, "The Quest for Water, from Biblical Times to the Present," *Environmental Science and Technology* 27, no. 7 (1993): 1265.

174. David Shipler, *Arab and Jew, Wounded Spirits in a Promised Land* (New York: Penguin, 1987), p. 57.

175. Interview with Eilon Schwartz, Tel Aviv, January 12, 1998.

176. Interview with Mahmoud Gazawi, Tel Aviv, January 13, 1998.

177. "The general inclination by far is the consumer society and an overwhelming desire to move from a Third World society to a modern one. The general permissiveness you find among us is a phenomenon similar to other places." Interview with Dr. Basel Ghattas, Shfaram, December 9, 1997.

178. Interview with Dr. Basel Ghattas, Shfaram, December 9, 1997.

179. Interview with Farkhan Shlebe, Sfinat Midbar, January 6, 1998.

180. Mahmud Ka'abiah, personal communication, July 12, 1993.

181. Interview with Dr. Basel Ghattas, Shfaram, December 9, 1997.

182. Giselle Hazzan, personal communication, January 21, 2002.

183. Philip Warburg, "Investing in Israel's Environmental Future," report submitted to the Israel Cooperative Program, Jerusalem, 1996, p. 15.

184. Liat Collins, "Up in Smoke," *Jerusalem Post Magazine*, August 16, 1996, pp. 16–18.

185. Netty Gross, "Smoke Signals," *Jerusalem Report*, September 19, 1996, p. 18.

186. Interview with Dr. Basel Ghattas, Shfaram, December 9, 1997.

187. Life and Environment, informational brochure, K'far Saba, 2001.

188. Arab historians frequently quote Don Peretz, who reported in 1958 that of 370 Jewish settlements established between 1948 and 1953, 350 were built on Arab absentee property. Don Peretz, *Israel and the Palestine Arabs* (Washington, D.C.: Middle East Institute, 1958).

189. Lustick, *op. cit.*, p. 246.

190. Hirschberg, *op. cit.*

191. El-Haj, *op. cit.*, p. 9.

192. Izhak Schnell and Amin Fares, *Towards High-Density Housing in Arab Localities in Israel* (Jerusalem: Floersheimer Institute for Policy Studies, 1996), p. 9.

193. Ibid., p. 68.

194. Interview with Dr. Basel Ghattas, Shfaram, December 9, 1997.

195. Yehuda Golan, "A Rise in the Number of Calories Consumed Is Recorded," *Maariv, Ha-Yom Supplement*, September 30, 1997, p. 2.

CHAPTER 11: ENVIRONMENTAL ACTIVISM HITS ITS STRIDE

1. Barbara Sofer, "A Movement Sweeps Israel," *Hadassah Magazine*, May 1991, pp. 18–20; see also Micha Odenheimer, "Retrieving the Garden of Eden," *The Melton Journal*, no. 24 (Spring 1991): 1–6.

2. David Suzuki, "A New Millennium," in *The Sacred Balance* (London: Allen and Unwin, 1997), pp. 207–240; see also Robert Gottlieb, *Forcing the Spring: The Transformation of the American Environmental Movement* (Washington, D.C.: Island Press, 1993).

3. Stuart Schoenfeld, "Negotiating the Environmental Social Dilemma: The Case of Greenpeace Canada," paper presented at the Jerusalem Conference on Canadian Studies, Hebrew University, July 1998.

4. Alon Tal and Eilon Schwartz, "The Environmental Crises as a Function of Relationships within Human Society," *Skira Chodshit* (June 1993): 38–39.

5. Eric Silver, "The New Pioneers," *Jerusalem Report*, April 21, 1994, pp. 12–15.

6. Shirli Bar-David and Alon Tal, *Harnessing Activism to Protect Israel's Environment: A Survey of Public Interest Activity and Potential* (Tel Aviv: Adam Teva V'din, 1996).

7. Gila Silverman, "Environmental Groups in Israel," internal memo prepared for the New Israel Fund, March 1994.

8. Orr Karassin, "Organizations Working for Quality of Life and the Environment in Israel," *National Priorities for the Environment in Israel, Second Policy Paper* (Haifa: Neeman Institute, 2001), p. 127.

9. Bar-David and Tal, *op. cit.*

10. Anat Tal-Shir and Zadok Yehezkeli, "The Greenpeace Commandos Conquered the Kishon," *Yediot Ahronot*, June 15, 2000.

11. Bar-David and Tal, *op. cit.*, pp. 14–15.

12. Karassin, *op. cit.*

13. Tal-Shir and Yehezkeli, *op. cit.*, p. 15.

14. Richard Laster, "Reading D: Planning and Building or Building and Then Planning," *Israel Law Review* 8 (1973): 492.

15. Yosef Charish, personal communication, July 1986.

16. *Israel Electric Company v. Avisar and Others*, Civil Appeal, 190/69, *Piskei Din* 23, II, 1969, pp. 318–321.

17. For a full description of the court battle see Earnest Katin and Mordechai Virshubski, "Environmental Law and Administration in Israel," *Tel Aviv University Studies in Law* 1 (1975): 213–215.

18. Interview with Yedidyah Be'eri, Tel Aviv, September 4, 1997.

19. *Israel Electric Company v. Fersht et al.*, Criminal Appeal, 151/84, *Piskei Din* 39, III, 1984, pp. 1–12.

20. D. Sivan, "Malraz—the Public Council for Prevention of Noise and Air Pollution," *Biosphera* 70, no. 3 (December 1970): 4.

21. To emphasize the nonpartisan nature of the organization, Malraz also recruited politicians from the left to its board, such as Knesset member and biology professor Boaz Moav.

22. Malraz, *Milestones in the Struggle against Nuisances* (Tel Aviv: Malraz, 1979).

23. "Malraz in the Year 1973," *Biosphera* 9 (1974): 3.

24. Malraz, *op. cit.*

25. Interview with Yedidyah Be'eri, Tel Aviv, September 4, 1997.

26. Josef Tamir, "Why I Established the Council for a Beautiful Israel and Life and Environment," in *Haver Knesset* (Jerusalem: Ahiabar, 1985), pp. 395–400.

27. Council for a Beautiful Israel, *See under Cleanliness* (Tel Aviv: Council for a Beautiful Israel, 1988).

28. Interview with Aura Herzog, Tel Aviv, March 25, 1996; see also *The Council for a Beautiful Israel News*, twenty-fifth anniversary edition, 1995.

29. Shirley Benyamin, personal communication, Karkur, September 3, 1997.

30. See debate over atomic work at the Dimona Reactor in *Divrei ha-Knesset*, December 7, 1965.

31. Herschell Benyamin, personal communication, Karkur, September 3, 1997.

32. Abraham Fund, "EcoNet Israel: The Ecological Network," *The Abraham Fund Directory* (New York: Abraham Fund, 1992), p. 240.

33. *Israel Nuclear News* 2, no. 1 (1987): 3.

34. *EcoNet News* 5, no. 4 (1990). Around the same time, unbeknownst to the group and to most Israelis, one of the first and largest Green Internet servers in San Francisco selected the same name.

35. Shirley Benyamin, personal communication, Karkur, September 3, 1997.

36. Shirley Benyamin, open letter to EcoNet Friends, May 1994.

37. John Goldsmith, "Counseling, Evaluation, and Research at Beer Sheva University to Immigrants Exposed to the Chernobyl Nuclear Reactor," *EcoNet News* 8, no. 2 (September 1993): 1–3.

38. "Health Risks from Low Level Radiation," *Israel Nuclear News* 4, no. 1 (1989): 1.

39. *EcoNet News* 8, no. 3 (December 1993): 1; Benyamin, open letter to EcoNet Friends, May 1994.

40. "Does Israel Really Need or Even Want a Nuclear Power Station?" *EcoNet News* 7, no. 3 (1992): 3.

41. Steve Rodan, "Clear and Present Option," *Jerusalem Post Magazine*, June 7, 1996, pp. 18–20.

42. Liat Collins, "Sarid: Country's Nuclear Reactors Pose No Threat," *Jerusalem Post*, March 31, 1993; Elli Elad, "Water, Air, Soil and Flora Samples from the Dimona Reactor to Be Passed Regularly to Minister Sarid," *Ha-Aretz*, March 31, 1993.

43. Benyamin, open letter to EcoNet Friends, May 1994.

44. Silver, *op. cit.*, pp. 14–16.

45. Yossi Moskowitz as quoted in "Storm at the Carmel Coast," *Maariv, Esekim* (Business), March 30, 1997, p. 4.

46. Interview with Lynn Golumbic, Haifa, November 24, 1997.

47. Penina Migdal Glazer and Myron Peretz Glazer, *The Environmental Crusaders: Confronting Disaster and Mobilizing Community* (University Park, Pa.: Pennsylvania State University Press, 1998), pp. 78–85.

48. "Treatment of Air Quality in the Haifa Union of Cities," in *Environmental Quality in Israel,* no. 16 (Jerusalem: Ministry of the Environment, 1990), pp. 123–125.

49. Interview with Lynn Golumbic, Haifa, November 24, 1997.

50. Glazer and Glazer, *op. cit.*

51. Interview with Lynn Golumbic, Haifa, November 24, 1997.

52. Ayana Goren, Shmuel Brenner, and Sara Helman, *Health Survey among School Children in Haifa Bay, 1989* (Tel Aviv: Ministry of the Environment, 1989).

53. Moshe Shachal, "The Ministry of Energy and Air Quality in the Haifa Region," speech by the Minister of Energy in *Biosphera* (October 1987).

54. Michael Eilan, "Fighting for Breath," *Jerusalem Post International Edition,* July 29, 1989, p. 9.

55. Ibid. p. 10.

56. Golumbic recalls: "I spoke at the National Planning Council meeting before they voted on the power plant. I had come with SPNI's local Director, Shoshi Perry (and she went crazy when the SPNI chairman, Yoav Sagi, who sat on the Council, abstained). But it didn't really matter because there was a strong majority to site the station in Hadera." Interview with Lynn Golumbic, Haifa, November 24, 1997.

57. Eilan, *op. cit.,* p. 10.

58. Efraim Orni and Elisha Efrat, *The Geography of Israel* (Jerusalem: Israel Universities Press, 1973), p. 65.

59. Ayana Goren, Shmuel Brenner, Sara Helman, *Health Survey among School Children in Beit Shemesh, 1994* (Tel Aviv: Ministry of the Environment, 1995), pp. 12–13.

60. Interview with Elli Vanunu, Beit Shemesh, February 22, 1999.

61. "Personal Decree for the Nesher Har-Tuv Cement Factory, Beit Shemesh," summary in *Environmental Quality in Israel* nos. 17–18 (Jerusalem: Ministry of the Environment, 1992), p. 64.

62. Ibid.

63. "Ecological Disaster and Health Risk," promotional material, Beit Shemesh Action Committee, October 31, 1991.

64. Zvi Greenberg, personal communication, January 3, 1999. Environmentalists were dumbfounded when Environmental Minister Yossi Sarid granted the factory a prize in 1994, citing its capital investment in pollution control technology.

65. Interview with Elli Vanunu, Beit Shemesh, February 22, 1999.

66. Ibid.

67. Chico Mendes, *Fight for the Forest: Chico Mendes in His Own Words* (New York: Inland Books, 1990).

68. Interview with Benny Shalmon, Kibbutz Ketura, January 14, 1998.

69. Sue Fishkoff, "For the Birds," *Jerusalem Post International Edition*, July 6, 1996, p. 16.

70. Reuven Yosef, "Is the Salt Marsh of Eilat (a Critically Important Habitat for Palearctic Migrants) Dead or Alive?" *Living Bird* (Winter 1998): 22–28.

71. Ibid., pp. 22–28.

72. H. Shirihai and D. Christie, *The Birds of Israel* (London: Academic Press, 1996).

73. Y. Waisal, "Vegetation of Israel," in *Encyclopedia of Plants and Animals of the Land of Israel*, vol. 8 (Tel Aviv: Ministry of Defense Press, 1984).

74. Yosef, *op. cit.*

75. Fishkoff, *op. cit.*

76. Interview with Reuven Yosef, Eilat, January 9, 1998.

77. Ibid.; see also Daniel Hillel, *Negev: Land, Water and Life in a Desert Environment* (New York: Praeger, 1982), pp. 45–46.

78. Reuven Yosef, "Physical Distances among Individuals in Flocks of Greater Flamingoes (*Phoenicopterus ruber*) Are Affected by Human Disturbance," *Israel Journal of Zoology* 43 (1997): 79–85.

79. Reuven Yosef, "Reactions of Gray Herons (*Ardea cinerea*) to Seismic Tremors," *Journal für Ornithologie* 138 (1997): 543–546.

80. Reuven Yosef, "On Habitat-specific Nutritional Condition in Graceful Warblers (*Prinia gracilis*); Evidence from Ptilochronology," *Journal für Ornithologie* 138 (1997): 309–313.

81. Reuven Yosef, "Clues to Migratory Routes of the Eastern Flyway of the Western Palearctics: Ringing Recoveries at Eilat, Israel," *Die Vogelwarte* 39 (1998): 203–208.

82. Fishkoff, *op. cit.*

83. Reuven Yosef, personal communication, August 29, 1998.

84. Interview with Reuven Yosef, Eilat, January 9, 1998.

85. Ibid.

86. Literally translated, Adam Teva V'din means "Man, Nature, and Law." This somewhat awkward name came about when the Ministry of the Interior discouraged use of the more conventional "Israel Union for Environmental Defense" on the grounds that it was too bombastic for a new organization. The latter name has begun to appear on the organization's English-language documents.

87. Responding to the public-interest action, the Ministry of the Environment eventually filed suit itself, and after a long legal struggle, the city capitulated and submitted a sewage plan. Adam Teva V'din, *Annual Report* (Tel Aviv: Adam Teva V'din, 1991).

88. Interview with Ruth Yaffe, Tel Aviv, January 12, 1998.

89. Prevention of Environmental Nuisances Law (Citizens' Suits), 1992, *Sefer ha-Hokim*, no. 1394, p. 184.

90. Water Law (Amendment No. 10) 1995, *Sefer ha-Hokim*, no. 1533, p. 362.

91. Prevention of Environmental Nuisances Law (Citizens' Suits), 1992, *Sefer ha-Hokim*, no. 1394, p. 55.

92. Environmental Law (Methods of Punishment) 1997, *Sefer ha-Hokim*, no. 1622, April 10, 1997 p. 132.

93. Adam Teva V'din, *Annual Report* (Tel Aviv: Adam Teva V'din, 1992).

94. Interview with Ruth Yaffe, Tel Aviv, January 12, 1998.

95. Ibid.

96. Adam Teva V'din, "UED Introduces a Comprehensive Sewage Strategy," in *Annual Report* (Tel Aviv: Adam Teva V'din, 1992), p. 9.

97. Ruth Yaffe, "The Public's Right to Participate in Environmental Decision-Making in Israel: A Progress Report on Issues of Law and Policy," *Tel Aviv University Studies in Law* 14 (1998): 13–15.

98. Daniel Fisch, "Israel's Environmental Problems," *Palestine-Israel Journal* 5, no. 1 (1998): 24.

99. Yaakov Garb, "The Road to Peace? The Trans-Israel Highway and the Fight for Sustainable Alternatives," *Sustainable Transport* (1996).

100. *Adam Teva V'din and Others v. The National Planning Council and Others*, Bagatz, 2924/94, *Piskei Din* 50 (III), 446 (1996).

101. Valerie Brachya and Uri Marinov, "Coastal Zone Management in Israel and Prospects for Regional Cooperation," in *Protecting the Gulf of Aqaba: A Regional Environmental Challenge*, ed. Philip Warburg et al. (Washington, D.C.: Environmental Law Institute, 1993), p. 198.

102. National Master Plan No. 13 (Jerusalem: Israel Ministry of the Interior), 1981.

103. Valerie Brachya, "Protecting Coastal Resources," *Environmental Science and Technology* 27, no. 7 (1993): 1269.

104. Robert Abel, "Cooperative Marine Technology for the Middle East," *Environmental Science and Technology* 27, no. 7 (1993): 1273.

105. Alon Tal, "Whose Coast Is It Anyway?" *Jerusalem Report*, June 1996.

106. Gil Doron, "The Marina of the Rich," *Ha-Ir*, July 29, 1995.

107. Daniel Aviv, "The Marina in Ashdod: The Municipality Will Discuss the Implications," *Kahn ha-Darom*, October 13, 1995.

108. Adam Teva V'din, *Annual Report* (Tel Aviv: Adam Teva V'din, 1995), p. 7.

109. Alan Roberts, "Tel Aviv Seeking Approval for Yarkon Project," *City Lights* 5, no. 181 (August 1, 1997): 1; see also Yizhak Bar-Yosef, "There Is a Need to Leave Something to the Children," *Yediot Ahronot*.

110. "Victory of the 'Greens' over the Developers," *Maariv, Esekim*, March 30, 1997, p. 4.

111. "Marina Plan Not to Be Submitted before Court Hearing," *City Lights*, February 14, 1997.

112. Lior El-Hai and Yizhak Bar-Yosef, "The Battle for the Coast," *Yediot Ahronot—24 Hour Supplement*, June 1, 1998, p. 8.

113. "Storm on Haifa Coast," *Maariv, Esekim*, March 30, 1997, p. 4.

114. Sagit Lampret, "There Is Law and There Is a Tower," *Kol Bo Haifa,* January 23, 1998, p. 62.

115. "Storm on Haifa Coast," p. 4.

116. Ibid., p. 4.

117. Janine Zacharia, "Haifa Builders Hit Back at Green Group Blocking Their Beachfront Project," *Jerusalem Report,* July 10, 1997.

118. "Storm on Haifa Coast," p. 4.

119. Lampret, *op. cit.*, p. 64.

120. Avner De-Shalit, "Where Do Environmentalists Hide?" in *Our Shared Environment—The Conference 1994,* ed. Robin Twite and Robin Menczel (Jerusalem: Israel-Palestine Center for Research and Information, 1995), pp. 274–276.

121. Interview with Azariah Alon, Beit ha-Shita, September 15, 1997.

122. Shmuel Gilbert, personal communication, March 17, 1999.

123. Shmuel Chen, "Green Ballots at the Ballot Box," *Green, Blue and White* (December 1998): 4–5, 12–13.

124. Menahem Rahat, "1,500 Die Each Year from Environmental Pollution, Does Anyone Care? Nehama Ronen, Leader of 'Kol ha-Sviva' in the Inaugural Ceremony," *Maariv,* January 18, 1998, p. 12.

125. Zafrir Rinat, "Dedi and Greens Vow Quality of Life," *Ha-Aretz,* March 29, 1999.

126. Veronique Boquelle, *Transportation in Israel* (Tel Aviv: Adam Teva V'din, 1997).

127. Ibid.

128. Alon Tal, "A Reform in Air Pollution from Motor Vehicles: Towards the Era of the Catalytic Converter," *Biosphera* 22 (1993): 4.

129. Zafrir Rinat, "A Third of the Cars in Israel Pollute the Air above What Is Allowed by Law," *Ha-Aretz,* January 30, 1998.

130. *Improving Enforcement for Air Pollution by Transportation—A Position Paper,* ed. Alon Tal (Jerusalem: Jerusalem Institute for Israel Studies, 2002).

131. Yizhak Bar-Yosef, "Increase in the Tel Aviv Air Pollution—A City Out of Breath," *Yediot Ahronot,* December 1, 1997, p. 6.

132. M. Luria, G. Sharf, and M. Peleg, "Forecast of Photochemical Pollution for the Year 2010," *Proceedings of the 24th Conference of the Israel Society for Ecology and Environmental Quality Sciences* (Jerusalem: Israel Society for Ecology and Environmental Quality Sciences: 1993).

133. Boquelle, *op. cit.*

134. Ibid.

135. Ibid.

136. Peter Hirschberg, "Road Rage," *Jerusalem Report,* May 8, 2000, p. 24.

137. Dalia Tal, "An Association against the Trans-Israel Highway," *Al ha-Sharon,* October 9, 1992.

138. Irit Sappir-Gildor, "Environmental Attitudes among Visitors to JNF Forests," master's thesis, Tel Aviv University, Department of Geography, 2001.

139. Palestine Israel Environmental Secretariat, "Nature Knows No Boundaries," promotional material, 1998.

140. Zafrir Rinat, "The Dining Room as an Ecological System," *Ha-Aretz,* February 20, 1997; see also http://www.arava.org.

141. *Desert Dreams* 7 (Kibbutz Ketura: The Arava Institute for Environmental Studies, Spring 2001).

142. International Center for the Study of Bird Migration at Latrun, "Migrating Birds Know No Boundaries," promotional material, 1997; http://www.birds.org.il.

143. "We have to discuss the environmental crisis from the perspective of culture; and part of culture, part of the way I look at things, for me as a Jew, is through Jewish tradition." Eilon Schwartz, "Are We as Trees of the Field? Jewish Perspectives on Environmental Ethics," in *Our Shared Environment—The Conference 1994,* ed. Robin Twite and Robin Menczel (Jerusalem: Israel-Palestine Center for Research and Information, 1995), pp. 4–5.

144. Lisa Beyer, "Trashing the Holy Land," *Time,* September 7, 1998, p. 54.

145. Asaf Levitan, personal communication, November 3, 1998.

146. Larry Derfner, "Clearing the Air," *Jerusalem Post Magazine,* October 30, 1998, p. 17.

147. Limiting of Smoking in Public Areas Law, 1983, *Sefer ha-Hokim,* 148 p. 658. All the same, passive smoking in Israel is thought to cause eight hundred deaths per year. This is in addition to the five thousand Israelis who die every year of smoking-related cancer and heart disease.

148. *The Greening of Urban Transport: Planning for Walking and Cycling in Western Cities,* ed. R. S. Tolley (New York: John Wiley and Sons, 1997).

149. Eilon Schwartz, Jeremy Benstein, and Nili Peri, *From Nature Protection to Environmental Practices* (Jerusalem: Melitz, August 1997), pp. 12, 16.

150. Yehuda Golan, "A Rise in the Number of Calories Consumed Is Recorded," *Maariv, Ha-Yom Supplement,* September 30, 1997, pp. 2–3.

151. Numbers 35:2–5; Leviticus 25:34.

152. Esther Zanberg, "2020 Vision," *Ha-Aretz Magazine,* January 1, 1998, pp. 14–16.

153. Naomi Tsur, "Are Developers Killing the Green Belt around Jerusalem?" *Jerusalem Report,* October 23, 2000, p. 96.

154. Tal and Schwartz, *op. cit.,* pp. 34–39.

155. Ibid., p. 39.

CHAPTER 12: TOWARD A SUSTAINABLE FUTURE?

1. Shoshana Gabbay, "The Environment in the Jordanian Peace Talks," *Israel Environment Bulletin* 18, no. 1, *Special Peace Issue* (1995): 3.

2. Interview with Yisrael Peleg, Tel Aviv, September 30, 1997.

3. Three events during the summer of 1992 are indicative of the frequency of oil spills. See Jackie Post and Adi Katz, "For the Second Time in a Month, an Oil Tanker Pollutes the Gulf of Aqaba," *Hadashot,* August 5, 1992;

Liat Collins, "Eilat Oil Spill Damage Assessed," *Jerusalem Post,* September 2, 1992.

4. Interview with Elli Varberg, Eilat, November 10, 1998.

5. Ibid.

6. Gabbay, *op. cit.,* in particular the description of the October 1994 Bahrain Environmental Code of Conduct, p. 12; see also The Gaza Strip and Jericho Agreement, May 4, 1994 (article 31 of Annex II); Moshe Hirsch, "Environmental Aspects of the Cairo Agreement on the Gaza Strip and the Jericho Area," *Israel Law Review* 28, no. 2–3 (1994): 374–401.

7. Gabbay, *op. cit.,* p. 10.

8. Alon Tal, "Preventing Pollution from Ports and Maritime Activity in the Gulf of Aqaba: Current Practices in Israel," *Protecting the Gulf of Aqaba: A Regional Environmental Challenge,* ed. Philip Warburg (Washington, D.C.: Environmental Law Institute, 1993), p. 265.

9. EcoPeace, *Symposium on the Environmental Heritage of the Gulf of Aqaba and Its Sustainable Development, Summary Report,* 1995, p. 6.

10. Alon Tal, Shoshana Lopatin, and Gidon Bromberg, *Sustainability of Energy-related Development Projects in the Middle East Peace Region* (Washington, D.C.: US AID's Energy Project Development Fund, March 1995).

11. Marlin Atkinson, Yehudith Birk, and Harold Rosenthal, *Evaluation of Pollution in the Gulf of Eilat: Report for the Ministries of Infrastructure, Environment, and Agriculture,* December 10, 2001.

12. EcoPeace, *An Updated Inventory of New Development Projects, Compiled from the Reports Presented by the Palestinian National Authority, the Hashemite Kingdom of Jordan, the State of Israel, and the Republic of Egypt to the Casablanca, Amman, and Cairo Middle East–North Africa Economic Summits* (Jerusalem: EcoPeace, March 1997), pp. 14–18.

13. Nanette Furman, "Changes in the Coral Reproductive Rate and Sea Anemone Survival at Eilat," *Ecology and Environment,* 2, no. 6 (1990): 87–92.

14. Nanette Chadwick-Furman, personal communication, December 1998.

15. Municipality of Eilat, promotional information, 1997.

16. Revital Levy-Stein, "Sewage Swamps Eilat," *Ha-Aretz,* August 17, 1998; or Revital Levy-Stein, "The Big Stink: Sewage Spill Closes More Eilat Beaches," *Ha-Aretz,* August 18, 1998.

17. Prime Minister Netanyahu's 1998 pledge to build thousands of new hotel rooms around the Dead Sea, despite Israeli and Jordanian environmentalists' calls for modest development, reflects the same dynamic. Irit Rosenblum, "Prime Minister Promises New Hotel Rooms at the Dead Sea," Ha-Aretz, August 26, 1998; see also EcoPeace, *Dead Sea Challenges, Final Report* (Jerusalem: EcoPeace, 1996).

18. Batiah Kurus, Spokesperson, Israel Employment Service, personal communication, January 28, 2002.

19. Ari Shavit, "Three Sins and a Miracle," *Ha-Aretz,* August 29, 1997, p. 1.

20. Gregg Easterbrook, *A Moment on the Earth: The Coming Age of Environmental Optimism* (New York: Viking, 1995).

21. Ministry of the Environment, *Environmental Quality in Israel*, no. 21 (1997): p. 96.

22. Highest estimates reach into the thousands (Gabi Zohar, "A Researcher in the Technion Estimates: About 1,000 Deaths per Year in Israel Resulting from Air Pollution," *Ha-Aretz*, January 20, 1995). One recent study projects 293 deaths each year in Tel Aviv alone as a result of tailpipe emissions (Gary Ginsberg, Aharon Serri, Elaine Fletcher, Joshua Shemer, Dani Koutik, and Eric Karsenty, "Mortality from Vehicular Particulate Emissions in Tel Aviv-Jafo," *World Transport and Policy and Practice* 4, no. 2 (1998): 27–31). Other researchers believe there may be only hundreds of fatalities (Lisa Pearlman, "Threat from the Air," *Jerusalem Post*, February 24, 1989, p. 4).

23. A. I. Goren and S. Helman, "Changing Prevalence of Asthma among Schoolchildren in Israel," *European Respiratory Journal* 10 (1997): 2279–2284.

24. Alon Tal, "The Air Is Free," *Eretz v'Teva* (September-October 1999).

25. Arie Laor, Leon Cohen, and Yehuda L. Danon, "Effects of Time, Sex, Ethnic Origin and Area of Residence on Prevalence of Asthma in Israeli Adolescents," *British Medical Journal* 307 (October 2, 1993): 841.

26. Netty C. Gross, "Malignant Neglect," *Jerusalem Report*, April 12, 1999, pp. 24–25.

27. Asaf Haim, "The Breast Cancer Rate in the Haifa Region Is Tens of Percent Higher Than the National Average," *Maariv*, October 17, 2001, p. 19.

28. Shoshana Gabbay, *The Environment in Israel* (Jerusalem: Ministry of the Environment, 1998), pp. 24–28.

29. Ministry of the Environment, *The Jordan River: Restoration, Conservation and Development* (November, 1995).

30. Shmuel Chen, "Ramat Hovav: You Could Explode from It," *Green, Blue and White* (August-September 1998): 4–5.

31. Yoram Yom-Tov and Heinrich Mendelssohn, "Altered Landscapes," *Eretz*, December 1995, pp. 51–54.

32. R. Nathan, U. Safriel, and H. Shirihai, "Extinction and Vulnerability to Extinction at Distribution Peripheries: An Analysis of the Israeli Breeding Fauna," *Israel Journal of Zoology* 42 (1996): 361–383.

33. Iris Millner, "The Rabbit's Foot," *Ha-Aretz*, August, 1998.

34. Arieh Neiger, lecture at Conference on Utilization of the Courts by Green Organizations, Netanya College, May 29, 2000.

35. Yehudah Gat, "Activities of the Electric Company for Environmental Protection at the Haifa Power Station: Present and Future Plans," *Eichut ha-Sviva*, July 1994, pp. 6–12.

36. Yossi Beier, "A System for Treatment and Reuse of Wastes in the Haifa Oil Refineries," *Eichut ha-Sviva*, July 1994, pp. 14–17.

37. Alon Tal, "The Economic Benefits of Noncompliance with Environmental Laws: The Role of Economic Analysis in Assessing Penalties for Polluters in Israel," *Ecology and Environment*, 6, no. 1 (March 2000): 3.

38. Alon Tal and Dorit Talitman, *The Enforcement System of the Ministry of the Environment*, internal report, Jerusalem, December 2000, pp. 10–15.

39. Amory Lovins and L. Hunter Lovins, "Reinventing the Wheels," *Atlantic Monthly*, April 1996.

40. Ginsberg et al., *op. cit.*

41. U.S. Office of Air Quality Planning and Standards, *National Air Quality and Emissions Trends Report*, 1997, EPA 454/R-97-013. The decrease in ambient concentrations probably reflects the fact that most ambient monitors are in urban areas, so the 10 percent drop reflects a decrease in mobile source NO_x emissions by about 6 percent from 1987 to 1996. Further decreases in mobile source NO_x emissions are scheduled within the next several years. Joe Kruger, U.S. EPA, personal communication, September 1, 1998.

42. Yom-Tov and Mendelssohn, *op. cit.*, p. 51.

43. Jonathan Katz, personal communication, August 9, 1998. These are measured at purchasing power parity (PPP) dollars, not straight dollars, which would amount to $18,500.

44. Ibid.

45. Alon Tal, "The Role of the Environmental Movement in Promoting ISO 14000—Changing the Payoff Scheme," in *Proceedings of the Workshop held by the Israel Engineers and Architects Association*, 1997.

46. Dorit Talitman, personal communication, November 14, 2001.

47. Nadav Aaronson, personal communication, April 14, 1999.

48. Alon Tal, "Of Protected Values and Environmental Violations," *Ha-Praklit* 40, no. 3 (1992): 413–421.

49. Marcia Gelpe, "The Goals of Enforcement and the Range of Enforcement Methods in Israel and the United States," *Tel Aviv University Studies in Law* 14 (1998): 135–177; see also Orit Marom-Albeck and Alon Tal, "Upgrading Citizen Suits as a Tool for Environmental Enforcement in Israel: A Comparative Evaluation." *Israel Law Review*, 2002, in press.

50. Environmental Quality Law (Methods of Punishment) (Amendments), 1997, *Sefer Ha-Hokim*, no. 1622, p. 132.

51. Zohar Shkalim, Director of Enforcement Coordination, Ministry of the Environment, personal communication, January 27, 2002.

52. Janine Zacharia, "The Wasteland," *Jerusalem Report*, August 21, 1997, pp. 16–18.

53. In 1764 he wrote, "The certainty of a punishment, even if it be moderate, will always make a stronger impression than the fear of another which is more terrible but combined with the hope of impunity." Cesare Beccaria, *On Crimes and Punishment* (London: Cambridge University Press, 1995), p. 58. This theme was taken up by the utilitarian philosopher Jeremy Bentham in his treatise *The Principles of Morals and Legislation* (London: Oxford, 1996) (see footnote in chapter 14, rule 1), p. 143.

54. Dr. Gilad Gilov, Chief Chemist, Ramat Hovav, oral presentation, November 11, 2001.

55. Mining Regulations (The Fund for Restoring Quarries), 1978, *Kovetz Takanot*, no. 3865, p. 1626.

56. Yair Shoshani, "Mining Mark or Question Mark?" *Green, Blue and White* (August-September 1998): 20–21.

57. Oded Shlomot, "Quarries That Fell between the Chairs," *Green, Blue and White* (August-September 1998): 16–17.

58. Aviva Lori, "The Whole Country Is Wounds," *Ha-Aretz, Musaf*, March 17, 1995, p. 86.

59. Efraim Orni and Elisha Efrat, *Geography of Israel* (Jerusalem: Israel Universities Press, 1973), p. 269.

60. Ephraim Shlain and Eran Feitelson, *The Formation, Institutionalization and Decline of Farmland Protection Policies in Israel* (Jerusalem: Floersheimer Institute for Policy Studies, 1996), pp. 10–11.

61. *Remote Sensing Support for Analysis of Coasts*, RESSAC Consortium, Palermo, Italy—April 1999; available at http://www.ctmnet.it/ressac.

62. Michal Korakh, Asaf-Shapira, "Open Spaces in Israel" in *Israel: Vital Signs 2001* (Tel Aviv: Heschel Center for Environmental Learning and Leadership: 2001), pp. 184–186.

63. Eran Feitelson, "Protection of Open Spaces in Israel at a Turning Point," *Horizons in Geography* 42–43 (1995), p. 20.

64. Lori, *op. cit.*

65. Pliah Albeck, in *Land Management Policy Alternatives in Israel* (Jerusalem: Land Policy Discussions Group, May 1996).

66. Tara Wolfson, "Former Israeli Prime Minister Shimon Peres Talks Peace and the Environment with AIES Students and Alumni," *Desert Dreams* (Kibbutz Ketura: Arava Institute for Environmental Studies, Spring 1998), p. 1.

67. "Shochet: The Management of Israel Chemicals Needs to Sit in the Negev, But It Can't Be Done by Force," *Telegraph*, September 23, 1994.

68. "Contract Signed for Leasing Land to Joint Israeli-Jordanian Industrial Park at the Sheich Hussein Bridge," *Globes*, June 2, 1998, p. 8.

69. Michael McCloskey, "The Emporer Has No Clothes: The Conundrum of Sustainable Development," *Duke Environmental Law and Policy Forum* 9, no. 2 (1999): 153.

70. Presumably the type of "sustainable" growth called for by the Director General of the Ministry of Finance, Ben Zion Zelberfarb, departs from even the excessively flexible traditional definitions of sustainability. See "New Budget Shifts Priority to Growth," *Ha-Aretz*, August 1998.

71. Adam Mazor, *Israel in the Year 2020* (Haifa: Technion Press, 1994).

72. Zafrir Rinat, "Go Forth, but Do Not Multiply," *Ha-Aretz*, November 8, 1998, p. B-4.

73. Yoav Sagi, "Escape from Megalopolis," *Eretz*, November 1996, p. 57.

74. Eran Feitelson, *The Pattern of Environmental Conflicts in the Metropolitan Development Process and Its Planning Implications* (Jerusalem: Jerusalem Institute for Israel Studies, 1996).

75. Uzi Paz, the first Director of Israel's Nature Reserves Authority, and a leading nature historian, was representative of the environmental community in 1981 when he wrote, "One outstanding change in the landscape of the land—and entirely for the good—is the tremendous growth in the population. Since the foundation of the State, the number of residents has grown fivefold and roughly six hundred new settlements arose, from the heights of the Galilee to the southern Arava. The number of motorcars has also increased on highways and on the dirt roads, taking hitchhikers who are thirsty for the landscape with them." Uzi Paz, *Eretz ha-Tsvi v'ha-Yael* (Land of the Gazelle and the Ibex), *I* (Givataim: Masada, 1981), p. 25.

76. Rinat, *op. cit.*

77. Philip Warburg, "Taboo That Needs Breaking," *Jerusalem Post,* January 1, 1997. Dan Perry also broke the silence: "It is incumbent on planners to propose a lowering of the natural rate of growth in Israel from the level of a developing country to the accepted level in developed countries. . . . In these days when the entire world recognizes the fact that population growth is the central problem of the human species, Israel cannot act as if it is not part of it." Dan Perry, "Rejecting the Essential," *Teva v'Aretz,* January 1995.

78. Netty C. Gross, "IVF Fever," *Jerusalem Report,* July 3, 2000, pp. 13–14.

79. Yehuda Golan, "A Rise in the Number of Calories Consumed Is Recorded," *Maariv, Ha-Yom Supplement,* September 30, 1997, p. 2.

80. Netty C. Gross, "Mum's the Word," *Jerusalem Report,* August 21, 1997, pp. 14–15.

81. "And so it is, that a Zionist family, whose two children will in the future be serving in the army and contributing to the economy of the state and its human capital, receives 338 shekels per month. An ultra-Orthodox family, whose ten children will evade military service and who will be a burden on the economy of the state will receive from the State of Israel 4,930 shekels for their children each month." Avraham Poraz, "Report to the Voter," February 1999.

82. Efraim Ya'ar, "More Children, Less Blessing," *Ha-Aretz,* February 24, 1999, p. B-2; see also "Research: Intelligence Quotients among Children from Large Families in Israel Are Below Average," *Ha-Aretz,* February 24, 1999, p. 1.

83. Mishnah, *Yevamot* 6:6.

84. Philip Warburg, letter to the Editor, *Jerusalem Post,* March 23, 1998.

85. "There will be a tendency for such people to have rather more children than the rest, and these children will inherit the wish to an enhanced extent and these will contribute a still greater proportion of the population." Charles Galton Darwin, "Can Man Control His Numbers?" in *Evolution after Darwin* 2, ed. Sol Tax (Chicago: University of Chicago Press, 1960), p. 464.

86. Dan Perry, personal communication, Tel Aviv, November 8, 1998.

87. Rabbi Yerachmyel Barkia, *The People of Israel's Links with the Negev* (Jerusalem: Jewish National Fund, 1999): 33–34.

88. Interview with Nehama Ronen, Tel Aviv, November 10, 1998.

89. Yaakov Garb and Meira Hanson, "Consumption and Population: Trends and What Lies between Them," in *Israel Vital Signs 2001* (Tel Aviv: Heschel Center for Environmental Learning and Leadership, 2001).

90. Erik Schechter, "Scientist Offers Clean Power. . . for Just 1.3 Billion," *Jerusalem Report*, April 12, 1999, p. 12. See also http://magnet.consortia.org.il/ConSolar/Sabin/Zas/ZasTOC.html.

91. Dan Zaslavsky, *Sustainable Energy in Israel: Assessment and Program*, report to Israel Ministry of the Environment, December 1997, pp. 22–24.

92. Other problems that appear readily given to technical solutions are disposal of the saline brine, the effect of the tower on migrating birds, and elevated humidity in the vicinity of the tower. Michael Zwirn, "*Questions in Planning, Environmental Impacts and Social Need in a Major Alternative Energy Proposal: The Israeli Energy Towers*," paper presented to Arava Institute for Environmental Studies, May 20, 1996.

93. Uri Shitrit Architects, et al., "Development Plan for Eilat-Eilot," Israel Lands Authority, June 1998.

94. Saul Arlozorov, "Water in Israel: Present Problems and Future Goals," *Towards Sustainable Development* (Jerusalem: Ministry of the Environment, April 1999), pp. 54–55.

95. Meir Ben-Meir, personal interview, Tel Aviv, November 19, 1998.

96. "A Sea Water Desalinization Device Fueled by Solar Energy," www.solarproject.com/desalframe.htm.

97. Ministry of Energy, *Energy* 95 (Jerusalem: Ministry of Energy, 1995).

98. Curtis Moore and Alan Miller, *Green Gold: Japan, Germany, the United States, and the Race for Environmental Technology* (Boston: Beacon Press, 1994).

99. Tal, Lopatin, and Bromberg, *op. cit.*

100. Mazor, *op. cit.*

101. Indeed, this subject, which has been so central to many environmental agendas, has only recently begun to attract attention in Israel, with a conference on brownfields held at the Technion on the subject in the winter of 2000.

102. "Your Guidebook to Nagasaki," www.shs.kyushu-u.ac.jp/ohki/tourist/dejima.html.

103. Ze'ev Goldberg, "A Green Island in the Sea," *Blue, Green and White* 20 (July 1998), pp. 6–7.

104. "No Fishing Zone Reduced around Artificial Reef," *Fisheries Western Australia* (July 1988).

105. Ze'ev Goldberg, "Driving Tel Aviv into Hong Kong Harbor: Traffic Congestion Is Forcing Tel Aviv to Consider a Seaward Migration," *Ha-Aretz*, August 21, 1998.

106. David Harris, "Israeli Holiday Islands Move a Step Closer," *Jerusalem Post*, Internet edition, January 9, 1997.

107. David Harris, "Report Recommends Coastal Islands," *Jerusalem Post* Internet edition, October 7, 1998.

108. Interview with Nehama Ronen, Tel Aviv, November 10, 1998.

109. Philip Warburg, "Sink the Artificial Islands," *Jerusalem Report*, April 12, 1999, p. 54.

110. Stuart Schoffman, "The Best Country on Earth," *Jerusalem Report*, June 26, 1997.

111. *The Climate Convention: A Workshop about Israel Greenhouse Gas Policy* (Jerusalem: Ministry of the Environment, April 2000).

112. Interview with Yossi Sarid, Jerusalem, December 30, 1997.

113. Ibid.

114. Law Limiting Smoking in Public Places, 1983, *Sefer ha-Hokim*, no. 148, p. 658.

115. Noga Morag-Levine, "The Politics of Imported Rights, Transplantation and Transformation in an Israeli Environmental Cause Lawyering Organization," in *Cause Lawyering and the State in a Global Era*, ed. Austin Sarat and Stuart Scheingold (London: Oxford University Press, 2000), p. 334.

116. Deuteronomy 11:10–14.

117. Deuteronomy 29:22–23.

Index

Map Rendering: Don Waller, Publication Services, Inc.
Compositor: Publication Services, Inc.
Text: 10/13 Aldus
Display: Aldus
Printer: Friesens Corporation
Binder: Friesens Corporation